# Micellization, Solubilization, and Microemulsions

Volume 1

# Micellization, Solubilization, and Microemulsions

## Volume 1

Edited by

### K.L. Mittal
*IBM Corporation*
*East Fishkill Facility*
*Hopewell Junction, New York*

Plenum Press · New York and London

Library of Congress Cataloging in Publication Data

Symposium on Micellization, Solubilization, and Microemulsions, Albany, 1976.
  Micellization, solubilization, and microemulsions.
  Symposium held at the Seventh Northeast regional meeting of the American Chemical Society.
  Includes indexes.
  1. Micelles—Congresses. 2. Solubilization—Congresses. 3. Emulsions—Congresses. I. Mittal, K. L., 1945-    II. Title. III. Title: Microemulsions.
QD549.S976 1976              541'.3451                              77-1126
ISBN 0-306-31023-6 (v. 1)

Proceedings of the first half of the International Symposium on Micellization,
Solubilization, and Microemulsions held at the Seventh Northeast Regional Meeting of the
American Chemical Society in Albany, New York, August 8—11, 1976

© 1977 Plenum Press, New York
A Division of Plenum Publishing Corporation
227 West 17th Street, New York, N.Y. 10011

All rights reserved

No part of this book may be reproduced, stored in a retrieval system, or transmitted,
in any form or by any means, electronic, mechanical, photocopying, microfilming,
recording, or otherwise, without written permission from the Publisher

Printed in the United States of America

PREFACE

This volume and its companion Volume 2 chronicle the proceedings of the International Symposium on Micellization, Solubilization, and Microemulsions held under the auspices of the Northeast Regional Meeting of the American Chemical Society in Albany, New York, August 8-11, 1976. The technical program consisted of 48 papers by 91 authors from twelve countries. The program was divided into six sessions, and Dr. Hartley delivered the Concluding Remarks. Subsequently, six more papers were contributed for the proceedings volumes with the result that these volumes contain 51 papers (three papers are not included for certain reasons) and are divided into seven parts. The first three parts are embodied in Volume 1 and the remaining four parts constitute Volume 2; each part is followed by a Discussion Section. Dr. Hartley's Concluding Remarks are included in both volumes.

· When the idea of arranging a symposium on micelles was broached to me, I accepted it without an iota of hesitation. I had two options: either to make it a one- or two-sessions symposium, or bring together the global community actively carrying out research in this area. I opted for the latter as the subject was very timely and I knew that it would elicit very enthusiastic response. In order to broaden the scope of the symposium, I suggested that the theme of the symposium should be Micellization, Solubilization and Microemulsions.

Two salient features of this symposium should be mentioned: (i) a truly international symposium of this magnitude rarely occurs at a Regional Meeting, and (ii) I do not know whether there was ever a symposium of this quality, magnitude, and breadth of coverage on this topic.

Micelles are colloidal species in solution that are produced by aggregation of a relatively large number (from 20 up to thousands) of small amphiphilic molecules or ions. Micellar systems are usually described as association colloids. The fundamental characteristics of micelle-forming monomers is their amphiphilicity - the presence of both polar and nonpolar portions in the same molecule. The great variety of possible monomers produces micelles of widely differing surface composition and interior. The applications of amphiphilic substances ranges from A to Z (anesthesiology to zoology) and their micelle formation has important ramifications. Micelle formation occurs by cooperative association of monomers over a narrow concentration range known as the critical micellization concentration (c.m.c.).

Their special structural characteristics and their ability to solubilize otherwise insoluble substances render micelles both important and useful. Among other things, micelles act as good model systems and are excellent catalysts for a host of reactions. The many and varied applications of micelles and microemulsions are summarized in the opening chapter of these proceedings volumes.

The earlier research activity in the area of micelles was primarily carried out by colloid chemists but a glance at the literature on this topic attests that presently researchers from many disciplines are actively pursuing research in this area. The energy crisis has given an impetus for increased work on application of microemulsions and micellar solutions for tertiary oil recovery. Also, permanent storage of energy through light-driven redox reactions is shown to be feasible using micellar surfactant systems. It should be added that the availability of sophisticated instrumentation has been a boon in micellar research.

These proceedings volumes contain a comprehensive coverage of both theoretical and practical aspects of micellization, solubilization, and microemulsions. These volumes bring together the latest theoretical and experimental research activities being carried out by scientists from various disciplines, bring out clearly the interdisciplinary and multidisciplinary aspects and importance of the subject symposium. The topics covered include: history, applications, and prospects of micelles; thermodynamics and kinetics of micellization, application of fast reaction kinetics; theoretical developments in understanding monomer-micelle equilibria and stepwise aggregation; stepwise aggregation and the concept of c.m.c.; micelles of ionic and nonionic surfactants in aqueous and nonaqueous media; micelles as model systems; micelles and oil recovery; mixed micelles; application of spectroscopic techniques to understand mechanisms of reactions and interactions in micellar media; micellar catalysis of a variety of reactions; solubilization of polar and nonpolar substances; formation and structure of microemulsions and reactions in microemulsion media.

# PREFACE

These volumes - a product of the efforts of more than one hundred scientists representing many countries - attest to the brisk research activity taking place in the realm of micelles and microemulsions, and all signals indicate that this tempo will be continued. As we probe more into this fascinating area of micelles and microemulsions, more new vistas will emerge.

Acknowledgments. First of all I am thankful to Dr. G. S. Hartley for his presence at the Symposium. His presentation, "Micelles: Retrospect and Prospect" was the highlight of the program. Thanks are due to the officials of the Northeast Regional Meeting for sponsoring the event, to IBM Corporation for permitting me to organize the symposium and to edit these volumes. Thanks are due to our secretary, Ms. Julie Hrib, for helping with the correspondence typing. I am thankful to my wife, Usha, for sacrificing many hours which rightfully belonged to her, and to my daughter, Anita, and son, Rajesh, for being very nice kids and letting their Daddy work at home. The reviewers should be thanked for their many valuable comments on the manuscripts. The enthusiasm and cooperation of the contributors, particularly those from overseas, is sincerely acknowledged. The Petroleum Research Fund of the American Chemical Society should be thanked for partial travel assistance to three speakers.

K. L. Mittal

IBM Corporation
East Fishkill Facility
Hopewell Junction, New York  12533

CONTENTS OF VOLUME 1

### PART I: GENERAL PAPERS

The Wide World of Micelles
    K. L. Mittal and P. Mukerjee . . . . . . . . . . . . . . . 1

Micelles - Retrospect and Prospect
    G. S. Hartley . . . . . . . . . . . . . . . . . . . . . . 23

Micellization, Solubilization, and Microemulsions
    L. M. Prince . . . . . . . . . . . . . . . . . . . . . . 45

Biological Implications of Micelle Formation
    A. T. Florence . . . . . . . . . . . . . . . . . . . . . 55

Fluorescent Probes for Micellar Systems
    N. J. Turro, M. W. Geiger, R. R. Hautala and
    N. E. Schore . . . . . . . . . . . . . . . . . . . . . . 75

Micellar Solutions for Improved Oil Recovery
    V. K. Bansal and D. O. Shah . . . . . . . . . . . . . . . 87

Discussion . . . . . . . . . . . . . . . . . . . . . . . . . . 115

### PART II. THERMODYNAMICS AND KINETICS OF MICELLIZATION IN AQUEOUS MEDIA

Thermodynamics of Micellization of Simple Amphiphiles
in Aqueous Media
    C. Tanford . . . . . . . . . . . . . . . . . . . . . . . 119

Thermodynamics of Amphiphilar Aggregation into Micelles
and Vesicles
    E. Ruckenstein and R. Nagarajan . . . . . . . . . . . . . 133

Thermodynamics of Micelle Formation
    K. S. Birdi . . . . . . . . . . . . . . . . . . . . . . . 151

Size Distribution of Micelles: Monomer-Micelle Equilibrium, Treatment of Experimental Molecular Weight Data, the Sphere-to-Rod Transition and a General Association Model
    P. Mukerjee . . . . . . . . . . . . . . . . . . . . 171

Ionic Interactions in Amphiphilic Systems Studied by NMR
    B. Lindman, G. Lindblom, H. Wennerström and
    H. Gustavsson . . . . . . . . . . . . . . . . . . . 195

Errors in Micellization Enthalpies from Temperature Dependence of Critical Micelle Concentrations
    N. Muller . . . . . . . . . . . . . . . . . . . . . 229

The Nature of the Local Microenvironments in Aqueous Micellar Systems
    P. Mukerjee, J. R. Cardinal, and N. R. Desai. . . . . 241

The Influence of Hydrophobic Counterions on the Thermodynamics and Kinetics of Ionic Micelles
    H. Hoffmann, H. Nüsslein and W. Ulbricht. . . . . . . 263

On the Use of Chemical Relaxation Methods to Distinguish between True Micellization and Continuous Association
    R. Zana, J. Lang, S. H. Yiv, A. Djavanbakt and
    C. Abad . . . . . . . . . . . . . . . . . . . . . . 291

Kinetic Study on Micellization of Ionic Surfactants by Means of Relaxation Methods
    K. Takeda, T. Yasunaga and H. Uehara. . . . . . . . . 305

On the Kinetics of Redistribution of Micellar Sizes
    M. Almgren, E. A. G. Aniansson, S. N. Wall and
    K. Holmåker . . . . . . . . . . . . . . . . . . . . 329

Ultrasonic Relaxation Studies Associated with Monomer-Micelle Exchange Processes
    W. J. Gettins, J. E. Rassing and E. Wyn-Jones . . . . 347

The Size, Shape and Thermodynamics of Sodium Dodecyl Sulfate (SDS) Micelles using Quasielastic Light Scattering Spectroscopy
    N. A. Mazer, M. C. Carey and G. B. Benedek. . . . . . 359

Quasielastic Light Scattering Spectroscopic Studies of Aqueous Bile Salt, Bile Salt-Lecithin and Bile Salt-Lecithin-Cholesterol Solutions
    N. A. Mazer, R. F. Kwasnick, M. C. Carey and
    G. B. Benedek . . . . . . . . . . . . . . . . . . . 383

## CONTENTS OF VOLUME 1

Quasielastic Laser Spectrometry Studies of Pure Bile Salt
and Bile Salt-Mixed Lipid Micellar Systems
    R. T. Holzbach, S. Y. Oh, M. E. McDonnell, and
    A. M. Jamieson. . . . . . . . . . . . . . . . . . . . . 403

Discussion . . . . . . . . . . . . . . . . . . . . . . . . . . 419

### PART III. MICELLES IN NONAQUEOUS MEDIA

Micelles in Apolar Media
    H. F. Eicke . . . . . . . . . . . . . . . . . . . . . . 429

Aggregation of Surfactants in Hydrocarbons. Incompatibility
of the Critical Micelle Concentration Concept with Experimental Data
    A. S. Kertes. . . . . . . . . . . . . . . . . . . . . . 445

Mixed Non-Ionic Detergent Systems in Aqueous and Non-Aqueous
Solvents
    I. Lo, F. Madsen, A. T. Florence, J.-P. Treguier,
    M. Seiller and F. Puisieux. . . . . . . . . . . . . . . 455

Calorimetric Studies of the Micelle Formation in Solutions
of Sodium Octanoate and Water in Aliphatic Alcohols
    J. B. Rosenholm, P. J. Stenius and M.-R. Hakala . . . . 467

Discussion . . . . . . . . . . . . . . . . . . . . . . . . . . 479

Concluding Remarks . . . . . . . . . . . . . . . . . . . . . . 485

About the Contributors . . . . . . . . . . . . . . . . . . . . xvii

Subject Index. . . . . . . . . . . . . . . . . . . . . . . . . xxv

CONTENTS OF VOLUME 2

PART IV: REACTIONS IN MICELLES AND MICELLAR CATALYSIS IN AQUEOUS MEDIA

The Kinetic Theory and the Mechanisms of Micellar Effects on Chemical Reactions
    K. Martinek, A. K. Yatsimirski, A. V. Levashov and I. V. Berezin . . . . . . . . . . . . . . . . . . . . . 489

A General Kinetic Theory of Rate Enhancements for Reactions between Organic Substrates and Hydrophilic Ions in Micellar Systems
    L. S. Romsted . . . . . . . . . . . . . . . . . . . . . 509

Laser Photolysis Studies of Photo Redox Processes in Micellar Solution
    M. Grätzel . . . . . . . . . . . . . . . . . . . . . . . 531

Radiation-Induced Redox Reactions in Micellar Solutions
    A. J. Frank . . . . . . . . . . . . . . . . . . . . . . 549

Radiation-Induced Processes in Nonionic Micelles
    K. Kalyanasundaram and J. K. Thomas . . . . . . . . . . 569

Radiation-Induced Peroxidation in Fatty Acid Soap Micelles
    L. K. Patterson and J. L. Redpath . . . . . . . . . . . 589

Bifunctional Micellar Catalysis
    R. A. Moss, R. C. Nahas, and S. Ramaswami . . . . . . . 603

The Use of Phase Transfer Catalysts with Emulsion and Micelle Systems in Electro-Organic Synthesis
    T. C. Franklin and T. Honda . . . . . . . . . . . . . . 617

The Catalytic Role of Micelle-Bisulfite Complexation in Vinyl Polymerization
    O.-K. Kim . . . . . . . . . . . . . . . . . . . . . . . 627

Discussion . . . . . . . . . . . . . . . . . . . . . . . . . . 645

## PART V: REACTIONS IN MICELLES AND MICELLAR CATALYSIS IN NONAQUEOUS MEDIA

Some Kinetic Studies in the Reversed Micellar System—
Aerosol OT (Diisooctyl Sulfosuccinate)/$H_2O$/
Heptane Solution
    M. Wong and J. K. Thomas. . . . . . . . . . . . . . . . . . 647

Catalysis by Cations in Cores of Non-Aqueous Micelles
    F. M. Fowkes, D. Z. Becher, M. Marmo, C. Silebi
    and C. C. Chao. . . . . . . . . . . . . . . . . . . . . . . . 665

Solubilization and Catalysis of Polar Substances in
Nonaqueous Surfactant Solutions
    A. Kitahara and K. Kon-no . . . . . . . . . . . . . . . . . 675

Ligand Exchange Reactions of Hemin and Vitamin $B_{12a}$ in
the Presence of Surfactants in Water and in
Nonpolar Solvents
    J. H. Fendler . . . . . . . . . . . . . . . . . . . . . . . . 695

Discussion . . . . . . . . . . . . . . . . . . . . . . . . . . . . 711

## PART VI: MICROEMULSIONS

Theory for the Phase Behavior of Microemulsions
    M. L. Robbins . . . . . . . . . . . . . . . . . . . . . . . . 713

Stability, Phase Equilibria, and Interfacial Free
Energy in Microemulsions
    E. Ruckenstein. . . . . . . . . . . . . . . . . . . . . . . . 755

Light Scattering of a Concentrated W/O Micro Emulsion;
Application of Modern Fluid Theories
    A. A. Caljé, W. G. M. Agterof and A. Vrij . . . . . . . . 779

Microemulsions Containing Ionic Surfactants
    S. Friberg and I. Buraczewska . . . . . . . . . . . . . . 791

Interactions and Reactions in Microemulsions
    R. A. Mackay, K. Letts, and C. Jones. . . . . . . . . . . 807

Discussion . . . . . . . . . . . . . . . . . . . . . . . . . . . . 817

# CONTENTS OF VOLUME 2

## PART VII. GENERAL PAPERS

Mixed Micelles of Methyl Orange Dye and Cationic Surfactants
  R. L. Reeves and S. A. Harkaway . . . . . . . . . . . . 819

Anionic Surfactant Complexes with Charged and Uncharged
Cellulose Ethers
  E. D. Goddard and R. B. Hannan. . . . . . . . . . . . . 835

Proposal for a New Theory of Molecular Transport across
Membranes: Implications for Lung Gas Transference
  B. Ecanow, B. H. Gold, R. Balagot, and R. S. Levinson . 847

Interfacial Tension Minima in Two-Phase Micellar Systems
  E. Franses, M. S. Bidner and L. E. Scriven. . . . . . . 855

Equilibrium Bicontinuous Structures
  L. E. Scriven . . . . . . . . . . . . . . . . . . . . . 877

Intramacromolecular Micelles
  U. P. Strauss . . . . . . . . . . . . . . . . . . . . . 895

Solubilization by Nonionic Surfactants in the HLB-
Temperature Range
  S. Friberg, I. Buraczewska, and J. C. Ravey . . . . . . 901

The Effect of Lysoplasmalogen on Some Physical Properties
of Dipalmitoyllecithin Bilayers: A Fluorescent Probe
Study
  D. A. N. Morris and J. K. Thomas. . . . . . . . . . . . 913

Concluding Remarks . . . . . . . . . . . . . . . . . . . . 927

About the Contributors . . . . . . . . . . . . . . . . . . 931

Subject Index . . . . . . . . . . . . . . . . . . . . . . 939

# Part I

## General Papers

# THE WIDE WORLD OF MICELLES

K. L. Mittal and P. Mukerjee
International Business Machines Corporation
Poughkeepsie, New York  12602  and
School of Pharmacy, University of Wisconsin
Madison, Wisconsin  53706

Following a brief introduction to the field of micelles, the relation between the role of amphipathic (amphiphilic) detergent-like monomers in aqueous solutions in surface-chemical and adsorption (binding) processes, and in self-association processes leading to the formation of micelles is pointed out. The meaning of the critical micellization concentration (c.m.c.) and the significance of the cooperativity of self-association when monomers show different patterns of self-association are discussed. A general method is indicated by which the extent of cooperativity at an apparent c.m.c. value and its validity can be tested. Examples are shown from some fluorine nuclear magnetic resonance data. The involvement of micelles in numerous technical and biological processes, as also their use as 'models' for diverse colloidal, interfacial, and membrane systems in fundamental research are reviewed with special attention to recent activity. The nature of microemulsions and their use in technical processes are discussed briefly. Several controversial issues and areas of fruitful research involving micelles are pointed out.

## INTRODUCTION

The multi- and interdisciplinary character of the subject of micelles is well attested by the great variety in the backgrounds and interests of researchers actively engaged in the area of micelles. A glance at the Tables of Contents of these Symposium volumes and the Biographical Data of the participants would provide sufficient justification for our title: The world of micelles is indeed wide.

The present article is in the form of a brief introduction to the field of micelles and an overview of current interests. However, the field is extremely large and diverse and no attempt has therefore been made to be comprehensive in a short article such as this. We have used several contributions to this Symposium as readily available examples of ongoing current activity. The field of micelles has been developing rapidly and branching out in unexpected directions in recent years. As might be expected for such an area, serious differences of opinion exist on many topics, including the precise meaning to be attached to the word *micelle*.

The word *micelle* is used in a variety of ways. According to the Webster Dictionary,[1] for example, a micelle is a unit of structure built up from polymeric molecules or ions as (a) an ordered region in a natural or synthetic fiber (as of cellulose, silk or viscose rayon), (b) a highly associated particle of a colloidal solution (colloidal micelles of soaps and detergents - J. W. McBain), (c) an organic colloidal particle ranging in size from one micron to one millimicron and found in coal and some shales. In this paper and in this Symposium, we are concerned exclusively with the second definition, which was used for the first time in 1913 by J. W. McBain[2] to represent aggregates of amphiphilic electrolytes in aqueous solutions.

It is interesting to note that in the early stages the concept of micelles met with some resistance before acceptance, as can be realized from the following quotation from McBain:[3]

> So novel was this finding that when in 1925 some of the evidence for it was presented to the Colloid Committee for the Advancement of Science in London, it was dismissed by the chairman, a leading international authority, with the words, 'Nonsense, McBain.'

Shortly thereafter, Hartley commenced his classical researches which laid the foundation for the field as it exists today. His 1936 monograph,[4] the first in the field, still continues to be a mine of ideas, and merits repeated and careful reading by all investigators in the field. Besides those of McBain and Hartley, the names of Ekwall, Harkins, and Debye must be added to the list of pioneers who were followed by Mysels, Stigter, Overbeek and many others still active in the field. A measure of the growth of the field can be obtained from the existing critical micellization concentration (c.m.c.) data. While the concept of the c.m.c. was new and unfamiliar some 45 years ago, a critical compilation[5] up to 1966 resulted in a listing of some 4,600 c.m.c. values on 721 compounds, even after the exclusion of certain classes of systems from consideration.

The involvement of micelles in practical problems such as detergency has been known for a long time. However, the fundamental investigations on micelles in the early days were concerned mainly with the structure and properties of micelles as association colloids, the monomer-micelle equilibrium and solubilization by micelles.[6-11] In recent years, the use of micelles as "model" systems[12] for a variety of problems related to monolayers, colloidal systems in general, proteins, enzymes and membranes as also the monomer-micelle equilibrium system as a "model" system for hydrophobic interactions[12] which are of central interest in biology have expanded the range of interest in micellar phenomena greatly. An additional impetus has been the recognition of the important role of micelles and mixed micelles in physiological systems.[13] The scope and range of current activity and interest in micellar systems can be gauged by a sampling of some monographs, review articles, and symposia proceedings of the last ten years or so dealing with micellar systems.[12-31]

## SURFACE ACTIVITY AND MICELLE FORMATION IN AQUEOUS MEDIA

The fundamental characteristic of micelle forming monomers is their amphiphilicity or amphipathicity, i.e. the presence in the same molecule or ion of polar and nonpolar moieties. This characteristic of amphipaths has been aptly described by Hartley[32] as their schizophrenia, and he has commented upon the use of the words amphiphilicity and ampipathicity. For aqueous systems, the polar portion is hydrophilic, and the nonpolar is hydrophobic.

An immense variety of amphipathic monomers are known.[5] The usual classification of non-ionic, anionic, cationic or zwitterionic surfactants depends upon the nature of the head groups. The ionic surfactants are, of course, always associated with counterions and their properties are often modified significantly by different counterions. The hydrophobic part can be of different lengths, can contain unsaturated portions or aromatic moieties, can be partly or completely halogenated as in fluorocarbon surfactants, and can be branched or consist of two or more chains. Many surfactants, such as the physiologically important bile salts, have rigid structures, as opposed to the flexible chains of soaps and detergents.

The amphipathic molecules have a tendency to collect at any interface where the hydrophobic groups can be partially or completely removed from the contact of water and the hydrophilic gorups can remain wetted. These general tendencies account for their surface activity (Figure 1), i.e. the ability to adsorb to air-water or oil-water interfaces and to surfaces of hydrophobic solids such as carbon or to macromolecules such as proteins. The same dual tendencies and the built-in asymmetry of the molecules allows them to form organized structures such as soap films and bi-layers (Figure 1). The formation of lipid membranes by relatively insoluble amphipaths involves similar forces and structural features also.

Micelle formation by amphipaths is the result of the same dual tendencies of the molecules mentioned above. These cause them to undergo self-association in solution to produce aggregates in which the organic moieties are put together in close juxtaposition so that the total contact area of the hydrophobic groups of the solute molecules with water is reduced. For detergents and other flexible chain surfactants, the process leads to the formation of micelles with the schematic structure shown in Figure 1: The hydrocarbon chains form a liquid-like core with the polar groups remaining exposed to the water at the surface.[4]

## CRITICAL MICELLIZATION CONCENTRATION

The c.m.c. is of central significance in micellar solutions.[5] Because of its importance and occasional misuse,[33] and because of numerous problems connected with the interpretation of data regarding the self-association of hydrophobic molecules of different structures and self-association in non-aqueous media, the c.m.c. is discussed here in some detail, and a general recommendation is made about the treatment of data based on an early suggestion by Hartley[4] which should settle some controversies. Numerous investigations have shown that micelle formation begins or becomes appreciable over a narrow range of concentrations. For such systems, the IUPAC Manual of Symbols and Terminology, Appendix II, Part I[35] makes the following statements about micelles and c.m.c.'s:

> Surfactants in solution are often association colloids, that is, they tend to form <u>micelles</u>, meaning aggregates of colloidal dimensions existing in equilibrium with the molecules or ions from which they are formed.

> There is a relatively small range of concentrations separating the limit below which virtually no micelles are detected and the limit above which virtually all additional surfactant forms micelles. Many properties of surfactant solutions, if plotted against the concentration, appear to change at a different rate above and below this range. By extrapolating the loci of such a property above and below this range until they intersect, a value may be obtained known as the <u>critical</u>

micellization concentration (critical micelle concentration), symbol $c_m$, abbreviation c.m.c. As values obtained using different properties are not quite identical, the method by which the c.m.c. is determined should be clearly stated.

The "relatively small range of concentration" mentioned above which includes the c.m.c. is narrow enough, however, to allow c.m.c. values to be determined quite precisely in many cases.[5] This sharpness of the c.m.c. region in flexible chain systems is due to a pronounced cooperativity of self-association which makes large aggregates containing many monomers (roughly 20 or more) much more stable than smaller species.[33,34,36,37]

The c.m.c. is one of the most easily obtainable and useful quantitative results about aqueous flexible chain surfactant systems.[5] It has numerous applications for the thermodynamics of micelle formation and for characterizing micellar solutions.[5,33] Nearly all of these depend upon attaching the proper significance and weight to the critical nature of the c.m.c. Indeed, the c.m.c. is sharp enough to form the basis of a two-phase model of micellar systems, which seems to overemphasize the critical nature of the c.m.c.[12]

One of the most practical uses of the c.m.c. is in the calculation of monomeric and micellar concentrations.[5,33] In many cases, the approximations that no micelles are present below the c.m.c. and above the c.m.c., the change in monomer concentration is small enough to be negligible, are reasonably valid.[33,37] As micelles are generally not very surface active, a corresponding practical approximate significance of the c.m.c. is that it is the equilibrium concentration where surface chemistry ends and colloid chemistry begins. Figure 1 shows how, from this point of view, the formation of micelles is a process that competes with any other adsorption or binding process in which the monomer participates.

Self-association of hydrophobic molecules is not, however, limited to that of flexible chain surfactants. Rigid, aromatic, heteroaromatic (including many dyes and drugs) as also alicyclic molecules (including bile salts) show self-association also. However, the patterns of self-association for such substances may be quite different from that of flexible chain surfactants,[33,34,36] depending upon the relative values of the stepwise association constants. When the pattern of self-association does not include much cooperativity, a c.m.c. which can be determined precisely and which can be used for the purposes mentioned earlier in the case of flexible chain surfactants is not available. Unfortunately, the operational procedure of determining the c.m.c. from a change in curvature of some solution properly plotted against some function of concentration can give rise to apparent c.m.c. values, even for associating systems involving only dimerization.[33] Such c.m.c. values can lead to seriously wrong conclusions regarding monomer concentration (activity) or the thermodynamics of monomer-micelle equilibrium. The

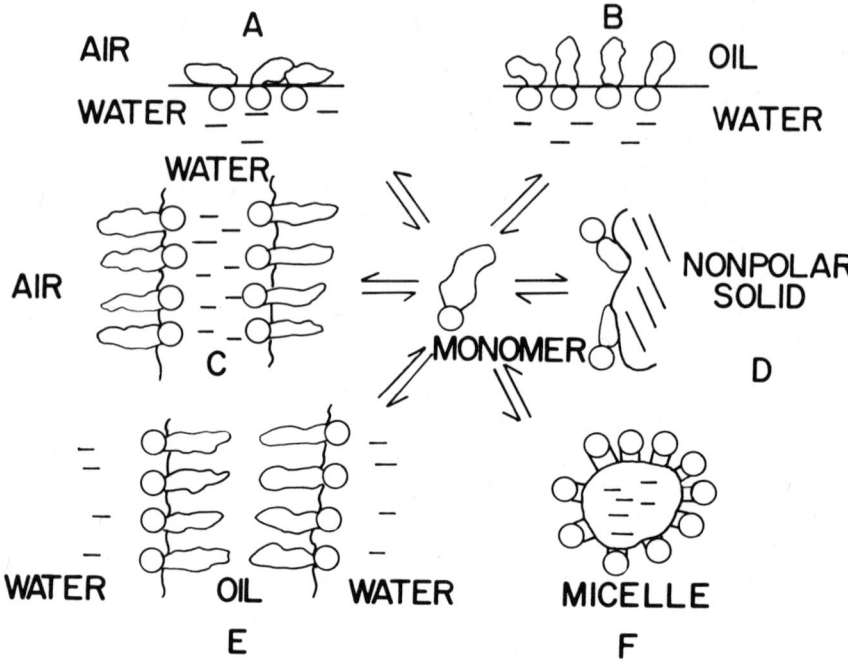

Figure 1. Schematic representation of surface activity and micelle formation. (a) air-water interface; (b) oil-water interface; (c) soap films, formation of organized structures); (d) adsorption to nonpolar solids (as also to polymers and proteins); (e) formation of a bilayer (a model for membranes); and (f) micelle.

problems of interpretation are sometimes compounded by an extension of the definition of the micelle to include small aggregates and even the dimer.[38]

To investigate the significance of an apparent c.m.c., a general procedure originally suggested by Hartley[4] and used by him in demonstrating how conductance data can show the sharp changes occurring at the c.m.c. in plots of differential conductance against concentration, appears to be extremely useful. The procedure has been adapted for solubilization data to indicate the difference between the highly cooperative self-association of sodium decane sulfonate and much less cooperative self-association in sodium cholate, a bile salt.[36] This method can be adapted for many types of data including surface tension,[39] rates of dialysis,[40] and NMR chemical shifts.[41] In the last case, the usual shift ($\delta$) data are multiplied by the total concentration, C, and the C$\delta$ data are then differentiated. Figure 2 shows an example of such a treatment of the $^{19}$F chemical shift data of the flexible chain compound sodium perfluorooctanoate in water and trifluoropromazine hydrochloride in 0.05 M NaCl. For the former system $\Delta(C\delta)/\Delta C$ from consecutive data points are plotted against the mean C. For the latter, a graphical differentiation of C$\delta$ values was made. Both systems show evidence of considerable self-association by this and other methods. This analysis, however, brings out clearly the cooperative nature of the self-association of the flexible chain surfactant reflected in the sharp transition region with the midpoint at about 0.032 M, which is close to the c.m.c. obtained by other methods. For the trifluoropromazine system, however, although apparent c.m.c. values at 0.01 to 0.02 M can be obtained from the chemical shift data and other methods, the present treatment shows the absence of any strong cooperativity of self-association, as is to be expected from the rigid structure of the molecule.[33] Treatment of NMR and other kinds of data in this manner should resolve any question about the significance of the c.m.c. in any aqueous or non-aqueous system.

It is interesting to note that certain patterns of self-association, e.g. the open-ended, non-cooperative, stepwise self-association,[33] can lead to the formation of aggregates with high aggregation numbers which will qualify as micelles according to the IUPAC definition,[35] without, however, the system showing a c.m.c. Dye aggregation, for example, is sometimes extensive enough to cause gel formation.[8]

## STRUCTURE OF MICELLES

The basic features of the roughly spherical micelle in aqueous solutions were worked out by G. S. Hartley in his pioneering work in the 1930's and 1940's. Figure 3 shows a schematic sketch of an ionic micelle including some features discussed at greater lengths elsewhere in this Symposium.[42] The micelle is a compact roughly

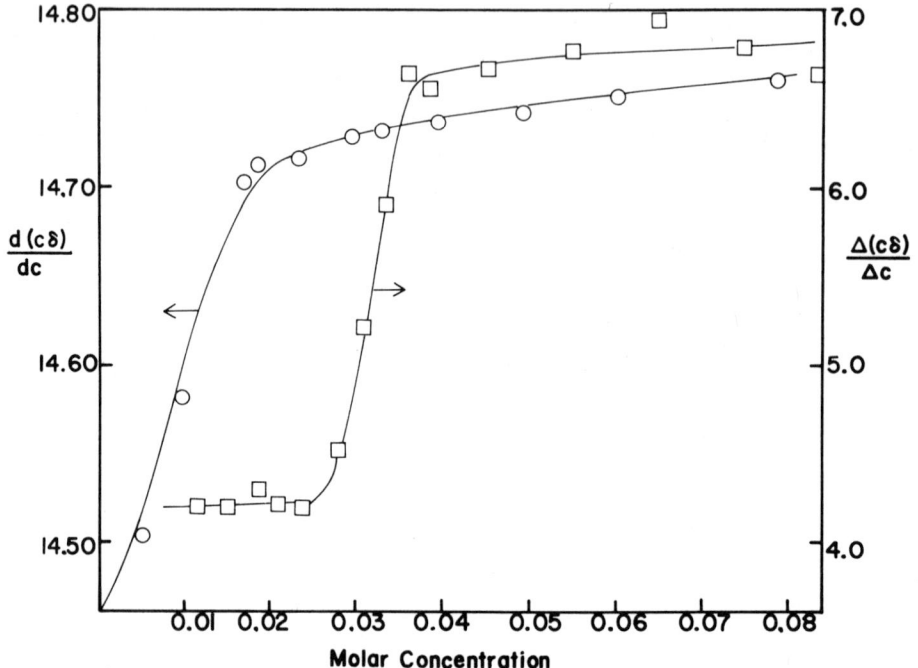

Figure 2.  $^{19}$F NMR data on (□) sodium perfluorooctanoate in water, and (O) trifluoropromazine hydrochloride in 0.05 M NaCl.

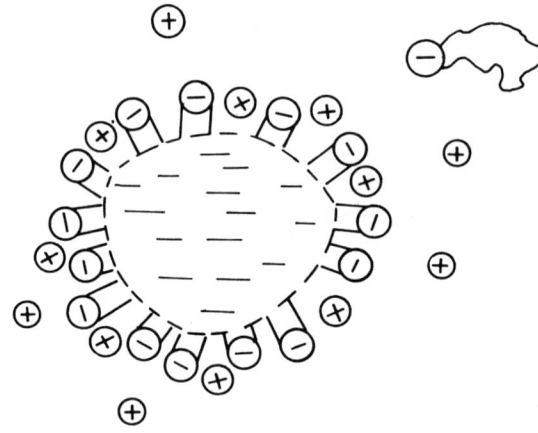

Figure 3.  Schematic representation of an aqueous ionic micelle, including representation of a liquid-like core, a dynamic structure, rough interface, counterion penetration and binding, possible wetting of methylene groups attached to the polar groups, and the diffuse double layer.

spherical body with a liquid-like hydrocarbon core. The polar head groups are at the surface which is rough.[43] A large fraction of the counterions is bound to the surface and forms part of the kinetic micelle. The remaining counterions form the diffuse double layer. The equilibrium between micelles and monomers is generally rapid. One or more methylene groups attached to the polar groups may be wet. These and many other properties of micelles have been extensively reviewed.[4,14-17,29,42]

In many surfactant solutions, even in dilute systems, very large, anisotropic micelles form. These appear to be worm-like, i.e. they are cylindrical with some flexibility. These systems are being studied with a variety of methods.[44-47]

In concentrated solutions of surfactants, a wide variety of structures and mesophases have been identified. Figure 4 shows schematically a number of these structures.[25b]

## MICELLE FORMATION IN NON-AQUEOUS MEDIA

Compared to aqueous solutions, solvophobic interactions of surfactants in non-aqueous media of even relatively high polarity such as glycerol or formamide tend to be much weaker.[48] In highly non-polar solvents[49-58] the polar groups of the amphipathic molecules become solvophobic and, in such media, aggregates form in which the polar groups form the core. Such species are often referred to as inverse, reverse, or reverted micelles. A possible structure of a reverse micelle in a nonpolar medium is shown in Figure 5. Theoreies and models for aggregation in nonpolar media were based largely on models derived from aqueous micelles until recently in terms of a monomer-micelle equilibrium and a c.m.c. However, the recognition of the low degree of aggregation observed in many cases and the application of step-wise self-association models of various kinds have now raised serious questions about the usefulness of a micellar model for such systems. The aggregation properties of surfactants in non-polar media are often altered markedly by the presence of traces of water or other additives. It should be noted that the general problems of the different patterns of self-association and the need for cooperativity for the existence of a c.m.c. are the same for non-aqueous media as for aqueous media,[33,34] and the previously proposed test of the extent of cooperativity at the c.m.c. should be as useful for nonpolar media as for aqueous media.

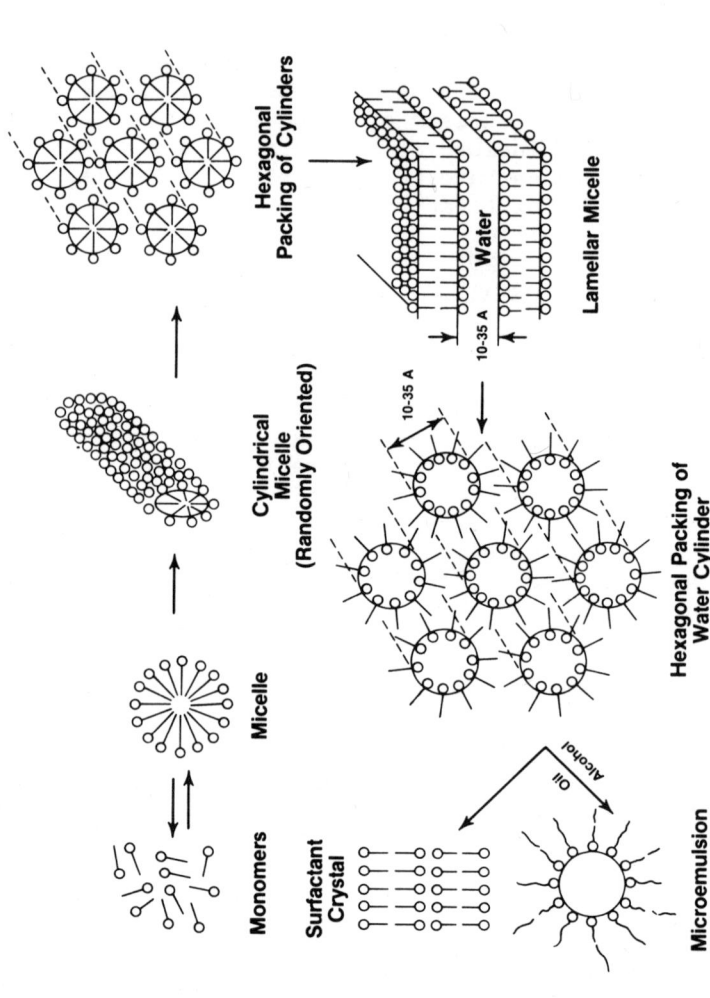

Figure 4. Different structures of micelles and mesophase (from reference 25b).

## SOLUBILIZATION

One of the most important properties of micellar systems is their ability to solubilize a variety of species. For aqueous micelles, solubilization is closely related to the hydrophobic and amphipathic properties of the solubilizate. Different sites of solubilization and orientations may be involved depending upon the structure of the solubilizate.[11,14,17,42] Figure 6 shows some possible sites for aqueous micellar systems of highly hydrophobic (A), amphipathic (B) and slightly polar (C) solubilizates as also relatively polar solubilizates in non-ionic micelles for which the voluminous polyethylene oxide mantle (shell) acts as a locus for solubilization (D). It should be noted that solubilization sites are not fixed: There is a rapid equilibrium between various possible sites as also between the solubilized state and the free state in the aqueous medium of the solubilized species. In non-polar media, polar substances tend to be solubilized by the aggregates.

## APPLICATION OF MICELLES

The role of micelles in practical systems derives from (a) their competition with monomers in adsorption processes, (b) their use as a "reservoir" of monomers, (c) their ability to solubilize materials and affect their physical and chemical properties, and (d) their usefulness as "model" systems.

### Micelles as Model Systems

The usefulness of the micelle as a "model" colloid for studying many problems of general interest in colloid science has been pointed out.[12] The primary reasons are the thermodynamic stability and reproducibility of many micellar systems, the simplicity of the structure of the micelle and of its surface composition. Ionic micelles are especially attractive for studying electrical interactions.[43,59,60,61] The compact micelles allow detailed examination of many theories of the behavior of spherical colloids. Because of their small radius, the characteristic Debye thickness of the double layer is often of the same order as the micelle size, so that electrical interactions are "felt" over large volumes.

Theformation of micelles is one of the most spectacular examples of "hydrophobic bonding" or hydrophobic interactions generally. Micellar systems have proved to be very useful for studying the factors involved in hydrophobic interactions.[12,29]

The micelle-solution interface incorporates the dielectric asymmetry and microenvironments typical of many interfacial systems such as monolayers, membrane surfaces, adsorbed layers (Figure 1). The micelle can be looked upon as a monolayer curved on itself.

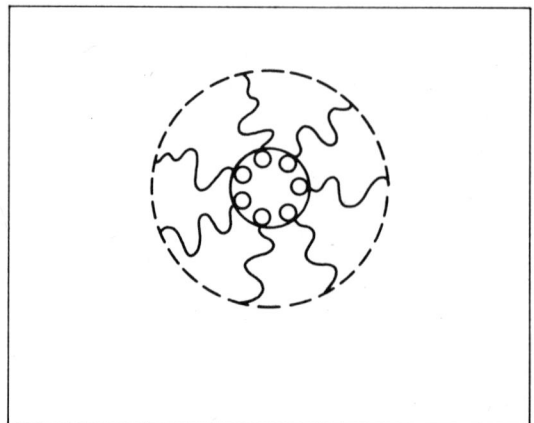

Figure 5. Schematic representation of a possible reverse micelle in nonpolar media.

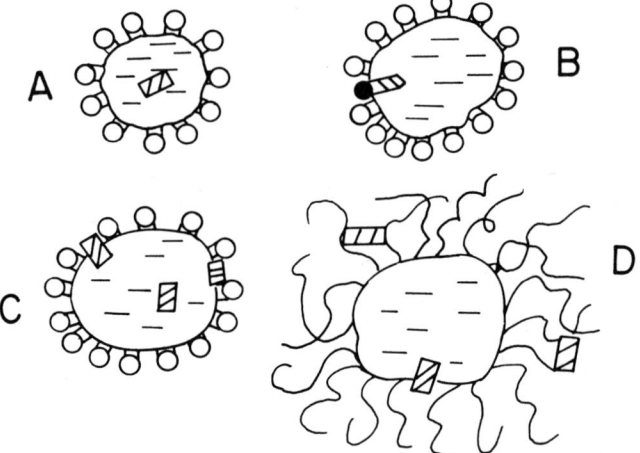

Figure 6. Schematic representation of solubilization in aqueous micelles. (a) nonpolar solubilizate, (b) amphipathic solubilizate, (c) slightly polar solubilizate, and (d) polar solubilizate in nonionic micelle showing the polyoxyethylene shell as a locus of solubilization.

Micellar systems are thus extremely useful for posing and answering many questions regarding the microenvironments encountered in interfacial systems of other kinds and membranes.[42] The major advantage of micellar systems in this respect,[12,63] is the relatively easy application of a variety of experimental techniques useful for bulk solutions but difficult to apply to actual bulk surfaces or membranes. Various spectroscopic methods are particularly useful.

## Biological or Physiological Implications of Micelles

The use of micelles as model systems for membranes or some aspects of enzymes has very important biological implications. The basic structural similarity of membranes and micelles has been referred to (see Figure 1). In recent years the importance of micelles and mixed micelles in the transport and absorption of lipids and as solubilizers for cholesterol in physiological systems has been recognized.[13,26,38] Self-association and mutual association of hydrophobic molecules generally are of great importance for many physiological molecules, food constituents and drugs.

## Reaction Catalysis in Micelles

Micellar media have been extensively used to affect rates of numerous organic and inorganic reactions. There are many reviews available on this subject.[54,64-68] It should be noted that first studies of micellar effects upon chemical reactions involved equilibria. Ionic micelles were found to have striking effects upon the protonation of indicator dyes.[62,70] The study of the kinetic effects of micellar solubilization began at a later date.[71] The catalysis or inhibition of solubilized species involve many kinds of interactions and may vary with the nature of the surfactant. Table 1 gives some examples of micellar catalysis and inhibition.[67] A proper choice of surfactant can lead to rate increases of 5- to 1,000-fold (in some cases much higher), compared with the same reactions in the absence of the surfactant. The motivation for studying chemical reactions in micellar systems is based on:[64] (a) the possibility of a better understanding of the factors which influence the rates and mechanisms of organic reactions; (b) the possibility of obtaining better insight into mechanisms of enzymatic catatlysts and (c) the opssible uses of micellar systems for the purpose of organic synthesis.

Table I. Micellar Catalysis and Inhibition (from reference 67).

| | | SURFACTANT | | |
| Substrate | Reagent | Cationic | Anionic | Nonionic |
|---|---|---|---|---|
| | | CTABr | NaLS | ethylene oxide |
| $RCO.O\text{-}C_6H_4\text{-}NO_2$ | $OH^-$ | catalyze | inhibit. | no effect |
| $(PhO)_2PO.O\text{-}C_6H_4\text{-}NO_2$ | $OH^-, F^-$ | catalyze | inhibit. | inhibit. |
| 1-Cl-2,4-dinitrobenzene | $OH^-$ | catalyze | inhibit. | no effect |
| $PhC(OMe)_3$ | $H_3O^+$ | inhibit | catalyze | slightly inhibit |
| $CH_2=CHPh \cdot OPO(OEt)_2$ | $H_3O^+$ | | catalyze | |

The kinetic studies performed so far in solutions of micelle forming surfactants can be grouped into two categories: The first includes those cases in which a surfactant micelle acts only as a medium for the reaction. The second class includes reactions in which the surfactant participates directly either as a catalyst or as a substrate.[64]

Efforts to enhance the effectiveness of micellar catalysis have included investigation of multi-charged[72] and functional micelle forming surfactants[69,73-76a] as well as attachment to macromolecular backbones.[76b-78]

Although most of the work on micellar catalysis has involved organic reactions, recently Chao and Morawetz[79] found that the reaction of mercuric ions with $CO(NH_3)_2Cl_2^{+2}$ is catalyzed up to 140,000-fold by micelles of sodium hexadecyl sulfate. Such studies may help unravel the catalytic role of metal ions in metalloenzymes.

Micellar catalysis in nonpolar media involves the solubilization of reactants (polar) in the hydrophilic portions of reversed micelles.[54] Properties of the polar cavity may be strongly affected by small amounts of water which is expected to behave very differently from ordinary water because of extensive binding and orientation effects.

One of the most spectacular catalytic effects observed so far has been for the aquation of the tris (oxalato) chromium (III) anion.[80] The aquation is up to 5.4 million times faster in a nonpolar medium containing octadecyltrimethyl ammonium tetradecanoate than in bulk water.

The cationic core of inverse micelles of metal sulfonates in hydrocarbons has a catalytic microenvironment like a molten salt but it is easily available to hydrocarbon soluble reactants. Fowkes et al.[81] have studied a few acid-catalyzed reactions and the rate enhancement seems to correlate well with the acidity of the cation. Water has a pronounced effect on the catalytic activity.

## Industrial, Technological, and Other Applications

A complete catalog of all the industrial and technological applications of micelles will be prohibitively large, and only a few selected examples are given. Most of the applications of micelles depend upon their ability to solubilize. Detergency is an example where micelles are quite important. In dry cleaning, solubilization in reverse micelles plays a vital role in removing polar dirt from clothes. Micelles in nonaqueous media are used in motor oils to solubilize corrosive oxidation products and to prevent them from reacting with engine parts. Solubilized systems are used in many agricultural sprays and dyeing media. Micellar solutions have been found very effective in removal of odor causing molecules from factories of food packaging plants. Micellar solubilization may be important for "deinking" in the paper industry. Micellar solubilization may be involved in photographic processes, and microencapsulation via interfacial polymerization. In the emulsion polymerization process, the polymerization process involves the solubilization of the monomers in the micellar core. Micellar media provide opportunity for practical organic electro-synthesis.[82] In electro-organic synthesis, it is vital to maintain a high conductivity of the medium as also a good solubility of organic species; micellar systems allow this combination.

The potential use of micellar solubilization in many commercial applications appears to be wide. It should be noted that micelles may play a very important role in the case of surfactant type corrosion inhibitors.[83] It has been found that certain inhibitors are effective only below the c.m.c., and others are effective only above their c.m.c., whereas some surfactants show no inhibition below or above the c.m.c. Micellar adsorption has been implied for those inhibitors which are effective only above the c.m.c. However, the concept of micellar adsorption is open to debate.[84]

## Energy Related Applications of Micelles

Because of the energy crisis, there has been increasing work on applications of micellar solutions as injection fluids for enhanced oil recovery. There are four principal enhanced oil recovery techniques now considered feasible for the production of crude oil:[85] thermal recovery methods, hydrocarbon miscible fluids, carbon dioxide floods, and water miscible micellar floods. Water miscible flooding appears to be the most generally applicable enhanced recovery technique. A variety of names including micellar flooding, surfactant flooding, chemical flooding, and proprietary names have been applied to this technique. The basic aim is to eliminate (reduce to less than $10^{-3}$ dyne/cm) the interfacial tension between the injected and reservoir fluids so that the remaining oil is immobilized and collected toward the oil wells. There are certain other requirements which the micellar solutions should fulfill for these to be efficient in enhanced oil recovery. These have been reviewed.[86] Microemulsion (microemulsions are discussed later) flooding is another technique for enhanced or tertiary oil recovery. Healy and Reed[87] have discussed the physicochemical aspects of microemulsion flooding, which is another technique for enhanced oil recovery.

Recently,[88] micellar systems have been employed to facilitate photochemical conversion of light energy into chemical energy which is potentially applicable for storage of solar energy in a useful form. For instance, a system containing dodecyltrimethylammonium chloride and sodium carbonate with duroquinone solubilized in the micelles has been used to obtain photoinduced evolution of oxygen from water. The micelles solubilize duroquinone and prevent a back reaction, thus facilitating the reaction.

## MICROEMULSIONS AND THEIR APPLICATIONS

Although the title of this paper is "The Wide World of Micelles", a short discussion of microemulsions is in order, for microemulsions are also known as micellar emulsions, swollen micelles, and micellar solutions. A microemulsion is a clear, transparent, and stable system consisting of essentially monodisperse oil in water (o/w) or water in oil (w/o) droplets with diameters generally in the range of 10-60 nm. The volume fraction of the disperse phase can generally be varied over a fairly wide range (e.g. 20 to 80%). Microemulsions are transparent because of their small particle size. A microemulsion contains four components: two fluid phases, surfactant, and a cosurfactant; whereas micelles form in two-component systems. The formation, stabilization, structure and properties of microemulsions have been reviewed.[89,90] Prince has recently compared microemulsions with micelles.[91] Microemulsions have also been used for studying reactions at interfaces[92,93] including metal ligand interactions.[94]

Microemulsions undergo structural changes when the water to oil ratio is increased. The phase inversion region of the microemulsion consists of the lamellar liquid-crystalline structure which can be used as a model membrane to study the effect of local anesthetics on membranes.[95]

Friberg[96] has recently pointed out the potential industrial applications of microemulsions, which include: tertiary oil recovery; octane improvement; pollution abatement; chemical processing; lubricants, surface coatings, and pesticide and other bioactive formulations. The use of microemulsions in cosmetic formulations is important for their esthetic appeal.

## SUMMARY AND CONCLUSIONS

A number of the areas under investigation have been referred to. Novel synthesis of surfactants of specified properties, the use of methods for following fast reactions to micellar equilibria[97,98] and many others can be added to this list. Some of the controversial areas have also been referred to such as the concept of the c.m.c. in nonpolar media. It appears that the field of micelles is going to remain active for a long time, and the combination of the interdisciplinary interests and the interactions between fundamental investigations and diverse applications should produce synergistic effects for the better understanding of micellar systems as also new applications.

## REFERENCES

1. "Webster's Third New International Dictionary", G. & C. Merriam Company, Springfield, Massachusetts, 1965.
2. J. W. McBain, Trans. Faraday Soc. 9, 99 (1913).
3. J. W. McBain in "Colloid Chemistry", J. Alexander, Editor, Vol. 5, pp. 102-103, Reinhold Publishing Corp., New York, 1944.
4. G. S. Hartley, "Aqueous Solutions of Paraffin Chain Salts", Hermann, Paris, 1936.
5. P. Mukerjee and K. J. Mysels, "Critical Micelle Concentrations of Aqueous Surfactant Systems", NSRDS-NBS-36, Superintendent of Documents, U.S. Government Printing Office, Washington, D.C. 20402, 1971.
6. J. W. McBain, "Colloid Science", D. C. Heath, Boston, 1950.
7. A. E. Alexander and P. Johnson, "Colloid Science", Vol. II, Clarendon, Oxford, 1949.
8. H. R. Kruyt, Editor, "Colloid Science", Vol. II, Elsevier, New York, 1949.
9. P. A. Winsor, "Solvent Properties of Amphiphilic Compounds", Butterworths, London, 1954.

10. M. E. L. McBain and E. Hutchinson, "Solubilization and Related Phenomena", Academic Press, New York, 1955.
11. K. Shinoda, T. Nakagawa, B. Tamamushi, and T. Isemura, Editors, "Colloidal Surfactants", Academic Press, New York, 1963.
12. P. Mukerjee, Adv. Colloid Interface Sci., $\underline{1}$, 241 (1967).
13. A. F. Hofmann and D. M. Small, Ann. Rev. Medicine, $\underline{18}$, 333 (1967).
14. M. J. Schick, Editor, "Nonionic Surfactants", Marcel Dekker, New York, 1967.
15. E. Jungermann, Editor, "Cationic Surfactants", Marcel Dekker, New York, 1970.
16. K. Shinoda, Editor, "Solvent Properties of Surfactant Solutions", Marcel Dekker, New York, 1967.
17. P. H. Elworthy, A. T. Florence, and C. B. McFarlane, "Solubilization by Surface Active Agents and Its Application in Chemistry and the Biological Sciences", Chapman and Hall, London, 1968.
18. "Molecular Association in Biological and Related Phenomena", ACS Symposium Series No. 84, American Chemical Society, Washington, D.C., 1968.
19. C. Tanford, "The Hydrophobic Effect", Wiley, New York, 1973.
20. E. H. Cordes, Editor, "Reaction Kinetics in Micelles", Plenum Press, New York, 1973.
21. J. Fendler and E. Fendler, "Catalysis in Micellar and Macromolecular Systems", Academic Press, New York, 1975.
22. L. M. Prince and D. F. Sears, Editors, "Horizons in Surface Science", Academic Press, New York, 1975.
23. K. L. Mittal, Editor, "Colloidal Dispersions and Micellar Behavior", ACS Symposium Series No. 9, American Chemical Society, Washington, D.C., 1975.
24. E. Wyn-Jones, Editor, "Chemical and Biological Applications of Relaxation Spectroscopy", Reidel Publications, Holland, 1975.
25. a. L. M. Prince, Editor, "Microemulsions: Theory and Practice", Academic Press, New York, in press.
    b. D. O. Shah, Chem. Eng. Education, in press.
26. M. C. Carey and D. M. Small, Amer. J. Med., $\underline{49}$, 590 (1970).
27. Symposium on Micellar Systems as Models for Radiation Biology, held at the 23rd Annual Scientific Meeting of the Radiation Research Society, Miami Beach, Florida, May 11-15, 1975.
28. Symposium on Micellization held at the Brunel University, England, September 23-24, 1976.
29. G. C. Kresheck, in "Water, a Comprehensive Treatise", F. Franks, Editor, Plenum Press, New York, 1975.
30. J. E. Gordon, "The Organic Chemistry of Electrolyte Solutions", John Wiley, New York, 1975.
31. M. N. Jones, "Biological Interfaces", Elsevier, New York, 1975.
32. G. S. Hartley, this Symposium volume.
33. P. Mukerjee, J. Pharm. Sci., $\underline{63}$, 972 (1974).

34. P. Mukerjee, in "Physical Chemistry: Enriching Topics from Colloid and Surface Science", IUPAC, H. van Olphen and K. J. Mysels, Editors, Theorex, La Jolla, California, 1975.
35. Manual of Symbols and Terminology, Appendix II, Part I, International Union of Pure and Applied Chemistry, Pure Appl. Chem., $31$, No. 4, 612 (1972).
36. P. Mukerjee and J. R. Cardinal, J. Pharm. Sci., $65$, 882 (1976).
37. P. Mukerjee, this Symposium volume.
38. D. M. Small in "The Bile Acids", P.P. Nair and D. Kritchevsky, Editors, Chapter 8, Plenum Press, New York, 1971.
39. P. H. Elworthy and K. J. Mysels, J. Colloid Interface Sci., $21$, 331 (1966).
40. M. Abu-Hamdiyyah and K. J. Mysels, J. Phys. Chem., $71$, 418 (1967).
41. P. Mukerjee and A. Y. S. Yang, unpublished work (1975).
42. P. Mukerjee, J. R. Cardinal and N. Desai, this Symposium volume.
43. D. Stigter and K. J. Mysels, J. Phys. Chem., $59$, 45 (1955).
44. P. Mukerjee, J. Phys. Chem., $76$, 565 (1972).
45. N. A. Mazer, G. B. Benedek and M. Carey, J. Phys. Chem., $80$, 1075 (1976).
46. R. J. M. Tausk, C. Oudshoorn and J. Th. G. Overbeek, Biophys. Chem., $2$, 53 (1974).
47. E. W. Anacker and H. M. Ghose, J. Am. Chem. Soc., $90$, 3161 (1968).
48. A. Ray, Nature, $231$, 313 (1971).
49. C. R. Singleterry, J. Amer. Oil Chem. Soc. $32$, 446 (1955).
50. N. Pilpel, Chem. Rev. $63$, 221 (1963).
51. F. M. Fowkes in "Solvent Properties of Surfactant Solutions", K. Shinoda, Editor, p. 65, Marcel Dekker, New York, 1967.
52. A. Kitahara in "Cationic Surfactants", E. Jungermann, Editor, p. 289, Marcel Dekker, New York, 1970.
53. A. Kitahara, in "Colloidal Dispersions and Micellar Behavior", K. L. Mittal, Editor, ACS Symposium Series No. 9, pp. 225-232, American Chemical Society, Washington, D.C., 1975.
54. J. H. Fendler, Accounts Chem. Res. $9$, 153 (1976).
55. A. S. Kertes and H. Gutman, in "Surface and Colloid Science", E. Matijevic, Editor, Vol. 8, pp. 193-295, Wiley, New York, 1975.
56. A. S. Kertes, this Symposium volume.
57. H. F. Eicke and H. Christen, J. Colloid Interface Sci., $46$, 417 (1974).
58. H. Christen and H. F. Eicke, J. Phys. Chem., $78$, 1423 (1974).
59. D. Stigter, J. Colloid Interface Sci., $47$, 473 (1974).
60. D. Stigter, J. Phys. Chem., $79$, 1008, 1015 (1975).
61. F. Huisman, Proc. Kon. Ned. Akad. Wetensch., Ser. B, $67$, 388, 407 (1964).
62. P. Mukerjee and K. Banerjee, J. Phys. Chem., $68$, 3567 (1964).
63. A. T. Florence, this Symposium volume.

64. E. H. Cordes and R. B. Dunlap, Accounts Chem. Res., 2, 329 (1969).
65. E. J. Fendler and J. H. Fendler, Adv. Phys. Org. Chem., 8, 271 (1970).
66. E. H. Cordes and C. Gitler, Prog. Bioorganic Chem., 2, 1 (1973).
67. C. A. Bunton, Prog. Solid State Chem., 8, 239 (1973).
68. I. V. Berezin, K. Martinek, and A. K. Yatsimirski, Russ. Chem. Rev. 42, 787 (1973).
69. R. A. Moss, R. Nahas and S. Ramaswami, in "Micellization, Solubilization, and Microemulsion", K. L. Mittal, Editor, Vol. 2, Plenum Press, New York, 1977.
70. G. S. Hartley, Trans. Faraday Soc., 30, 444 (1934).
71. E. F. J. Duynstee and E. Grunwald, J. Amer. Chem. Soc., 81, 4540, 4542 (1959).
72. C. A. Bunton, L. Robinson, J. Schaak, and M. F. Stam, J. Org. Chem., 36, 2346 (1971).
73. W. Tagaki, M. Chigira, T. Amada, and Y. Yano, J. Chem. Soc. Chem. Comm., 219 (1972).
74. D. G. Oakenfull and D. E. Fenwick, Aust. J. Chem. 27, 2149 (1974).
75. I. Tabushi and Y. Kuroda, Tetrahedron Lett., 3613 (1975).
76. a. K. Martinek, A. P. Osipov, A. K. Yatsimirski, and I. V. Berezin, Tetrahedron Lett., 709 (1975).
    b. H. Morawetz, Adv. Catalysis Related Subjects, 20, 341 (1969).
77. H. Morawetz, Accounts Chem. Res., 3, 354 (1970).
78. N. Ise, Adv. Polymer Sci., 7, 536 (1971).
79. J. R. Chao and H. Morawetz, J. Amer. Chem. Soc., 94, 375 (1972).
80. C. J. O'Connor, E. J. Fendler, and J. H. Fendler, J. Amer. Chem. Soc., 95, 600 (1973); and J. Chem. Soc. Dalton Trans., 625 (1974).
81. F. M. Fowkes, "Micellization, Solubilization, and Microemulsions", K. L. Mittal, Editor, Vol. 2, Plenum Press, New York, 1977.
82. T. C. Franklin and T. Honda, ibid.
83. A. Weisstuch and K. R. Lange, Materials Protection, 10, 29 (December 1971).
84. P. Mukerjee and A. Anavil, in "Adsorption at Interfaces", K. L. Mittal, Editor, A.C.S. Symposium Series No. 8, pp. 107-128, American Chemical Society, Washington, D.C., 1975.
85. T. M. Geffen, World Oil, March 1975.
86. V. K. Bansal and D. O. Shah, this Symposium volume.
87. R. L. Healy and R. L. Reed, Soc. Pet. Eng. J., 14, 491 (1974).
88. S. A. Alkaitis, G. Beck, and M. Grätzel, J. Amer. Chem. Soc. 97, 5723 (1975); S. A. Alkaitis, A. Henglein and M. Grätzel, Ber. Bunsenges. Phys. Chem. 79, 541 (1975); S. A. Alkaitis and

M. Grätzel, J. Amer. Chem. Soc. 98, 3549 (1976); R. Scheerer and M. Grätzel, J. Amer. Chem. Soc., Jan. 1977; R. Scheerer and M. Grätzel, Ber. Bunsenges. Phys. Chem., in print.
89. H. L. Rosano, J. Soc. Cosmetic Chem. 25, 609 (1974).
90. K. Shinoda and S. Friberg, Adv. Colloid Interface Sci., 4, 281 (1975).
91. L. M. Prince, J. Colloid Interface Sci., 52, 182 (1975).
92. R. A. Mackay, K. Letts, and C. Jones, in "Micellization, Solubilization, and Microemulsions", K. L. Mittal, Editor, Vol. 2, Plenum Press, New York, 1977.
93. K. Letts and R. A. Mackay, Inorg. Chem., 14, 2990, 2993 (1975).
94. G. D. Smith, B. B. Garrett, S. L. Holt, and R. E. Barden, J. Phys. Chem., 80, 1708 (1976).
95. D. O. Shah and R. M. Hamlin, Science 171, 483 (1971); D. O. Shah, Ann. N. Y. Acad. Sci., 204, 125 (1973).
96. S. Friberg, Chem. Tech., 6, No. 2, 124 (1976).
97. E. A. G. Aniansson, this Symposium volume.
98. H. Hoffmann, this Symposium volume.

# MICELLES - RETROSPECT AND PROSPECT

G. S. Hartley

57 Aurora Terrace, Hillcrest

Hamilton, New Zealand

The word "micelle" as now understood in physical chemistry, is an aggregate of amphiphilic molecules formed under the influence of rather simple non-vital forces. In its most simple form it is liquid, spherical and held together by the powerful self-attraction of the surrounding water molecules. At the opposite extreme it may be indefinitely extensive in two dimensions with properties considerably modified by specific internal forces. In this form the structure goes back to biology, not as an organelle but as the quasi-spontaneous membrane which surrounds it. My retrospect is of the evidence for the simplest form and of the struggle we had to escape from misconceptions and to establish (1) that separation of small, discrete units is not to be interpreted in terms of forces of attraction within the units and (2) that valid theory must explain not only aggregation but reversible limitation of it. We could, of course, have arrived more clearly and smoothly at the present position. The major prospect in this subject is undoubtedly the better understanding, assisted by important new techniques, of the more elaborate hydrophobic structures of direct biological interest, but I must leave this to others. My speculations are limited to what I understand to be still unsolved problems in the transition from spherical simplicity and perfect reversibility to anisometric structures responsive to external factors.

## INTRODUCTION

I find it first necessary to make some apology for the grasping way in which physical chemists of our kind took over a biologists' word "micelle" and used it for something quite different. In my case, I offered some feeble resistance, but, when my late professor, F. G. Donnan, asked me to write my little book on these units, he strongly urged me to use the "in" word. I added a new one to describe not the aggregate, but the kind of molecule which could build it. This was not strictly my word. It was created at my request by a friendly professor of Greek, the late T. M. Smilie.

I mention this for three reasons. Firstly, to conform to our usual professional practice of acknowledgment. Secondly to justify the choice of "amphipathic" rather than "amphiphilic" because I held that "love" implies a positive force of attraction (I was young then: I should be less sure now even of that) and a word indicating possession of both "feelings" ("pathos" in Greek, despite the present use of its Anglicized form, has a more general meaning than "philos") would best describe the molecular schizophrenia on which the important properties of these substances depend. I may add, but dare not use more parentheses, that I chose a Greek rather than Latin root because the professor of Latin was much less helpful. He thought science vulgar and that scientists could not use even their own language properly.

This is all very much in retrospect because, on the enormous modern university campus the scientist would not even know the name of his language professors unless they shared a club vice-presidency or a garden fence. Also because my third reason is to admit my acceptance of the inevitable advance of the more popular "amphiphilic". Pedantically, they are neither satisfactory, because, as I was at pains to point out, the operative forces arise in the water rather than in the amphiphile. Micelles exist because water is intensely cohesive.

So, when we use the word "micelle" in this conference we are talking about aggregates formed spontaneously in solutions of amphiphiles. They are turning out to have important significance for the units of living matter to which Nageli first gave the name. That subject, however, belongs more in prospect than retrospect.

It is quite a long time since I have done any active research in this field. I was very pleased when Dr. K. L. Mittal, Chairman of the Symposium, invited me to contribute but there has been such a good response from other invitees that my contribution could easily have been dispensed with. This leaves me free to attempt a rather broad perspective view over a field where I know I am out of date on many important developments of detail.

## THE IMPORTANCE OF AMPHIPHILES

It may, I think, be useful to examine why this subject does arouse the wide and active interest which is here represented. I would list four factors but they react on one another and I put them in no order of history or importance.

(1) Amphiphiles provide the easiest way of getting matter thinned down in one dimension to molecularly significant thickness. "Soap" films have interested many philosophers, whether it be for the interference colors at sub-micron level, the deeper significance of the much thinner blacks, their "weightless" representation of liquid form as developed by Plateau or the monolayers, invisible directly, formed on a water surface. There was a small pond on Clapham common in London - I don't know if it still exists - by which your great Benjamin Franklin sat and contemplated. Why Clapham common, I don't know. James Michener might and would tell us of the other animals, pre-historic and human, that had been there before, but it was then that the study of monolayers, later so much developed by Pockels, Langmuir, Adam and others, began.

There should be a plaque or a statue but, although Franklin would come out well in marble or bronze, representation of a monolayer would surpass the genius of Epstein or Moore. With most forms of matter we need devices of great sophistication to get down to molecular dimensions. With a solution of a suitable amphiphile, the careless sweep of a child's arm can create scores of envelopes each weighing much less than the air it contains.

(2) The opposition of properties within an amphiphilic molecule naturally arouses the interest of the enquiring researcher to see how nature resolves the conflict in various situations.

(3) Amphiphiles permit controlled separation of fatty compounds from aqueous solution in which they are produced by living chemistry and play in many cases a controlling part in the synthesis. While stereo-regular polymers of water soluble molecules can also separate out of the local environment, and play an essential role in the skeletal tissue of higher forms of life, it is only the amphiphiles which form molecularly selective barriers and are therefore more necessary to biochemical elaboration.

(4) Amphiphiles permit the manipulation of oils in many vital and industrial processes using cheap, harmless and abundant water - otherwise most unsuitable - as a diluent or carrier. By their means, oils are displaced from fabrics into water (detergency), air or oil is displaced by water (wetting out and oil production), oily substances are spread very thinly (paints, cosmetics, pesticides) and are brought into intimate contact with the high heat capacity

of water and the chemical reactivity of many reagents which can be dissolved in it (emulsion polymerization, saponification, digestion).

## THE MICELLE AS COMPETITORS

With one exception, these important biological and technical behaviors are not directly assisted by micelle formation - rather the reverse. The micelle forms as an alternative to adsorption at a surface or interface. Water must retain the "head" - the hydrophilic part of the amphiphile - but reject the "tail" - the what? There is the linguistic problem. Neither "lipophilic" nor "hydrophilic" is good. Paraffin is no more attracted to itself than it is to water and there is no positive repulsion from the latter. The water rejects the intruding paraffin just as it would reject a vacuum if it could not destroy it by compression. It does not want its strongly self-attracting molecules forced apart by a substance which can offer nothing in exchange. So we have a hydrophilic head and a tail unwanted by water.

The easiest solution to the problem is found when there is another phase present which can accommodate the tails without detriment while the water retains the heads. The amphiphiles crowd into the interface and reduce its tension. Whether, in extreme cases, they can cause spontaneous extension of an interface between liquids is an interesting question. If there is no - or insufficient - suitable interface and the concentration of amphiphile in water is increased, it comes to a rather critical value at which a new paraffinic phase forms spontaneously in a very fine state of division - the micelle. The interior of the micelle is populated entirely by tails rejected by water, but all the heads are held in the surface and one dimension is therefore limited to the length of a tail. Beyond this critical concentration there is very little further increase of chemical potential with concentration. Formation of the micelle provides relief for the strain of separation of water molecules. The crowding of existing interfaces, which increased with concentration below the critical value, increases very little further above it.

Since the formation of the micelle is an alternative to crowding of an interface one might expect greater interfacial activity if micelle formation could be made more difficult. It was a source of some surprise and disappointment to me that a considerable effort that I put into exploring this possibility - doing the synthesis and purification[1] as well as the physical measurements[2] - attracted very much less attention than my earlier work on the nature of micelles in solution of orthodox amphiphiles. 1939-40 were, of course, bad years to publish academic work. I prepared and examined sulphonates of resorcinol ethers in which two paraffin chains of any length could easily be included. Industrial production of sulphonates of succinic esters, more cheaply made, were a better commercial proposition and, at the time, no one seemed very interested in the reason for their remarkable properties.

One dimension of the micelle is limited by the distance between the head group and the end of a tail. Replace a single tail by two of half length and we reduce this distance to about half. If the aggregate, formed with a liquid interior by the extruding action of water, is also near spherical - and a large amount of evidence indicated that this was so - the micelle formed from a compound with two 8 C tails would have only about one eighth of the volume of that from a similar compound with one 16 C tail. Its surface would be inadequately populated with heads and its formation would provide a much less satisfactory solution to the energy problem. An existing oil phase could be just as ready to receive two 8 C tails as one 16 C tail. Moreover the awkward-shaped two tail compound would pack less satisfacotrily in crystal form. So, one would expect the two tail compound to be somewhat less active in an existing interface than the single tail compound at the same very low concentration, but that a higher chemical potential would be reached as the concentration increased and so the surface activity would become greater.

The facts gave better support to the theory than I dared hope. The interfacial tension - oil, water - ceased to fall when a single chain 16 C compound reached a concentration of about N/1000, below which a two chain compound was rather less active. The surface activity went on increasing, however, with the latter until it became immeasurably (by my method) small at about 4 times this concentration. 1.1 dynes/cm was the lower value reached between water and a carbon tetrachloridecyclohexane mixture in the presence of a compound with a single 16 C chain. With a compound with two 8 C chains, 0.04 dynes/cm was measured before the meniscus at the capillary tip became unstable and convection and emulsification occurred in the capillary bore.

There cannot be so great an advantage at gas-water interface because the surface tension cannot get below the value of about 24 dynes/cm for the surface of a rather low-volatile paraffin. With perfluoro compounds one may get rather less. One did not expect either that the two short chains could form so stable a surface layer as the single double-length chains. The two tail compounds would therefore lower the oil-water interfacial tension much more but they would not form films in air so easily or of such stability. But they did. The 8-8 compound produced larger bubbles which blackened more quickly, but lived longer than any I had handled from a soap-type solution. The 10-6 compound was even better, but the behavior is so sensitive to impurities that I would attach no significance to this without more experiment. Winsor[3] later reported the same behavior with straight paraffin chains bearing a sulphonate ester group near the middle. The sulpho-succinates also form stable foams. In depths of more than a few mm all these two-chain foams drain quickly to the "ghost" stage - i.e. the film areas are black and only the network of plateau borders is clearly visible.

One other prediction was not quite right. The two-chain compounds, whose complex geometry would not so easily fit into a crystalline arrangement, were, despite less easy micelle formation, sufficiently soluble to be very surface-active. When crystals of the K salt of the 8-8 compound grew slowly from an evaporating aqueous glycerol solution they formed massive pyramids unlike the very thin platelets of the single chain compounds. Why, I do not know. In water they dispersed spontaneously to a turbid solution when the concentration was above about 0.1%.

I confess I was too pleased with the success of the main prediction (although few others seemed to be) to consider well the wrongness of the prediction of poor foam stability or the significance of the spontaneous turbidity. It is sometimes a misfortune for a scientific theory that the first experiments should clearly verify it. Probing by exception should always be encouraged.

## SOLVENT ACTION OF MICELLES

I mentioned that one behavior of amphiphiles, important in biology and technology, is not antagonized by micelle formation. This is the phenomenon known as "solubilization", a word which I have never been happy about. The liquid interior of the micelle can dissolve paraffin-soluble compounds. If these compounds - e.g. naphthalene and azobenzene - have their solubility sufficiently limited by their crystalline nature, the micelle is not grossly swollen and the solubility in the paraffin dispersed as micelles is not greatly different from that in bulk liquid paraffin.[4]

If the dissolved non-polar compound is itself liquid, there must be indefinite growth and no clear distinction between a micellar solution of a small additional volume of solute and an emulsion of a large volume, unless some different factor limits the growth. There is no doubt at all that, in simple systems such as cetyl pyridinium chloride - water - benzene, there is a quite definite limit to the amount of solute oil which can be brought into a solution which satisfies all the tests of equilibration. Excess forms an emulsion which is not particularly stable and in which the wide range of globule size is dependent on conditions of agitation. The globule size, in other words, is either around 4-5 nm with less than half the volume benzene and strictly in equilibrium, spontaneously formed - or, it is in the 2-3 μm range and above with no clear limit on proportion of benzene and growing spontaneously by several mechanisms towards eventual separation when left at rest.

If the solute oil is not strictly non-polar but itself amphiphilic, but perhaps not strongly enough to form micelles on its own, it can be "solubilized" in the micelle in much greater volume. If solubility is restricted by crystallinity, the total solubility in the aqueous solution of the primary amphiphile may be greater than it would be in the water and paraffin if separated. The reason, of course, is that it can find a place in the surface as well as in the interior of the micelle, and, making the surface of the paraffin-interior more polar, it permits some submersion of head groups and expansion of the micelle. There is still a definite limit to the amount of a semipolar oil which can be held in clear spontaneous solution and the emulsion of excess is in this case very unstable.

The semipolar solute oil - middle range primary alcohols, about C 6 to C 12, are preeminent - can further increase very greatly the solvent power of the aqueous solution of the primary amphiphile. Solution can form spontaneously containing oils in several times the volume of primary amphiphile. Their globular micelles are in the 5-10 nm range giving strong light scattering. Such systems were called "micro-emulsions" by Schulman[5] but they do not differ in principle from very similar "solubilized" oils described earlier by Winsor.[6]

The thermodynamics of these complex systems and the factors limiting their range of stability are considered in several papers at this conference. A point I wish to emphasize here is that the balance of factors in a large multi-component micelle is necessarily much more delicate than in the small micelle of a single amphiphile. It takes longer to determine equilibration and there is necessarily more dependence on external factors. Sedimentation would be significant if convection were arrested and Winsor describes systems where the light scattering is changed by gentle stirring.

My own main interest in the solvent action of micellar solutions was to demonstrate that the inner paraffinic contents of the micelle was essentially liquid.[7] Indeed the solvent action for a crystalline non-polar compound (I chose azobenzene on account of its color and stability but, when I came to measure the concentration of dilute solutions in daylight, I had to make an interesting diversion into photo-induced cis-trans isomerism[8]) provides some of the best evidence both for the existence of micelles and for their fluidity. A good solvent, molecularly dispersed in a poor solvent, loses most of its solvent power. The solubility of a third substance in the mixture is more nearly a logarithmic than linear function of composition. At a concentration of 1% or less the good solvent can

usually be ignored. Many amphiphiles in aqueous solution can be
diluted much further before they lose much of their solvent action
and, when they do lose it, the loss is catastrophic. One really
needs no other evidence for micelles, representing molecularly undi-
luted solvent, for their sudden formation when the concentration
scale is ascended and for their fluidity.

## SOME OTHER CHANGES CAUSED BY MICELLE FORMATION

Another measurement which could have provided all the evidence
needed but requiring very great exactness and avoidance of dissolved
air would have been density. It was explored early by Bury[9] in his
work on very short-chain compounds - butyric acid, sodium caproate,
etc. - which could be examined in high concentration and which originated the conception of a cirtical concentration. Tartar,[10]
Harkins[11] and Benjamin[12] later measured partial volumes in systems
in which the micellar state had already been established by other
methods. The most widely measured property in the early work was
electrical conductivity.

The ionic head group, however, is not essential. In some
respects it complicates the behavior but at the same time introduces
a set of properties easily and exactly measurable. The early compounds were all ionic (<u>pace</u> Bury's butyric acid) because no other
single small group was sufficiently hydrophilic. Most of the important technical non-ionic amphiphiles have, of course, a multiplicity
of rather feebly hydrophilic groups. We divide into tail and head
a compound with say, 12 $CH_2$ groups and 10 $-O-CH_2-CH_2-$ groups by cus-
tom and courtesy rather than logic. Micelles of these compounds
have a predominantly paraffinic core, outside of which is a diffuse
region grading off from poly-ether to water. This naturally considerably modifies the solvent properties. I used to complain of the
ionic compounds that their behavior would have been more easily
studied had the electronic charge had only half its value. It is
not only in political and commercial affairs that pure scientific
theory meets irrelevant restrictions! The most compact, powerful
non-ionic hydrophilic group is the amine oxide and had the amine
oxide[13] amphiphiles been the first available the structure of solutions of amphiphiles might have advanced more quickly.

## HYPERSTRUCTURE IN PARAFFIN-CHAIN SALT SOLUTIONS

There are two main effects of ionic head groups on the structure of amphiphile solutions. Firstly, repulsion between them is a
factor limiting growth. I will return to this again under the
heading of size limitation. It is, I think, less important in this
respect than is often thought. Secondly, repulsion between micelles

# MICELLES - RETROSPECT AND PROSPECT

is greater and more extensive when they are highly charged. This gives rise to a hyperstructural effect in the whole solution.

The X-ray diffraction evidence for a dominant long-spacing in these solutions was first held to be evidence of laminar or platelet micelles.[14,15,16] I won't say that it was held to disprove the existence of spherical liquid micelles because the first, and incorrect, interpretors were unaware of the conclusive evidence from other lines of work with less sophisitcated equipment. The error, and its correction, are quite an interesting example of the danger of not thinking far enough back. X-ray diffraction patterns arise from a spatially regular disposition of diffracting centers. This regularity is found pre-eminently in crystalline solids where centers are regularly spaced in planes and planes regularly spaced from one another. From single crystals a spot pattern is developed: from a powdered specimen, a ring pattern. The method had made such enormous advances in our knowledge of e.g. crystal structure and, through this, of molecular geometry that the association of ring pattern with parallel planes had become automatic. The concentrated solution of a paraffin chain salt produced a primitive type of crystal powder diagram - therefore it must consist of a chaotic assembly of flat platelets of regular thickness. The experts at the time were so certain their technique could speak with full authority that the evidence from more primitive types of measurement was often not even read.

The regular spacing, however, need not be between planes. Curved surfaces with constant parallel spacing would suffice although, if the parallel areas were very small, the diagram would be less intense and sharp - as indeed the solution diagrams were in comparison with true crystal powder diagrams. Schulman and Riley[17] were the first to consider the most clearly defined spacing found in the solution to be that between regularly spaced globules. This interpretation was applied to micro-emulsions near to the high concentration ($\pi/3\sqrt{2}$ volume fraction) needed for close packing of spheres. They found evidence of distortion when this volume was exceeded. I pointed out at about the same time[18] that mutual electrostatic repulsion would tend to set highly charged spheres in the same regular spacing at much lower concentrations. The long spacing, $\ell$, if it were the distance between a point on one micelle and a corresponding point on any of its 12 closest neighbors, would be related to the radius, $r$, of the micelle (assumed spherical) and the volume fraction $\phi$ by

$$\ell^3 = \frac{8\pi}{3\sqrt{2}} \phi r^3 \tag{1}$$

The observed long spacing and its variation with concentration fitted well with this equation, taking r to be a little more than the length of a fully extended chain, an assumption consistent with the theory of the spherical micelle and with diffusion measurements.

Solutions containing laminar micelles should show two long spacings, one being the thickness of the lamina and the other the distance between corresponding surfaces. The latter only would increase with dilution. Assuming this to be the dominant one observed, a major difficulty in interpretation was that water between micelles accounted for only a part of the total,[19] but varied with it. Even more difficult to explain was the disproportionate increase of long spacing when an oil was solubilized:[20] it had to be assumed that swelling of the paraffin layer increased the spacing, already tens of water molecules thick, between micelles. The regularly spaced spheres accounted simply for these variations. Solubilized oil increases both r and $\phi$ and a disproportionate increase of $\ell$ directly follows.

There was at first some reluctance to accept that the X-ray findings were consistent with the spherical micelle. Harkins et al.[21] were doubtful whether a rather ill-defined short spacing was consistent with a purely unordered structure within the micelle and preferred an oblate spheroid in which many chains near the minor diameter would have to be parallel to the minor axis. There is, however, some order in a liquid straight-chain paraffin and one cannot expect the micelle interior to be any more chaotic. Corrin[22] accepted the view that the long spacing arose from a hyperstructure among the micelles and was not indicative of shape except that shape determines size and therefore density of population. Shortly before he died I had a most constructive talk with Harkins which levelled out any real differences. The later more exhaustive work of Reiss-Husson and Luzatti[23] confirms these views.

## LIMITATION OF GROWTH

The low viscosity, optically clear, dilute aqueous solutions of many paraffin-chain salts, on which I concentrated perhaps a disproportionate amount of attention during my active research in this field, passed all the classical tests for systems in complete equilibrium. Nevertheless they showed clear evidence of extensive aggregation at concentrations much lower than it had been possible for McBain to investigate in the hydrolyzable soaps when laying the foundation of this subject. A crystalline phase could form a very fine dispersion and liquid-crystalline phases could separate in concentrated systems. I therefore felt it necessary to propose an aggregate of very different structure which could form in dilute solutions and to show how this aggregate could be clearly limited in growth as the evidence for equilibrium required.

An internal core of paraffin-chains with the ionic head groups in its surface and therefore in contact with the surrounding water was a basic necessity in any postulated structure. A laminar form seemed to have no limit to its growth in the plane and to be too like an element of the crystalline solid. Moreover the chains, in parallel array, would not be expected to have the observed solvent action on non-polar crystalline solids. High partial volume of the paraffin, solvent action and absence of birefringence all indicated a liquid and isodimensional form. The radius of the inner paraffin core would be limited by the length of a fully extended chain, since the interior must be filled and the head group be retained by the water. If it were to expand in all directions, some ionic heads would have to leave their hydrating environment. Two factors would operate against contraction - more naked paraffin would be exposed to water, against the free energy of association of water with itself or other polar groups, and the disposition of chains anchored in the surface would be restricted. The sphere, if formed at all, would at once expand to its full radius and the initial concentration phenomenon would be explained.

Studies by sophisticated spectroscopic probing techniques,[24,25] have confirmed the fluidity of the core and shown that it is less viscous in the center than near the surface where restriction of the heads will have greatest effect. Other studies, notably elaboration of X-ray diffraction technique and its interpretation and the extensive exploitation of light scattering measurement have shown that the mean size of micelles in isotropic, mobile solution is rather greater than predicted by the spherical model and that it does generally increase with increase of concentration. Distortion from the spherical form must be assumed and this has been particularly expounded by Tartar[27] and Tanford.[28] I would not wish to dispute the interpretation of the evidence or the validity of theory and I must admit that my spherical micelle was an over-simplification. It should be remembered that it was prepared in opposition to - or rather, in necessary distinction from - the laminar aggregates which were almost universally assumed at the time. Also, as far as micelle formation is concerned, it is much more important that an aggregate of, say, 50 paraffin chain ions is formed without significant appearance of much smaller aggregates than that the real number should be nearer 100 and rise with concentration to 200 instead of remaining constant.

If a macroscopic liquid sphere with a zero-thickness boundary expands in one axis only (that of revolution, becoming a prolate ellipsoid) the ratio of surface to volume falls from $3/b$ for the sphere towards half this value for an infinite rod ($a = \infty$, $b$ being the constant minor axis). Intermediate factors are 0.85 for $a = 2b$ and 0.80 for $a = 4b$. If the micelle expands in two axes (rotation about the fixed minor axis, $b$, i.e. an oblate ellipsoid), the surface to volume ratio decreases at first twice as rapidly and approaches 1/b, i.e. 1/3 of the sphere value, as a → ∞ (the indefinite lamina of thickness 2b) intermediate factors for a = 2b, 4b being 0.69 and 0.57. For small deviation from the sphere the ratio of surface to volume is a linear function of axial ratio.

Since the core is made up of paraffin tails the ratio of surface to volume is proportional to the area of surface available to each head group. For the indefinite lamina, or bi-layer, this would be the area in a close-packed monolayer having a minimum value of about 2 nm for a straight paraffin with a polar, but not enlarged, head such as octadecyl alcohol. The corresponding values for the cylinder and sphere are 0.3 and 0.6 $nm^2$ in which very small polar groups such as -OH would not be close packed.

Tanford emphasizes the importance of mutual repulsion of ionic heads which would tend to favor the large surface/volume ratio of the sphere. In the presence of excess electrolyte, reducing repulsion, or when the head group is non-ionic there will be a tendency towards the lower surface-volume ratio of some extended form. This will be particularly favored when the amphiphile has two tails (or a tail otherwise increased in cross section) so that, even in the bi-layer, the heads will not be close-packed. Opposing this tendency to expansion is the higher degree of order necessary in the posture and mutual fit of chains necessary to allow it. Tanford[29] has examined the thermodynamics of this balance rather fully in view of the biological importance of spontaneous bilayer formation. He makes the effect of ellipsoidal distortion on the surface available per head rather greater than the figures I have quoted for a macroscopic ellipsoid. His very reasonable argument is that the first, and perhaps second, $CH_2$ group in the "neck" of the molecule is subordinate to the water-interaction of the ionic head and is not free to form part of the separated paraffin interior of the micelle. The ionic centers should not, in other words, be considered as sessile on a sphere of radius 0.254 × n/2 nm (where n is the number of C atoms in the chain) but rather as extending 0.1-0.2 nm out from a correspondingly smaller sphere containing a disproportionately smaller number of chains. This correction increases considerably the available area per head outside the spherical micelle but makes no difference to that outside the bilayer. The cylinder is intermediate.

## TRANSIENT IRREGULAR GROWTH

While I accept that extensive growth of the micelle must be through a prolate ellipsoid towards a rod or through an oblate ellipsoid towards a plate, I think it is unnecessary, and indeed erroneous, to assume that the early stage of expansion must be considered this way. A small macroscopic ellipsoid with a smooth contour is not a true picture. The surface must be subject to irregular deviations from smoothness. The Brownian translation of the micelle as a whole and its Brownian rotation about a variable axis must be accompanied by what one can call amoeba-like Brownian movements of the surface. A liquid paraffin core of mean radius 2 nm but effective chain length only 1.85 nm need only have somewhere a small flat area depressed 0.15 nm below the mean surface to keep the central cavity filled by a terminal $-CH_3$ group. The identity of this methyl group and the location of the flat are, of course, frequently changing. The increase of area when a liquid sphere of constant volume is flattened at one place so as to depress the center of the flat by a distance $\delta$ below the mean radius a can be shown to be $\pi\delta^2$ when $\delta \ll a$. Note the quadratic form of dependence on $\delta$, compared with the linear form for ellipsoidal distortion. The increase of interfacial energy for the 0.15 nm depression in the example needs an effective interfacial tension as high as 7.5 dynes/cm to reach kT at ordinary temperature. I am not competent to take this sort of statistical mechanics any further but it seems to me obvious that amoeboid movement in the micelle surface can permit the mean radius of a micelle of generally spherical form to exceed quite considerably the true hydrocarbon chain length.

One may even question the non-existence of a central cavity. Is a cavity of 0.15 nm radius not a common but fluctuating occurrence in a liquid paraffin? It would have a disproportionate effect on the content of the core. The volume of this cavity is 0.014 $nm^3$. Its existence at the center would increase the paraffin volume from 26.5 to 33.5 $nm^3$ accommodating 68 and 86 chains. Its permanent existence, additional to any transient cavities created by thermal motion, could decrease the density by less than 1 in 2000 and I doubt if any existing methods would detect this difference.

I think that a tendency towards a central vacuum is closely connected with the expansion of the micelle by solubilized non-polar liquid and the definite limitation of this process. The expanded spherical micelle can be permanently populated only by the solubilized liquid which must attain its full vapor pressure when intrusion by terminal groups of the chain, very frequent at small expansion, becomes impossible.

Very great expansion of the micelle either of the amphiphile itself or with solubilizate, must occur anisodimensionally. When either the prolate or oblate form has become so unsymmetrical that the population of heads can no longer be evened out by preferential inclination of chains away from the minor axes, there can be no barrier to indefinite extension. There will result a separation of an anisotropic phase and, as more than one such is known to exist in several systems, they must be of different form. It is I think generally agreed that indefinite rods in parallel alignment, one with the aqueous liquid continuous, the other with the paraffin continuous, form two such phases. Alternate aqueous and paraffinic laminae form a third.

As Tanford emphasizes, the tendency towards indefinite laminae will be greater in two-chain amphiphiles because even in the lamina array the area available per head is excessive. I should not have been surprised when my two-chain compounds gave aqueous solutions forming films in air which were both exceptionally stable and exceptionally fluid. The film array, of course, is inverted – paraffin outside – but the same argument applies, unless an external oil phase favors a water-in-oil type of micelle. This is a reintroduction of the old "oriented wedge" theory of emulsion stability: while farcically inapt for emulsion globules in the μm order it is highly significant for the micelle of 1-2 nm order.

Explanation of my earlier reluctance to accept laminar forms for the micelle illustrates a mistake which I think many of us have made from time to time but without positively attributing it to anyone else. There is no doubt that the first micelles formed in dilute solution are globular rather than laminar – this is not in dispute. I felt, however, that when the evidence indicated growth to much larger units these must not have the parallel chain arrangement necessary for the thickest laminae consistent with all heads in the water – because this is a description of the crystal and the compound would not be soluble. For laminae to exist in solution, the crystal must cleave along the ionic planes and one could not have an equilibrium state with micelles of limited size. Two assumptions were involved. Firstly, that, in solutions of laminar micelles, the strict equilibrium state, demonstrated for dilute solutions (of globular micelles), was still valid: this requires more investigation. Secondly, that the parallel array of chains would be solid. This involved an error that I had been at pains to avoid and expose. The whole drive towards micelle formation comes from the self-attraction of water. Whether the paraffin chains have a disposition towards regular organization is irrelevant to this main cause. The laminar arrangement, for some compounds produces significantly more separation of the paraffin from water, than does the globular arrangement. The extra separation is much less than that resulting from the primary aggregation in globules but may still be important. The parallel array of chains may therefore be forced on the amphiphile at a temperature above that at which array in fixed relative positions is dictated by the chains themselves. Because aggregation is forced by major forces operating in the water the shape and internal organization of the extruded aggregates is left to a more delicate balance of energetic and entropic factors operating in a locally concentrated system.

# MICELLES - RETROSPECT AND PROSPECT

The existence of spontaneous bilayers in some systems is undeniable and their importance, via vesicle formation, for vital membrane structure is immense. I have contributed nothing in this field and it would be presumptuous for me to comment. I would like to conclude with some questions about areas of experimental data where explanatory theory seems at present to be poorly developed and further investigation of which could, I feel, throw some useful light on the formation of more elaborate structures from expanding globular micelles. These areas may be interrelated.

## SEPARATE ISOTROPIC PHASES

The first is the, to me, remarkable formation of a second isotropic phase at rather low concentrations of many nonionic amphiphiles in water and probably of some two-tail ionic amphiphiles. Shinoda's[30] investigation of phase diagrams and cloud point in solutions of pure synthesized ethylene oxide amphiphiles has established clearly that, above a critical temperature, solutions which may contain as little as 1% of amphiphile separate into a solution which may contain as little as 10% on the one hand and nearly pure water on the other. At the critical point the concentration must of course be more nearly equal but there are formidable difficulties in experimentation. The phenomenon is not restricted to the ethylene oxide compounds, which have heads longer than their tails, since the amphiphiles with the compact powerful hydrophilic groups, amine oxide and phosphine oxide, have been shown by Hermann[31,32] to have similar behavior.

Both groups of workers draw a parallel with the Flory-Huggins theory of highly swollen polymers in equilibrium with an almost pure liquid which is just not quite a complete solvent. Shinoda considers that large micelles, probably elongaged prolate, are behaving as "poly-soaps" in which the hydrophobic portion is a covalent macro chain. A small degree of preference of molecules for company of their own kind, multiplied up on the Brønsted principle in the macromolecule, can produce this behavior but swelling does not usually go to this extreme without complete solution. Moreover, the micelles, formed by extrusion of the paraffin chains from water, have extremely hydrophilic surfaces which would not be expected to have long-range DVLO type attraction. This is particularly true of the amine oxides and phosphine oxides the solubility of which indicates a far greater attraction of these small groups to water than to one another.

I would like to know what geometrical structure is envisaged in these rather dilute solutions which form a phase boundary with almost pure water. One might imagine a network of bi-layers rather than filaments forming a micro version of a compact, stable foam in contact with excess air, but the composition of the bilayer network in water would seem necessarily to depend on its mode of formation. The phases, however, have equilibrium compositions and moreover both are isotropic. One would expect either network to show some birefringence on shearing.

## SHEAR-INDUCED STRUCTURE

The second area where, to my knowledge, explanation has failed may not be unrelated. It is the elastic, and associated, behavior of very dilute aqueous solutions of some amphiphiles. It is highly specific with regard to head structure (including counterions) of the salt and, although first observed in rather concentrated ammonium oleate solutions, it shows up very dramatically in some solutions as dilute as 1 in 10,000. These solutions, after shaking to introduce small air bubbles to mark the flow, always appear heterogeneous in the sense that the bubbles are collected together in non-random groups. This seemed contradictory to their apparent equilibration in other respects.

I published a short note[33] on a closer look at the related phenomena but with no real attempt at molecular-level explanation, but have not followed it up, nor have I seen other discussion by researchers in micelle structure. I have heard that some researchers in the displacement of oil from porous earths have been interested in the phenomenon under the name of "positive non-simple behavior". The structure-creation is similar to that underlying the work-hardening of metals but is spontaneously reversible. The reason for the apparent heterogeneity of solutions in which air bubbles have been entrained is that a resistance to stress has been built up by shear which itself has irregular incidence. A slow succession of small air bubbles will rise normally in a quiescent volume of liquid but if a region in their path is stirred, they hesitate and cluster. Correspondingly, if the flow of liquid through a porous plug is examined, we have the following sequence of behavior. Under sufficient pressure difference, the flow is as of water. After a period of rest, flow under any suddenly created pressure difference is initially as of water, but, if the pressure difference is small enough, the flow rapidly decreases to a small fraction of the water value. When this stage has been reached, higher pressure differences, in either direction, will not release the "blockage" unless they are considerably greater than some values which would not by themselves produce blockage. A period of complete rest - few to many minutes - will restore original behavior.

Clearly, moderate shear creates or organizes a framework which is not effective in the resting solution. The framework is destroyed by excessive shear or will destroy itself spontaneously in the solution at rest.  The associated elasticity has probably therefore a rubber-like mechanism - thermal randomization tends to restore the pre-strain posture.  A few experiments I made with elastic solutions of wholly different substances - saliva and seed mucilages - suggest that the reversible framework production and the elasticity are rather generally associated.  Irreversible coagulation by shear was an important defect in the early days of emulsion paints before manufacturers learnt how to control it by suitable formulation. This exemplifies how shear can force together particles of the 100 nm order over an energy barrier which thermal movement cannot surmount.  In the case of emulsion paints, the effect was irreversible, but, if the energy trough, at close approach, were not many times more than kT below the crest but considerably higher than the level at full separation, it would be possible for shear to produce a clumping of a temporary nature.

It seems at first unlikely that adhesion, short of complete fusion, could occur between micelles formed so as to leave a fully hydrophilic exterior, but the delicate balance between spheroidal and elongated forms may provide the answer.  Let us suppose that, for a certain head and counterion configuration, the thermodynamic balance at rest just favors the near-spherical form (with amoeboid expansion).  Forcing of micelles together by shear would then tip the balance in favor of rod-like extension, but, at rest, the rods would segment into globular micelles.  These, taking up a random distribution and even tending, by repulsion, to a honeycomb lattice hyperstructure, would cause the whole environment of the temporary rod to become more spherical.  I think the phenomenon is worthy of more experiment and, with statistical-mechanical interpretation of which I am not competent, it might throw valuable light on the factors tending to produce spontaneous extension of micelles.

## THE EXTRUSION FORCE ON SINGLE CHAINS

The primary force producing aggregation of paraffin chains arises from cohesive energy of the surrounding water, or rather from the excess of cohesive energy over that of adhesion to paraffin. Langmuir[34] and Harkins[35] explored the idea that the free energy of transfer of a paraffin chain from a water environment to one of paraffin in bulk could be calculated from the interfacial tension in bulk between paraffin and water and the area of the intruding molecule.  There is some difficulty of course in defining this area and doubt about the validity of assuming the bulk value of interfacial tension to apply to so small an area, but it is instructive to see where this idea leads us.

The volume occupied by a chain can be estimated from Avogadro's number and the density of the paraffin formed by cutting off the heads. Assuming the crystallographic distance 0.25 nm between penadjacent C atoms we can then calculate the area of this volume, if it is cylindrical. The higher paraffins give us a cylindrical surface area of about 0.02 nm$^2$ per $CH_2$ group. The interfacial tension is about 40 dynes/cm, whence the interfacial energy of displacement is about $8 \times 10^{-14}$ ergs per $CH_2$. This is about 2 kT at ordinary temperature and we should expect an increment in $\log_{10}$ (partition coefficient) of 0.87 per $CH_2$. The value agreed in much experimental[36,37,38] work is about 0.5. One can say that the simple calculation is a serious overestimate or that, in view of its obvious crudity, it gives a surprisingly close prediction. Two things are wrong with the simple calculation. Firstly the chain is flexible and even *in vacuo* would not be cylindrical but randomly crumpled. In water, as Langmuir pointed out, it should be forcibly crumpled to a near-spherical form. A 16C chain of volume 0.42 nm$^3$ would have an area in cylindrical form about 1 nm$^2$ greater than in spherical form, having a difference in interfacial energy of about 10 kT. This is far too great for crumpling to be seriously resisted. When the spherically crumpled chain is extruded from water the free energy dictating the partition into paraffin should be proportional to $N^{2/3}$ where N is the number of C atoms.

Measurements of solubility of hydrocarbons and partition of fatty acids into paraffin from water indicate a linear relationship of log (partition) against N. Ward[39] has explained this for the lower fatty acids by a counteraction of the basic proportionality to $N^{2/3}$ and the lower density of the lower members but for the higher members one could expect linearity against $N^{2/3}$ to show up. Smith and Tanford[40] found linearity against N for fatty acids as high as N = 22. On the other hand, Baker[41] and Franks,[42] attempting to extend to high N values McAuliffe's[43] accurate measurements of solubility of hydrocarbons up to N = 8, found a serious flattening off above N = 10. So Smith and Tanford find solubilities for N = 18 which are less than Langmuir would have predicted, but Baker and Franks find solubilities unaccountably much greater.

Baker particularly mentions possible micelle formation and says that the solubility (by Tritium estimation) was reduced by millipore filtration. I think the errors are in experiment but the micelle suggestion is worth mention here to illustrate the dissimilarity of an amphiphile, where there is a good mechanism of limitation of growth, and a hydrocarbon where growth would be inevitable. Although inevitable, slow. Coalescence of very small globules of oil in pure water at great dilution is extremely slow: growth will mostly occur through the aqueous solution by the Ostwald ripening process. I calculate that such growth in a dispersion of octane would be measurable over a few hours but in octadecane, if $N^{2/3}$ theory is right, in years. If the N theory is right, in centuries.

Honour where it is due. I have made serious criticism of much physical chemistry appearing in the pesticide literature, but I can recommend one paper[44] in an entomological journal to physical chemists who want to measure very low solubilities. Do it quietly and allow a long time, not for the molecular concentration to rise to saturation but for the early values, made erroneously high by dispersed particles, to fall.

I would like to see these solubilities much more carefully studied by diffusion methods involving no agitation and in dilute salt solutions to reduce electro-repulsive effects. The larger the molecule, the more nearly right should the interfacial theory be - but with $N^{2/3}$, not N.

The usefulness of interfacial tension should not be reduced by the interesting discoveries that separation of water molecules decreases entropy rather than increases[45,46,47] energy - as many H bonds are formed but the water needs a more structured arrangement around a "neutral hole" to exploit them. The structuring effect near a hydrophobic surface should to a first approximation be independent of the size of the intruding object. It is part of the mechanism of the high interfacial tension. In contrast to the "destructuring" effect of an ionic charge it has some interesting consequences. To me the most dramatic is the finding that the Ar atom diffuses[48] in water more slowly than the hydrated $K^+$ and $Cl^-$ ions. This however is not relevant to the $N^{2/3}$ vs. N difficulty I have raised - or only so if the interplay of structure and bonding is such that there is a major effect of curvature so that the effective interfacial tension increases with increasing radius of the intruding hydrophobic object. Linearity of log (partition) with N at a slope only about 2/3 that predicted by the simple theory would then be due to an accident of compensating factors.

This is not really a satisfactory foundation for our knowledge of the primary driving force towards micelle formation. I have said nothing of the deeper studies of the thermodynamics of micelle formation. They are discussed at this symposium by workers more competent in this subject than I ever was. They may throw light on these questions, but I hope there will also be an improved mechanical picture for this light to shine on.

## REFERENCES

1. G. S. Hartley, J. Chem. Soc., 1828 (1939).
2. G. S. Hartley, Trans. Faraday Soc., 37, 130 (1941).
3. P. A. Winsor, Trans. Faraday Soc., 44, 376 (1948).
4. M. E. L. McBain and E. Hutchinson, "Solubilization and Related Phenomenon", Academic Press, New York, 1955.

5.  J. H. Schulman and T. S. McRoberts, Trans. Faraday Soc., 428, 165 (1946).
6.  P. A. Winsor, "Solvent Properties of Amphiphilic Compounds", Butterworth, London, 1954.
7.  G. S. Hartley, J. Chem. Soc., 1968 (1938).
8.  G. S. Hartley, J. Chem. Soc., 633 (1938).
9.  D. G. Davies and C. R. Bury, J. Chem. Soc., 2263 (1930).
10. A. B. Scott and H. V. Tartar, J. Amer. Chem. Soc., 65, 692 (1943).
11. W. D. Harkins, R. W. Mattoon and M. L. Corrin, J. Colloid Sci., 1, 105 (1946).
12. L. Benjamin, J. Phys. Chem., 70, 3790 (1966).
13. K. W. Hermann, J. Phys. Chem., 68, 3504 (1964).
14. K. Hess and J. Gundermann, Ber. 70, 1800 (1946).
15. H. Kiessig and W. Philipoff, Naturwiss., 27, 593 (1939).
16. J. Stauff, Koll. Zeit., 89, 224 (1939).
17. J. H. Schulman and D. P. Riley, J. Colloid Sci., 3, 383 (1948).
18. G. S. Hartley, Nature, 163, 787 (1949).
19. J. D. Bernal, Trans. Faraday Soc., 42B, 180 (1946).
20. R. W. Mattoon, R. S. Stearns and W. D. Harkins, J. Chem. Phys., 15, 209 (1947).
21. R. W. Mattoon, R. S. Stearns and W. D. Harkins, J. Chem. Phys., 16, 156 (1948).
22. M. L. Corrin, J. Chem. Phys., 16, 844 (1948).
23. F. Reiss-Husson and V. Luzatti, J. Phys. Chem., 68, 3504 (1964).
24. M. Shinitzky, A. C. Dianaux, C. Gitler and G. Weber, Biochemistry, 10, 2106 (1971).
25. A. S. Waggoner, O. H. Griffith and C. R. Christensen, Proc. Nat. Acad. Sci. U.S.A., 57, 1198 (1967).
26. O. H. Griffith and A. S. Waggoner, Accounts Chem. Res., 1, 17 (1969).
27. H. V. Tartar, J. Phys. Chem., 59, 1195 (1955).
28. C. Tanford, J. Phys. Chem., 76, 3020 (1972).
29. C. Tanford, "The Hydrophobic Effect", Wiley, New York, 1973.
30. K. Shinoda in "Solvent Properties of Surfactant Solutions", K. Shinoda, Editor, p. 27, Dekker, New York, 1967.
31. K. W. Hermann, J. G. Bushmiller and W. L. Coerchene, J. Phys. Chem. 70, 2909 (1966).
32. K. W. Hermann, J. Phys. Chem., 66, 295 (1962).
33. G. S. Hartley, Nature, 142, 161 (1938).
34. I. Langmuir, in Alexander's Colloid Chemistry, 1, 525 (1926).
35. W. D. Harkins, in Alexander's Colloid Chemistry, 1, 192 (1926).
36. C. Tanford, Ref. 29, p. 14.
37. F. Irmann, Zeit. Chemi. Eng. Tech., 37, 789 (1965).
38. C. Hansch and S. M. Anderson, J. Med. Chem., 10, 745 and refs. therein (1967).
39. A. F. H. Ward, Trans. Faraday Soc., 42, 399 (1946).
40. R. Smith and C. Tanford, Proc. Nat. Acad. Sci. U.S.A., 70, 289 (1973).

41. E. G. Baker, Science, 129, 871 (1959).
42. F. Franks, Nature, 210, 87 (1966).
43. C. McAuliffe, J. Phys. Chem. 70, 1267 (1966).
44. K. S. Park and W. N. Bruce, J. Econ. Ent. 61, 770 (1968).
45. H. S. Frank and M. W. Evans, J. Chem. Phys., 13, 507 (1945).
46. G. Némethy and H. A. Scheraga, J. Chem. Phys., 36, 3382 (1962).
47. H. S. Frank, Science, 169, 635 (1970).
48. R. E. Smith, E. T. Fries and M. F. Morales, J. Phys. Chem., 59, 382 (1955).

# MICELLIZATION, SOLUBILIZATION, AND MICROEMULSIONS

Leon M. Prince
Consulting Surface Chemist
7 Plymouth Road
Westfield, New Jersey   07090


A typical commercial microemulsion of oil, water, surfactant, and cosurfactant was studied to distinguish among the relevant types of colloidal aggregates present.  A phase map and a plot of the ratio of the interphase volume to the core volume vs. aggregate diameters were drawn to aid in the interpretation of the results.  It was proposed that microemulsions may occur in the droplet diameter range of 10 to 200 nm as an alternative to micelle formation providing the chemical type of the oil and emulsifier match.  When droplet diameters are below 7.5 nm only ion pairs and micelles form because oil molecules outnumber surfactant molecules in the aggregate moiety.  Above this limit, surfactant species outnumber oil molecules so that surfactant and cosurfactant species may adsorb to the oil/water interface and there form an interphase capable of enveloping oil or water in microdroplets.  This would be a two phase emulsion system as opposed to a one phase micellar system.  In this context, the term solubilization embraces stable, transparent, or translucent systems of all kinds.

## INTRODUCTION

The terms micellization, solubilization, and microemulsions are considered in this presentation as types of aggregation in which a system of water, oil, and surfactant may exist. In this sense, solubilization and micellization convey meanings other than those originally ascribed to them but because of their recent wide usage in this context have become familiar terms to investigators in this field, and to this extent are unambiguous. It is the object of this work to try and differentiate among these terms in a pragmatic way.

To this end, a phase map has been drawn of a typical commercial microemulsion system which is characterized by water-in-oil (w/o) dispersions and oil-in-water (o/w) dispersions adjacent to and separated by a viscoelastic gel stage. The interpretation of these data is made on the basis of interactions among molecular species in a mixed film and disregards some of the newer thermodynamic approaches to systems of this type.

## EXPERIMENTAL

### Chemicals

The two oils, alpha-pinene and alpha-terpineol, were commercial samples supplied by the Organics Department of Hercules, Inc. The alpha-terpineol consisted of 95-97% alpha-terpineol and 3-5% other tertiary terpene alcohols. The alpha-pinene consisted of 88-93% alpha-pinene, 6-10% camphene, and 3-5% related terpene hydrocarbons.

AMP, 2-amino-2-methyl-1-propanol, was a commercial sample furnished by Commercial Solvents Corporation. This material is reputed to have a purity in the 90% range. The grade used in these experiments contained 5% water.

The Oleic acid was U.S.P.-F.C.C. Food Grade (J. T. Baker Chemical Co.) and the water was distilled tap water.

### Procedure

Three parts by weight of alpha-pinene were blended with one part of alpha-terpineol to make the _oil_ phase. The _soap_ phase consisted of 4 g Oleic acid and 5 g 95% AMP. Fifteen separate blends of soap with oil were studied using amounts of oil ranging from 0 to 140 g. To each of these blends, small aliquots of water were successively added and carefully mixed after each addition. The addition of water was continued until the system inverted to a fluid o/w dispersion. The pH of these dispersions was above 10.5.

Visual and optical streaming birefringent observations with polarizing plastic sheets were made after each water addition. The compositional points of systems were identified by different colors and shapes of points: (a) transparent (clear), isotropic w/o and o/w systems; (b) translucent (opalescent), isotropic w/o and o/w systems; (c) opaque and unstable, isotropic w/o and o/w systems; (d) visually hazy and birefringent thin gels of a cylindrical liquid crystalline phase; (e) a thick, clear birefringent gel of a lamellar liquid crystalline phase; and (f) a viscous, non-birefringent macroemulsion inversion stage.

The oil and soap were blended in a 250 ml or 600 ml beaker using a hot plate magnetic stirrer. To facilitate the blending with water, the temperature of the mixture was on occasion raised as high as 60°C. This helped to bring the blends homogeneously through the gel stage.

## RESULTS AND DISCUSSION

Figure 1 presents the experimental data in a phase map. Such a map is one showing the room temperature equivalents of the phases on the phase diagram.[1,2] Figure 1 differs from the usual Friberg School phase equilibria diagrams in three other important respects. First of all, the cosurfactant, alpha-terpineol, is included in the oil phase. Secondly, the oil phase blend was carefully selected to yield results similar to those obtained with the Carnauba wax, the natural and synthetic "emulsifiable" waxes, Chlordane, mineral oil, alkyds, and other commercial systems.[3,4] Thirdly, only phases relevant to the objectives of this study were identified.

The inclusion of the alpha-terpineol with the hydrocarbon, alpha-pinene, as a single component was an empirical carryover from the original Carnauba wax emulsion systems in which the alcohol component was part of the wax.[5] In a sense, the 3:1 mixture of alpha-pinene and alpha-terpineol emulates the Carnauba wax from an emulsifiable point of view. From a scientific viewpoint, this blending was not too unreasonable since the bulk of the alcohol resides in the oil phase where it lowers the original interfacial tension before addition of the soap from $\gamma_{o/w}$ to $(\gamma_{o/w})_a$.[6] Moreover, the fraction of alpha-terpineol in the interphase changes with the ratio of oil to soap and the water content of the system.[7] Thus, Figure 1 is less than perfect as a phase map. However, this is a matter of consequence only in the upper portions of the diagram, which, fortunately, is not important for the purposes of these arguments.

What is of paramount importance is the kind of pattern that is obtained by using the selected components. This is the pattern that is exhibited by commercial manufacturing processes. It is noteworthy that after 38 years in the field, only about 35 to 40 microemulsifiable combinations of oils and emulsifiers (including

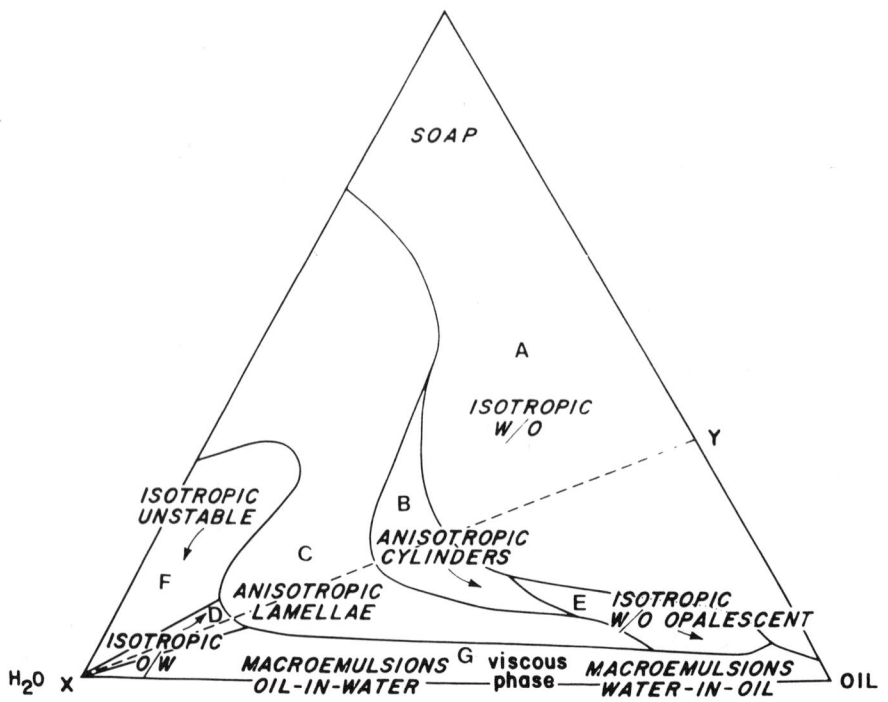

Figure 1. Phase map of water, oil (3 parts alpha-pinene: 1 part alpha-terpineol) and soap (4 g oleic acid plus 5 g AMP). Nine regions are shown.

nonionics) are known to the author. The reason for this is found in the interpretation of the data and forms the basis for the differentiation between micellar solutions and microemulsions.

In practice, microemulsions are found by slowly adding water to a mixture of a given oil and a high level of emulsifier, 30-100% on the weight of the oil. If the emulsifier matches the oil, a fluid transparent w/o dispersion first forms which, upon subsequent addition of small aliquots of water, undergoes inversion to a translucent or opalescent (more rarely, a transparent, clear) o/w microemulsion via the viscoelastic stage. Improper matching results in failure to achieve any one or all of these stages.

Experience has also shown that an oil can be matched to a given emulsifier by judicious blending with another oil. In the present work translucent or transparent o/w microemulsions were first made of alpha-pinene and alpha-terpineol alone. It was found, however, that a 3:1 blend of hydrocarbon to alcohol gave systems which passed through the viscoelastic gel stage with less difficulty.

In Figure 1, the addition of water to a given blend of oil and soap was traced along a straight line beginning on the oil-soap base line and extending to the water apex, e.g., line XY. Fifteen such lines were traced. The result of classifying each point on each line effectively outlined the regions A through G. These are discussed in terms of the molecular interactions believed to be taking place among the several species.

In Region A, since there was 5% water in the amine, the data points did not begin on the oil-soap base line. It is of interest, however, that in <u>all</u> cases, the first mixtures of oil, soap, and this small amount of water were clear.

Three kinds of aggregates probably exist in Region A but were not distinguishable by the techniques used in this work. The first type of aggregate to be formed at very low levels of water is one in which the water hydrogen bonds to the polar carboxylate, alcoholic, and amino groups on the soap and alpha-terpineol. This water is not free to form a core of bulk water until all the hydrogen bonding sites are occupied.[8,9] Such dispersions are considered to be ion pairs. Thereafter, and just prior to the formation of the liquid crystalline phases, inverted micelles form in which the water exists as a bulk phase in increasingly larger cores. These are called micellar solutions or $L_2$ phases. In proximity to the gel stage, it is submitted that because of the larger size of the droplets (core plus interphase), the aggregates are w/o microemulsions. To put this contention in better perspective, the arguments to support it are reserved until later in the discussion.

Region B is a liquid crystalline phase identified as optically hazy and birefringent. It is a thin gel in these systems and probably consists of an hexagonal array of cylinders of water.[7,8] Its disappearance above 50% soap is undoubtedly associated with the lowering of the amount of alcohol in the system with the concomitant lowering of the ratio of alcohol to soap in the interphase.

Region C is also a liquid crystalline phase but is made up of lamellar micelles. This heavier gel stage exists in low, medium, and high oil content systems adjacent to o/w microemulsions in Region D as well as adjacent to unstable o/w emulsions in Regions F and G. It also exists in the soap-water system at zero oil at intermediate water contents. Its ubiquitousness serves to illustrate the point that, although a given oil and emulsifier system may form a clear, transparent w/o dispersion and form a viscoelastic gel stage, this is not assurance that an o/w microemulsion will form.

Region D is obviously a very select region, occurring only as a result of careful matching of emulsifier in the compositional areas of both amount and kind. Microemulsions in this range were photographed in the electron microscope using special techniques.[10,11] Droplets ranging from 120 nm down to 7.5 nm were observed. It is of interest that the 7.5 nm droplets were considered at that time (1960) to approximate the dimensions of swollen micelles and myelin figures, dimensions which were independently determined by x-ray measurements. Such small aggregates were not believed to be in the microemulsion range. Thus, in Region D there may be two types of aggregates, o/w microemulsions and normal micelles, i.e. an $L_1$ phase or micellar solution phase.

The appearance of opalescent or translucent w/o dispersions at higher levels of oil in Region E came as a surprise. They seemed to coincide with the disappearance of cylindrical micelles in the viscoelastic gel stage. It may not be just a coincidence that this situation occurs in the lower part of the map in which lamellar micelles invert to o/w macroemulsions.

Region F is bounded by lamellar micelles, translucent dispersions, and the water-soap base line. Normal micelles ($L_1$ phase) exist on the base line at zero oil and opalescent o/w microemulsions are adjacent to it in Region D. The dispersions in the center of this region separate into a lower, translucent (aqueous) phase and an upper, opalescent, o/w microemulsion phase. This behavior is ascribed to the shortage of alcohol needed for the interphase to achieve microemulsification of all the oil.

Region G is the macroemulsion range. In it very unstable w/o macroemulsions on the right invert through a viscous or creamy stage[12] to remarkably stable o/w macroemulsions on the left. This behavior differs markedly from that of the systems above it in the map. It suggests that, at least in this range, the Hydrophile-Lipophile Balance (HLB) is playing its normal role, forming more stable w/o emulsions in the emulsifier HLB range of 3-8 and more stable o/w emulsions in the emulsifier range of 10-18. (The HLB of the amine soap and terpineol emulsifier system is in the higher range.) This may be a clue to differentiating microemulsions from micellar solutions.

It is based upon the way in which oil molecules in a system of oil, water, and surfactant may possibly interact with different sized aggregates. Figure 2, drawn from previous data,[7] puts the issue in a geometric perspective.

The curve is a plot of the ratio of the volume of the interphase, $V_i$, to the volume of the core, $V_c$, versus the diameter of spherical aggregates having a 2.5 nm thick interphase (the length of the oleate radical). The sharp bend in the curve may signal a change in aggregate form from swollen micelle to microdroplet to macrodroplet. When the film thickness is only 0.5 nm, the bend occurs at slightly lower (ca. 6 nm) droplet diameters.

Geometrical considerations are the key to this viewpoint. As $V_i/V_c$ decreases, the number of molecules of surfactant and cosurfactant in the interphase simultaneously increases in proportion to the square of the diameter of the core of the bulk internal phase. This is because the number of surfactant species that can orient themselves on the surface of the core is proportional to the area of the polar group in the case of w/o dispersions and to the area of the tail in the case of o/w dispersions. The result is that as the diameter of an aggregate increases, the volume of the interphase unoccupied by surfactant species, and therefore available to

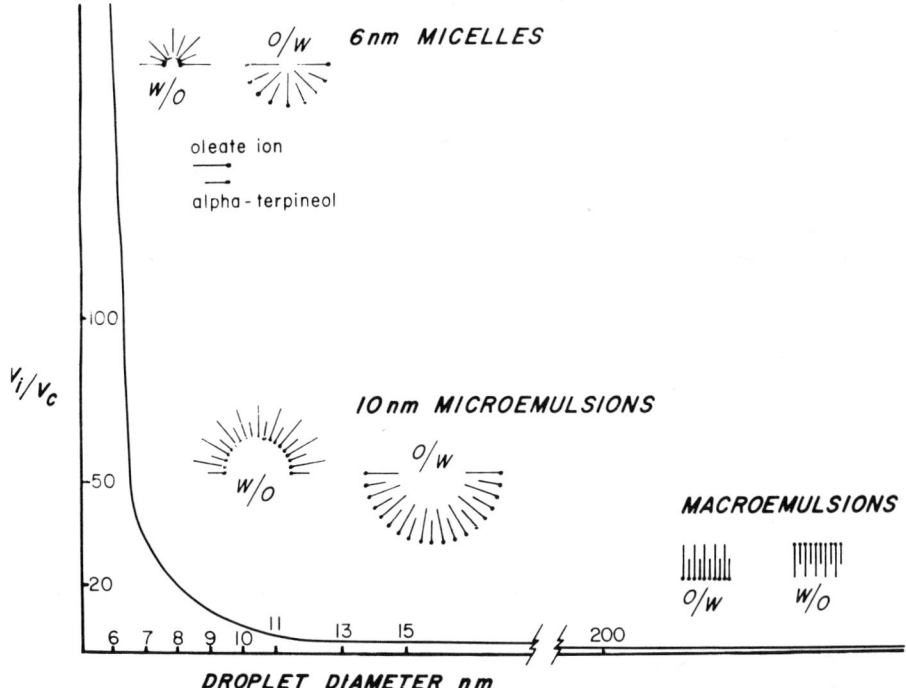

Figure 2. Curve of the ratio of interphase volume, $V_i$, to core volume, $V_c$, versus droplet diameters having an interphase thickness of 2.5 nm. Models of a change in aggregate character from micelle to microemulsion to macroemulsion are shown corresponding to changes in $V_i/V_c$.

hydrocarbon molecules, decreases. These are the only molecules that can fill the interfacial void volume since water molecules cannot penetrate the interphase beyond the polar heads and the alpha-terpineol distribution is fixed.[6]

It is understood that the showing of the interphase species as rigid sticks in an interface that is exactly 25 Å thick is for illustrative purposes only. The molecules are, of course, in a continuous state of motion. This, however, only affects the volume relationships shown in the models in a minor way.

Thus the changes in void volume with aggregate size may be divided into three distinct stages. In the first, below an aggregate diameter of 7.5 nm, alpha-pinene molecules greatly outnumber the surfactant species in the interphase. In the second, in the droplet range of about 10 to 200 nm, the surfactant species outnumber the hydrocarbon molecules. In the third, the packing of the surfactant species is so tight and they are adlineated so parallel that few hydrocarbon molecules penetrate the interphase. These circumstances are illustrated by models drawn on the graph for both w/o and o/w dispersions.

A feature of these models is the difference between the positions that the hydrocarbon molecules occupy in the w/o as opposed to the o/w systems. This is due to the inversion of the taper of the wedge formed by the surfactants in the interphase. When oil is the continuous phase, hydrocarbons penetrate the open volume among the tails up to the polar heads. When water is the continuous phase, however, the bulk of the hydrocarbon molecules are situated closer to the polar heads where the larger void volume now exists. There is probably some dissolution of surfactant tails in the bulk oil phase but it is suggested that this is minimal. This aspect of the interphase has not received the attention it deserves. It explains, among other things, why a large chaotropic cation like AMP favors o/w dispersion and why a mole-ratio of cation to oleate ion in excess of 1 abets o/w microemulsification. In the present study this ratio is about 4.

In the perspective of the models of Figure 2, a differentiation between micellar solution and microemulsion may be made providing one also takes into consideration the matching of emulsifier and oil molecules. This matching is used in the same context as is "chemical type" in the HLB concept. When droplet diameters are below 7.5 nm and oil molecules outnumber surfactant molecules in the interphase, it would appear that surfactant and cosurfactant species can only aggregate in the form of ion pairs or micelles, spherical, cylindrical, lamellar, etc. But above this 7.5 nm limit, when surfactant species outnumber oil molecules in the interphase, there is a possibility that as an alternative to micelle formation, the surfactant and cosurfactant species may adsorb to the oil/water interface and there form an interphase capable of enveloping oil or water in microdroplets. This would be a two phase system as opposed to a one phase micellar system. In the droplet diameter range of 10 to 200 nm, the two phase systems are called microemulsions. The responsibility for their formation lies with the matching of oil to emulsifier, whatever this may entail. Failure to match results in micellar solutions in this same aggregate size range. Reasons for and the consequences of this matching have been discussed previously.[6,7,13,14,15]

Some support for the validity of these inferences may be found in Figure 1. On line XY, w/o and o/w microemulsions appear to exist on either side of the viscoelastic gel stage at the same emulsifier HLB. This is contrary to the HLB concept. The premise that addition of water to these systems can change emulsifier HLB is a mistaken one. Schulman and cockbain[12] inverted their macroemulsions by the use of polyvalent cations. Sears and Schulman[16] found that electrostatic repulsion between adjacent soap molecules was not an important factor in determining the expansion of a monolayer. And finally, even drastic changes in phase volume ratios do not necessarily effect inversion.[17] At this time, one can only associate this unusual behavior with the high volume ratio of interphase to core of these very small droplets. In such a situation, the effect of water molecules on the polar groups of the tenants of the interphase and its penetration by oil phase molecules play an as yet undetermined role.

## DEFINITIONS

In conclusion, three definitions are offered:

<u>Micelles</u>. Micelles are aggregates of colloidal dimensions existing in equilibrium with the molecules or ions from which they are formed.[18] No solvent, oil, or water need be included.

<u>Microemulsions</u>. Microemulsions are spherical aggregates of oil or water dispersed in the other liquid and stabilized by an interfacial film of one or more surfactants such that the droplets range in diameter from 10 to 200 nm.

<u>Solubilization</u>. In the context of this study, solubilized systems are stable, fluid, isotropic, transparent or opalescent dispersions of oil, water, or surfactant. Solubilized systems are generally water-continuous dispersions but may be oil-continuous. Cosolubilized systems,[19,20] in which the components are molecularly dispersed, are also included in this category.

## ACKNOWLEDGMENTS

The advice and constructive suggestions of Dr. Marjorie Vold are deeply appreciated. The author wishes to thank Commercial Solvents Corporation and Hercules, Inc., for their cooperation in furnishing samples.

## REFERENCES

1. M. J. Buerger, L. B. Smith, F. V. Ryer, and J. E. Spike, Jr., Proc. Nat. Acad. Sci. <u>31</u>, 226 (1945).
2. A. W. Adamson, "Physical Chemistry of Surfaces", p. 377, Interscience Publishers, Inc., 1966.
3. L. M. Prince, in "Emulsions and Emulsion Technology", K. J. Lissant, Editor, pp. 134-139, Marcel Dekker, Inc., New York, 1974.
4. L. M. Prince, "Microemulsions, Theory and Practice", Chapter 2, Academic Press, Inc., New York, in press.

5. A. H. Warth, "The Chemistry and Technology of Waxes", 2nd ed., Reinhold Publishing Corp., New York, 1956.
6. L. M. Prince, J. Colloid Interface Sci., 23, 165 (1967).
7. L. M. Prince, J. Colloid Interface Sci., 52, 182 (1975).
8. D. O. Shah and R. M. Hamlin, Jr., Science, 171, 483 (1971).
9. P. Ekwall, L. Mandell, and K. Fontell, J. Colloid Interface Sci., 33, 215 (1970).
10. W. Stoeckenius, J. H. Schulman, and L. M. Prince, Kolloid Z., 169, 170 (1960).
11. L. M. Prince, Soap Chem. Specialties, 36, No. 9, 103 (1960); ibid., No. 10, 99 (1960).
12. J. H. Schulman and E. G. Cockbain, Trans. Faraday Soc., 36, 661 (1940).
13. J. H. Schulman, W. Stoeckenius and L. M. Prince, J. Phys. Chem. 63, 1677 (1959).
14. L. M. Prince, J. Colloid Interface Sci., 29, 216 (1969).
15. L. M. Prince, J. Soc. Cosmet. Chem., 21, 193 (1970).
16. D. F. Sears and J. H. Schulman, J. Phys. Chem. 68, 3529 (1964).
17. A. W. Adamson, "Physical Chemistry of Surfaces", p. 403, Interscience Publishers, Inc., 1966.
18. Pure Applied Chem., 31, 612 (1972).
19. D. O. Shah, "On Distinguishing Microemulsions from Cosolubilized Systems", presented at 48th National Colloid Symposium under the auspices of the American Chemical Society, Austin, Texas, June 24-26, 1974.
20. D. O. Shah, R. D. Walker, W. C. Hsieh, N. J. Shah, S. Dwivedi, J. Nelander, R. Pepinsky and D. W. Deamer, "Some Structural Aspects of Microemulsions and Co-Solubilized Systems", Paper No. SPE 5815, presented at Improved Oil Recovery Symposium of the Society of Petroleum Engineers of AIME, Tulsa, Oklahoma, March 22-24, 1976.

# BIOLOGICAL IMPLICATIONS OF MICELLE FORMATION

Alexander T. Florence

Department of Pharmaceutical Technology
University of Strathclyde
Glasgow, G1 1XW, United Kingdom

There are several well-known examples of biologically important micellar behaviour in living organisms. Probably the best known is the involvement of the bile salts in the absorption of fats, in which a micellar phase can be identified. There are many other processes in which the aggregated or micellar phase of naturally occurring molecules, drug molecules or added surfactant can possibly influence biological evens. Self-association or mixed association may be one method by which it is possible for the body to control and store bioactive molecules. Aggregate formation between ATP and 5-hydroxytryptamine is thought to allow these molecules to be stored in permeable storage granules and subsequently to be released on deaggregation in response to changes in calcium levels. The action of cyclic polypeptides on membrane function has been considered to be due to reversible molecular aggregation. Mixed micelle formation between amphipathic drug molecules and other amphipaths may result in decreased biological activity. The propensity of many drug molecules to aggregate is not in itself of great biological importance but can be of use in controlling the transport of these drugs across polymer membranes in the form of microcapsules or implant capsules. Although surfactants used pharmaceutically may reduce the amount of free drug by solubilisation and can thereby reduce biological availability, surfactant micelles are too labile to allow utilisation of solubilised drug systems in the control of drug release. Some attempts have recently been made to form more stable micellar systems.

## INTRODUCTION

The work of Alexander and Trim[1] on bactericide-surfactant mixtures showed 30 years ago that surface active molecules, especially those with the ability to form micelles, are not biologically inert substances - that they can act synergistically with bactericides and at micellar concentrations inactivate active species by solubilisation. Since that time there has been much work on the nature of biomembranes which aids in our understanding of surfactant effects in biological systems, but there are still several gaps in our knowledge. In this paper some biological implications of micelle formation, whether by naturally occurring biologically active species or by supposedly inactive detergent molecules, will be discussed. It is probably best to separate two distinct events that occur on micelle formation: the change in the properties of the <u>monomer</u> and the change in the properties of the <u>solution</u>. Both have biological significance. To this can be added the importance of the presence of a micellar "phase" in relation to solubilisation of active molecules or membrane components and to the direct involvement of the micellar phase in catalysis of reactions[2].

When amphipathic molecules aggregate, several biologically significant changes occur in the system of which they constitute a part: the concentration of monomeric species may increase only slowly or may decrease with increase in total concentration and the transport and colligative properties of the system are changed. If the amphipathic molecules have an intrinsic biological activity, then the formation of aggregates might well lead directly to an alteration in biological activity. One can envisage the aggregation of naturally occurring molecules leading directly to a change in biological activity due to decreased transport rates or decreased ability to pass through biological barriers. Or the ability of the aggregated species to interact with other biological species may change and there may be a physical alteration in the environment caused by micelle formation. To neglect the possibility of aggregation of biological molecules is to neglect what is undoubtedly in some cases the mechanism of controlling the thermodynamic activity and hence biological activity of the compound <u>in vivo</u>.

Biological activity and especially the specificity of many biochemical processes demands organisation. Aggregation provides one level of organisation of molecules and it is reversible. The monomer $\rightleftharpoons$ micelle transition may well be one of the many control mechanisms in the body. Certainly, the bilayer lamellar $\rightleftharpoons$ micellar equilibrium in biological membranes postulated by Lucy and others[3] seems to be widely accepted as a means of controlling membrane properties. A great effort has been applied to the study of biological membranes but less attention has been paid to control mechanisms which are not membrane mediated. While many biological activities are understood at a pharmacological or biochemical level, the molecular nature of the activity is often less well understood. One of the areas of lack of knowledge is in the field of control and regulation of activity of natural hormones and the molecular changes that occur in the sequence of events that leads to biological activity.

Macromolecules are frequently implicated as receptors and biological barriers but the body has evolved systems in which assemblies of molecules arise through their amphipathic properties; thsse assemblages frequently behave like macromolecules in colloidal terms, but have the distinct advantage that they are rapidly reversible systems. This is an essential attribute to maintain the rapidity of many biological events. Because biological membranes have evolved as complicated bilayers with hydrophobic interiors and hydrophilic exteriors, foreign surfactant molecules readily interact with them. Many drug molecules, such as local anaesthetics and tranquillisers, whose activity depends on the interaction with membranes are, not suprisingly, amphipathic in character[4].

These observations provide the three themes of this paper, namely a discussion of aggregation of biological molecules, the interaction of foreign surfactants in the micellar state with biological systems and some observation on the significance of micelle formation in systems containing drugs or other biologically active substances. The paper will attempt to define some areas in which the aggregation of molecules of natural origin might be of biological importance especially in a biological control or regulatory function, and will discuss better known natural surfactant systems such as the bile salts and their involvement in fat and drug absorption pathways. As surfactant molecules are ingested in foodstuffs and medicines[5], an account is given of the significance of the micellisation of some of these adjuvant molecules. This is based on work in our laboratories on the biological action of drug-surfactant mixtures[6].

Mukerjee[7] has pointed out the many ways in which amphipathic molecules can associate. It is not the purpose of this paper to discuss modes of association or micellisation, nor to quantify the properties of the resulting aggregates except insofar as these draw attention to the possible implications of the process in the biological milieu. The terms "micelle" and "micellisation" will be used freely in spite of the fact that the particular systems may not always meet Mukerjee's criteria. The micellisation process or processes akin to it, which result from hydrophobic interactions in solutions and elimination of hydrocarbon-water contacts, is so widespread in nature that the evolution of so many systems with the ability to form ordered arrays must have resulted from some biological advantage.

## AGGREGATION OF NATURALLY OCCURRING MOLECULES

The critical concentration for the association of many naturally occurring compounds is high and often is greater than circulating levels of the compound. This does not diminish the significance of association, for many hormones and other substances are concentrated in cells or special organelles. For example, the surface active steroidal antibiotic, fusidic acid, is concentrated approximately fifty times inside E. coli in solutions of $2 \times 10^{-6}$g ml$^{-1}$[8] and adenosine triphosphate and biogenic amines attain concentrations of the order of 20% in storage granules[9]. While it has been recognised that the quaternary structure of many proteins is crucial for their activity and function, it has also been recognised that many protein molecules aggregate[10]. Frieden[11] has suggested that the reversible association of the enzyme bovine liver glutamate dehydrogenase (GDH-ase) may serve a critical function in the control of certain metabolic pathways in vivo. The early observation that effector molecules known to modify GDH-ase activity also dramatically altered the state of aggregation led to this hypothesis that association was probably linked to a metabolic control mechanism, a proposal which is realistic as GDH-ase reaches concentrations in liver cell mitochondria (several milligrams per litre) above the concentrations at which Eisenberg[12] found it to aggregate in vitro. Because of aggregation, the enzyme undergoes an allosteric transition at a lower level of activation than the monomeric form. The tendency for aggregation of the active form is much greater than for the inactive form. As active polymers are formed, the concentration of active monomeric is depleted since the equilibrium between active and inactive forms is maintained. This pulls some inactive monomers into the active form[13]. There seems to be little reason then that aggregation of other active molecules will not induce biological changes. In the case of the enzyme, it is the change in conformation and masking of reactive groups on association that causes change in activity.

## MEMBRANES AND MICELLISATION

One of the regulatory structures in the body living organisms is the cell membrane. The simple bimolecular leaflet structure for the lipid portion of the membrane probably still is considered to form the main bulk of the mammalian membrane, but the concept of a more flexible and mobile structure has been suggested by Lucy and others[3]. While the current views on membrane morphology are constantly shifting and recent texts should be consulted, it is salutory to consider work which has discussed the micellar nature of membrane lipids. Much of this is speculative but illustrates one approach to the problem.

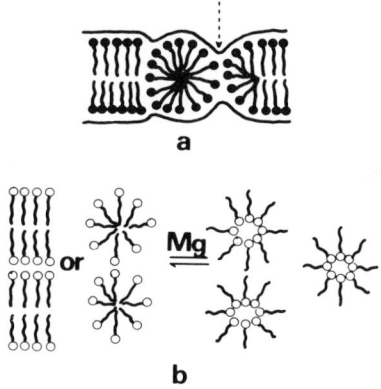

Figure 1.a) A representation of the association of lipids in a biological membrane and the formation of 'polar discontinuities' caused by the juxtaposition of polar head groups, after Watkins[14].
   b) Inverse micelle formation in membrane in bilayer or micellar state induced by Magnesium ion, after Maas and Coleman[16].

Association of the hydrophobic tails of membrane lipids leads, as depicted in Figure 1.a, to what Watkins[14] has described as polar discontinuities. Transition to the micellar state is considered to be essential to allow cell fusion to occur as the biomolecular leaflet is thermodynamically a stable system which would resist coalescence with similar structures. External influences can, however, induce phospholipid aggregation and can thus alter the permeability of cell membranes to water-soluble and oil-soluble species. Calcium ions for example induce inverse micelle formation in phospholipid systems[15]. Other metal ions also result in this transformation (Figure 1.b); addition of adenosine triphosphate (ATP) removes the metal and leads to a reversion to the normal micellar pattern. Maas and Coleman[16] have postulated that such transitions may have significance in nerve membrane operation. Metal-ATP-phospholipid complexes can be seen as forming lamellar sheets; on removal of ATP, for example by ATP-ase, the metal-phospholipid complex will assume a quite different orientation. Replacement of ATP from some metabolic source will then cause a reversion to the original configuration. Changes in the solution bathing cells lead to shifts in membrane properties.

The glycoprotein components of cell membranes are amphipathic and it has been suggested that increases in cation, protein or polycation content in the aqueous phase cause surface aggregation of these glycoprotein molecules[17]. Cations bind to the surface negative groups and reduce surface negative charge. This allows aggregation of mobile glycoprotein units to occur and membrane resistance decreases immediately on aggregation[18]. The change in

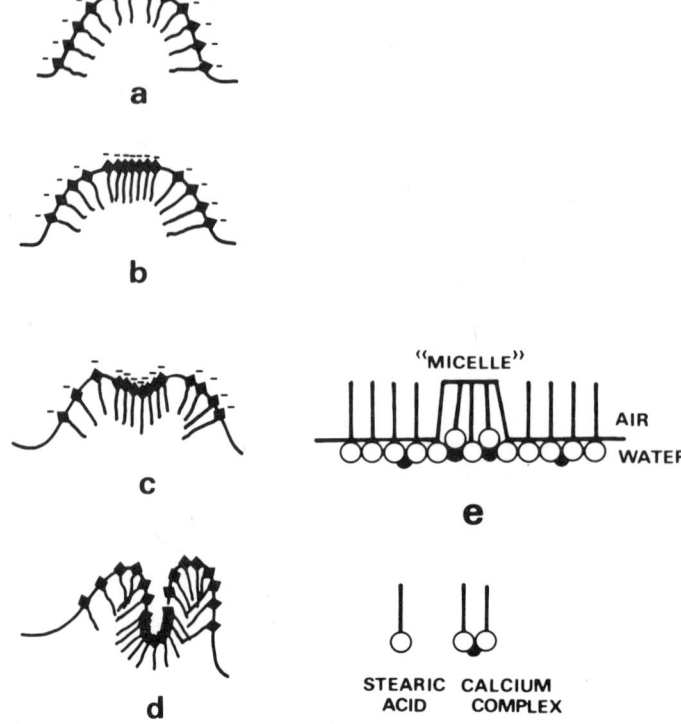

Figure 2.a-d) Aggregation of amphipathic glycoprotein as a trigger for pinocytosis after Gingell in Mammalian Cell Membrane, Butterworths, 1976, see reference 20. a) normal state of cell surface, b) salt or protein in medium aggregates glycoproteins, c) filaments contract and changed ionic milieu, d) channel formation.
 e) Surface "micelle" formation by stearic acid in the presence of calcium ion from Neumann[21], by permission of Academic Press.

the composition of the surface layer has been thought to provide a trigger for pinocytosis (Figure 2a - d). The increase in surface pressure resulting from aggregation would be relieved by the expansion and folding of the membrane. Elworthy's suggestion that changes in local dielectric constant can cause micellar changes in membranes (based on work on the effect of dielectric constant on the transition of micellar states of lecithins)[19] predates the globular micelle membrane model.

Calcium ions are vital to the functioning of cell membranes and to many biological processes. The surface micelles that Bray has postulated[20] have analogues in simpler surfactant systems. At around physiological pH for example, monolayers of stearic acid molecules associate in the presence of Calcium chloride into surface micelles approximately 6 nm in diameter[21] (Figure 2e). The formation of surface micelles in membranes may be closely related to the structural changes and entropy decrease reported to occur during membrane excitation[21].

Molecules added to biological systems may find their way into membranes and associate in the lipid system. Several compounds of fungal or bacterial origin (such as alamethacin, which forms micelles in aqueous systems[22]) readily incorporate into planar lipid bilayers and form channels by self-association, channels which allow the translocation of ions and other hydrophilic species through the otherwise hydrophobic membranes. Based on calculations Mueller[23] has made on the rates of association possible for channel formation by lateral diffusion of random molecules of alamethacin located in the membrane surface, he concluded that the molecules are probably already aggregated in the local structures at the membrane surface.

The stability of flat arrays of globular micelles may require interactions with protein or mucoprotein molecules because of the proximity and concentration of charged head groups: on the other hand, the ability of phospholipid-cholesterol dispersions to form neat and middle phases in which there is a similar juxtaposition of head groups suggests that this might not be a problem. Labilisation and stabilisation of membranes may be connected respectively with induction and reduction of the micellar state. On increasing the micellar fraction, the permeability to water will rise dramatically and lead to eventual osmotic damage to the membrane system. The volume of the membrane must change as the micellar phase is much less economical in the use of space. Surface active drugs such as chlorpromazine, SKF 525 A[24], and local anaesthetics and surfactant molecules may well induce micelle or mixed micelle formation in membranes. Chlorpromazine hydrochloride has been shown to efficiently solubilise pure membrane lipids[25]. Certainly the penetration theory of their action does not lead to an easy explanation of the ability of these drugs to increase membrane transport whereas penetration and induction of a micellar phase might.

## STORAGE AND RELEASE OF NATURAL SUBSTANCES IN THE BODY

Hormones and transmitter substances are frequently stored in the body in relatively high concentrations awaiting signals for their release. There are three possible explanations of the ability of high concentrations of organic electrolytes to be stored within an organelle enclosed by a lipid membrane: i) the free drug concentration is reduced to negligible proportions by complexation with a macromolecule, or ii) by self-association or mixed micelle formation, or iii) the permeability of the membrane is only

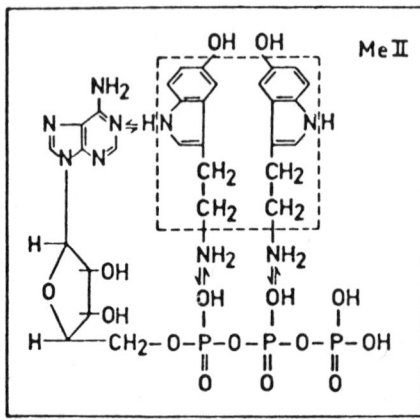
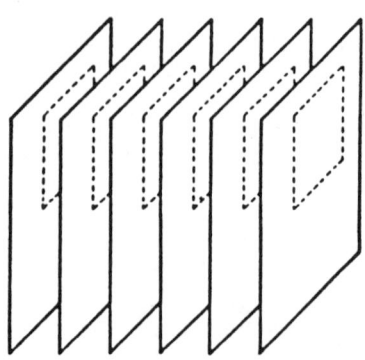

Figure 3. Hypothetical structure of aggregates formed by 5-hydroxytryptamine and ATP reproduced with permission from A. Pletscher et al., Biochem. Psychopharmacol. Vol. 2, 205. Raven Press, New York.

increased when release is required. The association explanations have been suggested for the storage of biogenic amines in storage vessels within various sites in vivo[26]. One of the striking observations about norepinephrine (NE) storage vessels is that they contain considerable quantities of ATP nearly always in a stoichiometric ratio to NE of 1:4. An important means of storing amines in the adrenal and neuronal vesicles may therefore involve interactions between NE and ATP, possibly facilitated by divalent cations (Figure 3). The micellar aggregates postulated by Pletscher[27] are large and non-diffusable. Berneis[28] has found that the size of the aggregates increases with increasing concentration of the solutes and is dependent on calcium ion concentrations.

Micelle formation can also explain the osmotic stability of the organelles despite their high levels of ATP (20% w/v) and amines such as 5-hydroxytryptamine (25%). The rapidity of release of the biogenic amines is of the order of 1 ms, judged by synaptic delay times and much of this time must be accounted for by the time required for molecules to traverse the neuronal gap which is 100 nm wide[9]. Rates of micellar breakdown of ∼0.1 ms have been observed for surfactant micelles[29] while rates of exocytosis (the alternative mode of extracting the amines from inside the storage vesicles) are too long - of the order of 100 to 200 ms[9].

## OTHER AGGREGATING SYSTEMS AND THEIR SIGNIFICANCE

Chlorophyll associates in non-aqueous solvents, the aggregated species possessing different spectral properties. It has been suggested that the shift in the emission maximum of chlorophyll which occurs in the greening of Euglena results from the formation of aggregates containing about 4 molecules of chlorophyll[30].

Protoporphyrin is highly aggregated in aqueous systems above 4 µM[31] and although the possible biological role of these aggregates has not yet been evaluated, if reservoirs such as the postulated haem pools exist, aggregation undoubtedly takes place in vivo and must be taken into account in biochemical studies on these systems[32]. The realisation of the possibility of association in biochemical systems in vitro is important. "Micellar" aggregation in solutions of purified Forssman hapten results in a loss in activity[33], and similar complications arise in systems containing caseins. Although not a typical small amphiphile K-casein is amphipathic and associates to give rise to structures similar to the surfactant-like micelles proposed by Waugh et al[34] for α and β-casein[35]. Hydrodynamic measurements suggest spherical micelles containing 20 - 30 monomers with radii of 10 nm. The non-polar part of the molecule is oriented inwards while the charged acidic peptides and the hydrophilic carbohydrate containing portions are near the micelle surface (Figure 4). Neutrophils can recognise casein molecules; but this chemotactic activity is very dependent on the molecular form of the casein[36]. It is found that proteins which excite a locomotor response in cells do so by virtue of having externally positioned non-polar groups[36]. Therefore chemotactic activity is reduced when the casein is in a micellar form.

◇ charged groups    ◀ non-polar groups

Figure 4. Diagrammatical representation of a K-casein micelle redrawn from Wilkinson[36] by permission. This shows the mode of association in aqueous media with non-polar groups in the centre of the aggregate and charged groups to the exterior, as in a conventional micelle.

Calcium is again implicated. At low $Ca^{++}$ concentrations the casein is in monomeric state and chemotactic activity is maximal.

## BILE SALTS

The bile salts represent a more typical range of micelle forming surface-active molecules. Their micelle-forming capacity has been comprehensively reviewed[37,38]. They have a well-known role in fat absorption processes. It is in fact generally agreed that one of their principal physiological functions is aiding fat absorption[39]. They render fat absorption more efficient although in the total absence of bile about 70% dietary fat can be absorbed. Hofmann and Borgström [40] have shown that mixed micelles of bile salts, fatty acids and monoglycerides can act as vehicles of fat transport. The role of the bile salts in lipolysis is not an obligatory one as bile salts can be replaced by non-surfactant molecules or synthetic surfactants[39]. Below their cmc's bile salts prevent lipase from inactivation by preventing unfolding at the fat-water interface. An inhibitory effect of bile salts above their cmc's has been reported recently but only in the absence of the amphipathic peptide "co-lipase". Naturally in the intestinal contents the bile is above the cmc and co-lipase is a necessary factor[41,42]. At the cmc Brogström and Erlanson[42] have suggested that a monolayer of bile salts prevents the lipase reaching the substrate. As it is surface-active, co-lipase can penetrate this bile salt barrier.

Bile salts above their cmc displace lipase from hydrophobic interfaces, and this parallels the loss of lipase activity[43,44]. Anionic detergents of the acyltaurine, decanoyl taurine and dodeconylsarcosyltaurine type also inhibit lipase when present above their cmc's, a process reversed by co-lipase. These detergents behave as bile salts[45] and it may be of interest that analogues have been isolated from the gastric contents of the crab[46]. A number of steroidal antibiotics, 3-acetylfusidic acid, cephalosporin $P_1$ and helvolic acid have recently been found by Carey et al[47] to have micellar properties very similar to the bile salts and it has been suggested that they may serve as model compounds for detergent replacement in bile salt deficiency syndromes.

The biliary route is a major route of drug elimination, but the importance of bile salt micelles in drug transport, absorption and metabolism is not as well understood as their role in fat absorption. Whether binding to micelles has importance in hepatic transport and biliary excretion of organic anions has been discussed by Vonk and his colleagues[48]. Most authors agree that there is not a simple relationship between stimulation of bile flow and stimulation of biliary output of organic anions, but taurocholate (TC), which promotes formation of biliary micelles, stimulates biliary excretion of certain compounds more than agents like theophylline and dehydrocholate and hydrocortisone which induce biliary excretion without micelle formation. Several suggestions implicating micelles have been put forward[49], including facilitation of transport from liver to bile by direct effect on canicular membranes; stimulation of micelle formation inside the liver cells; binding of the drug anions to micelles and subsequent exocytosis of these aggregates into bile canaliculae and binding of anions to micelles in the canaliculae with consequent reduction of free drug concentration leading to decreased transfer of drug from bile back into the liver.

Taurocholate is able to form micelles; infusion of TC has been shown[48] to increase output of phospholipid and cholesterol with which it forms mixed micelles. Dehydrocholate has a low tendency to micelle formation and has little influence on the biliary excretion of cholesterol and phospholipid. However, Hardison and Apter[49] have found that while micelle formation is an important factor in the biliary excretion of lipids it cannot alone explain the results they have obtained. Dehydrocholate, a synthetic agent, is a potent choleretic presumably because its osmotic activity in bile is not diminished by micelle formation and it may therefore exert an osmotic force in biliary canaliculae approximately equal to its concentration. As a result, bile flow is increased more by this substance than by micelle-forming bile salts. Some of the metabolites of dehydrocholate are, however, capable of micelle formation[50].

Hypersection of acid leads to precipitation of bile salts in the intestine and malabsorption syndromes. Neomycin and kanamycin precipitate bile salts and among other adverse effects also lead to malabsorption[51].

## INTERACTIONS OF FOREIGN MICELLAR SYSTEMS

The ability of surfactants to solubilise membrane components has been used to advantage in studies of membrane structure and function, but despite extensive literature describing surfactant treatment of membranes there is no clear account of mechanisms of membrane-surfactant interactions[52]. Part of the reason is the diversity of membrane structure and composition and the wide range of surfactant types studied. Sodium dodecyl sulphate binds to all components uniformly and the release of components from the detergent-membrane aggregate is roughly in proportion to the water solubility of the components[53].

Onset of lipid solubilisation coincides with the cmc. This is the level at which conformational changes in membranes begin and rodlike NaDS-protein complexes form after incorporation of all the lipid into NaDS micelles. These complexes can also solubilise cholesterol, the main component responsible for the integrity of the membranes. Above the cmc, phospholipids are removed from proteins by deoxycholate micelles which permits solubilisation of cholesterol. Similar sequences are observed with non-ionic detergents such as Triton X-100[54].

It is little wonder then that incorporation of surfactants into biological systems readily alters transport across membranes but in no simple way. Surfactants are added to medicines as emulsifiers, suspending and wetting agents therefore the biological implications must be considered[53]. The biphasic influence of many surfactants on biological membranes has been emphasised: at low concentrations polysorbate 80 and similar non-ionic surfactants enhance the permeability of biological membranes (Figure 5). At the cmc this effect is maximal and activity determined by transport decreases due to solubilisation of the free drug molecules, and we believe physical blocking of the membrane surface, perhaps

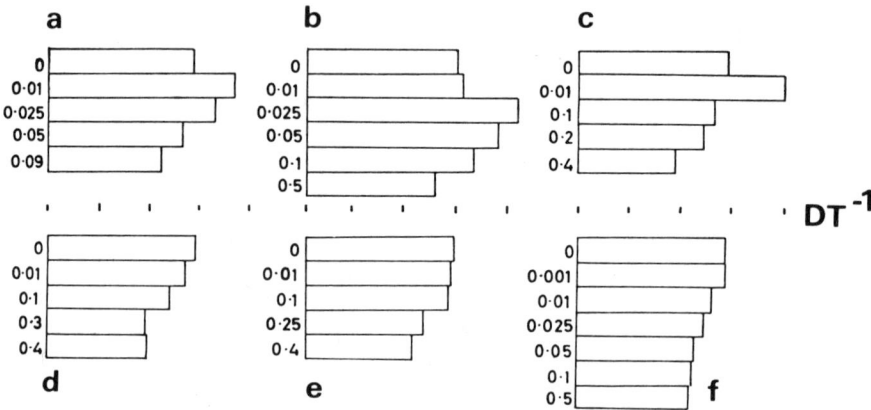

Figure 5. Absorption of thioridazine in goldfish in the presence of increasing concentrations of various non-ionic detergents, the rate of absorption being proportional to the reciprocal of the death time of the fish, reciprocal death time is plotted on the ordinate concentrations of surfactants (% w/v) are marked. Lack of enhancement of absorption by some surfactants probably due to poor ability to penetrate lipid membranes because of shape factors. Decrease in absorption due to non-ionic micelle formation. From Florence & Gillan[6] by permission of the Journal of Pharmacy and Pharmacology. The surfactants are all Atlas products (Honeywill-Atlas, U.K.).
a) Atlas G2162    b) Renex 650    c) Atlas G1790
d) G1298          e) G1300        f) Cremophor EL

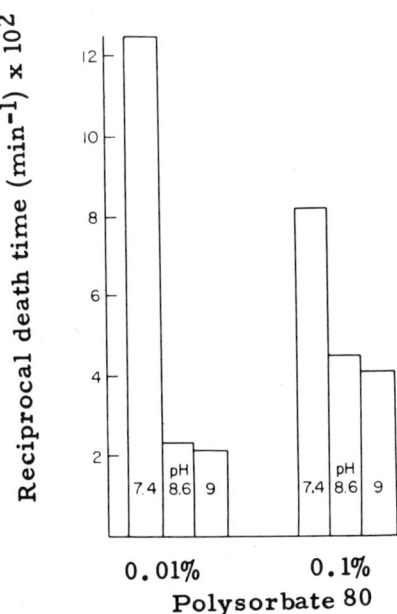

Figure 6. The influence of pH and polysorbate 80 concentrations of the absorption from solutions or suspensions containing 0.08% thioridazine as shown by reciprocal death times of goldfish from Florence & Gillan[7] by permission of Pesticide Science.

by formation of hemi-micelles, or by simple competition with drug molecules for surface sites.

In the presence of saturated solutions with solid drug, surfactant micelles will alter dissolution controlled rates of absorption by increasing the rate of solution of drugs. The effect of two concentrations of polysorbate 80 on the activity of thioridazine in goldfish is shown in Figure 6 illustrating the relative importance of drug solubilisation in increasing solubility (i.e. the concentration gradient) and decreasing the free drug concentration. The relative affinity of the drug for the micelles and for the membrane is of course a relevant factor to be considered. The change in partitioning and transport properties of the drug at the cmc in non-ionic surfactant systems is shown in Figure 7.

Solubilised systems therefore have to be used with caution in medicine. Too high concentrations of surfactant may cause tissue damage and also decrease the thermodynamic activity of the drug substance[55]. The main problem with such systems as drug carriers is their lability - on dilution _in vivo_ drug is precipitated albeit as fine particles. Experiments at an early stage in our laborat-

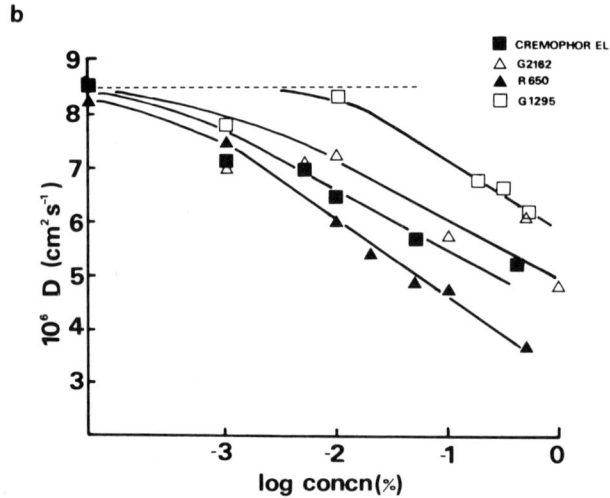

Figure 7. a) The apparent partition coefficient of thioridazine between decane and phosphate buffer solution as a function of polysorbate 80 concentration.
b) Diffusion coefficients of thioridazine (0.4%) as a function of concentration of various non-ionic surfactants.
■ Cremophor EL   △ Atlas G2162   ▲ Renex 650   and □ Atlas G1295*.
Florence A.T. & Gillan J.M.N. (unpublished work).

ories have been aimed at polymerising micelles by gamma irradiation of polyoxyethylated non-ionic micelles[56]. Cross-linked micelles should have some interesting detergent, biological and pharmaceutical properties.

The polymerised micelles prepared by Elias[57] perhaps have some use in pharmacy although they were synthesised for electron microscopy and the related 'nanocapsules' of Speiser[58] also have interesting properties. If liposomes can be considered to be micelle of sorts, these convoluted amphipathic dispersions which have the ability to solubilise drug substances and whose surface charge can be altered for specialised chemotactical exercises[59] (to direct the drug-loaded liposome to specific cells) should be included in a comprehensive review of this subject.

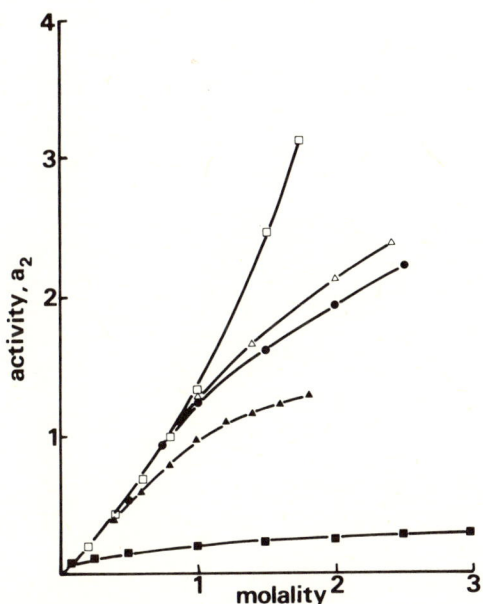

Figure 8. Solute thermodynamic activity of alkyl polyoxyethylene ethers in aqueous solutions as a function of molality.
☐ $CH_3CH_2(OCH_2CH_2)_6OH$ - does not form micelles.
△ $CH_3(CH_2)_3(OCH_2CH_2)_6OH$ at 20° and ▲ at 30° and ● $(CH_3)_2CH.CH_2(OCH_2CH_2)_6OH$ at 20°; ■ $(CH_3CH_2)_2CH.CH_2(OCH_2CH_2)_6OH$ at 20°C.

## ASSOCIATION OF DRUG MOLECULES

A large number of drug molecules are amphipathic and associate, usually at fairly nonphysiological high concentrations to form aggregates of varying size[4,60]. It is most likely that the surface-activity of these molecules rather than the ability to micellize is of significance biologically, although the propensity of the molecules to form associations by hydrophobic bonding will manifest itself in interactions with proteins, with other drugs and with hydrophobic receptor sites. While accumulation of drug molecules in certain sites in the body is likely and while this may possibly as with the biogenic amines and porphyrins lead to aggregate formation, it is probably only in pharmaceutical systems that micellar concentrations are reached. The non-ionic polyether surfactants have local anaesthetic properties[61]. The change in thermodynamic activity[62] of these compounds at their cmc, shown in Figure 8 obviously complicates their behaviour when an homologous series is under study. It is in homologous series of anaesthetics (and many other drugs) that a decrease in biological activity is seen at higher hydrocarbon chain lengths[63]. While some have ascribed this

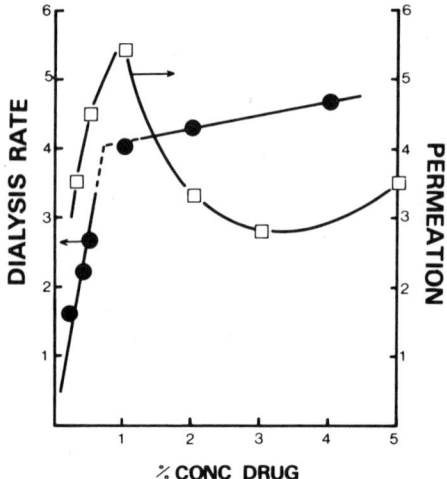

Figure 9. Rates of dialysis of chlorpromazine, a micelle forming drug with a cmc ~ 0.7% in water, through cellophane membranes in aqueous solutions above and below its critical micelle concentration. This may be compared with the rate of permeation of the drug through Silastic[R] membranes.

# BIOLOGICAL IMPLICATIONS OF MICELLE FORMATION

decreased activity to limitations in aqueous solubility it may also arise from micelle formation.

There is in medicine a growing need for systems which control the release of drugs, systems such as polymer (Silastic$^R$) implants and microcapsules[64]. Use of concentrated solutions of surface active drugs or of solubilised systems leads to a degree of control like that exercised in storage granules. Figure 9 shows the variation in penetration rate of chlorpromazine from aqueous solutions above and below its cmc across cellophane and polymethylsiloxane membranes. In the former a simple filtration of monomers occurs. The movement across the lipophilic silicone membrane, which probably resembles more a biological barrier, results from diffusion of the drug into the polymer; the changed partition characteristic matters here as the permeability coefficient P (a measure of drug transfer from bulk solution to the solution on the other side of the membrane) is the product of the drug partition coefficient (Kp) and the drug difference coefficient (Dm) within the membrane, provided that diffusion through the aqueous diffusion layers is not rate-limiting. Under steady state conditions P can be calculated from[65]

$$P = \frac{(dq/dt)l}{AC}$$

where $dq/dt$ is the rate of permeation at steady state, l and A are the membrane thickness and area respectively and C is the concentration of unionised drug in the bulk fluid.

$$\frac{dq}{dt} = (constant)(KpDm)C$$

Dm will be unaltered by micelle formation but both Kp and C are altered, Kp decreasing on aggregation and the effective concentration C being affected both directly by micellization and by an alteration in the ionisation of the total system following on aggregation (Figure 10). This would qualitatively explain the results with silicone membranes and micellar drug. As biological membranes are more akin to lipophilic polymer membranes than cellophane membranes these measurements have some relevance in understanding the biological activity of micellar solutions of drug molecules.

## CONCLUSIONS

The properties of micellar systems in biological experiments cannot be viewed in isolation. In the laboratory concentration of surfactant at surfaces does not significantly alter the state of the system but accumulation of surfactant molecules at surfaces are

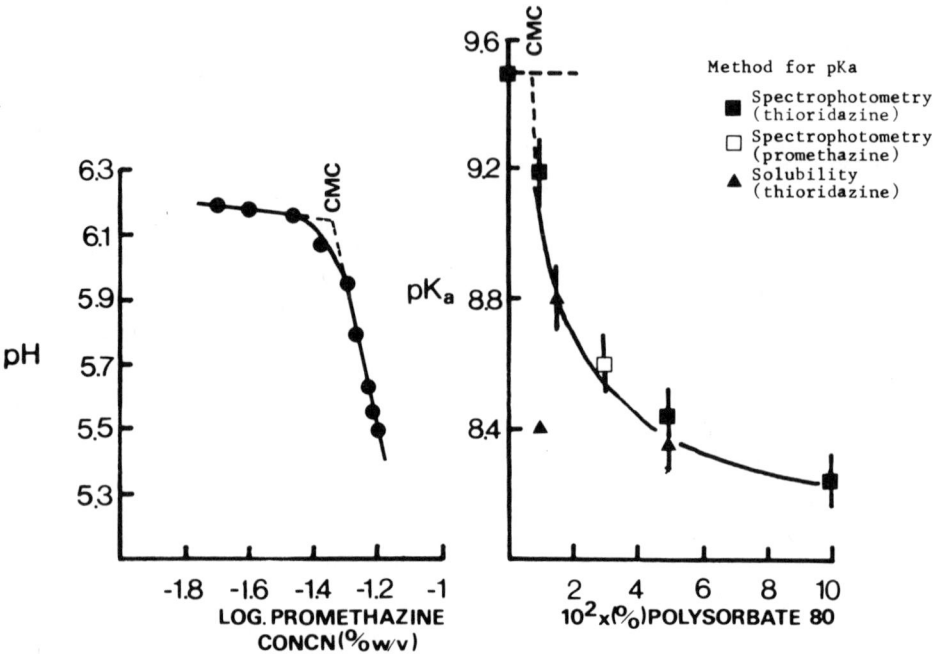

Figure 10. Change in effective pKa on micelle formation by the drug molecules is indicated by pH decrease on micelle formation by cationic drug. Association of these drugs with non-ionic detergent micelles also causes a reduction in apparent $pK_a$ values. (Florence & Gillan, unpublished).

of vital importance in living systems. The micellar phase observed in simple solutions cannot be held to be the same as in a biological environment where there are many possibilities for solubilisation or an interaction of membrane components and other materials in the environment.

## ACKNOWLEDGEMENTS

The author thanks the following for permission to use published material: Academic Press for Figure 1a, the Editor of Nature for Figure 1b and Churchill-Livingstone for Figure 4. Other sources are listed in the legends to the figures.

## REFERENCES

1. A.E. Alexander and A.R. Trim, Proc. Roy. Soc. [B], <u>113</u>, 220 (1946)

2. J.H. Fendler, Acc. Chem. Res., 9, 153 (1976)
   J.H. Fendler and E.J. Fendler, in "Catalysis in Micellar and Macromolecular Systems", Academic Press, New York, N.Y. (1975)
3. J.A. Lucy, J. Theoret. Biol., 7, 360 (1964)
4. A.T. Florence, Adv. Colloid. Interface, Sci., 2, 115 (1968)
5. P.H. Elworthy, A.T. Florence and C.B. Macfarlane in "Solubilisation by Surface-Active Agents", Chapman and Hall, London, 1968
6. A.T. Florence and J.M.N. Gillan, J. Pharm. Pharmac., 27, 152 (1975)
   A.T. Florence and J.M.N. Gillan, Pestic. Sci., 6, 429 (1975)
7. P. Mukerjee, J. Pharm. Sci., 63, 972 (1974)
8. P.M. Bennett and O. Maale, J. Mol. Biol., 90, 541 (1974)
9. R.J. Baldessarini in "Handbook of Psychopharmacology" Volume 3, S.D. Iverson and S.H. Snyder, Editors, Plenum Press, New York 1975
   H.G. Weder and U.W. Wiegard, FEBS Lett., 38, 64 (1973)
10. I.M. Klotz, Science, 155, 697 (1967)
11. C. Frieden, J. Mol. Biol., 238, 3286 (1963)
12. H. Eisenberg, Acc. Chem. Res., 4, 379 (1970)
13. R.J. Cohen, J.A. Jedziniak and G.B. Benedek, Proc. Roy. Soc. Lond., A345, 73 (1975)
14. J.C. Watkins, J. Theoret.Biol., 9, 37 (1965)
15. M. Wolman and H. Weiner, Nature, 200, 486 (1963)
16. J.W. Maas and R.W. Coleman, Nature, 208, 41 (1965)
17. D. Gingell, J. Theoret. Biol., 38, 677 (1973)
    D. Gingell in "Mammalian Cell Membranes" Volume 1, G.A. Jamieson and D.M. Robinson, Editors, Butterworths 1976
18. P.W. Brandt and A.R. Freeman, Science, 155, 582 (1967)
19. P.H. Elworthy and D.S. McIntosh, Kolloid Z., 195, 27 (1964)
20. D. Bray, Nature, 244, 93 (1973)
21. R.D. Neuman, J. Colloid Interface Sci., 53, 161 (1975)
22. A.I. McMullen and J.A. Stirrup, Biochim. Biophys. Acta, 241, 807 (1971)
23. P. Mueller, Ann. N.Y. Acad. Sci., 264, 247 (1975)
24. A.T. Florence, J. Pharm. Pharmacol., 22, 1 (1970)
25. M.C. Corey, P. Hirom and D.M. Small, Biochem. J., 153, 519 (1976)
26. P. O'Brien, M. Da Prada et al., Life Sci., 11, 749 (1972)
27. A. Pletscher et al., Brain Res., 62, 317 (1973)
28. K.H. Berneis, U. Goetz, M. Da Prada and A. Pletscher, Naunyn-Schmiedeberg's Arch. Pharmacol., 277, 291 (1973)
29. N. Muller, J. Phys. Chem., 76, 3016 (1972)
30. S.S. Brody, J. Theoret. Biol., 7, 352 (1964)
31. S.B. Brown, M. Shillock and P. Jones, Biochem. J., 153, 279 (1976)
32. S.B. Brown, and R.F.G.J. King, Biochem. J., 153, 479 (1976)
33. R.L. Rush and M.F. Mallette, Immunochemistry, 7, 115 (1970)
34. D.F. Waugh, L.K. Creamer, C.W. Slattery and G.W. Dresdner, Biochemistry, 9, 786 (1970)
35. P.C. Wilkinson, Nature, 251, 58 (1974)
36. P.C. Wilkinson in "Chemotaxis and Inflammation", Churchill-

Livingstone, Edinburgh, 1974
37. A.F. Hofmann and D.M. Small, Ann. Rev. Med., 18, 333 (1967)
38. M.C. Carey and D.M. Small, Arch. Intern. Med., 130, 506 (1972)
39. H. Brockerhoff and R.G. Jensen in "Lipolytic Enzymes", Academic Press, New York, 1974
40. A.F. Hofmann and B. Borgström, J. Clin. Invest., 43, 247 (1964)
41. B. Borgström and C. Erlandson, Eur. J. Biochem., 37, 60 (1973)
42. M.F. Maylie, M. Charles, M. Astier and P. Desneulle, Biochem. Biophys. Res. Comm., 52, 291 (1973)
43. B. Borgström, J. Lipid Res., 16, 411 (1975)
44. B. Borgström, and J. Donner, J. Lipid Res., 16, 287 (1975)
45. R. Lester, M.C. Carey, J.M. Little, L.A. Cooperstein and S.R. Dowd, Science 189, 1098 (1975)
46. A. Van den OOrd et al., J. Biol. Chem., 240, 2242 (1965)
47. M.C. Carey, J-C. Montet and D.M. Small, Biochemistry, 14, 4896 (1975)
48. R.J. Vonk, P. Jekel and D.K.F. Meijer, Naunyn Schmiedeberg's Arch. Pharmacol., 290, 375 (1975)
49. W.G. Hardison and J.T. Apter, Amer. J. Physiol., 222, 61 (1972) G.E. Gibson and E.L. Forker, Gastroenterology, 66, 1046 (1974)
50. W.G. Hardison, J. Lab. Clin. Med., 77, 811 (1971)
51. W.W. Faloon, I.C. Paes, D. Woolfolk, H. Nankin, K. Wallace and E.N. Haro, Ann. N.Y. Acad. Sci., 132, 879 (1966)
52. A. Helenius and K. Simons, Biochim. Biophys. Acta, 415, 29 (1975)
53. F.H. Kirkpatrick, S.E. Gordesky and G.V. Marretti, Biochim. Biophys. Acta, 345, 154 (1974)
54. K. Simons, A. Helenius and H. Garoff, J. Mol. Biol., 80, 119 (1973)
55. A.T. Florence, J. Pharm. Pharmacol., 22, 265 (1970)
56. A.T. Florence, T.L. Whateley and A. Al Saaden, unpublished work
57. U. Kammer and H. Elias, Kolloid Z., 250, 344 (1972)
58. P. Speiser, Progr. Colloid Polymer Sci., 59, 48 (1976)
59. G. Gregoriadis, N. Engl. J. Med., 292, 215 (1975)
60. D. Attwood, A.T. Florence, R. Grieg and G.A. Smail, J. Pharm. Pharmac., 26, 847 (1974); A.T. Florence, J. Pharm. Pharmac., 22, 1 (1970)
61. A.T. Florence, A. Bowman and W.E. Sneader, unpublished results
62. A.T. Florence, Ph.D. Thesis, University of Glasgow (1965) P.H. Elworthy and A.T. Florence, Kolloid Z., 208, 157 (1966)
63. T. Eckert, E. Kilb and H. Hoffmann, Arch. Pharm., 297, 31 (1964)
64. "Microencapsulation", J.R. Nixon, Editor, Marcel Dekker, New York, 1975
65. E.G. Lovering and D.B. Black, J. Pharm. Sci., 63, 671 (1974)

# FLUORESCENT PROBES FOR MICELLAR SYSTEMS

Nicholas J. Turro, Margaret W. Geiger,
Richard R. Hautala and Neil E. Schore
Chemistry Department
Columbia University
New York, New York 10027

Naphthalene and pyrene are useful fluorescent probes for the study of micellar systems. The fluorescence lifetime and intensity of these molecules are sensitive to variation which depends on micellar structure and dynamics. It is shown that oxygen effects may result from the effect of oxygen in micelles. The probes provide information concerning local oxygen concentration in the micellar phase. The processes of micellar catalysis and inhibition may be directly observed by using the naphthalene and pyrene probes. A direct measurement of the partitioning of naphthalene between the micelle and aqueous phases has been made. The fluorescent properties of the indole group have also been employed to probe micellar structure. An indole capable of making its own micelles has been prepared. Wavelength correlated single photon counting has been used to resolve the fluorescence of the compound into aqueous and micellar components.

## FLUORESCENT PROBES

A chromophore is an atom or group of atoms that may be conveniently considered as a unit which absorbs visible or ultraviolet light. A lumophore is an atom or group of atoms that may be conveniently viewed as a unit which emits visible or ultraviolet light. For example, aromatic hydrocarbons behave as lumophores, as do ketone groups. There are two commonly encountered luminescences: fluorescence and phosphorescence. Fluorescence is generally associated with the emission of light accompanying the transition of excited singlet states to the ground state and phosphorescence is generally associated with the emission of light from the lowest triplet state to the ground state (Figure 1).

A luminescent probe is any molecule or group of atoms whose luminescent properties may serve to explore and examine the structural and dynamic features of systems of interest. Thus there exist fluorescence probes and phosphorescence probes. In order to be effective, a property of a luminescent probe must change and be measurable when the probe is added to the test system. The general idea behind a luminescence probe[2] is that luminescence parameters are sensitive to changes in the microenvironment of the probe and that a luminescence probe in different microenvironments will display experimentally distinct luminescence properties which will characterize uniquely each environment. The readily measurable properties of lumophores[1] are

(1) Emission spectrum (intensity of emission as a function of wave length for a fixed absorbed intensity).

(2) Excitation spectrum (intensity of emission at fixed wavelength as a function of absorbed intensity).

(3) Luminescence decay (fall off of the intensity of emission spectrum as a function of time after an excitation pulse).

(4) Quantum yield of luminescence (the number of photons emitted relative to the number of photons absorbed).

(5) Polarization of luminescence (the orientation of the emitted light vector relative to the orientation of the absorbed light vector).

(6) Quenching (a decrease in luminescence intensity or increase in decay rate as a function of added solute or changing environment).

(7) Sensitization (an increase in luminescence intensity or decrease in decay rate as a function of added solute or changing environment).

The "dynamic range" of luminescence probes spans 13 orders of magnitude! Radiative lifetimes of fluorescence commonly range from $10^5$ sec$^{-1}$ to $10^9$ sec$^{-1}$ and radiative lifetimes of phosphorescence range from $10^{-1}$ sec$^{-1}$ to $10^3$ sec$^{-1}$. Since quantum

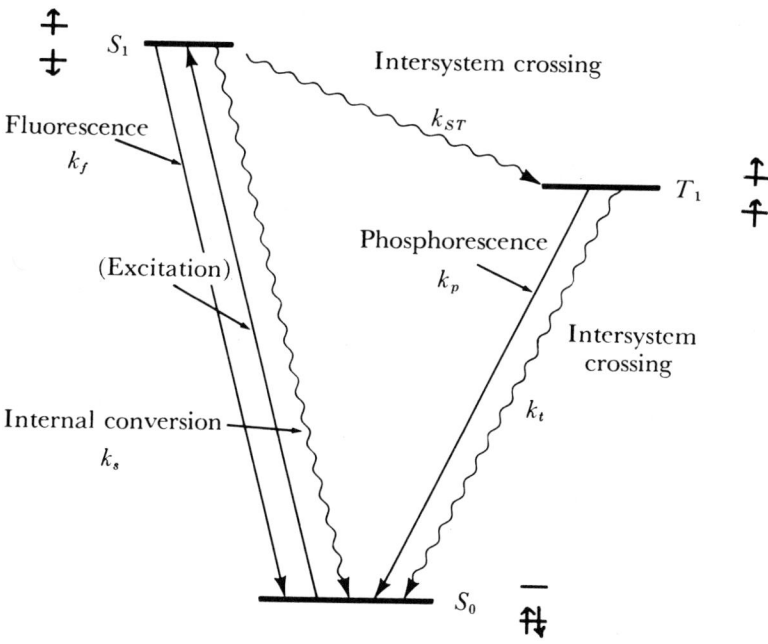

Figure 1. Typical energy diagram. The diagram shows schematically the major radiative and radiationless transitions commonly encountered in organic molecules.

yields of emission of $10^{-3}$ are readily measured, the effective experimental dynamic range of fluorescence is $10^6$ $sec^{-1}$ to $10^{12}$ $sec^{-1}$ and that of phosphorescence is $10^{-1}$ to $10^6$ $sec^{-1}$. Thus there is a continuum for dynamic measurements for the range $10^{-1}$ $sec^{-1}$ to $10^{12}$ $sec^{-1}$ (Figure 2).

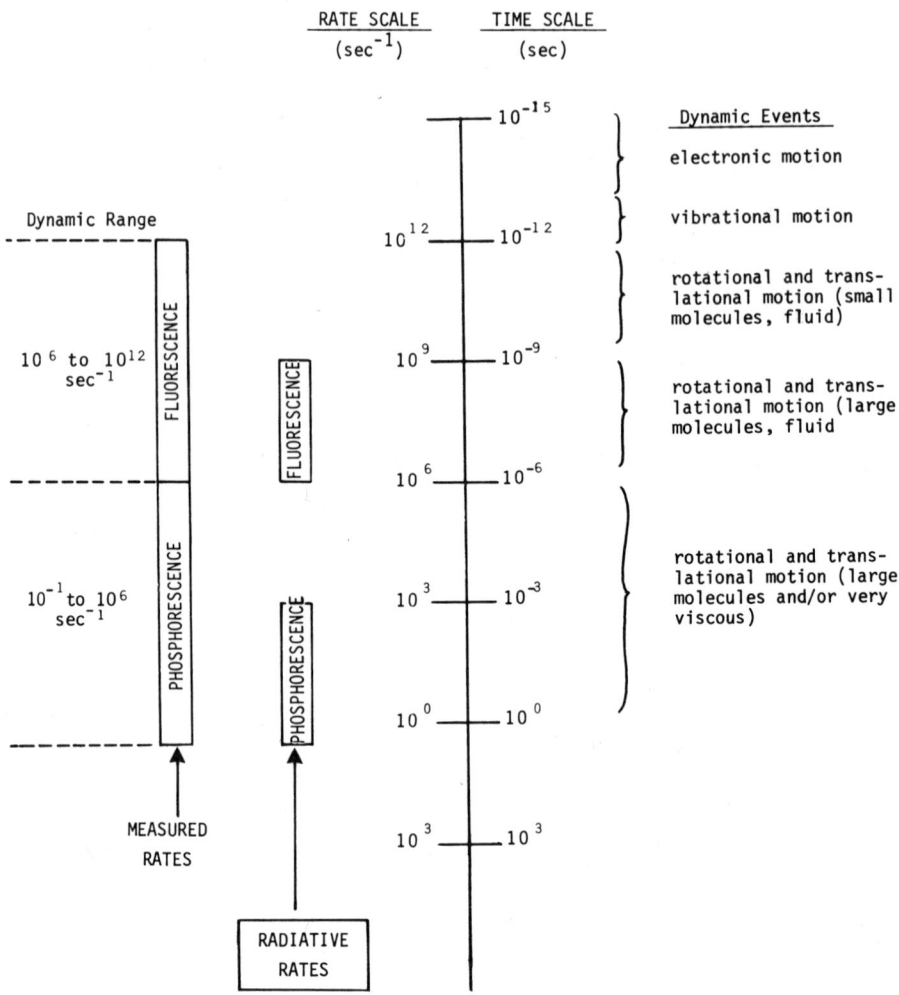

Figure 2.  The dynamic range of luminescence probes.

The measured changes in a luminescent probe have been related to molecular properties of the environment of the probe such as:
 (1) Polarity
 (2) Viscosity and diffusion

(3) Concentration of quenchers
(4) Distance between groups

Luminescence probes may be thought of as "sensors" of microenvironments. The photons emitted by such probes are "reporters" of the local structure in the vicinity of the probe at the time of emission. The emitted photons may be employed to "read back" the map of the microenvironment of the probe. Use of pulsed excitation adds a temporal dimension to the probe. In this mode, the probe serves as a clock of molecular dynamics of its environment.

Luminescent probes provide the greatest degree of information for characterization of microenvironments when the probe exhibits a strong and selective affinity for a given environmental site or if, when it is distributed among distinct sites (e.g., micellar environment and bulk aqueous solution) the measureable luminescent properties are distinctly different. For example, if the emission spectrum, the excitation spectrum, the decay time, the quantum yield of emission, the polarization of luminescence, or the quenching (or sensitization) of emission is environment dependent, the luminescence probe can be employed to characterize the environment. The effective use of luminescence probes requires a detailed knowledge of the nature of radiative and radiationless transitions and of the mechanisms of excited state interactions with different environments.

It should be stated that one difficulty must always be considered when interpreting results derived from the use of luminescence probes : the probe of necessity is a "foreign body" in the environment that is being probed. It is therefore crucial to provide independent evidence that the probe does not change the inherent quality of the measured property. In the case of micellar systems, it must be established whether the associations of the probe in the micelle causes a fundamental change in the nature of the micelles. In the work reported here, it was found that the inclusion of naphthalene, pyrene or indoles in detergent solutions did not change the CMC and that the change in fluorescence phenomena observed only occurred at the CMC or at higher concentrations of detergent.

In this work, we have employed only fluorescence probes.

Many fluorescence probes which have been employed in the literature[2] suffer from the disadvantage that the probe is a relatively large and/or highly polar molecule. Such probes are inherently likely to be of little use in studying the properties of the hydrophobic regions of micelles, since they may cause significant perturbations of the very environment they are intended to explore.

With respect to the study of micellar systems, fluorescent probes have been applied by various research groups. We have employed napthalene (N), pyrene (Py), and the indole (In) lumophore (Scheme I) as fluorescent probes to examine the structure and dynamics of solutions of the common micelle forming detergents.[3-5] The experimental techniques employed were standard fluorescence intensities and time correlated single photo counting.

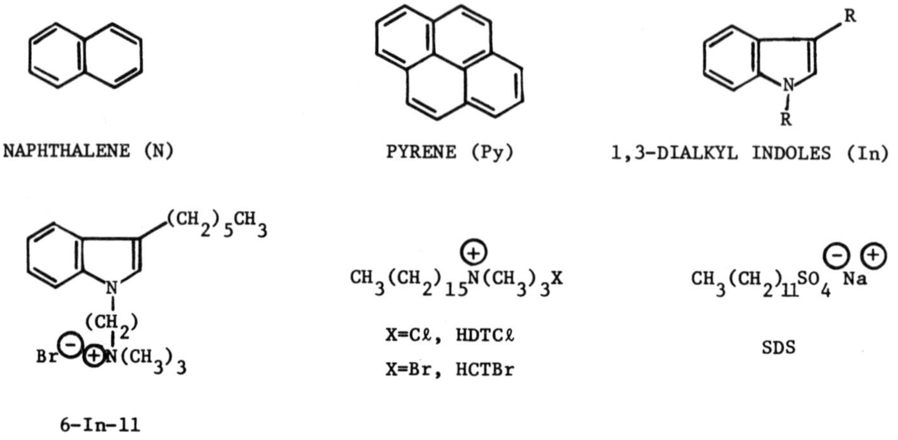

Scheme I. Structures of Molecules Employed in This Study.

## Naphthalene in HDTBr, HDTCℓ and SDS

The spectral distribution of the fluorescence spectrum of naphthalene is essentially identical in aqueous and in non-polar solvents. It is therefore expected and found that there is little difference between naphthalene's fluorescence spectrum in aqueous solution or in micelle containing detergent solutions.[3] However, the lifetime of naphthalene fluorescence ($\tau_F$ is very different in micellar versus aqueous solutions). The reason for this difference is the differing effectiveness of quenching of naphthalene fluorescence in different environments.

For example, the lifetime of naphthalene[3] fluorescence in (air saturated) water is $\sim 40 \times 10^{-9}$ sec, whereas (above the CMC) in HDTCℓ, HDTBr and SDS (Scheme I) $\tau_F$ is $\sim 23 \times 10^{-9}$ sec, $\sim 10 \times 10^{-9}$ sec and $\sim 70 \times 10^{-9}$ sec, respectively. How can we explain these very wide differences in $\tau_F$? A clue derives from the observation that upon pruging these solutions with nitrogen gas, $\tau_F$ is found to increase substantially in all cases except for HDTBr.

We interpret the results as follows:
(1) Quenching by oxygen determines the observed $\tau_F$ of naphthalene in aerated water and in micellar solutions of HDTCℓ and SDS.
(2) Quenching by $\mathrm{Br}^-$, the gegen ions of the cationic micelle HDTBr determines $\tau_F$ in micellar solutions of HDTBr.
(3) Oxygen is more soluble in a micellar environment than in an aqueous environment.
(4) Oxygen molecules may be transported across the micellar bulk aqueous interface.

In the case of naphthalene in micellar solutions of HDTCℓ and in HDTBr, time correlated single photon counting revealed (Figure 3) <u>two</u> distinct values of $\tau_F$, indicating that naphthalene is located in <u>both</u> a micellar and an aqueous environment. The observation of two distinct values of $\tau_F$ requires that the dynamics of movement of the fluorescence probe are slow relative to lifetime limiting processes. These results represent simultaneous direct observation of processes occuring in <u>both</u> a micellar and aqueous environment.

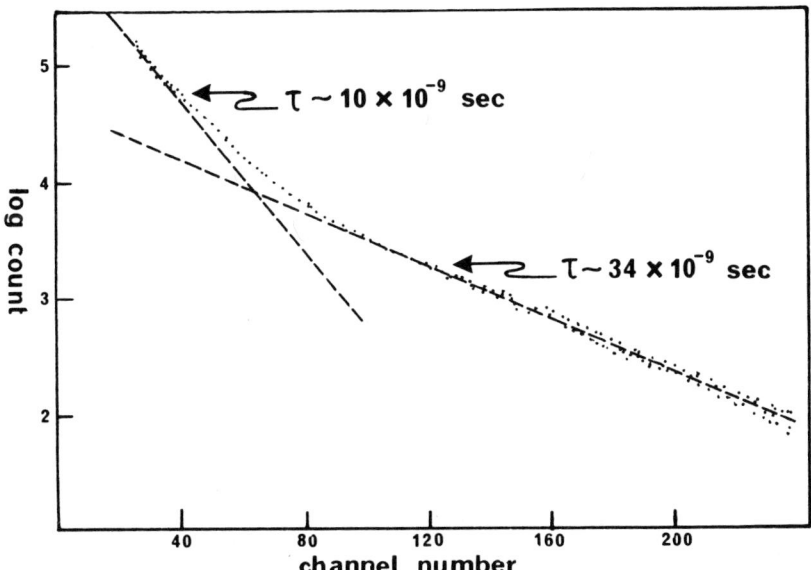

Figure 3. Fluorescence decay curve of naphthalene ($2 \times 10^{-4}$M) in HDTBr ($2 \times 10^{-2}$M) measured using the single photon counting technique. Extrapolation of the long component ($\tau_b$ = 34 nsec) to zero excitation time (channel 21) yields a value of 0.11 for the coefficient of the component. The short component (a = 0.89) corresponds to a lifetime of 10 nsec. The multichannel analyzer was calibrated at $9.25 \times 10^{-10}$ sec per channel for this experiment.

The fact that bromide (but not chloride) quenches naphthalene fluorescence allows a study of the factors influencing interactions between a substrate "dissolved" in a micelle, counterions of the micelle and other ions in the aqueous phase. This type of situation is typical of micellar catalysis chemical reactions. The figure shows examples of micellar "catalysis" and micellar "inhibition" of the quenching of naphthalene fluorescence. Thus, $\tau_F$ is smaller in HDTBr micelles than in the aqueous phase because the bromide ions are associated as counterions with the positively charged surface of the micelles, and therefore possess a high local concentration with respect to quenching naphthalene fluorescence. On the other hand, in solutions containing anionic detergents (e.g., SDS) the excited naphthalene is protected from quenching by bromide ions. In this case, the negatively charged surface of the micelle causes <u>inhibition</u> or <u>protection</u> from quenching by negative ions. The "catalysis" and "inhibition" are shown schematically in Figure 4.

## Pyrene in HDTCℓ, HDTBr and SDS

Pyrene (Py) is a relatively large non-polar molecule that possesses an unusually long inherently fluorescent lifetime. For example, in $N_2$ saturated water $\tau_F \sim 225 \times 10^{-9}$ sec (cf. naphthalene, $\tau_F \sim 60 \times 10^{-9}$ sec). The value of $\tau_F$ for Py ($\sim 10^{-6}$M) is lower in aerated aqueous and micellar solution than in $N_2$ purged solution.[5] For example, $\tau_F$ is $\sim 125 \times 10^{-9}$ sec, $\sim 160 \times 10^{-9}$ sec and $\sim 160 \times 10^{-9}$ sec in aerated water, HDTCℓ and SDS, respectively. Figure 5 shows the effect of detergent concentration of $\tau_F$ of Py. Clearly, micellar effects are operating.

It is found that oxygen may be added or removed from the solutions and $\tau_F$ is thereby affected reversibly. This is taken to mean that oxygen molecules move freely in and out of micelles. From quantitative study of the effect of oxygen quenching on $\tau_F$ of Py, it may also be concluded that oxygen is at least as or more soluble in the micellar environment than in bulk aqueous solution. Comparison of the lifetimes of pyrene and naphthalene in nitrogen purged and aerated HDTCℓ allows calculation of $k_q^{O_2}[O_2]$, i.e., the oxygen quenching constant of fluorescence times the oxygen concentration. The results are $1.4 \times 10^7$ sec$^{-1}$ and $1.2 \times 10^7$ sec$^{-1}$ (these values are experimentally indistinguishable) for napthalene and pyrene respectively. We take the experimental identity of the $k_q^{O_2}[O_2]$ measurements to imply that the HDTCℓ micelles are "kinetically identical" and therefore structurally identical when naphthalene or pyrene are guest solutes.

Although $Eu^{3+}$ is an excellent quencher of Py fluorescence ($k_q \sim 2 \times 10^9$ M$^{-1}$ sec$^{-1}$ in water), it is found that Py dissolved in HDTCℓ is completely protected from quenching by $Eu^{3+}$ ($k_q < 1$

# FLUORESCENT PROBES FOR MICELLAR SYSTEMS

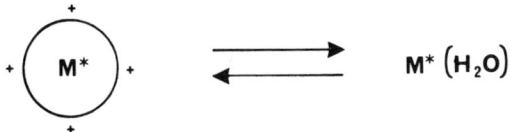

**Catalysis**

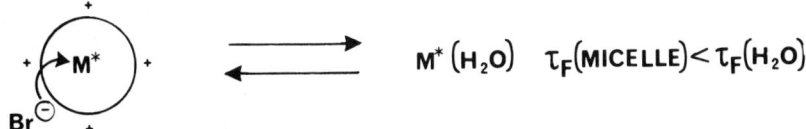

**Inhibition**

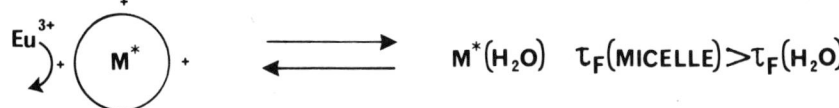

**Oxygen sequestering**

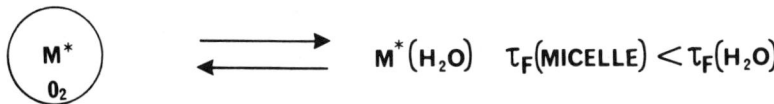

Figure 4. Schematic representation of the "catalysis" of quenching of naphthalene fluorescence in HDTBr and the "inhibition" of naphthalene fluorescence in SDS.

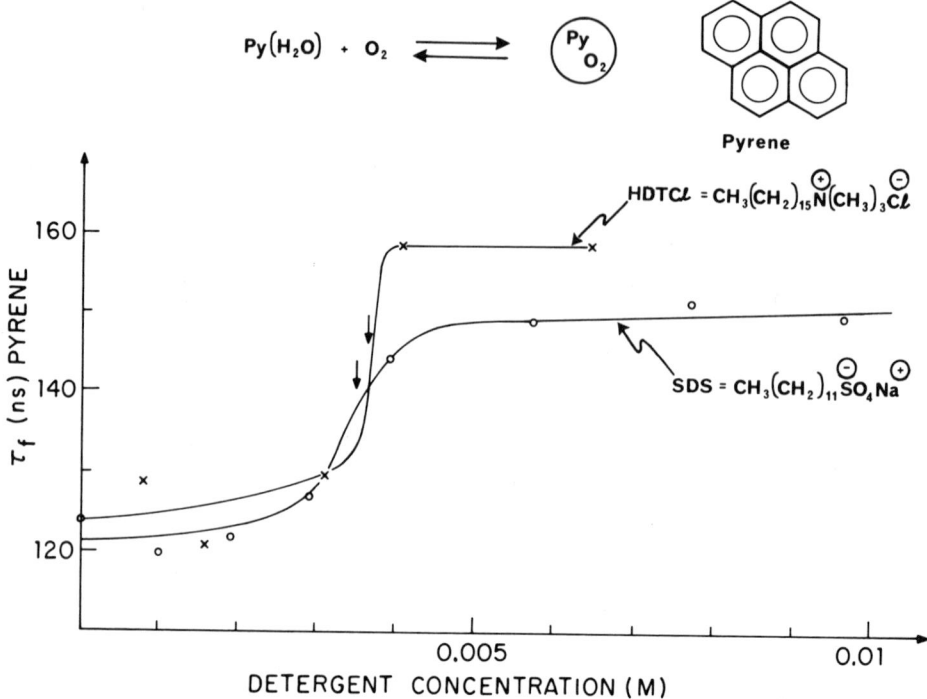

Figure 5. The effect of detergent concentration and micelle formation on pyrene lifetime. The arrows indicate approximate CMC's. The x refers to HDTCℓ and O refers to SDS.

x $10^6$ $M^{-1}$ $sec^{-1}$). The positively charged micelle inhibits quenching by the positively charged quencher.

1,3-Dialkylindoles in HDTBr. A Fluorescent Probe that Forms Its Own Micelles.

In contrast to aromatic hydrocarbons (e.g., naphthalene and pyrene), indoles are found to have highly solvent dependent fluorescent properties. For example, 1,3-dimethylindole[4] shows fluorescent maxima at 305 nm and 370 nm in cyclohexane and water, respectively. Its fluorescence lifetime is also strongly solvent dependent (4 x $10^{-9}$ sec and 16 x $10^{-9}$ sec in cyclohexane and water, respectively).

A survey of 1,3-dialkyl indoles indicated that these compounds are associated with micelles from HDTBr and HDTCℓ but that they tend to remain close to the ionic interface. However, the indole detergent 6-In-11 was found to be incorporated in micelles of HDTBr and <u>also form its own micelles</u>. Furthermore, 6-In-11 shows

# FLUORESCENT PROBES FOR MICELLAR SYSTEMS

a dramatic shift in its fluorescence maximum as it passes through the fluorescence lifetime of 6-In-11 in both a micellar environment and in bulk aqueous solution (Table I). It is found that:

(a) The flourescence maximum of 6-In-11 in an aqueous environment occurs at ∿370 nm.
(b) The fluorescence maximum of 6-In-11 in a micellar environment occurs at ∿350 nm.
(c) The fluorescence lifetime of 6-In-11 is ∿6 x $10^{-9}$ sec and ∿18 x $10^{-9}$ sec in a micellar and in an aqueous environment, respectively.

Table I. Fluorescence Decay of 6-In-11 in Aqueous Solution

| [6-In-11] in M | $\tau_F$ (micelle) in ns | $\tau_F$ (water) in ns |
|---|---|---|
| 1 x $10^{-5}$ | – | 20 |
| 1 x $10^{-4}$ | 6 | 18 |
| 1 x $10^{-3}$ | 6 | 15 |

These data provide evidence that indole emission takes place at a much more rapid rate than does exchange between the micellar environment and the aqueous environment.

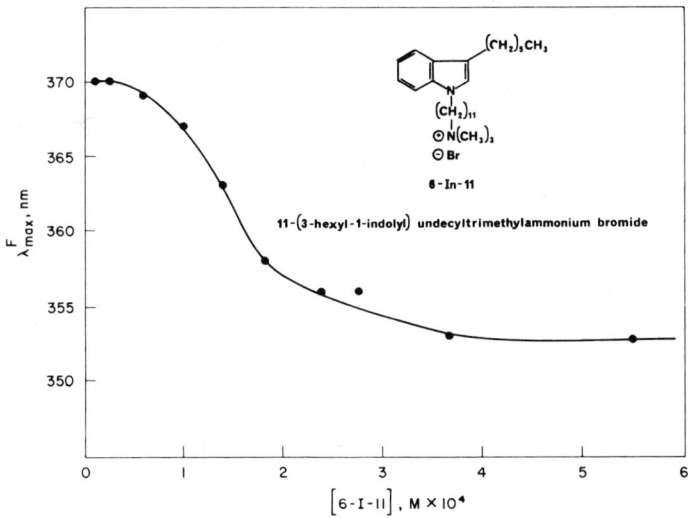

Figure 6. Fluorescence maximum of 6-In-11 as a function of concentration in water.

## CONCLUSIONS

The results of our investigations have provided information concerning (1) partitioning of fluorescence probes between a micellar environment and bulk aqueous solution; (2) an organizing of fluorescence probes in the micellar environment which results in specific "catalysis" or "inhibition" of quenching; (3) sequestering of oxygen into the micellar environment; (4) the free penetrating of the micelle-water inferface by oxygen molecules.

It is desirable to extend these studies using <u>phosphorescence probes</u>. Such probes will extend the dynamic range of investigation into the microsec and millisec domain. It may be possible with such probes to measure directly the dynamics of movement in and out of the micellar environment.

## ACKNOWLEDGEMENT

The authors at Columbia University wish to thank the Air Force Office of Scientific Research (Grant AFOSR-74-2589) and the National Science Foundation (Grants NSF-CHE70-02165 and NSF-MPS73-04672) for their generous support of this research.

## REFERENCES

1. (a) For a review of photochemistry see N.J. Turro, "Molecular Photochemistry," W.A. Benjamin, San Francisco, 1965.
   (b) For a review of luminescence parameters see Parker, "Photoluminescence," Elsevier, London, 1968.
2. For a review of luminescence probes see G.M. Edelman and W.O. McClure, Acc. Chem. Res., $\underline{1}$, 65 (1967); L. Brand and J.R. Gohlke, Annu. Rev. Biochem., $\underline{41}$, 843 (1972).
3. R.R. Hautala, N.E. Schore and N.J. Turro, J. Am. Chem. Soc., $\underline{95}$, 5508 (1973).
4. N.E. Schore and N.J. Turro, ibid., $\underline{97}$, 2488 (1975).
5. M.W. Geiger and N.J. Turro, Photochem. Photobio., $\underline{22}$, 273 (1975).
6. See ref. 5 for citations of recent work by other researchers on fluorescent probes of micelles.

# MICELLAR SOLUTIONS FOR IMPROVED OIL RECOVERY

V.K. Bansal and D.O. Shah

Departments of Chemical Engineering and Anesthesiology

University of Florida, Gainesville, Florida 32611

Micellar solutions of various surfactants have been employed as injection fluids to improve oil recovery during the tertiary oil recovery process. These solutions produce ultra low interfacial tensions ($10^{-2}$ to $10^{-4}$ dynes/cm) at the oil/micellar solution interface.

Petroleum sulfonates have been used as surfactants in improved oil recovery studies. Various physico-chemical techniques, such as electrical conductivity, light-scattering, surface tension, viscometry, and high resolution NMR have been employed to elucidate the structure of such micellar solutions. It was observed that the minimum in interfacial tension and the maximum in electrophoretic mobility of oil droplets occur at the same surfactant concentration. The micellar size increased with an increase in salt concentration. The association between surfactant and polymer molecules strikingly alters the rheology and interfacial properties of micellar solutions. Several aspects of the micellar solutions, such as adsorption, the influence of electrolytes and cosolvents, phase equilibrium with oil and brine, rheology, and polymer-surfactant interaction have been discussed.

## INTRODUCTION

The recovery of oil from a reservoir can be divided into three stages. In the primary oil recovery process, oil is recovered due to the pressure of natural gases which force the oil out through production wells. When this pressure is reduced to a point where it is no longer capable of pushing the oil out, water is injected to build up the necessary pressure to force the oil out. This is generally called the secondary oil recovery or water flooding process. The average oil recovery during the primary and secondary stages is about 30% of oil-in-place. The purpose of the tertiary oil recovery process is to recover at least part of the remaining 70% oil-in-place. Various techniques proposed for this stage are carbon dioxide injection, steam flooding, or surfactant flooding by either micellar or microemulsion solutions. The use of surfactants for improved oil recovery is not a recent development in Petroleum technology. In 1930 De Groot[1,2] filed a patent describing the use of water-soluble surfactants as an aid to improve oil recovery. In 1958, Holbrook[3] filed for a patent which the use of surfactant (organic perfluoro compounds, fatty acids, soaps, poly-glycol ethers, salts of sulfonic acids) dissolved in water was proposed as an aid to improve the oil recovery process. The other patents issued to Holm and Bernard[4], Gogarty and Olson[5] and Jones[6,7] essentially described the use of high concentration of surfactant in the form of microemulsion for improved oil recovery.

The process of using surfactants for improved oil recovery can be divided into two groups. In the first, a solution containing a low concentration of a surfactant in the form of micellar solution is injected. In the second, the surfactant concentration is relatively high and the injected slug is formulated with three or more components and is known as microemulsion. The basic components of the microemulsion are hydrocarbon, surfactant, water, alcohol and salt. In the second process a relatively small pore volume (about 3 to 20%) as compared to the first (15-60%) is injected. This paper deals basically with the first process i.e., using a low concentration of a surfactant in the form of micellar solution as an aid to improve oil recovery.

The micellar solution flooding process is an immiscible-type displacement process. Two basic well configurations- the "five spot" pattern or the "line drive" pattern are used for the micellar flooding process. In the "five spot" pattern (Figure 1) four production wells are drilled at the corners of a square, and the injection well through which micellar solution is pumped, is at the

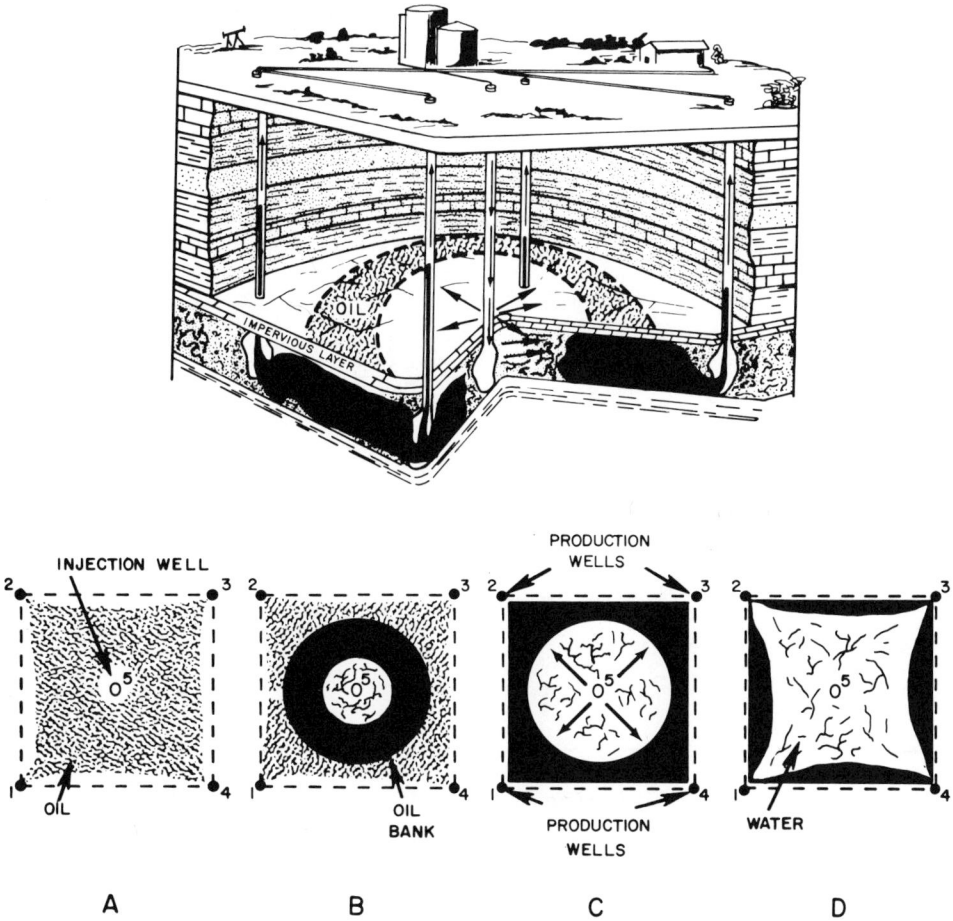

Figure 1. Five spot pattern for displacement of oil.

center of this square. In the "line drive" pattern, production and injection wells are drilled in alternate rows. In both cases, the injected micellar solution tends to displace the oil towards the producing wells. Although considerable work has been done on the oil recovery process under both laboratory and field conditions, our basic understanding of the process and mechanism of oil displacement is far from clear.

## MOLECULAR AGGREGATES IN SURFACTANT SOLUTIONS

When a surfactant is dissolved in water, it tends to adsorb at the gas-liquid interface. The adsorption of surfactant at the interface results in a greater concentration at the interface as compared to that in bulk solution. Above a critical concentration depending upon the structure of surfactant molecules as well as physicochemical conditions, the surfactant molecules form aggregates called micelles (Figure 2). This characteristic concentration is called

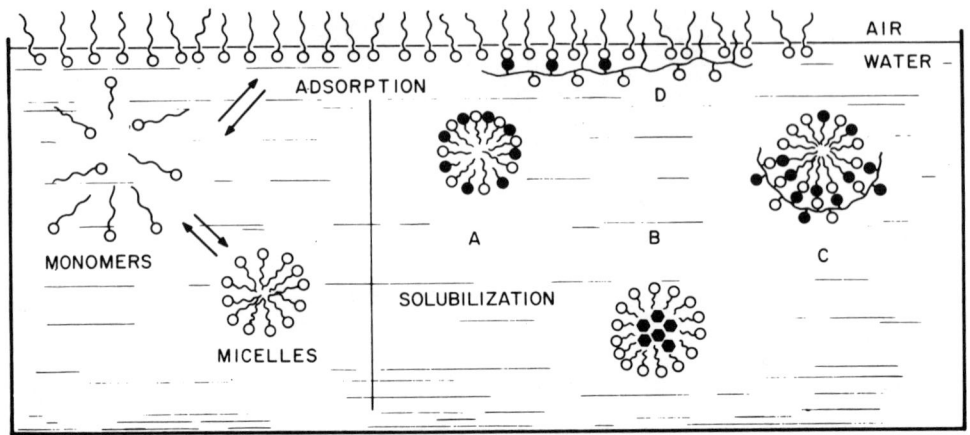

Figure 2. Adsorption, micelle formation, solubilization and interaction at the micelle surface.

the critical micelle concentration (CMC). Micelles are spherical aggregates of surfactant-molecules containing 20 to 100 molecules. The formation of micelles in aqueous solution creates local nonpolar environments within the aqueous phase. Any oil soluble materials such as dyes, pigments or nonpolar oils can dissolve within the micelles (Figure 2B). Using ionic and nonionic surfactants, one can produce mixed micelles which are often larger in size and in the number of molecules within a micelle (Figure 2A). If a surfactant solution contains a surface active polymer, then a mixed adsorbed film of polymer and surfactant occurs at the interface. The polymer surfactant interaction can also occur at the micellar surface (Figure 2C and 2D). The solubilization of oil within micelles can also occur when such micellar solutions are injected into the oil fields.

Surfactant molecules can be considered as building blocks. One can make various types of structures of surfactant molecules by simply increasing the concentration of surfactant in water. Figure 3 schematically shows various structures that are formed in the sur-

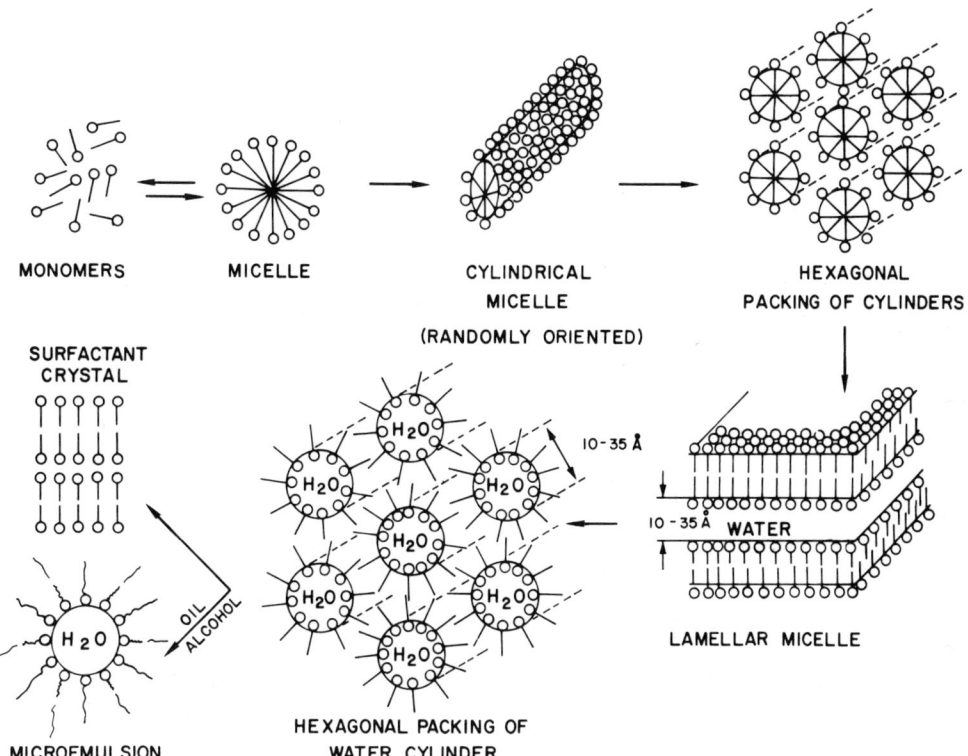

Figure 3. Structure formation in surfactant solution.

factant solution upon increasing the concentration of surfactant. The spherical micelles become cylindrical ones. Upon further increasing the concentration, there is a hexagonal packing of surfactant cylinders. If the concentration is still increased the lamellar structures are formed. Upon further addition of surfactant, the lamellar structures are converted to a hexagonal packing of water cylinders. Upon addition of oil and a short-chain alcohol, one can convert such water cylinders into water-in-oil microemulsions. It is possible to induce a transition from one structure to another by changing the physicochemical conditions such as temperature, pH, addition of mono- or divalent cations in the surfactant solution. It should be emphasized that the scheme shown in Figure 3 is a general scheme and a surfactant may skip several phases depending upon its structure and the physicochemical conditions. The flow behavior of surfactant formulations containing such structures through porous media is not explored in detail and a careful study on the effect of such structure on the oil displacement efficiency is desirable.

## THE ROLE OF VARIOUS INTERFACIAL PARAMETERS IN A CONCEPTUAL OIL DISPLACEMENT MECHANISM

As we visualize, four interfacial parameters are responsible for enhanced oil recovery by micellar flooding. These parameters are: (1) interfacial tension, (2) interfacial viscosity, (3) interfacial charge, (4) contact angle. It has been established[8-10] that for the success of a tertiary oil recovery process interfacial tension should be in the order of $10^{-3}$ dynes/cm. Foster[11] explained on the basis of capillary number ($N_{ca}$) that interfacial tension should be reduced 10,000 times to recover a larger amount of oil. Such low interfacial tensions reduce the work of deformation necessary for oil droplets to emerge from the narrow neck of pores (Figure 4). It is then necessary for these displaced oil droplets usually referred to as oil ganglia, to coalesce and thereby form an oil bank (Figure 5). For this coalescence to occur, a very low interfacial viscosity is desirable. The moving bank coalesces with more oil ganglia (Figure 6) and causes further displacement of residual oil toward the producing wells. For a hydrodynamically stable system (for mobility control) it is necessary that the micellar solution be followed by a mobility control polymer solution (Figure 7).

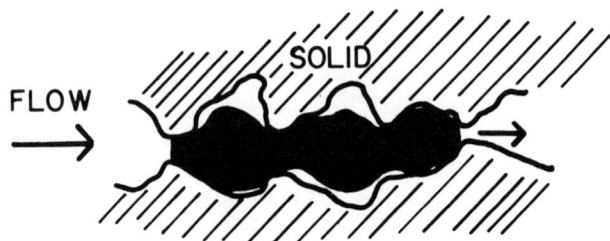

Figure 4. Movement of oil ganglia through a narrow neck of pores; a very low interfacial tension is desirable.

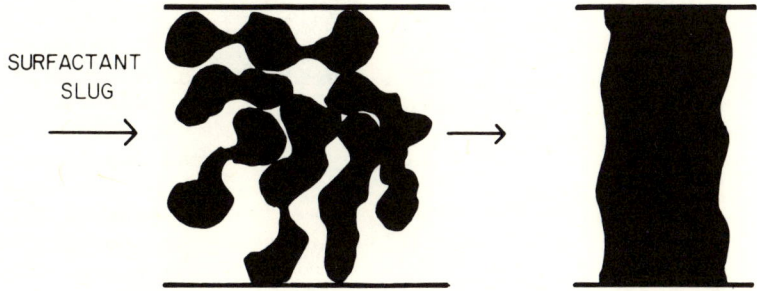

Figure 5. Formation of continuous oil bank through coalescence of displaced oil ganglia: For this a very low interfacial viscosity is desirable.

Figure 6. Coalescence of oil ganglia with oil bank.

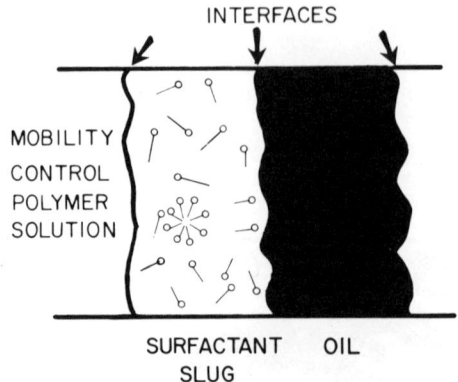

Figure 7. Formation of three interfaces in porous media during tertiary oil recovery by surfactant and polymer flooding.

The relationship between interfacial charge and interfacial tension has been established[22] for the brine/oil system and it has been found that the electrophoretic mobility and interfacial tension curves are inverse images of each other. The importance of interfacial charge at crude oil/aqueous phase system suggests that the nature and magnitude of the charges on solid surfaces are also important variables in determining the efficiency of the oil displacement process. Wagner and Leach[12] and Leach et al.[13] suggested that reversing wettability of a porous material (from oil-wet to water-wet) during a water flood will result in an increase in oil recovery. With a proper choice of surfactant, one can selectively alter the contact angle (wettability) of oil on solid surfaces (Figure 8) thus creating more favorable conditions for oil displacement. Melrose and Brandner[8] and Morrow[14] concluded that for optimal recovery of residual oil by a low interfacial tension flood the rock structure should be water wet. Slattery and Oh[15] suggested that for the most efficient displacement of residual oil, the porous structure should be water-wet. Intermediate wettability may be less desirable than either oil wet or water-wet behavior.

The success of the above approach for improved oil recovery will depend on the proper choice of chemicals in formulating the optimum micellar slug. The composition of the micellar slug is dependent upon the properties of the micellar solution itself (e.g. viscosity, salt tolerance, temperature stability,etc.) as well as on the conditions prevalent in the reservoir.

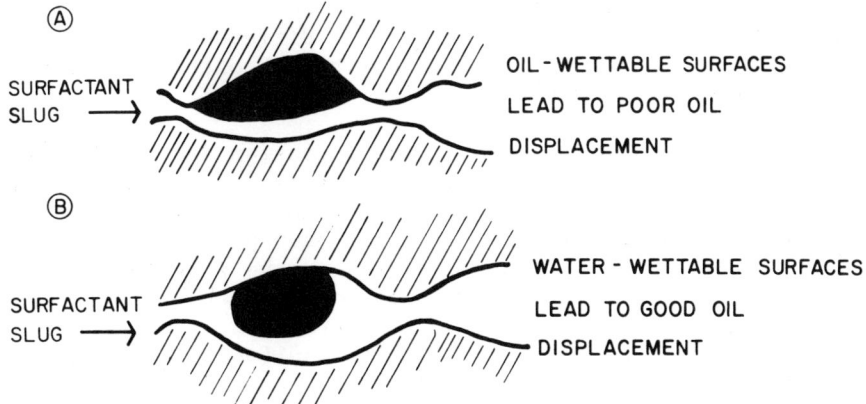

Figure 8. Effect of wettability on oil displacement.


## INTERFACIAL PROPERTIES OF MICELLAR SOLUTIONS

It has been established[11] that for the success of a tertiary oil recovery process the capillary number (ratio of viscous to interfacial forces) should be in the region of $10^{-2}$. During an ordinary water flooding process, the capillary number is of the order $10^{-6}$ and the residual oil at this capillary number is about 50% pore volume. To recover a larger amount of oil, i.e., to minimize the residual oil to near zero, requires a capillary number of four orders of magnitude larger than this. In practice, this can be accomplished only by reducing the interfacial tension at the oil/brine interface which has values of the order of 10 dynes/cm. Reducing this interfacial tension to a value of $10^{-3}$ dynes/cm will produce a capillary number in the desired range. Recently, studies have been conducted on aqueous solutions of petroleum sulfonate that demonstrate that such low tensions can be obtained with the use of relatively small concentrations of surfactants[11,16,17,18].

It has been shown[19] that a petroleum sulfonate with an equivalent weight distribution that is relatively narrow and/or symmetrical about the median is the most effective in lowering interfacial tension. A minimum in interfacial tension can also be obtained through an adjustment of the electrolyte content of the aqueous phase. Sodium chloride was shown to be more effective than sodium sulfate, carbonate or tripolyphosphate in increasing the interfacial activity of a petroleum sulfonate. It was pointed out that the production of low interfacial tensions for different types of oils might require different optimum electrolyte concentrations and a different equivalent weight of petroleum sulfonates.

Much of the work done on interfacial tension does not have a systematic approach for evaluating the effect of different variables on the interfacial tension at a micellar solution-oil interface. Recently, however, a systematic study has been performed[20,21] to evaluate the effects of several variables on the interfacial tensions of a homologous series of hydrocarbons. The variables examined were: (1) salt concentration, (2) surfactant concentration, (3) average equivalent weight of petroleum sulfonate, (4) surfactant concentration and (5) aging.

It has been found[20,21] that, for a given concentration of salt and surfactant, there is a definite affinity for a particular hydrocarbon and a pronounced minimum in interfacial tension at a particular value of R (number of carbon atoms in the oil). In some cases the interfacial tension of the two adjacent alkanes is one or two orders of magnitude larger than the minimum value. The preferred value of R varies inversely with the surfactant concentration and directly with the salt concentration. It has also been established[21] that production of a low interfacial tension requires an optimal electrolyte concentration for a given surfactant concentration, and an optimal surfactant concentration for a given electrolyte concentration. At a given salinity and surfactant concentration, the value of R corresponding to the minimum tension is the same for alkylbenzenes and alkane. However, the absolute value of the interfacial tension at the minimum is lower for alkylbenzenes as compared to alkanes. The correlations for surfactant and electrolyte concentrations are basically the same for both the alkylbenzenes and alkanes. However, the minimum tension achieved with the alkanes seems to occur at higher surfactant and salt concentrations than with the alkylbenzenes.

The effect of the equivalent weight of petroleum sulfonate on the position of the minimum has also been studied. The main conclusion reached was that as the equivalent weight of the surfactant increases, the preferred value of R (corresponding to the interfacial tension minimum) also increases. The actual factors responsible for this shift have not been identified, but prediction of the direction of the shift of the minimum by a mixture of two surfactants is possible if the general trends of each surfactant on a homologous series is established. It has also been found that the ammonium ion shifts the minimum to higher values of R for alkanes. As the molecular weight of the cation is increased, the minimum tension is again shifted to larger R values. As has been discussed, such a shift is also characteristic of an increase in the surfactant equivalent weight. This suggests a strong association and formation of ion pairs between the cation and surfactant.[21]

The effect of different alcohols on the value of R at the tension minimum has also been investigated[21]. It was found that an increase in the molecular weight of the alcohol shifted the tension minimum to larger values of R. Changes in interfacial tension with time has also been investigated[20] and the "aging" effect has been observed both in the presence and absence of co-surfactants (alcohols). The interfacial tension measured against a variety of hydrocarbons tends to change with time. The direction of change is most often toward higher interfacial tension, the magnitude of these changes being dependent on the surfactant concentration and the alcohols used as cosolvents.

## A CORRELATION OF INTERFACIAL TENSION WITH ELECTROPHORETIC MOBILITY

The salt or surfactant concentration in a micellar solution is one of the most important and effective variables in manipulating the interfacial tension between the micellar solution and oil, although the molecular mechanism of this process is not elucidated. It is known that the salt concentration in a micellar solution affects both the interfacial charge and micellar size. In our laboratory, the relationship between interfacial charge and interfacial tension[22] has been studied extensively by measurements of the interfacial tensions and electrophoretic mobilities of various systems. It has been established for the brine/oil system that the electrophoretic mobility and interfacial tension curves are inverse images of each other. The micellar system used for these measurements consisted of TRS-10-80 (surfactant) + 1% NaCl + n-octane (oil) and the results obtained are shown in Figure 9. The results of interfacial tension measurements obtained by K. Chan[23] are in agreement with those reported by Cash et al.[20]. The above system exhibits an unusual minimum in interfacial tension at 0.05% TRS-10-80. At the same concentration, the electrophoretic mobility exhibits a striking maximum. From the results reported previously[23] it is clear that the electrophoretic mobility, which is an indirect measurement of interfacial charge density, has a strong correlation with the interfacial tension for the oil/brine or oil/surfactant system. These results also demonstrate that the electrophoretic mobility and interfacial tension curves are almost inverse images of each other. We have shown[23] a similar correlation between interfacial tension and electrophoretic mobility for various crude oils in the presence of NaOH solutions. These results indicate that perhaps a detailed study of the electrochemistry of the oil/water interface in the presence of a surfactant may provide a better insight into how the ultra-low interfacial tension is achieved, rather than attacking this problem by looking at the more structural aspects of the surfactant molecule.

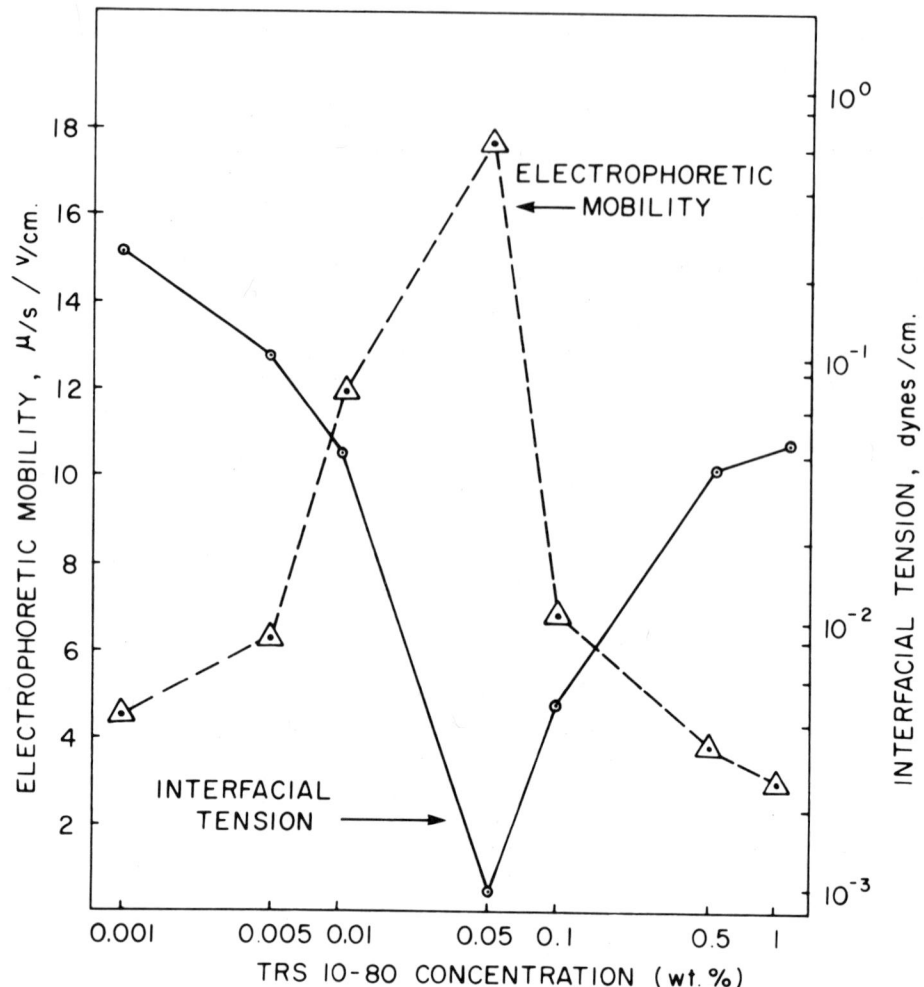

Figure 9. Effect of surfactant concentration on interfacial tension and electrophoretic mobility of oil droplets.

We believe that a greater emphasis on interfacial charge as compared to interfacial tension is highly desirable. If one considers only the ultra-low interfacial tension in the oil displacement process, then the nature of the charge on the solid surfaces (rocks, minerals and clays) is of no consequence. However, if one considers the interfacial charge at the crude oil/surfactant formulation interface, then the nature and magnitude of the charges on the solid surfaces become extremely important in determining the efficiency of the oil displacement process. A better understanding

## LIGHT SCATTERING STUDIES ON MICELLAR SOLUTIONS OF PETROLEUM SULFONATES

Light scattering methods have been used to calculate the molecular weight of a micelle for the system TRS-10-410 + isobutanol in different salt concentrations. The surfactant (TRS-10-410) concentration was 5% by weight and the alcohol (isobutanol) concentration 3% by volume. A Wood dual photometer, model 5000 was used to measure the light scattered at 90° by these solutions. The formulation was then diluted with a salt solution, thereby keeping the salt concentration constant, and varying the concentrations of TRS-10-410 and isobutanol, the ratio of surfactant to cosolvent remaining constant. Figure 10 illustrates the effect of NaCl concentration on the light scattered at 90° by the TRS-10-410 solutions. The average molecular weight of a micelle was calculated from the position of the maximum using the approach of Debye and Bueche[24] for concentrated solutions. According to their analysis, the concentration at which the maximum occurs depends on the ratio of the molar volumes of the solute and solvent and is given by the equation

$$\phi_{max} = (1 + \sqrt{n})^{-1}, \quad \phi_{max} \approx (\sqrt{n})^{-1} \qquad [1]$$

for large n (high molecular weight of solute) where n is the ratio of the molar volume of the solute to that of the solvent and $\phi_{max}$ is the volume fraction calculated at the maximum in light scattering. For the system when the solvent is water and assuming the density of micelle is one, Equation 1 reduces to

$$\text{Molecular weight of micelle} = \frac{18}{\phi_{max}^2} \qquad [2]$$

From Figure 10 it can clearly be seen that as the concentration of NaCl increases, the light scattered at 90° exhibits a sharper maximum that is shifted toward the left (lower concentrations). Calculated values of the molecular weight of a micelle by Equation [2] and the number of surfactant molecules per micelle are given in Table I. Our results, as shown in Figure 10, are very similar to those obtained by Tager and Andreeva[25] for polymer solutions of increasing molecular weight. We would like to emphasize that the molecular weight of a micelle should be accepted as an approximate value in view of the complex nature of the surfactant. We believe that this is as far as we can go quantitatively with this "impure" but practical system.

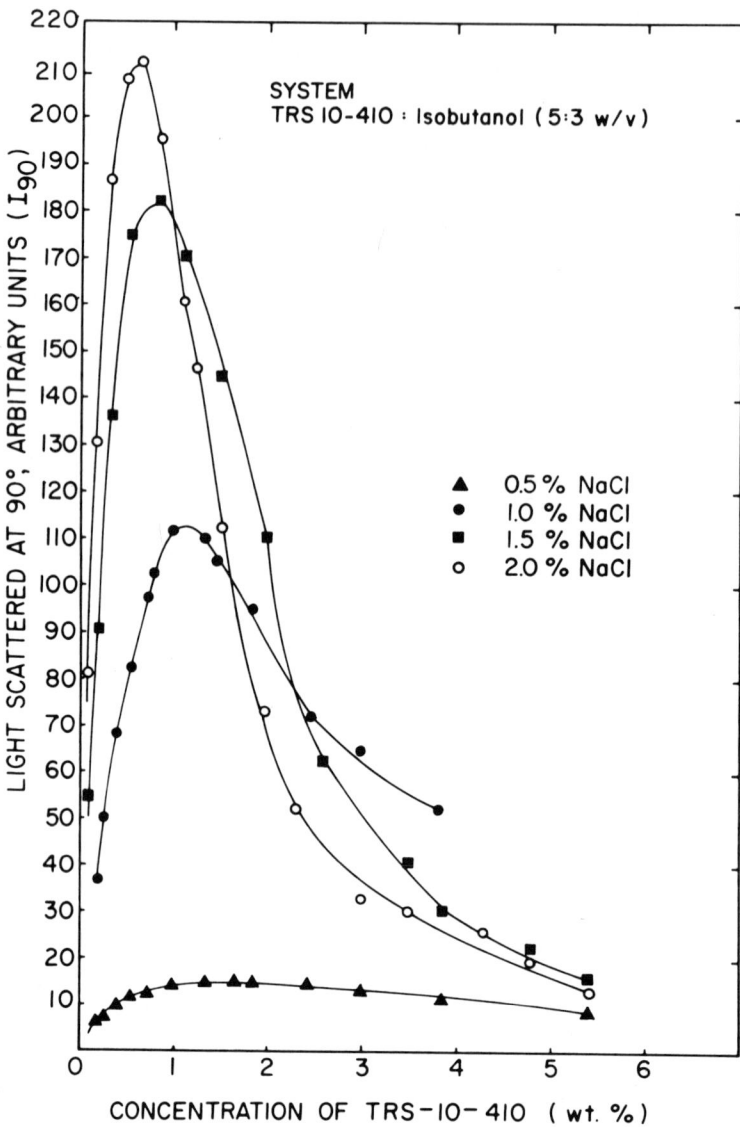

Figure 10. Effect of NaCl and surfactant concentration on light-scattered at 90° by a surfactant formulation.

Table I. The Effect of NaCl Concentration on Micellar Weight of a Surfactant Formulation

| Surfactant Formulation | | | | Number of Molecules in a Micelle | |
|---|---|---|---|---|---|
| System | NaCl Conc.% | Position of Maxima wt.% | Mol.Weight of micelle | TRS-10-410[1] | Isobutanol[2] |
| TRS-10-410 + IBA | 0.5 | 1.6 | $7.0 \times 10^4$ | 121 | 121 |
| TRS-10-410 + IBA | 1.0 | 1.0 | $1.8 \times 10^5$ | 364 | 364 |
| TRS-10-410 + IBA | 1.5 | 0.8 | $2.8 \times 10^5$ | 566 | 566 |
| TRS-10-410 + IBA | 2.0 | 0.6 | $4.9 \times 10^5$ | 992 | 992 |

1. The equivalent weight of 420 for TRS-10-410 has been assumed to be its molecular weight. The weight of surfactant is not corrected for its oil content (i.e., it is assumed that TRS-10-410 is 100% active surfactant).

2. It is assumed that the surfactant and Isobutanol molar ratio in a micelle is 1:1.

Interfacial tensions of the above systems, used for light scattering, were measured with a spinning drop interfacial tensiometer against hexadecane oil. Interfacial tensions at two different TRS-10-410 concentrations (0.5% and 1.0%) and various salt concentrations are shown in Figure 11. It is evident from this figure that there is a decrease in the interfacial tension with increasing salt concentration. This decrease in interfacial tension with increasing salt concentration is associated with an increase in the micellar size as shown by the light scattering results (Table I).

## PHASE EQUILIBRIUM AND SOLUBILIZATION PHENOMENA

In general, the oil displacement process involves the interaction of three components, namely, oil, surfactant and brine. It is therefore both convenient and instructive to employ a ternary representation for a phase equilibrium study. A simple ternary diagram for the three component system is shown in Figure 12.[26] In this Figure, the multiphase region is bounded by a continuous binodal curve. Everywhere above this binodal curve there exists a single phase that undergoes transitions among various structural states as the compositional point moves about the diagram.

In the multiphase region, the most simple three-component system involves only two phases throughout the region; an oil-external phase and a water-external phase. The actual micellar solutions used in tertiary oil recovery are more complex than this, always involving more than three components and three or more phases in equilibrium. Despite these complexities, the ternary diagram representing the system can be used to trace the possible events that can occur in a micellar solution that is injected into a porous medium.

It is evident that a large variety of phases can exist in equilibrium with each other. Each phase might involve a different micellar structure, and in equilibrium with each other. This intermicellar equilibrium concept was put forth by Winsor[27] and is illustrated in Figure 13. Spherical micelles consisting of oil cores in a water continuous medium are called an $S_1$ or water-external phase. The inverse of this is the $S_2$ or oil-external phase. An intermediate lamellar structure, which may be a gel or liquid crystal, is called the G phase. It has been proposed[18] that for the oil recovery process, the structural state of the single phase region is unimportant as long as its viscosity is not large. This requirement rules out the use of highly viscous lamellar structures and microgels for the tertiary oil recovery process.

The effect of important variables such as salt, surfactant and cosolvents concentrations have been studied on the phase behavior[28-30] and on other properties such as viscosity, electrical conductance and birefringence. It has been observed that the use of alcohols as cosolvents modifies the phase behavior of brine/oil/

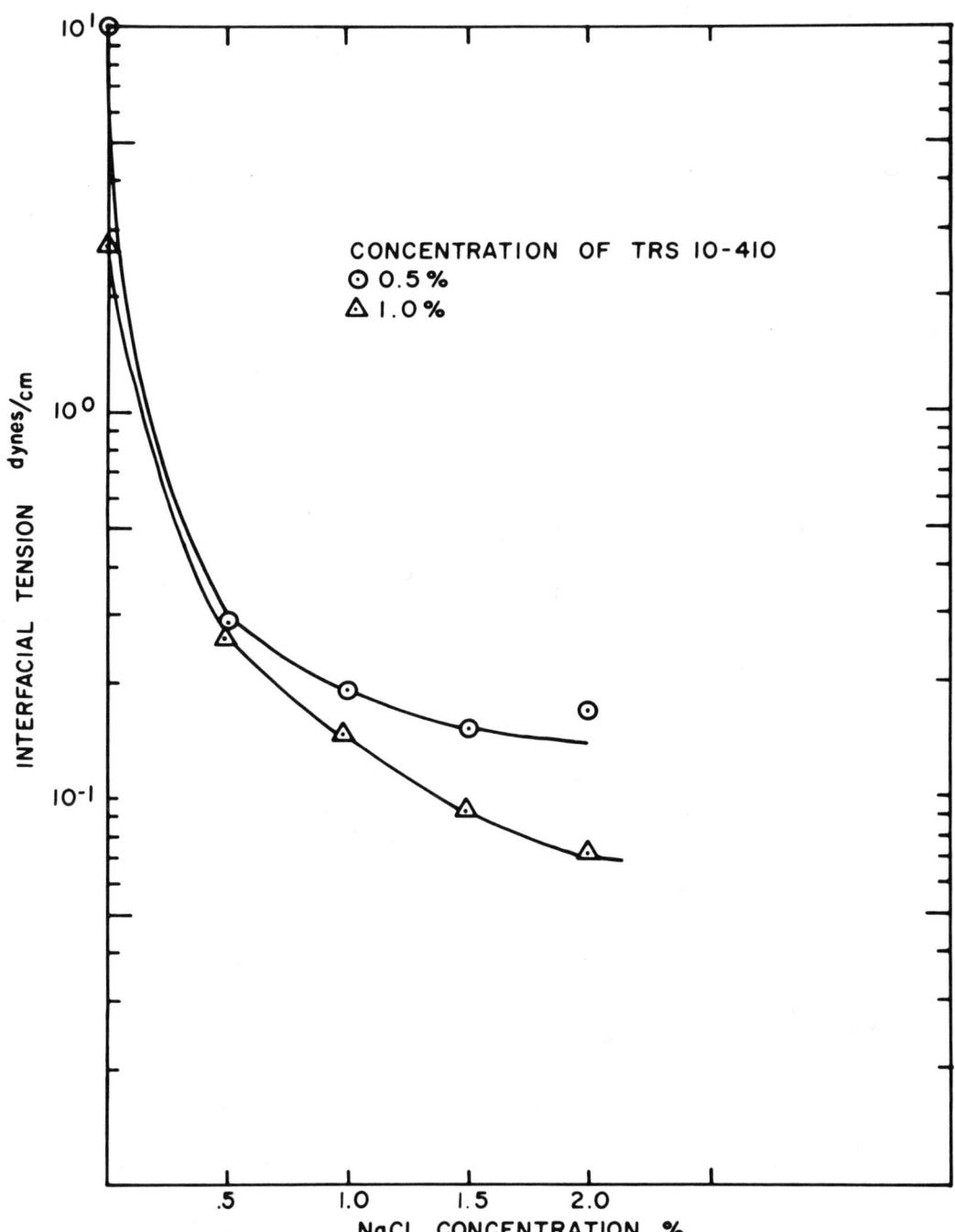

Figure 11. Effect of NaCl concentration on interfacial tension of hexadecane/micellar solution.

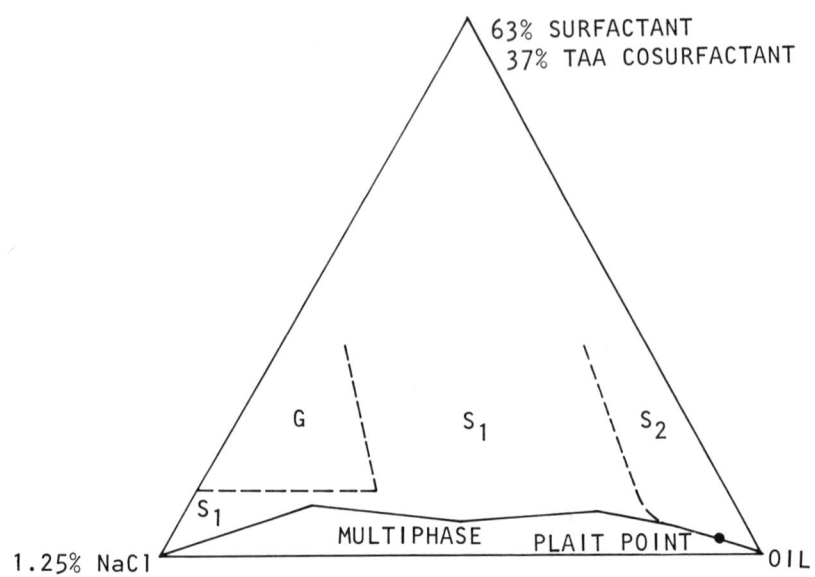

Figure 12. Phase equilibrium diagram (ref. 26).

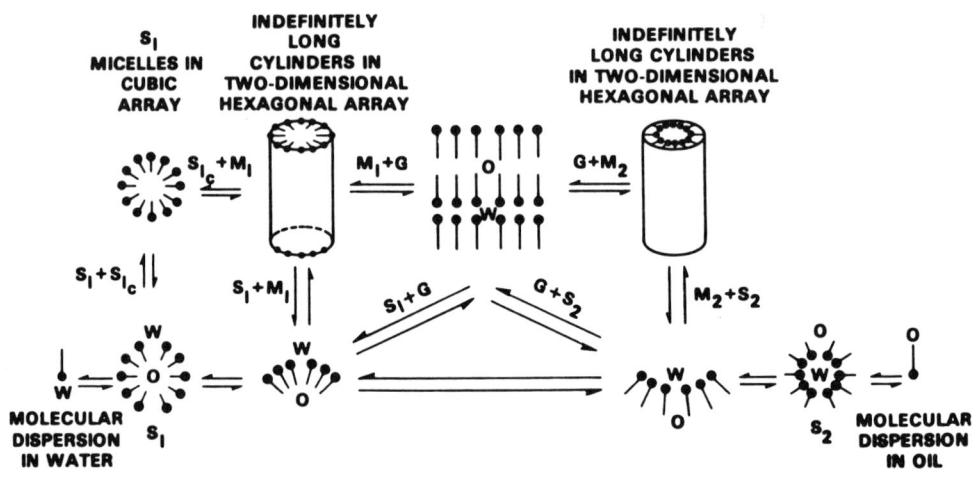

Figure 13. Winsor's concept of intermicellar equilibrium.

surfactant systems. A water soluble alcohol increases the solubility of surfactant in a brine while a water insoluble alcohol decreases the surfactant solubility. The concentration of NaCl changes the alcohol requirements. An increase in electrolyte concentration decreases the requirement of water insoluble alcohols and increases the requirement for water soluble alcohols.

Robbins[31,32] suggested a model which predicts quantitatively the phase behavior of micellar and microemulsion systems consisting of nonionic surfactants. The model relates the volume uptake of water and oil in the microemulsion phase with HLB value of surfactant. It was also shown that water uptake in microemulsion depends on the oil type used. The inverse correlation between $V_w/V_s$ or $V_o/V_s$ (where $V_w$ is volume of water, $V_o$ of oil and $V_s$ of surfactant in the microemulsion) with the interfacial tension was found. It was observed that for ethoxylated surfactant, the increase in temperature, salt concentration and oil aromaticity increase the oil uptake and decrease water uptake.[32] Reed and Healy[33] related interfacial tension $\gamma_{mo}$ and $\gamma_{mw}$ ($\gamma_{mo}$ interfacial tension between microemulsion phase and oil and $\gamma_{mw}$ interfacial tension between microemulsion phase and water) and solubilization parameter, $V_o/V_s$ and $V_w/V_s$ to salt concentration. It was observed that as salt concentration increases, $\gamma_{mo}$ decreases and $\gamma_{mw}$ increases. The salinity, $C_\gamma$, where $\gamma_{mo}$ intersects $\gamma_{mw}$ was defined as the interfacial tension optimal salinity. Similarly, $C_\phi$, the phase behavior optimal salinity, was defined by the intersection of $V_o/V_s$ with $V_w/V_s$. It was also observed that $C_\gamma$ value was very close to $C_\phi$ value. Puerto and Gale[34] developed the methods for estimation of optimal salinity and solubilization parameters for alkyl orthoxylene mixtures. It was found that as the amount of low molecular weight sulfonate ($C_n \leq 9$) in surfactant mixtures increases, $C_\phi$ increases, but solubilization parameter $V_o/V_s$ or $V_w/V_s$ decreases resulting in higher interfacial tensions. Higher molecular weight sulfonates ($C_n \geq 12$) and highly oil soluble alcohols increase the solubilization parameter at $C_\phi$ but the value of $C_\phi$ is considerably reduced. Correlation between interfacial tension and solubilization parameters suggests that phase volumes can replace the interfacial tension measurements as a preliminary measure of interfacial activity.

## THE EFFECT OF DILUTION ON THE RHEOLOGICAL PROPERTIES OF SURFACTANT FORMULATIONS

The effect of surfactant concentration on the bulk properties (bulk viscosity, screen factor, NMR half band width), of the system TRS-10-410 + isobutanol in brine has been investigated. The original formulation contained 5% by weight of TRS-10-410 and 3% by volume of isobutanol in a 2% NaCl brine. This was diluted with a solution of 2% NaCl. Bulk relative viscosities were measured with a Cannon-Fenske Capillary Viscometer. A screen viscometer was used to obtain screen factors and the half band width was calculated from the NMR spectra obtained at different surfactant concentrations.

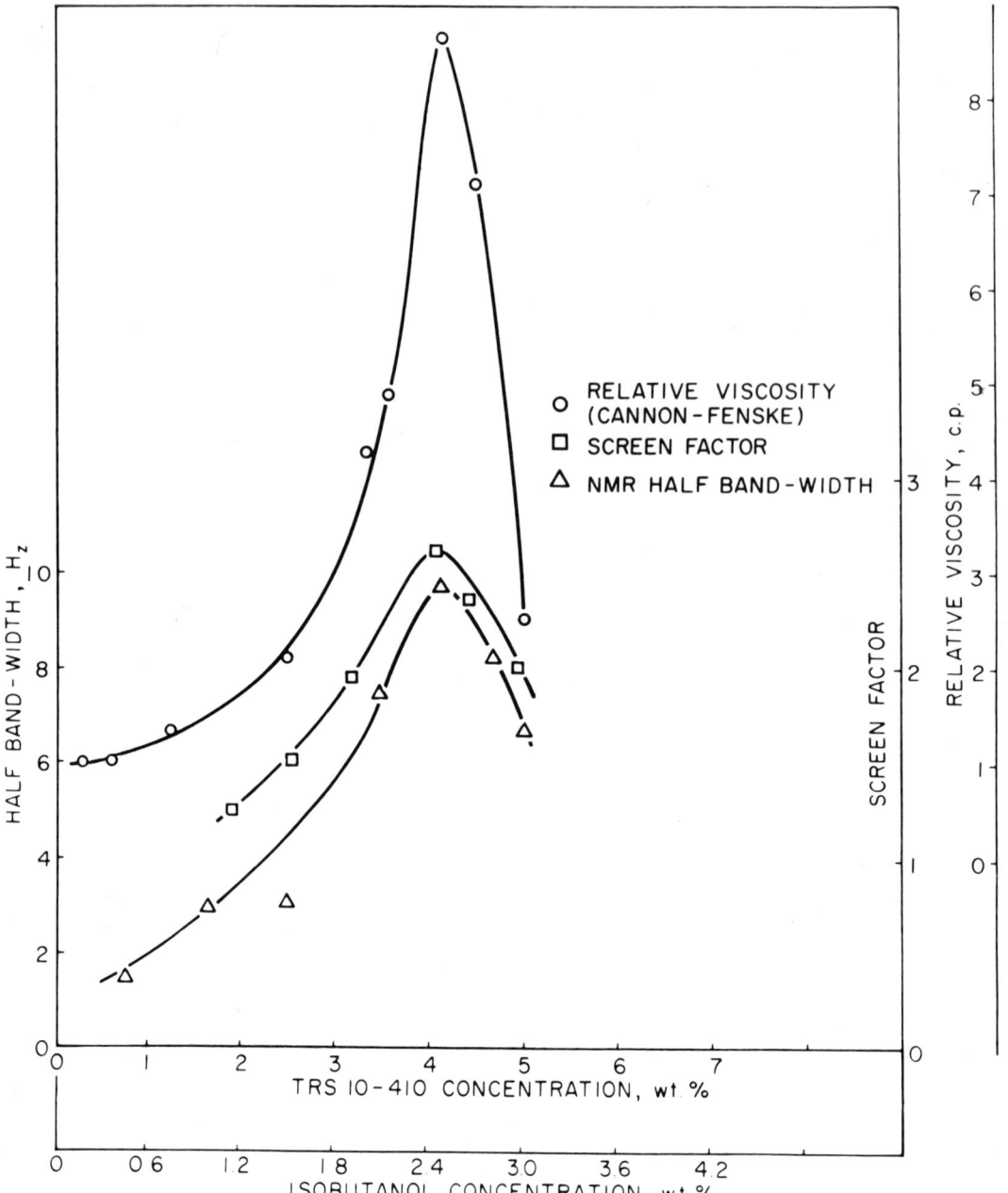

Figure 14. Effect of dilution of a surfactant formulation on relative viscosity, screen factor and NMR half band width.

The results shown in Figure 14 indicate that around a TRS-10-410 concentration of 4% there is a maximum in screen factor and relative viscosity. This indicates that the bulk viscosity is maximum at this point. The NMR half band width is related to the molecular packing in the solution. The broader the peak, i.e., the larger the half band width, the longer it takes for the excited proton to relax back to its original state, hence the closer is the packing of the surfactant and alcohol molecules. Thus the peak in the half band width at 4% TRS-10-410 indicates that the surfactant-alcohol molecular packing is closest at this concentration.

We believe that the above mentioned effect observed upon dilution is an indication of structural changes occuring in the surfactant formulation. This study has implications for injection of such a formulation in the reservoir. As the concentrated surfactant formulation is injected into the ground, there will not be any injectivity problem because of the low viscosity of the formulation. However, once into the ground, it will be diluted with the field brine and will exhibit a higher viscosity presumably more appropriate for the mobility control of the surfactant slug. Therefore, we believe that such a formulation which at the injection point exhibits a low viscosity and upon subsequent dilution increases the viscosity might have certain advantages. However, if the viscosity peak is very high or narrow then it may create a conformance problem.

## ADSORPTION FROM MICELLAR SOLUTIONS

It is well known that surface-active agents tend to concentrate at interfaces. In the tertiary oil recovery process, when the micellar slug comes into contact with the reservoir rocks and clays, there would be a loss of surfactant due to adsorption at the solid-liquid interface. The adsorption loss is taken into consideration for the selection of an optimum micellar slug size and its ability to lower the oil-water interfacial tension. Adsorption of various petroleum sulfonates on reservoir cores has been studied by various investigators.[11,16,17,35,36,37] The following factors are important in determining the adsorption loss of petroleum sulfonates:

1. Specific surface area and electrochemical characteristics of reservoir solids
2. Temperature of the reservoir
3. Composition and concentration of electrolytes in reservoir brine
4. Equivalent weight of the surfactant
5. pH of the reservoir brine and micellar solution
6. Structure and concentration of cosolvent
7. Microstructure of surfactant formulation, e.g. spherical, cylindrical, lamellar structures, or microemulsions (oil- or water-external).

It has been found that an increase in salt concentration increases the adsorption of petroleum sulfonates[11,17,35]. The adsorption of petroleum sulfonate was also found to be a strong function of its equivalent weight; a sharp increase in adsorption being observed at a petroleum sulfonate equivalent weight of about 450.[35] Some inorganic salts were found to decrease the adsorption of petroleum sulfonates.[11,16] Sodium tripolyphosphate effectively reduces the adsorption of petroleum sulfonate and apparently some inorganic salts can be used as sacrificial chemicals to reduce the adsorption of petroleum sulfonates.[11]

The adsorption isotherm for petroleum sulfonate does not correspond to any of the Langmuir types, since the isotherm exhibits a maximum[31,32] which is reported to occur near the critical micelle concentration. A minimum in adsorption is also observed to occur at higher surfactant concentration. At this concentration, the physical appearance of the micellar fluid changes from a turbid, caramel colored fluid to an amber, transparent solution, thus indicating a change in the micellar structure.[37]

The adsorption isotherms of TRS-10-410 in zero and 1% NaCl for crushed Berea sandstone at 30°C are shown in Figure 15.[38] Both these isotherms exhibit maxima. The clay fraction of the crushed Berea sandstone was separated and it was found that most of the adsorbed petroleum sulfonate was on the clay fraction and a negligibly small amount of adsorption occurred on the sand. The maximum observed in the isotherm near the CMC was explained on the basis of competition for the calcium-surfactant anion complex (formed due to adsorption of surfactant on clay) between the clay surface and micelles.

## SURFACTANT-POLYMER INTERACTION

In tertiary oil recovery, it is customary to follow the slug of micellar solutions with a mobility control polymer solution (Figure 7). Often the polymer is added also to the micellar solution to increase its viscosity for mobility control purposes. There is thus a large possibility of the polymer and micellar solutions coming in contact with each other and leading to surfactant-polymer interaction. Such interactions can affect the properties of the interface between the polymer and micellar banks, as well as the properties of the micellar solution due to mixing effects. It has been found[39] that there is a decrease in apparent viscosity at the micellar slug-polymer interface.

It has been found[40] that the polymer tends to move ahead of the surfactant solution. This phenomenon was explained by the concept of the "polymer inaccessible pore volume".[40] This arises from the fact that due to their greater molecular weight, polymer

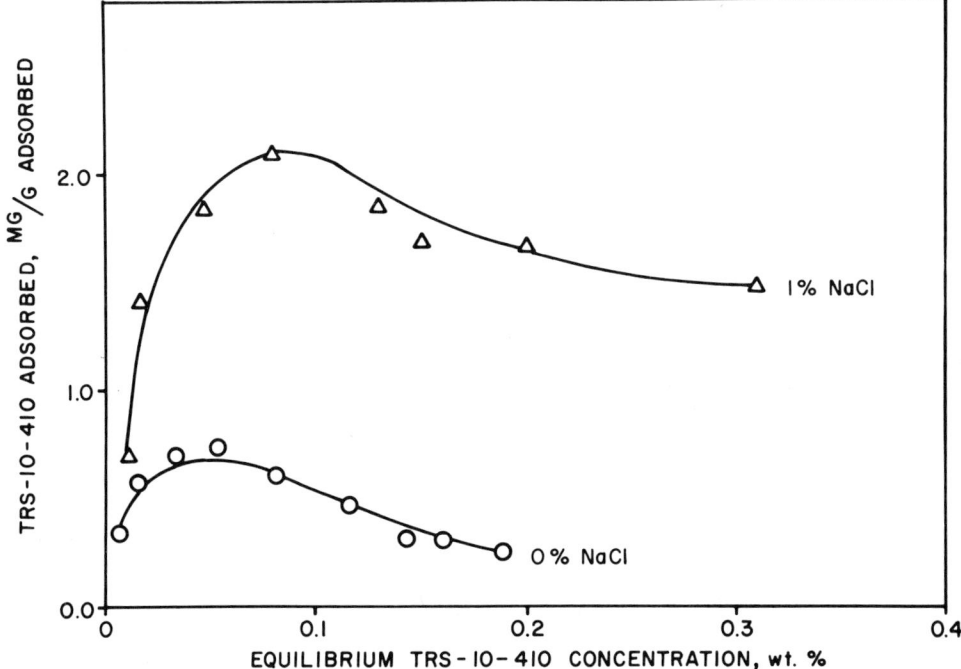

Figure 15. Adsorption of Deoiled TRS-10-410 on crushed Berea sandstone at 30°C.

molecules are excluded from the small pores and can propagate through only larger pores in the porous media. In contrast, water can travel through both the small and large pores. The pore volume available to the polymer molecules is hence less than that available to water. This phenomenon leads to polymer molecules moving faster than the carrier water, and thus can move ahead of the micellar-polymer interface leading to a surfactant-polymer interaction in the micellar fluid. It has been observed[41] that such interactions cause a separation of the micellar formulation into two phases, one of which is trapped in the porous medium. These polymer-surfactant interactions can also cause flocculation to take place in the micellar slug. Both phase separation and flocculation in the micellar slug lead to a substantial loss of surfactant from the micellar slug, with consequent impairment of the efficiency of the process. This surfactant-polymer interaction can be reduced[41] if the salinity of the mobility buffer bank is lower than the salinity of the micellar fluid. The phase separation can also be eliminated by the addition of cosolvents or cosurfactants to the mobility buffer or micellar solution.

In our laboratory, the effect of a surfactant (sodium pentadecyl benzene sulfonate) on the bulk and surface properties of a poly-

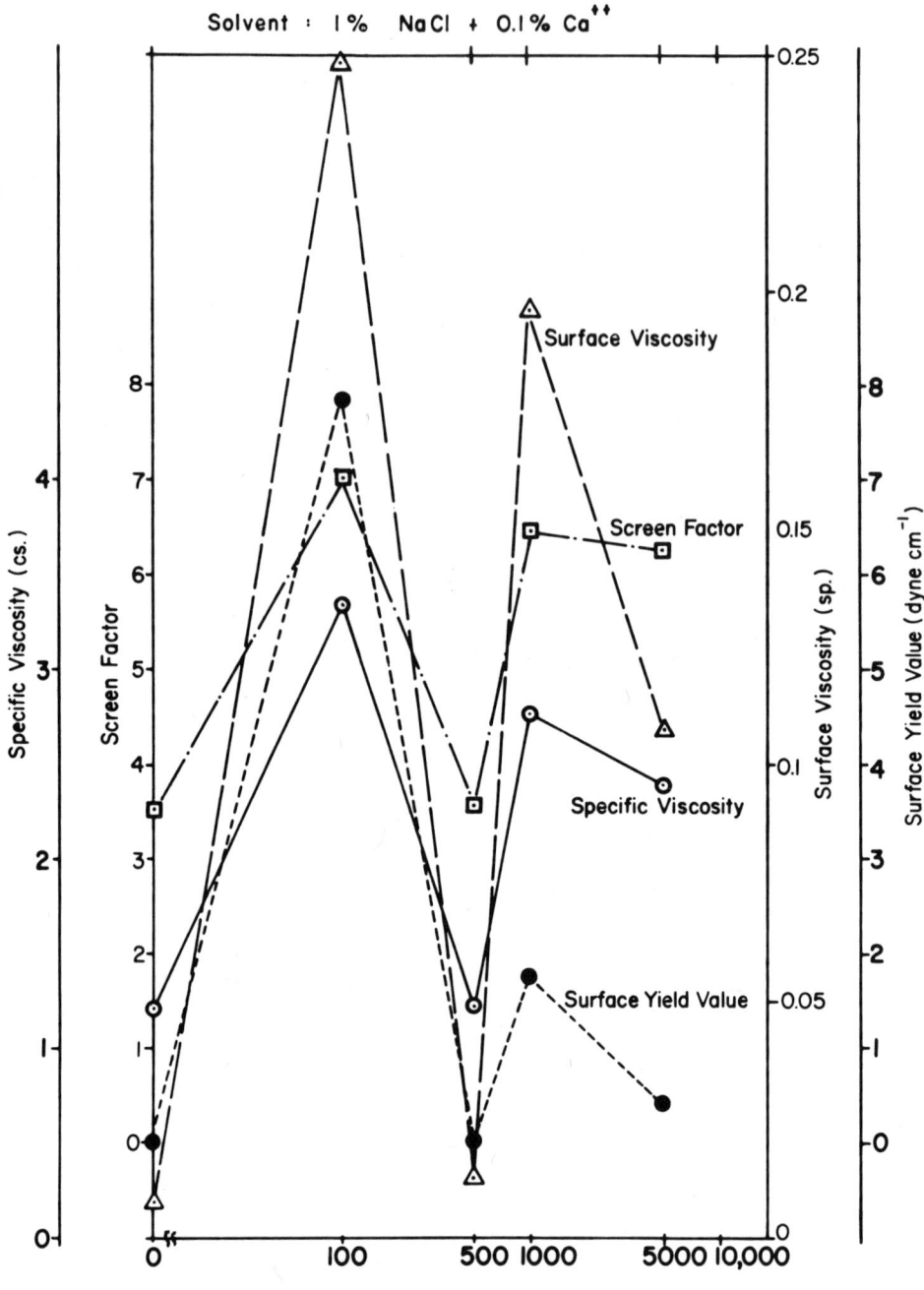

Figure 16. Effect of surfactant concentration on bulk and surface properties of a polyacrylamide solution.

acrylamide solution (Nalco VX 222) has been studied in three solvents, namely, distilled water, 1% NaCl, and 1% NaCl + 0.1% $CaCl_2$.[42] The measured parameters were screen factor and relative viscosity (both bulk properties) as well as surface viscosity and surface yield value (both surface properties). With use of either of the former two solvents (water and 1% NaCl), all the measured properties, with the exception of the screen factor, exhibited a smooth decrease with increasing surfactant concentration. This indicates a greater fluidity of both the bulk and surface phases with increasing surfactant concentration. The results obtained with 1% NaCl + 0.1% $CaCl_2$ as the solvent (Figure 16), however, were strikingly different. Unlike the previous cases, all the measured parameters exhibited a trend of maxima and minima. The addition of the divalent calcium ion has obviously caused tremendous changes in both the surface and bulk properties of the solutions. The evidence suggests possible aggregate formation at a specific surfactant concentration of 500 ppm. In summary, the results obtained on this polymer-surfactant system illustrate the importance of investigating and understanding the polymer-surfactant interaction for the development of an efficient tertiary oil recovery process.

## ACKNOWLEDGEMENTS

The authors wish to thank the National Science Foundation-RANN (Grant AER-75-13813) and the consortium of 20 major oil and chemical companies for their continuing support of the Improved Oil Recovery Research Program at the University of Florida.

## REFERENCES

1. M. De Groot, U.S. Patent No. 1,823,439, 1929.
2. M. De Groot, U.S. Patent No. 1,823,440, 1930.
3. O. C. Holbrook, U.S. Patent No. 3,006,411, 1958.
4. L. W. Holm and G. G. Bernard, U.S. Patent No. 3,082,822, 1959.
5. W. B. Gogarty and R. W. Olson, U.S. Patent No. 3,254,714, 1962.
6. S. C. Jones, U.S. Patent No. 3,497,006, 1967.
7. S. C. Jones, U.S. Patent No. 3,506,070, 1976.
8. J. C. Melrose and C. F. Brander, J. Can. Pet. Tech. 13, 54, 1974.
9. O. R. Wagner and R. O. Leach, Soc. Pet. J. 6, 335, 1966.
10. J. J. Taber, Soc. Pet. J. 9, 3, 1969.
11. W. R. Foster, J. Pet. Tech. 25, 205, 1973.
12. O. R. Wagner and R. O. Leach, Trans. AIME 216, 65, 1959.
13. R. O. Leach, O. R. Wagner, and H. D. Wood, J. Pet. Tech. 14, 206, 1962.
14. N. R. Morrow, presented at the 27th Annual Technical Meeting of Pet. Soc. of CIM, Calgary, Alberta, June 7-11, 1976.
15. J. C. Slattery and S. C. Oh, ERDA Symposium on Enhanced Oil and Gas Recovery, Tulsa, Oklahoma, September 9-10, 1976.

16. H.J. Hill, J. Reisberg and G. L. Stegemeier, J. Pet. Tech., 186, 1973.
17. W. W. Gale and E. C. Sandvik, Soc. Pet. J. $\underline{13}$, 131, 1973.
18. R. N. Healy and R. L. Reed, SPE paper No. 4583, presented at 48th Annual Fall Meeting, Las Vegas, Nevada, September 30-October 3, 1973.
19. M. P. Wilson, C. L. Murphy, and W. R. Foster, SPE paper No. 5812, presented in Symposium on Improved Oil Recovery, Tulsa, Oklahoma, March 22-24, 1976.
20. R. L. Cash, J. L. Cayias, M. Hayes, D. J. MacAllister, T. Shares, R. S. Schechter and W. H. Wade, SPE paper No. 5564, presented at 50th Annual Fall Meeting of SPE, Dallas, Texas, September 28-October 1, 1975.
21. R. L. Cash, J. L. Cayias, G. R. Fourneir, J. K. Jacobson, C. A. LeGear, T. Schares, R. S. Schechter, and W. H. Wade, preprint, J. Colloid Interface Sci., in press.
22. D. O. Shah, K. S. Chan and M. Chiang, paper presented at 50th Colloid and Surface Science Symposium, Puerto Rico, June 1976.
23. D. O. Shah and K. S. Chan, University of Florida "Improved Oil Recovery Research Program Semi-annual Report", page A-15, June 1976.
24. P. Debye and A. M. Bueche, J. Chem. Phys. $\underline{18}$, 1423, 1950.
25. A. A. Tager and V. M. Andreeva, J. Poly. Sci. $\underline{C10}$, 1145, 1967.
26. R. N. Healy and R. L. Reed, Soc. Pet. J. $\underline{14}$, 491, 1974.
27. P. A. Winsor, "Solvent Properties of Amphiphilic Compounds", Butterworth's Scientific Publication, 1954.
28. L. E. Scriven, D. R. Anderson, M. S. Bidner, H. T. Davis and C. D. Manning, SPE paper No. 5811, presented at Symposium on Improved Oil Recovery, Tulsa, Oklahoma, March 22-24, 1976.
29. P. B. Lorenz and A. F. Bayazeed, SPE paper no. 4751, presented at Symposium on Improved Oil Recovery, Tulsa, Oklahoma, April 22-24, 1974.
30. S. C. Jones and K. D. Dreher, SPE paper No. 5566, presented at 50th Annual Fall Meeting, Dallas, Texas, September 28-October 1, 1975.
31. M. L. Robbins, paper presented on Interfacial Phenomena in Oil Recovery, AIChE National Meeting, Tulsa, Oklahoma, March 1974.
32. M. L. Robbins, SPE paper No. 5839, presented on Improved Oil Recovery Symposium, Tulsa, Oklahoma, March 22-24, 1976.
33. M. C. Puerto and W. W. Gale, SPE paper no. 5814, presented on Improved Oil Recovery Symposium, Tulsa, Oklahoma, March 22-24, 1976.
34. R. L. Reed, R. N. Healy, paper presented at 81st AIChE Meeting, Kansas City, Mo., April 13-14, 1976.
35. B. C. Hurd, SPE paper No. 5818, presented on Improved Oil Recovery Symposium, Tulsa, Oklahoma, March 22-24, 1976.
36. J. H. Bae, C. B. Patrick and R. Ehrlich, SPE paper No. 4749, presented on Improved Oil Recovery Symposium, Tulsa, Oklahoma, April 22-24, 1974.

37. S. P. Trushenski, D. L. Dauben and D. R. Parrish, Soc. Pet. Eng. J. 14, 623, 1976.
38. R. D. Walker, Jr., University of Florida, "Improved Oil Recovery Research Program Semi-Annual Report", p. D-38, June 1976.
39. H. L. Wilchester, E. W. Malmberg, J. C. Shepard, E. F. Schultze, J. P. Parmley and D. W. Dycus, SPE paper No. 4742, presented on Improved Oil Recovery Symposium, Tulsa, Oklahoma, April 22-24, 1974.
40. R. Dawson and R. R. Lantz, Soc. Pet. Eng. J. 12, 448, 1972.
41. S. P. Trushenski, paper presented at 81st AIChE Meeting, Kansas City, Mo., April 13-14, 1976.
42. J. Noronha and D. O. Shah, University of Florida "Improved Oil Recovery Research Program Semi-annual Report", p. A54-61, June 1976.

DISCUSSION

On the paper by G. S. Hartley

L. E. Scriven, *University of Minnesota*: The same thermal activity of molecules which is responsible for Brownian motions, in these liquid systems, also causes fluctuations in surface zone density and in core density as well as the fluctuations in micelle shape which you mentioned. Shape fluctuations, for example, range all the way down to a wavelength of just one head-group spacing. At this scale individual amphipathic molecules are slipping inward and outward from their mean location in the micelle. One knows that sometimes they escape the micelle altogether; this is of course half of the story of the substantial exchange between micelle and monomers in solution. One expects that sometimes they hide all the way down in the micelle core (in principle there is a certain probable number of molecules in this state). Is your point that some or all of these fluctuations give rise to <u>larger</u> average volume of the micelle than one calculates from static packing of molecular models? Would you say that the whole spectrum of thermal fluctuations of <u>liquid</u> micelles is in need of consideration?

G. S. Hartley: I agree entirely with your comments. Occasional submersion of a whole amphipathic molecule in the interior will certainly enable the size to increase since the size is limited by the necessity to avoid a hole in the middle. If half of one $C_{16}$ chain from among the hundred or so present can always be located in the middle the increase in size is quite disproportionate. Evidence for this submergence comes from the increase of size when a counter-ion providing a more oil-soluble ion pair is substituted for one less oil soluble - e.g. the size increases among cetyl trimethyl ammonium salts in the order Cl Br I. The iodide is yellow in oil solution. It is also yellow in micellar solution in water. What I called amoeba-like distortion of the spherical shape provides another mechanism of size increase but I agree that these are causally related - probably not experimentally distinguishable.

I feel that the necessary "fogginess" of the micelle boundary makes exact deductions from light scattering difficult and I would like to hear comment on this from Dr. Mazer.

## On the paper by L. M. Prince

H. Chun, *General Mills*: I have found that sometimes a clear microemulsion turned "white" in color on the surface upon standing at 45°C for a prolonged period. Why is that? Sometimes I encounter this problem during the mixing process too. It always appears on the surface of the emulsion around the beaker (the beaker is kept covered). Secondly, in your opinion, what is the best way to determine whether a microemulsion is o/w or w/o?

L. M. Prince: The following, I hope, answers your questions.

The "white" ring on standing is probably due to poor blending of the ingredients if the system was made by the inversion process. Another possibility is that some of the droplets in your microemulsion are larger than 3000 Å in diameter. The rest of them are in the lower range. In time, these larger droplets will cream out. You need more emulsifier or a slight change in the emulsifier system you are using. Ordinarily, a good stable microemulsion requires many formulations before it remains stable under all the conditions you would like it to.

Your second question is well answered by Dreher and Syndansk, J. Petroleum Technology Forum, p. 1437, December 1971. They recommended both conductivity and miscibility measurements.

## On the paper by N. J. Turro, M. W. Geiger, R. Hautala, and N. Schore

J. H. Fendler, *Texas A&M University*: The two component decay for the naphthalene fluorescence became one upon degassing. This was interpreted as being due to differential solubilities of the oxygen in micellar and bulk aqueous phases. In the original explanation (Mol. Photochem. $\underline{4}$, 545, 1972) the two component decays were rationalized in terms of the probe being in different environments. Is the original explanation no longer sensible?

N. J. Turro: The degassing procedure removes the factor which discriminates the fluorescence lifetimes of naphthalene in the aqueous and micellar environments. The "natural" lifetime of naphthalene fluorescence is limited by intersystem crossing

## DISCUSSION

in degassed hydrocarbon or aqueous solution. When the detergent solutions are degassed $\tau_F = (k_F + k_{ST})^{-1}$. We presume that the latter two rate constants are the same in water and in micellar environment. When oxygen is present $\tau_F^{O_2} = (k_F + k_{ST} + k_q[O_2])^{-1}$ different measured values of $\tau_F^{O_2}$ occur in water and in the micelle because $k_q[O_2]$ is <u>different</u> in the two environments.

J. K. Thomas, *University of Notre Dame*: First let us indicate that pulseradiolysis date at N.D. show no evidence for significant oxygen solubility in CTAB micelles. Oxygen diffusion data in lipids and membranes show that $O_2$ diffusion is considerably slower in these systems compared to water. These two facts are in agreement with the observation that $O_2$ quenching of probes in micelles is slower than in homogeneous solution. However the interpretation of your data with naphthalene in micelles appears to be contrary to the above. Have you any comment?

Your data also indicate that a large portion of naphthalene resides in the aqueous phase compared to the micelle. This follows for the nonlinear decay of naphthalene. Did you check this by varying the surfactant concentration? The extent of naphthalene in water relative to micelle should cause the change.

N. J. Turro: The naphthalene and pyrene systems might of course behave differently with respect to oxygen solubility and quenching. However, we have measured $k_q[O_2]$ for <u>both</u> systems and find them to be the <u>same</u>. This suggests that there are not dynamical differences or that a fortuity exists. I have no explanation for the pulse radiolysis experiments.

# Part II

# Thermodynamics and Kinetics of Micellization in Aqueous Media

THERMODYNAMICS OF MICELLIZATION OF SIMPLE AMPHIPHILES IN AQUEOUS MEDIA

Charles Tanford

Department of Biochemistry, Duke University Medical Center, Durham, N.C. 27710

A completely rigorous thermodynamic treatment of micellization is obtained if one writes down the equilibrium constants for formation of micelles of all possible degrees of association. All observable parameters (critical micelle concentration, micelle size and distribution, etc.) can be expressed in terms of these equilibrium constants, and the latter can themselves be expressed in terms of the free energy of a monomer molecule as a function of the state of association. This free energy is the sum of a hydrophobic contribution (favoring association) and a size-limiting term resulting from repulsion between head groups. The hydrophobic contribution can be calculated quite accurately on the basis of existing knowledge, and in most cases demands a disk-like shape for micelles in aqueous media (so as to minimize the area of contact between hydrocarbon and water). Calculation of the repulsive term is more difficult, especially for nonionic micelles. If a micelle-forming amphiphile can form stable monolayers at a water-hydrocarbon interface, the repulsion factor can be estimated from monolayer compression data. Using this procedure, micelle characteristics have been calculated for sodium dodecyl sulfate in 0.1 M NaCl, and results in excellent agreement with experiment are readily obtained.

## INTRODUCTION

A completely rigorous thermodynamic treatment of micelle formation presents no conceptual difficulties at all. The problem consists simply of the evaluation of the equilibrium constants for the self-association of monomeric amphiphile (Z) into micelles ($Z_m$), for all values of m from 2 to $\infty$. In the treatment I shall present here it will be assumed that an amphiphile solution is not subject to thermodynamic non-ideality from interactions other than self-association, so that the equilibrium constants,

$$K_m = [Z_m]/[Z]^m \tag{1}$$

can be expressed in concentration units. Non-ideality of the monomer, as results for example from interaction of an ionic amphiphile with other ions in the solution, can be readily incorporated into the thermodynamic analysis by means of an activity coefficient. Interaction of micelles with solvent components is often automatically taken into account in terms of the parameter $W_m$ defined below.

For a variety of reasons it is simplest to express concentrations in mole fraction units ($X_i$). It is also convenient to replace the concentrations of micelles per se by the stoichiometric concentration of amphiphile monomers in micelles of a given size, which I shall designate by $X_m$. Thus $[Z_m]$ is replaced by $X_m/m$, and Equation (1) becomes

$$K_m = X_m/mX_1^m \tag{2}$$

where $X_1$ is the mole fraction of monomeric amphiphile.

## MICELLE SIZE AND SIZE DISTRIBUTION

A plot of $X_m$ versus m defines the distribution function for micelle size. We can define an optimal micelle size (m*) as that value of m for which $X_m$ is a maximum, i.e.,

$$(d \ln X_m/dm)_{m = m*} = 0 \tag{3}$$

The weight-average micelle size, as one would measure by light scattering, is

$$\bar{m}_w = \sum_{m=2}^{\infty} mX_m / \sum_{m=2}^{\infty} X_m$$

and the number average is

$$\bar{m}_n = \sum_{m=2}^{\infty} X_m / \sum_{m=2}^{\infty} (X_m/m) \qquad (5)$$

It follows from the principle of le Chatelier that $m^*$, $\bar{m}_n$, $\bar{m}_w$ and higher size averages must all increase with increasing amphiphile concentration. For the purpose of actual calculations it is simplest to consider $X_1$ as the independent variable, generating equations such as

$$\bar{m}_w = \sum_{m=2}^{\infty} m^2 K_m X_1^m / \sum_{m=2}^{\infty} m K_m X_1^m \qquad (6)$$

The overall concentration of amphiphile in micellar form ($X_{mic}$) is related to $X_1$ as

$$X_{mic} = \sum_{m=2}^{\infty} X_m = \sum_{m=2}^{\infty} m K_m X_1^m \qquad (7)$$

and the total stoichiometric concentration of amphiphile in the solution ($X_{tot}$) is

$$X_{tot} = \sum_{m=1}^{\infty} X_m = X_1 + X_{mic} \qquad (8)$$

The relation between micelle size and $X_{tot}$ is readily generated by evaluating both $X_{tot}$ and $\bar{m}_w$ or other size parameters as functions of $X_1$.

## CRITICAL MICELLE CONCENTRATION

Micelle formation does not represent the formation of a new macroscopic phase, and is therefore not characterized by a single critical transition point. The use of the term "critical micelle concentration" (hereafter called "cmc") can therefore be misleading, and in some system we would probably be better off not to use the term at all. However, when there is in fact only a narrow range of $X_1$ values within which an observable concentration of micelles can exist, then the use of the cmc as a characteristic of the system becomes convenient. The requisite conditions are (1) that micelle formation be cooperative, such that very small micelles do not contribute significantly to the equilibrium population; and (2) that the optimal micelle size be reasonably large (perferably $m^*$ should be >50). The second condition alone is not sufficient: random polymerization can lead to very large values of $m^*$, and the equations given above are formally applicable in relating size parameters to concentration, but the transition from monomer to aggregates is in this case not even moderately sharp.

When the cmc is used as a characteristic of a micelle-forming system, it has to be understood that no unique <u>experimental</u> definition exists, i.e., different experimental procedures should not be expected to yield precisely the same value. An arbitrary <u>formal</u> definition can of course be made in various ways, but this is not very useful if experiments designed to measure the cmc do not in fact determine the quantity so defined. The simplest practical procedure is to let the experiment define the cmc. I shall assume that one can interpret the experiment being done in terms of a critical ratio,

$$X_{mic}/X_{tot} = \rho_{crit}$$

and the cmc is then automatically defined in terms $K_m$ values by use of Equations (7) and (8).

## FREE ENERGY OF MICELLIZATION

The standard free energy of forming a micelle of size m is given by $-RT \ln K_m$, and if mole fraction units are used for concentrations the free energy will automatically be in <u>unitary</u> units[1]. It is convenient and customary to work in terms of the standard (unitary) free energy of a monomer molecule in a micelle rather than the free energy of formation of the whole micelle, i.e., where $\Delta G°_m$ represents the free energy of transfer of a monomer molecule from aqueous solution to a micelle of size m,

$$-RT \ln K_m = m \Delta G°_m \qquad (10)$$

For simple amphiphiles (single head groups joined to a hydrocarbon tail) it is appropriate to consider $\Delta G°_m$ as the sum of two parts

$$\Delta G°_m = \Delta U°_m + W_m \qquad (11)$$

where $\Delta U°_m$ represents the contribution of the hydrocarbon tail (predominantly ascribable to hydrophobic repulsion by the solvent) and $W_m$ represents the contribution of the head group (predominantly repulsion between head groups since the head groups remains in a solvent environment when the micelle is formed).

It is essential for a rigorous treatment of micelle formation that $\Delta G°_m$ and its component parts be evaluated for <u>all possible values</u> of m, and that realistic relations to represent the dependence on m be employed. Most previous treatments have used simplifying assumptions that are not valid. The availability of modern electronic computers (or even deskcalculators) makes the use of such assumptions unnecessary.

## GEOMETRICAL RESTRICTIONS

As has been known since Hartley's elegant treatise on the subject in 1936[2], the thermodynamic driving force for micelle formation is the hydrophobic effect, whereby hydrocarbon tails of amphiphile molecules are expelled from an aqueous medium so as to avoid, as far as possible, any hydrocarbon-water contact. For this reason the <u>core</u> of any simple micelle consists of pure liquid hydrocarbon, as compact as structural packing requirements permit. There is however one limitation on the achievable degree of compactness - one dimension of the hydrocarbon core clearly may not exceed the length of two hydrocarbon chains approaching the center of the core from opposite directions. When one is dealing with micelles having liquid cores, this limiting dimension must be appropriate for a hydrocarbon chain in the liquid state, i.e., it will be less than the length of two fully extended chains. Estimates can be made with the aid of experimental and theoretical data provided by Flory[3].

If one calculates the volume of a sphere with a diameter equal to the limiting dimension, one finds that it corresponds to the volume of a relatively small number of hydrocarbon chains, and this would be true even if the fully extended chain lengths were used as a basis for the calculation. Measured values for the micellar state of aggregation are nearly always much larger than can be accommodated by a micelle with a spherical core[4,5]. One can see the thermodynamic reason for this by calculating the surface area which a spherical micelle core would have. Even if this area is minimized by assuming it to be smooth, it is much larger than the area occupied by amphiphile head groups, so that much of the area remains in contact with the aqueous environment. The thermodynamic requirements of the hydrophobic effect therefore dictate an increase in the state of aggregation: regardless of the shape of the aggregate there will always be a decrease in the surface/volume ratio and a concomitant decrease in the exposed surface between head groups.

What is the actual shape? Keeping in mind the restriction on one dimension of the hydrocarbon core, there are two possibilities: a disk-like structure (a x a x b; where b is the restricted dimension) or a rod-like structure (a x b x b). The surface/volume ratio decreases much more rapidly with an increase in state of aggregation when the shape is disk-like, and one would therefore intuitively expect a disk-like shape to be preferred, at least at low amphiphile concentrations where there is thermodynamic pressure to keep the state of aggregation low. Similar conclusions have been reached previously by Klevens[6] and Tartar[4]. They are supported by hydrodynamic data for both ionic and nonionic micelles[7].

## MICELLE SHAPE IN DETAIL

The term "disk-like" used above allows for a variety of possible detailed specifications of shape. In a previous paper[7] I have used an oblate ellipsoid of revolution as a basis for calculation. The ellipsoidal shape can be modified by rounding off edges. An alternative possibility is a flat disk with a semi-toroidal ring around the edge. Because the hydrocarbon is liquid there are likely to be rapid fluctuations in shape that may blur the distinction between these possibilities. In addition, restrictions on molecular bond angles near the surface may tend to introduce corrugations into the core surface, as previously described[8].

From the point of view of thermodynamic calculations the important parameter is the area of contact between hydrocarbon and solvent at the surface of the core, and we shall use the symbol $A_{Hm}$ to designate this area per constituent monomer molecule in a micelle of size m. For small micelles the differences between possible detailed shapes is small. For example, for the hydrophobic core that is formed by straight alkyl chains containing 12 hydrophobic carbon atoms, with m = 100, and measuring $A_{Hm}$ at the closest distance of approach of water molecules (1.5 Å outside the actual core surface) we get $A_{Hm}$ = 66.2 Å$^2$ for an oblate ellipsoid and 63.5 Å$^2$ for a flat disk with a semi-toroidal edge. (For a prolate ellipsoid $A_{Hm}$ = 80.0 Å$^2$ and for a straight cylinder with hemispherical caps it is 74.2 Å$^2$). The limiting trans-micelle dimension (2b) was set at 24.36 Å for this calculation, which would be the value appropriate for completely liquid-like alkyl chains, but the effect of stiffening the chain is again relatively small: e.g., if we set 2b = 28 Å, $A_{Hm}$ becomes 63.3 Å$^2$ for an oblate ellipsoid and 61.7 Å$^2$ for a disk with a semi-toroidal edge. All calculations are based on a smooth surface: "corrugations" would increase $A_{Hm}$ (regardless of shape) by perhaps 10%.

The hydrophobic core of a micelle is surrounded by a solvent-permeated layer containing the head groups. The actual geometry is dictated by bond lengths and angles. In ionic micelles we are interested in the location of the centers of charge, and it is presumably correct to think of them as lying on a surface parallel to the core surface, with an area per head group ($A_{Rm}$) which will be larger than $A_{Hm}$ because of the curvature of the surface. Non-ionic amphiphiles with small head groups can be treated in the same way, but consideration of polyoxyethylene derivatives is more difficult because the head groups are here very long and it is not yet known whether they are randomly coiled or assume some definite (perhaps helical) structure.

## HYDROPHOBIC COMPONENT OF THE FREE ENERGY OF MICELLIZATION

Three steps are involved in the estimation of numerical values for the size-dependent hydrophobic component ($\Delta U_m°$) of the free energy of micellization. (1) We begin with the free energy of transfer of the hydrophobic portion of the amphiphile from water to the pure anhydrous liquid state. (2) We allow for the fact that the hydrophobic core of a micelle is more ordered than bulk liquid hydrocarbon, in that the amphiphile tails are constrained at one end, i.e., where they are attached to the polar group projecting from the core surface. (3) We allow for the incomplete removal of the amphiphile tail from the aqueous environment by introducing a positive contribution to $\Delta U_m°$ proportional to the excess area of contact with aqueous solvent at the core surface. To exemplify this approach, I shall restrict myself to amphiphiles with saturated linear paraffin chains, for which requisite experimental data are most readily available.

Thermodynamic data for transfer of hydrocarbons from water to pure liquid hydrocarbon are based on the extensive solubility studies of McAuliffe[9]. For all saturated hydrocarbons (linear, branched, and cyclic) the unitary free energy of transfer (in cal/mole) at 25°C is numerically equal to approximately 25 times the molecular area (in $Å^2$) at the position of centers of water molecules in contact with the hydrocarbon molecule[10]. There is some uncertainty in the measurement of molecular area, and a more precise representation of the data for n-alkanes is obtained by assigning -2100 cal/mole per $CH_3$ group and -884 cal/mole per $CH_2$ group[1]. The contribution per $CH_2$ group derived from solubility studies using fatty acids[11] or alcohols[12] is numerically somewhat smaller (-825 cal/mole), but the organic solvent in these experiments was always n-heptane, and there is bound to be some contribution from non-ideal mixing in the organic phase.

There is good evidence[1] suggesting that a $CH_2$ group immediately adjacent to a polar group contributes little to the hydrophobicity of an amphiphile, persumably because the solvent molecules in contact with this $CH_2$ group are pre-empted into the structural water around the polar group. We have assumed that this first $CH_2$ group of an alkyl chain does not in fact enter the hydrophobic core of a micelle: the number of hydrophobic C atoms per chain ($n_c$) is taken to be <u>one less</u> than the actual chain length. It has been established by magnetic resonance measurements (e.g., Seelig and Seelig[13]) that the first few carbon atoms of an alkyl chain within a fluid hydrophobic core are more ordered than those near the chain terminus, but even the rest of the chain has less freedom of motion than in a pure liquid because it is constrained to have an orientation more or less perpendicular to the core surface, and only the terminal $CH_3$ groups can be expected to be uninfluenced by this. The complete removal of the amphiphile alkyl chain from water thus makes a contribution to $\Delta U_m^\circ$ which may be expressed in terms of the contributions from $CH_3$ and $CH_2$ groups as $-[u_{CH_3} + su'_{CH_2} + (n_C-s-1)u_{CH_2}]$ where s is the number of "more ordered" $CH_2$ groups, likely to be in the range of s=5 to 8. The value of $u_{CH_3}$ is expected to be close to 2100 cal/mole, the value of $u_{CH_2}$ is expected to be significantly less than 885 cal/mole, and $u'_{CH_2} < u_{CH_2}$. The parameter $u_{CH_2}$ can in fact be obtained experimentally because it turns out that the effect of alkyl chain length on the cmc is almost exlusively determined by this term. All studies with homologous series of detergents (both ionic and nonionic) indicate that $u_{CH_2}$ is very close to 700 cal/mole. The difference between $u'_{CH_2}$ and $u_{CH_2}$ becomes an empirical parameter, but should have the same value for all amphiles with unbranched alkyl chains. (On the basis of the single calculation presented below the difference appears to be about 100 cal/mole).

The foregoing numerical values apply to complete removal of the alkyl chain from contact with water. If the chains were close-packed as in crystalline hydrocarbons, with a surface area of 21 Å² per chain (perpendicular to the chain orientation), there would be no contact with water. We allow for residual hydrocarbon-water contact by multiplying $A_{Hm}$ (expressed in Å²) by the factor 25 (cal/mole/Å²) experimentally observed for the hydrophobic interaction in pure hydrocarbon solutions.

The overall expression for $\Delta U_m^\circ$ (with $u_{CH_3}$ = 2100 and $u_{CH_2}$ = 700 cal/mole) becomes

$$\Delta U_m^\circ = -1400 - 700n_C + C' + 25(A_{Hm}-21) \qquad (12)$$

where C' represents the empirical parameter $s(u_{CH_2}-u'_{CH_2})$. It should be noted that $A_{Hm}$ is the only parameter in Equation (12) which depends on micelle size, so that the value of C' has no significant effect in the evaluation of optimal micelle size.

## HEAD GROUP REPULSION

An a priori theoretical calculation of the contribution of head group interactions to the free energy of micellization ($W_m$ of Equation (11)) is difficult. Coulombic repulsion is presumably the principal factor for ionic micelles, but calculations are complicated by the high charge density generated by the closely spaced head groups and by counterion binding that results from it. Steric repulsion is likely to be a major factor in some nonionic micelles, but attractive forces may contribute and lead to preferred head group separations.

It should be possible to evaluate $W_m$ at least approximately from force-area curves of amphiphile monolayers at a hydrocarbon-water interface. The hydrophobic portions of the amphiphile molecules are in a liquid hydrocarbon environment in this type of experiment, and the head groups extend into the aqueous phase. If we assume that the mixing of the hydrophobic tails with solvent molecules of the hydrocarbon phase is a thermodynamically ideal process, we can ascribe the excess work of compression (as measured experimentally) entirely to repulsion between head groups. There are some as yet unresolved questions in exactly how to extract the required parameter from the monolayer data. It appears probable that the interaction term is obtained correctly as the difference between the observed work of compression (which is $\int \Pi dA$, where $\Pi$ is surface pressure and A the area per head group), from infinite dilution to any given A, and the ideal work for the same process. In a previous attempt to evaluate the interaction[8], the calculation was based on $\int A d\Pi$ and incorrect results were obtained. (I am very grateful to Dr. J. Lang of the Procter and Gamble Co. for drawing my attention to this).

Monolayer compression curves (after correcting for the contribution from ideal mixing) yield values for W as a function of the area (A) per molecule: because the hydrocarbon-water interface is planar the numerical value of A is the same at the interface between the phases as it is at a parallel plane in the aqueous phase where one visualizes that the interaction is taking place. At the curved surface of a micelle the area per monomer increases as one moves into the aqueous phase from the surface of the hydrocarbon core. For ionic micelles at relatively high ionic strength (small thickness of ionic atmosphere) this is probably the major effect of surface curvature, i.e., it should be a good approximation to equate $W_m$ with the monolayer W value at an area A that corresponds to the area $A_{Rm}$ calculated at an appropriate distance from the micellar core surface.

## SAMPLE CALCULATION

Table I shows the results of a sample calculation. The system chosen is sodium dodecyl sulfate in 0.1 M NaCl at room temperature. The experimental value of $\bar{m}$ is about 90, and the experimental cmc is about $1.4 \times 10^{-3}$M[14,15]. It has been assumed that $\rho_{crit} = 0.05$ at the cmc. The monolayer compression data of Mingins[16,17] have been used to evaluate W, yielding

Table I. Calculations for Sodium Dodecyl Sulfate in 0.1 M NaCl

| Adjustable Parameters[a] | | | | Calculated Results[b] | | |
|---|---|---|---|---|---|---|
| $b_o/1_{max}$ | $\delta x_R$ (Å) | $\delta x_H$ (Å) | C' | m*(oblate) | m*(disk) | C'(correct) |
| 0.75 | 4 | 1.5 | 500 | 55 | 63 | |
| | | 2.0 | | 80 | 86 | |
| | | 2.25 | | 94 | 98 | |
| | | 2.5 | | 110 | 110 | |
| | | 3.0 | | 145 | 133 | |
| 0.75 | 4 | 2.25 | 500 | 94 | 98 | 520 |
| 0.85 | | | | 90 | 94 | 650 |
| 0.75 | 3 | 2.25 | 500 | 100 | 98 | 420 |
| | 3.5 | | | 97 | 98 | |
| | 4 | | | 94 | 98 | 520 |
| | 4.5 | | | 92 | 98 | |
| | 5 | | | 91 | 98 | 600 |
| 0.75 | 4 | 2.25 | 1000 | 98 | 101 | |
| | | | 500 | 94 | 98 | |
| | | | 0 | 91 | 96 | |

[a] $b_o/1_{max}$ is the ratio of the limiting micelle dimension to the length of fully extended alkyl chains; $\delta x_R$ is the distance from the surface of a hypothetically smooth hydrophobic core surface to the surface where head group interaction occurs; $\delta x_H$ is a similar distance for calculating the effective area of contact with solvent; C' is the value of the parameter used to calculate m*.

[b] m* values are given for two alternative geometrical models, an oblate ellipsoid of revolution and a flat disk with a semi-toroidal edge; C' is the value that would have to be used (without change in other parameters) to obtain a cmc value in agreement with experiment.

$$W = \frac{1.432 \times 10^5}{A} - \frac{4.23 \times 10^6}{A^2} + \frac{9.65 \times 10^7}{A^3} \quad (13)$$

with W in cal/mole and A in Å$^2$. I have used $n_C = 11$, allowing for the probability that the $CH_2$ group next to the sulfate head groups is outside the hydrophobic core. This choice dictates that the centers of charge of the sulfate head group be located at a distance ($\delta x_R$) of 4 Å outside the core surface. (If $n_C = 12$ is used, $\delta x_R$ has to be reduced to about 2.75 Å, and the major overall effect on the calculation is to diminish the value of C' required to obtain the correct cmc. This point will be discussed below). The limiting trans-micelle dimension (2b) was set at 0.75 times the length of two fully extended 11-carbon chains on the basis of the data of Flory[3] for liquid-like chains, but the results show that stiffening the chains (setting 2b = 0.85 times the fully extended length) has little effect.

The only variable parameter with a significant effect on the calculation of optimal micelle size is $A_{Hm}$. Assuming a disk-like micelle, one can use an oblate ellipsoid or a flat disk with a semi-toroidal edge as a geometrical model. The results show that this choice has almost no effect. The increase in surface area due to surface corrugations has been allowed for by calculating $A_{Hm}$ at a greater distance ($\delta x_H$) from the surface of a perfectly smooth hydrophobic core than the minimal distance corresponding to the closest approach of water molecules ($\delta x_H = 1.5$ Å). The data show that optimal results are obtained by setting $\delta x_H \approx 2.25$ Å, which corresponds (at m = 90) to a surface area about 7% larger than the minimal area at $\delta x_H = 1.5$ Å.

The results show that varying C' (ignoring its effect on the calculation of the cmc) has little effect on the calculation of m*. Similarly, varying $\delta x_R$ by as much as 1 Å has little effect, the reason for this being that varying $\delta x_R$ affects $W_m$ more than it affects $dW_m/dm$. The magnitude of W and its dependence on A do of course influence the results. If we calculate W on the basis of the pressure-area curves of Brooks and Pethica[18] (now known to be incorrect[17]) we would get m*$\approx$65 using $\delta x_H = 2.25$ Å.

The last column shows the value of C' that must be used to obtain a cmc value in agreement with experiment, using $\delta x_H = 2.25$ Å and several different choices for other parameters. The required value is seen to be about 500 cal/mole. Since C' = $s(u_{CH_2} - u'_{CH_2})$ and s is about 5, this indicates that the ordering of $CH_2$ groups near the core surface affects the free energy by about 100 cal/mole per $CH_2$ group, which is a reasonable result. As noted above, this result is influenced by the choice of $n_C$ = 11, and if $n_C$ = 12 were chosen instead the optimal C' value would have to be about 700 cal/mole larger, as is obvious from Equation (12). With the choice of $n_C$ = 12 one would in effect be required by the experimental data to distribute throughout the alkyl chain a positive free energy to compensate for the extra negative free energy that is introduced (unrealistically, I believe) when the $CH_2$ group adjacent to the polar head group is taken to be intrinsically equal to other $CH_2$ groups in hydrophobicity.

(A minor factor might be mentioned here. I have not made extensive calculations of the effect of alkyl chain length on the cmc of homologous amphiphiles, and the value of 700 cal/mole for $u_{CH_2}$ is only approximate. It is likely that a value of perhaps 720 cal/mole would be more accurate, and it is seen from Equation (12) that this would increase the C' derived from the present data to about 700 cal/mole).

An important aspect of these calculations is that they do not in fact require the <u>assumption</u> of a disk-like shape for sodium dodecyl sulfate micelles. Parallel calculations have been make for rod-like micelles using a prolate ellipsoid or a cylinder with hemispherical caps as geometrical model. The calculations show that the monomer concentration required to form rod-like micelles is considerably larger than the concentration needed to form disk-like micelles. In other words, at the cmc calculated for disk-like micelles the concentration of rod-like micelles is completely negligible. Experimental data from a variety of studies unambiguously support a disk-like shape for sodium dodecyl sulfate in solutions of moderate ionic strength[7].

The distribution functions calculated on the basis of parameters that yield results in agreement with experimental data are moderately broad and quite symmetrical, so that there is only a small difference between $\bar{m}_n$, $\bar{m}_w$ and m*. For example, for the parameters in the third line of Table I, at the cmc, the weight concentration falls to 50% of its maximal value at m* = 94 when m = 73 and 117. Average degrees of association are $\bar{m}_n$ = 92 and $\bar{m}_w$ = 96.

## DISCUSSION

The most important feature of the sample calculation given in Table I is that <u>no freely adjustable parameters were employed</u>. The uncertainty in the calculation of $A_{Hm}$ (as reflected in the choice of $\delta x_H$) is small. The value of $\delta x_H$ = 3 Å corresponds to $A_{Hm}$ 15% above its minimal value and this is a reasonable upper limit to the effect of surface corrugations. The value of $\delta x_H$ = 2.25 Å that would lead to agreement with the experimental micelle size corresponds to $A_{Hm}$ 7% above its minimal value, and this is a reasonable result. Similarly, the value of C' required for agreement with the experimental cmc is entirely reasonable. Questions remain about the validity of the direct use of the W expression from surface compression data (Equation(13)) for head group repulsion at the micelle surface, but modifications in this procedure could not affect the results greatly - at most they could change the optimal $\delta x_H$ by a few tenths of an Å and the value of C' by 100 or 200 cal/mole.

It should also be noted that even the small adjustments in these parameters that were possible in this attempt to account for the experimental data for one ionic detergent at one ionic strength are subject to experimental test. The physical meaning of these parameters is such that they should be the same regardless of alkyl chain length or of ionic strength, and in simple situations should also be independent of the amphiphile head group. The ultimate test of the theory presented here will come from extension of the calculations to micelles formed by other alkyl sulfates and over a range of ionic strength, and to other ionic amphiphiles for which appropriate monolayer studies have been or are being made. Essentially the same values of $\delta x_H$ and $C'$ should apply to all such data.

In more general terms, the equations presented here should provide a rigorous test for theoretical calculations of the head group interaction term (e.g., where monolayer data are not available) or for possible absolute calculations of the hydrophobic free energy term. The thermodynamic equations relating micelle size, cmc, and the effect of such variables as alkyl chain length, are rigorous and adaptable to alternative expressions for the terms $\Delta U_m^\circ$ and $W_m$ of Equations (11).

## ACKNOWLEDGMENT

This work has been supported by the National Science Foundation.

## REFERENCES

1. C. Tanford, "The Hydrophobic Effect", John Wiley & Sons, New York, 1973.
2. G.S. Hartley, "Aqueous Solutions of Paraffin-Chain Salts", Hermann & Cie., Paris, 1936.
3. P.J. Flory, "Statistical Mechanics of Chain Molecules", John Wiley & Sons, New York, 1969.
4. H.V. Tartar, J. Phys. Chem., 59, 1195 (1955).
5. C. Tanford, J. Phys. Chem., 76, 3020 (1972).
6. H.B. Klevens, J. Amer. Oil Chem. Soc., 30, 74 (1953).
7. C. Tanford, J. Phys. Chem., 78, 2469 (1974).
8. C. Tanford, Proc. Nat. Acad. Sci. USA, 71, 1811 (1974).
9. C. McAuliffe, J. Phys. Chem., 70, 1267.
10. J.A. Reynolds, D.B. Gilbert and C. Tanford, Proc. Nat. Acad. Sci. USA, 71, 2925 (1974).
11. R. Smith and C. Tanford, Proc. Nat. Acad. Sci. USA, 70, 289 (1973).
12. D.B. Gilbert, C. Tanford and J.A. Reynolds, Biochemistry, 14, 444 (1975).
13. A. Seelig and J. Seelig, Biochemistry, 13, 4839 (1974).
14. K. Mysels and L. Princen, J. Phys. Chem., 63, 1696 (1959).
15. M.F. Emerson and A. Holtzer, J. Phys. Chem., 71, 1898 (1967).
16. J. Mingins (1974), personal communication.
17. J.A.G. Taylor and J. Mingins, J. Chem. Soc. Faraday Trans. I, 71, 1161 (1975).
18. J.H. Brooks and B.A. Pethica, Trans. Faraday Soc., 61, 571 (1965).

THERMODYNAMICS OF AMPHIPHILAR AGGREGATION INTO MICELLES AND VESICLES

E. Ruckenstein and R. Nagarajan, Faculty of Engineering and Applied Sciences, State University of New York at Buffalo, Buffalo, NY 14214

A unified thermodynamic treatment of self-aggregation into micelles and/or vesicles of amphiphiles with one or two hydrocarbon tails in aqueous media is developed. Empirical expressions provided by Tanford are used for the free energy of the aggregates to obtain the size distribution function and the type of aggregation for the two kinds of amphiphiles. Calculations have been carried out for different tail lengths and different repulsive interaction strengths between the polar head groups. They show that, for the range of parameters considered, the amphiphiles with one hydrocarbon tail aggregate as micelles, while those with two hydrocarbon tails aggregate as vesicles. A critical vesicle concentration (the analog of CMC) is calculated on the basis of a sharp change in the dependence of the total concentration of aggregates on the total amphiphilar concentration.

## INTRODUCTION

Amphiphilar molecules in aqueous media achieve segregation of their hydrophobic parts by self-aggregation. The aggregates can be either in the form of micelles with an internal hydrocarbon core and a surface composed of polar groups or in the form of bilayers with two layers of amphiphiles in contact having the hydrocarbon tails inside, and the polar groups outside in contact with the aqueous medium. Micelles can be spherical, ellipsoidal or cylindrical, whereas bilayers can be either planar or spherical with an internal cavity filled with solvent. Bilayers of the latter structure are called vesicles. Amphiphiles with single hydrocarbon tails form micelles[1] which grow in size as the length of the tail increases or as the ionic strength of the aqueous medium increases. Lipid molecules which possess a single polar head group and two long hydrocarbon tails form bilayers, both planar as well as spherical[1].

Aggregation of amphiphiles into micelles has been treated either as a stepwise association phenomenon or as a phase transition process[2,3]. The formation of spherical vesicles from amphiphiles in solution has not yet been examined from a thermodynamic point of view. In an earlier paper[4] micellization was treated, assuming that micellar aggregates of all possible sizes coexist. At low amphiphilar concentrations, the size distribution of the aggregates was predicted to decrease monotonically with size. As the total concentration was increased, the size distribution was found to change from a monotonic decreasing function to one exhibiting two extrema, a minimum and a maximum. A critical concentration was defined as the total amphiphilar concentration at which the change in the shape of the size distribution function occurs. This critical concentration was shown to be a close lower bound of the critical micelle concentration (CMC)[5]. In the region close to the CMC, the physico-chemical properties of the system were shown to change rapidly but continuously and the concentration of single amphiphiles was shown to increase slowly. Calculations showed that addition of surfactant above the CMC leads not only to an increase in the number of aggregates, but also to larger average sizes of the aggregates. These predictions are in agreement with experimental observations. It may be pointed out that this treatment was not based on any particular model of micellization. However, the treatment considered the existence of a single type of aggregate, the micelles.

The goal of the present paper is to determine the conditions under which micelles and/or vesicles form. It will be shown that a unified thermodynamic framework can be provided to describe the aggregation of amphiphiles with one or two hydrocarbon tails in aqueous solution into micelles and/or spherical vesicles.

In the next section the size distribution model is formulated for aggregates of arbitrary shape and kind. Explicit expressions for the free energy are introduced in section III. The types of aggregation of the amphiphiles with one and two hydrocarbon tails are considered in sections IV and V.

The main theoretical conclusion is that single chain amphiphiles aggregate generally as micelles and double chain amphiphiles as vesicles. As mentioned above, this is in agreement with experiment.

## II. SIZE DISTRIBUTION MODEL

The amphiphilar system considered here is composed of $N_S$ solvent molecules, $N_A$ single amphiphiles, and $N_{gi}$ aggregates of type i (micelles or vesicles) and size g. Aggregates of different sizes are considered as distinct species, each characterized by its own standard chemical potential. The standard chemical potentials of the solvent molecule and of the single amphiphile are denoted by $\mu_S^o$ and $\mu_A^o$ respectively. The standard chemical potential per amphiphile of the aggregate of type i is separated into a size independent part denoted by $\mu_{Bi}^o$ and a size dependent part $\mu_{gi}^o/g$. As explained below the size independent part of the standard chemical potential per amphiphile $\mu_{Bi}^o$ is the same for all types of aggregates and hence will be denoted by $\mu_B^o$. It is assumed that the solution is sufficiently dilute in amphiphiles and, therefore, that the interaction forces between the aggregates are negligible. For the total thermodynamic potential $\Phi$ we use the expression:

$$\Phi = N_S \mu_S^o + N_A \mu_A^o + \sum_i \sum_g N_{gi} (\mu_B^o g + \mu_{gi}^o)$$
$$+ kT [N_S \ln v_S + N_A \ln v_A + \sum_i \sum_g N_{gi} \ln v_{gi}] \quad , \tag{1}$$

where the free energy of mixing has been written in terms of the volume fractions $v_S$, $v_A$ and $v_{gi}$ of the solvent molecule, single amphiphiles and aggregates of type i and size g to account for the large differences in sizes. Here k is Boltzmann's constant and T, the absolute temperature. The summation over g is carried out between 2 and $\infty$ for micelles and between the minimum size (as defined below) and $\infty$ for vesicles. The summation over i is carried out for all types of aggregates. Denoting by a the ratio of the volume of a single amphiphile to the volume of a solvent molecule, the total potential $\Phi$ can be rewritten as

$$\Phi = N_S \mu_S^o + N_A \mu_A^o + \sum_i \sum_g N_{gi} (\mu_B^o g + \mu_{gi}^o)$$
$$+ kT [N_S \ln \frac{N_S}{F} + N_A \ln a \frac{N_A}{F} + \sum_i \sum_g N_{gi} \ln ag \frac{N_{gi}}{F}] \quad , \tag{2}$$

where

$$F = N_S + a N_A + \sum_i \sum_g a g N_{gi} \quad . \tag{3}$$

Since the total number of amphiphiles is a constant, F is also a constant. The equilibrium condition, corresponding to the minimum of the total potential $\Phi$ subject to the constraint (3), yields

$$-g\,\mu^o_A + g\mu^o_B + \mu^o_{gi} + kT\,[-g\,\ln a\,\frac{N_A}{F} - g + 1$$

$$+ \ln a\,g\,\frac{N_{gi}}{F}] = 0 \quad . \tag{4}$$

Equation (4) can be rearranged to give the equilibrium size distribution for aggregates of type i

$$\frac{N_{gi}}{F} = \xi^g\,(aeg)^{-1}\,\exp\,(-\mu^o_{gi}/kT) \quad , \tag{5}$$

where

$$\xi = (\frac{N_A}{F})\,ae\,\exp\,-\,(\frac{\mu^o_B - \mu^o_A}{kT}) \tag{6}$$

and e is the base of the Naperian logarithm. It was shown[4] that such a size distribution function is monotonically decreasing with size below a critical concentration and exhibits a minimum and a maximum above that concentration. From the size distribution function (5) one can compute any size dependent property of the system such as the true number average aggregation number

$$(\bar{g}_n)_i = \sum_g g\,N_{gi} / \sum_g N_{gi} \quad , \tag{7}$$

and the true weight average aggregation number

$$(\bar{g}_w)_i = \sum_g g^2\,N_{gi} / \sum_g g\,N_{gi} \quad . \tag{8}$$

The dispersion in size of the aggregates of type i is measured by the variance $\sigma_i$ of the size distribution function

$$\sigma^2_i = \frac{\sum_g (g - (\bar{g}_n)_i)^2\,N_{gi}}{\sum_g N_{gi}} \quad , \tag{9}$$

or by the ratio

$$\frac{\sigma_i}{(\bar{g}_n)_i} = [(\bar{g}_w)_i / (\bar{g}_n)_i - 1]^{1/2} \quad . \tag{10}$$

At a critical value $(\frac{N_A}{F})^*$ of the concentration of non-aggregated amphiphiles, phase separation occurs. At this point the non-aggregated amphiphiles and the aggregates in aqueous solution are in thermodynamic equilibrium with the bulk amphiphilar phase. The phase equilibrium is

$$\mu_{bulk} = \mu_A = \mu_A^o + kT \left[ \ln a\left(\frac{N_A}{F}\right)^* + 1 - a\left(\frac{N_s + N_A}{F}\right)^* - \sum_i \sum_g a\left(\frac{N_{gi}}{F}\right)^* \right], \quad (11)$$

where $\mu_{bulk}$ is the chemical potential per amphiphile of the bulk amphiphilar phase and $\mu_A$ is the chemical potential of the non-aggregated amphiphiles in solution. The asterisk denotes phase separation conditions. The concentration $\left(\frac{N_A}{F}\right)^*$ of the non-aggregated amphiphiles is a limiting value below which aggregation of any added amphiphile is favored and above which phase separation into one phase containing aggregates and a bulk amphiphilar phase occurs. The equations derived so far can be applied to different kinds of amphiphiles and types of aggregation by using suitable expressions for the free energy terms $\mu^o$. In the next section we describe one such expression.

## III. EXPRESSIONS FOR THE FREE ENERGY

For illustrative purposes, Tanford's empirical expressions[1] are used. In that treatment the standard free energy of formation of an aggregate, per amphiphile, is separated into attractive and repulsive components. The attractive component arises from the hydrophobic effect which seeks to minimize the hydrocarbon-water contact. This component is assumed to be independent of the nature of the head group and contains both size independent and size dependent parts. The size independent part represents the free energy change of the hydrocarbon chain from an aqueous environment to complete immersion in the micellar core. Because the hydrocarbon environment and the translational and rotational constraints are similar for micelles and vesicles, this part can be assumed to be the same for all types of aggregates. However, it differs for different types of amphiphiles. The size dependent part is due to the interfacial interaction between the aqueous medium and the exposed hydrocarbon surface and is a function solely of the (size dependent) aggregate surface area A per head group. The repulsive component of the free energy depends both on the nature of the amphiphilar head group as well as on the separation between the head groups. The surface area A is used as a measure of this separation. In the present paper the standard free energy of formation of an aggregate is decomposed into size independent and size dependent terms. Since the size dependence occurs through A alone, the size dependent part of the free energies of different kinds of amphiphiles and aggregates can be expressed by the same function of A.

For amphiphiles consisting of a single hydrocarbon tail, the size independent part of the standard free energy of aggregation is

$$\frac{\mu^o_B - \mu^o_A}{kT} = \frac{-2000 - 700\,(n_c - 1)}{RT}, \qquad (12)$$

where $n_c$ is the total number of carbon atoms in the hydrocarbon tail and R is the gas constant in cal/mole°K. The size dependent part of the free energy is given by

$$\frac{\mu^o_g}{kT} = \frac{25\,(A_H - 21)\,g}{RT} + \frac{\alpha\,g}{A_R\,RT}, \qquad (13)$$

where $\alpha$ is a constant in cal Å$^2$/mole representing the repulsion between the polar head groups, $A_H$ and $A_R$ are the surface areas per amphiphile in Å$^2$ corresponding to an aggregate surface containing the bases or centers of the head groups respectively. For simplicity, $A_H$ and $A_R$ have been considered equal and denoted by A. The parameter $\alpha$ is dependent on the nature of the polar head group and on the ionic strength of the medium. The use of this expression implies that the surface area A per amphiphile is uniform throughout the aggregate surface. The area A per amphiphile is different for different types of aggregates and different types of amphiphiles and depends on g.

For amphiphiles with two hydrocarbon tails, the two tails have some degree of mutual association, thus decreasing the area of their exposure to the aqueous medium. Consequently, the free energy change of transfer of amphiphiles from an aqueous medium to an aggregate phase for a double chain amphiphile can be expected to be not quite twice that of the corresponding single chain amphiphile. This has, in fact, been observed experimentally[1]. For amphiphiles with two hydrocarbon tails, Tanford[1] has estimated that addition of a second tail increases the energy change by only 60%. Using this estimate, the size independent part of the free energy of transfer for such amphiphilar aggregates is given by

$$\frac{\mu^o_B - \mu^o_A}{kT} = 1.6\left[\frac{-2000 - 700\,(n_c - 1)}{RT}\right]. \qquad (14)$$

The size dependent part of the free energy is given by the same equation as for a single chain amphiphile, since as mentioned earlier the influence of the two tails is incorporated in the area A per amphiphile.

$$\frac{\mu^o_g}{kT} = \frac{25\,(A - 21)\,g}{RT} + \frac{\alpha g}{ART}. \qquad (15)$$

# THERMODYNAMICS OF AMPHIPHILAR AGGREGATION

The interactions between the two polar surfaces of vesicles makes a negligible contribution to the free energy[6].

It should be mentioned that the empirical equations of Tanford are used here because of their simplicity and their ability to predict reasonably the micellar aggregation[1,4,5]. Expressions with a clearer physical significance for various terms can be obtained[7] and a treatment based on them will be published elsewhere. In addition to the well established terms representing cohesive interactions and electrostatic repulsions, that treatment accounts for the interfacial free energy and for the translational and rotational constraints on amphiphiles following their incorporation into an aggregate.

For amphiphiles with a single hydrocarbon tail, the volume $v_o$ and extended length $\ell_o$ of the hydrocarbon chain are[1]:

$$v_o = 27.4 + 26.9 \, (n_c - 1) \text{ in } \text{Å}^3 \qquad (16)$$

and

$$\ell_o = 1.5 + 1.265 \, (n_c - 1) \text{ in } \text{Å} \, , \qquad (17)$$

respectively. For amphiphiles with two hydrocarbon tails, the corresponding equations are

$$v_o = 2 \, [27.4 + 26.9 \, (n_c - 1)] \text{ in } \text{Å}^3 \qquad (18)$$

and

$$\ell_o = 1.5 + 1.265 \, (n_c - 1) \text{ in } \text{Å} \, . \qquad (19)$$

For micellar aggregates of spherical shape and size g, the surface area per amphiphile is

$$A = \frac{1}{g} \, 4\pi \, [(\frac{3gv_o}{4\pi})^{1/3} + \delta]^2 \, , \qquad (20)$$

where $\delta$ accounts for the surface roughness. For micellar aggregates of cylindrical shape with hemispherical ends, the surface area per amphiphile is

$$A = \frac{1}{g} \, [2\pi L \, (\ell_o + \delta) + 4\pi \, (\ell_o + \delta)^2] \, , \qquad (21)$$

where $\ell_o$ is the core radius of the cylinder and the sphere and L is the length of the cylindrical section given by

$$L = (gv_o - \frac{4}{3}\pi \ell_o^3)/\pi \ell_o^2 \quad . \tag{22}$$

For aggregates in the form of spherical vesicles, the hydrocarbon shell volume is

$$gv_o = \frac{4\pi}{3}(R_o^3 - R_i^3) \tag{23}$$

where $R_o$ and $R_i$ are the outer and inner radii of the hydrocarbon shell. The thickness of the hydrocarbon shell is taken to be twice the length of the extended hydrocarbon tail, i.e.

$$R_o - R_i = 2\ell_o \quad . \tag{24}$$

Then the outer radius $R_o$ and the inner radius $R_i$ of the hydrocarbon surfaces are related to $g$, $v_o$ and $\ell_o$ through

$$R_o = [\frac{gv_o}{8\pi \ell_o} - \frac{\ell_o^2}{3}]^{1/2} + \ell_o \tag{25}$$

and

$$R_i = [\frac{gv_o}{8\pi \ell_o} - \frac{\ell_o^2}{3}]^{1/2} - \ell_o \quad . \tag{26}$$

The average surface area per amphiphile at a distance $\delta$ from the core surface is calculated from

$$A = \frac{1}{g} 4\pi [(R_o+\delta)^2 + (R_i-\delta)^2] \quad . \tag{27}$$

The computations have been carried out assuming $\delta = 3$ Å and a value of 30 Å$^3$ for the volume of a solvent molecule. Obviously the minimum size of the vesicles will correspond to $R_i = \delta$.

## IV. AGGREGATION OF AMPHIPHILES WITH SINGLE HYDROCARBON TAILS

The aggregation of single chain amphiphiles into micelles and vesicles was examined for hydrocarbon chain lengths varying from $n_c = 6$ to $n_c = 12$ and for values of the repulsive parameter $\alpha$ between $6 \times 10^4$ and $8 \times 10^4$ cal Å$^2$/mole. In the case of micelles, since the spherical shape cannot be maintained beyond a critical radius equal to the length $\ell_o$ of the extended hydrocarbon tail, the computations were carried out for spherical micelles up to this critical radius. Beyond this, the shape was assumed cylindrical with hemispherical ends. The concentrations of non-aggregated amphiphiles and the corresponding equilibrium concentration of aggregates in the form of micelles and of vesicles are presented in Table I for the four sets of values of $\alpha$ and $n_c$. The computations show that as the total concentration of amphiphiles is increased the number of micellar aggregates keeps increasing and the concentration of vesicles is almost zero in the whole range of parameters examined.

The size distribution of micellar aggregates formed from amphiphiles with a tail length of 12 carbon atoms is shown in Figure 1. As discussed earlier[4], the size distribution function undergoes a transition in shape from a monotonic decreasing one to one exhibiting extrema. The dispersion of sizes of the micellar aggregates at the CMC is found to be relatively small ($\sigma/\bar{g}_n = 0.19$). A similar behavior is exhibited when $n_c$ = 6 to 12 and $\alpha$ = 6 x $10^4$ to 8 x $10^4$ cal Å$^2$/mole.

For $\alpha$ = 4 x $10^4$ cal Å$^2$/mole a peculiar result was obtained. Although the calculated size distribution of vesicles was found to be a monotonic decreasing function of size, the number of vesicles was much larger than that of micelles. There are, however, no experimental results indicating the formation of vesicles from single

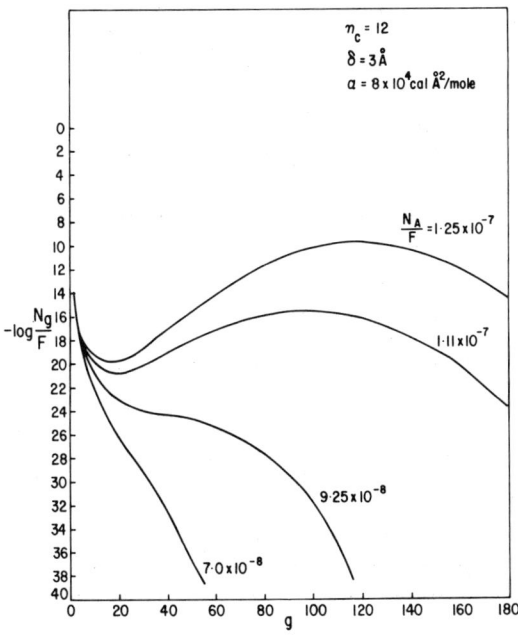

Figure 1. Micellar Size Distribution for Various Concentrations of Non-Aggregated Single Chain Amphiphile with $n_c$ = 12 and $\alpha$ = 8 x $10^4$ cal Å$^2$/mole.

Table I. Equilibrium Concentrations of Non-Aggregated Amphiphiles, Micelles and Vesicles for Single Chain Amphiphiles[a]

| | | $\alpha = 6 \times 10^4$ cal Å$^2$/mole | | | $\alpha = 8 \times 10^4$ cal Å$^2$/mole | |
|---|---|---|---|---|---|---|
| | $\left(\dfrac{N_A}{F}\right)$ | $\left(\Sigma g \dfrac{N_g}{F}\right)$ micelles | $\left(\Sigma g \dfrac{N_g}{F}\right)$ vesicles | $\left(\dfrac{N_A}{F}\right)$ | $\left(\Sigma g \dfrac{N_g}{F}\right)$ micelles | $\left(\Sigma g \dfrac{N_g}{F}\right)$ vesicles |
| $n_c = 6$ | $1.515 \times 10^{-4}$ | $2.076 \times 10^{-9}$ | $1.003 \times 10^{-24}$ | $2.10 \times 10^{-4}$ | $5.46 \times 10^{-10}$ | $9.82 \times 10^{-75}$ |
| | $1.710 \times 10^{-4}$ | $1.108 \times 10^{-6}$ | $1.002 \times 10^{-18}$ | $2.55 \times 10^{-4}$ | $5.23 \times 10^{-8}$ | $2.36 \times 10^{-60}$ |
| | $1.785 \times 10^{-4}$ | $1.481 \times 10^{-3}$ | $1.560 \times 10^{-15}$ | $3.00 \times 10^{-4}$ | $1.14 \times 10^{-4}$ | $2.59 \times 10^{-48}$ |
| $n_c = 12$ | $6.75 \times 10^{-8}$ | $2.01 \times 10^{-14}$ | $1.01 \times 10^{-90}$ | $1.17 \times 10^{-7}$ | $2.40 \times 10^{-12}$ | $\sim 0$ |
| | $7.05 \times 10^{-8}$ | $3.23 \times 10^{-11}$ | $6.98 \times 10^{-81}$ | $1.26 \times 10^{-7}$ | $1.10 \times 10^{-8}$ | $5.32 \times 10^{-176}$ |
| | $7.30 \times 10^{-8}$ | $7.42 \times 10^{-7}$ | $5.37 \times 10^{-73}$ | $1.35 \times 10^{-7}$ | $7.91 \times 10^{-4}$ | $2.11 \times 10^{-160}$ |

[a] The size distribution of vesicles is monotonically decreasing for all values of $\alpha$, $n_c$ and $\dfrac{N_A}{F}$ used in the Table.

chain amphiphiles. We feel that the expressions for the free energies $\mu^o$ used to obtain this anomaly are not adequate enough to draw definite conclusions.

## V. AGGREGATION OF AMPHIPHILES WITH TWO HYDROCARBON TAILS

The formation of micelles and of vesicles by double chain amphiphiles was examined for hydrocarbon tail lengths in the range $n_c = 6$ and $n_c = 12$ and for values of the repulsive parameter $\alpha$ between $4 \times 10^4$ cal $\text{Å}^2$/mole and $\alpha = 8 \times 10^4$ cal $\text{Å}^2$/mole. The equilibrium concentration of all aggregates in the form of micelles and of vesicles corresponding to given concentrations of non-aggregated amphiphiles is presented in Table II for four sets of values of $\alpha$ and $n_c$. The computations show that as the total concentration is increased, enormous numbers of vesicles are formed and the concentration of micellar aggregates is practically zero in the whole range of parameters studied.

The size distributions of spherical vesicles formed of amphiphiles with $n_c = 12$ and $n_c = 6$ are shown in Figures 2 and 3 for different values of $\alpha$. The minimum in the size distribution function

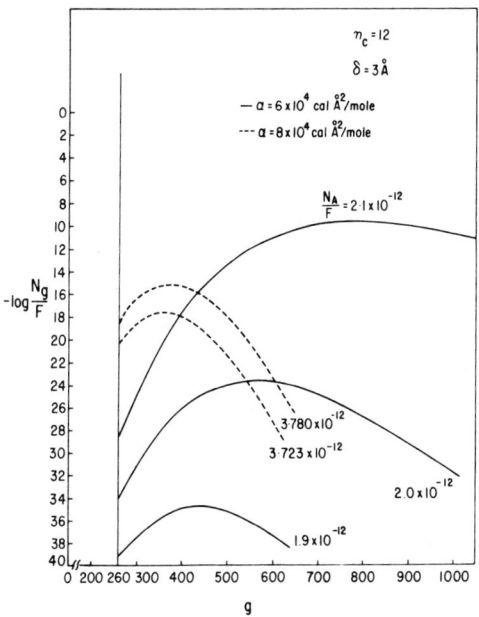

Figure 2. Vesicle Size Distribution for Various Concentrations of Non-Aggregated Double Chain Amphiphile with $n_c = 12$ and $\alpha = 6 \times 10^4$ and $8 \times 10^4$ cal $\text{Å}^2$/mole.

Table II. Equilibrium Concentrations of Non-Aggregated Amphiphiles, Micelles and Vesicles for Double Chain Amphiphiles[a]

| | $\alpha = 4\times10^4$ cal Å$^2$/mole | | | $\alpha = 8\times10^4$ cal Å$^2$/mole | | |
|---|---|---|---|---|---|---|
| | $\frac{N_A}{F}$ | $(\Sigma g \frac{N_g}{F})$ micelles | $(\Sigma g \frac{N_g}{F})$ vesicles | $\frac{N_A}{F}$ | $(\Sigma g \frac{N_g}{F})$ micelles | $(\Sigma g \frac{N_g}{F})$ vesicles |
| $n_c=6$ | $1.504\times10^{-7}$ | $4.193\times10^{-16}$ | $5.579\times10^{-16}$ | $5.46\times10^{-7}$ | $5.34\times10^{-15}$ | $3.75\times10^{-9}$ |
| | $1.540\times10^{-7}$ | $4.419\times10^{-16}$ | $2.773\times10^{-12}$ | $5.73\times10^{-7}$ | $5.98\times10^{-15}$ | $1.29\times10^{-6}$ |
| | $1.572\times10^{-7}$ | $4.624\times10^{-16}$ | $3.329\times10^{-6}$ | $6.00\times10^{-7}$ | $6.66\times10^{-15}$ | $6.17\times10^{-4}$ |
| $n_c=12$ | $1.0032\times10^{-12}$ | $2.233\times10^{-22}$ | $9.528\times10^{-16}$ | $3.62\times10^{-12}$ | $2.698\times10^{-21}$ | $3.65\times10^{-18}$ |
| | $1.0036\times10^{-12}$ | $2.235\times10^{-22}$ | $1.282\times10^{-14}$ | $3.71\times10^{-12}$ | $2.844\times10^{-21}$ | $2.00\times10^{-14}$ |
| | $1.0041\times10^{-12}$ | $2.237\times10^{-22}$ | $8.000\times10^{-13}$ | $3.80\times10^{-12}$ | $2.995\times10^{-21}$ | $1.50\times10^{-10}$ |

[a] The size distribution of micelles is monotonically decreasing for all values of $\alpha$, $n_c$ and $\frac{N_A}{F}$ used in the Table.

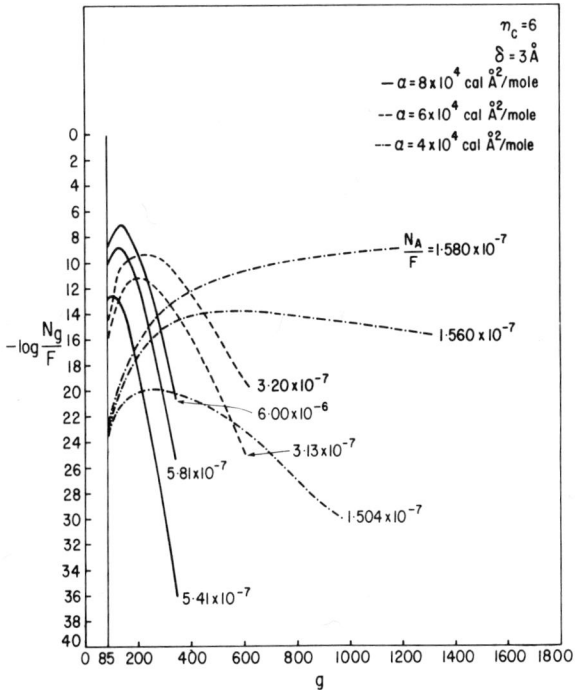

Figure 3. Vesicle Size Distribution for Various Concentrations of Non-Aggregated Double Chain Amphiphile with $n_c = 6$ and $\alpha = 4 \times 10^4$, $6 \times 10^4$ and $8 \times 10^4$ cal Å²/mole.

is not plotted in the figure since it occurs in a size range wherein vesicles cannot exist. The size distribution function representing the transition from a monotonic decreasing function to one exhibiting extrema is also not plotted in the figure, because the corresponding aggregate concentrations are almost zero. The shape of the size distribution functions in Figure 3 shows that as the value of $\alpha$ increases, the size distribution function tends to become a monotonic decreasing function of size, while as its value decreases, the size distribution function tends to become an increasing function of size leading to phase separation.

The dependence of the concentration of spherical vesicles on the single amphiphilar concentration $N_A/F$ is presented in Table III for amphiphiles with $n_c = 12$ and for $\alpha = 6 \times 10^4$ cal Å²/mole. The table also contains the number average and weight average aggregation numbers and the dispersion in size. Over a large range of concentrations of the aggregates the change in the degree of aggregation is small and the size dispersion is quite narrow.

Table III. Dependence of the Concentration of Aggregates on the Concentration of the Non-Aggregated Double Chain Amphiphile[a]

| $\frac{N_A}{F} \times 10^{12}$ | $\Sigma g \frac{N_g}{F}$ | $\bar{g}_n$[b] | $\bar{g}_w$[b] | $\bar{g}_w/\bar{g}_n$ | $\sigma/\bar{g}_n$ |
|---|---|---|---|---|---|
| 2.010 | 3.82x10$^{-18}$ | 585 | 591 | 1.0102 | 0.1010 |
| 2.020 | 7.51x10$^{-17}$ | 603 | 609 | 1.0104 | 0.1020 |
| 2.030 | 1.60x10$^{-15}$ | 622 | 629 | 1.0105 | 0.1024 |
| 2.040 | 3.69x10$^{-14}$ | 643 | 650 | 1.0105 | 0.1024 |
| 2.045 | 1.84x10$^{-13}$ | 654 | 661 | 1.0107 | 0.1034 |
| 2.050 | 9.36x10$^{-13}$ | 665 | 673 | 1.0108 | 0.1039 |
| 2.055 | 4.89x10$^{-12}$ | 677 | 685 | 1.0108 | 0.1039 |
| 2.060 | 2.62x10$^{-11}$ | 690 | 698 | 1.0110 | 0.1049 |
| 2.070 | 8.16x10$^{-10}$ | 716 | 724 | 1.0115 | 0.1072 |
| 2.100 | 5.27x10$^{-5}$ | 811 | 821 | 1.0123 | 0.1109 |

[a] $n_c = 12$, $\delta = 3$ Å, $\alpha = 6 \times 10^4$ cal Å$^2$/mole

[b] aggregation numbers rounded off to integral values

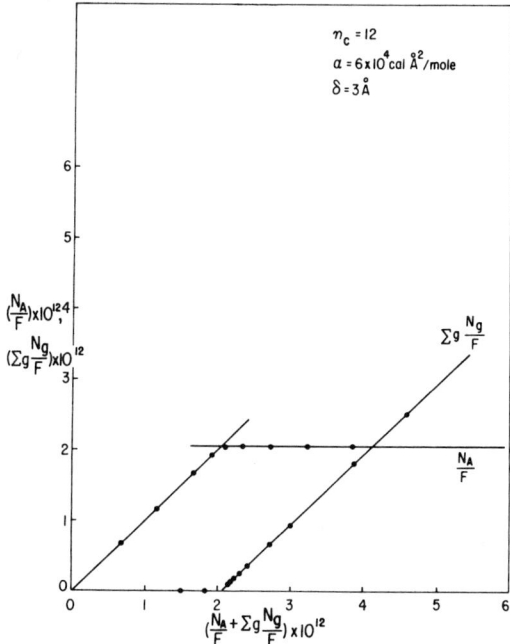

Figure 4. Dependence of Non-Aggregated Amphiphile and Vesicle Concentrations on the Total Concentration of Double Chain Amphiphiles with $n_c = 12$ and $\alpha = 6 \times 10^4$ cal Å$^2$/mole.

The concentration of amphiphiles present as single amphiphiles and as aggregates is plotted against the total amphiphilar concentration in Figure 4 for vesicles formed of amphiphiles with $n_c = 12$ and for $\alpha = 6 \times 10^4$ cal Å$^2$/mole. From the sharp change in the values of either of these two concentrations a CVC value (the analog of CMC) for formation of spherical vesicles can be determined. The plot shows that the changes are sharper than those in the case of micellar aggregates[5]. This indicates that spherical vesicles are almost monodispersed. Also the concentration of the single amphiphiles appears to remain almost constant beyond the CVC. An aggregation model assuming a single size of vesicles would be adequate to describe the system.

The CVC values, the average aggregation numbers and the variance in the size distribution of spherical vesicles are tabulated in Table IV for $n_c = 6$ and $n_c = 12$ and for three different values of $\alpha$. The CVC values are much lower than those generally measured for micellization, but as in micellization the aggregates grow drastically to very large sizes as $\alpha$ is decreased. The dispersion of sizes

Table IV. Dependence of the Characteristics of Vesicles from Double Chain Amphiphiles on Tail Length $n_c$ and Head Group Repulsion $\alpha$[a]

| $n_c$ | $\alpha$ cal Å$^2$/mole | CVC in molar fraction units | $\bar{g}_n$ | $\bar{g}_w$ | $\bar{g}_w/\bar{g}_n$ | $\sigma/\bar{g}_n$ |
|---|---|---|---|---|---|---|
| 6 | 4x10$^4$ | 1.644x10$^{-7}$ | 690 | 742 | 1.0756 | 0.2750 |
|   | 6x10$^4$ | 3.307x10$^{-7}$ | 190 | 195 | 1.0255 | 0.1596 |
|   | 8x10$^4$ | 5.804x10$^{-7}$ | 117 | 119 | 1.0165 | 0.1285 |
| 12 | 4x10$^4$ | 1.062x10$^{-12}$ | 7466 | 8057 | 1.0791 | 0.2812 |
|   | 6x10$^4$ | 2.14 x10$^{-12}$ | 650 | 656 | 1.0106 | 0.1030 |
|   | 8x10$^4$ | 3.90 x10$^{-12}$ | 363 | 366 | 1.0080 | 0.0894 |

[a] $\delta = 3$ Å, $\bar{g}_n$, $\bar{g}_w$ and $\sigma$ are calculated at the CVC.

of vesicles is small for larger values of α and increases as α decreases.

## VI. CONCLUSIONS

In this paper it is shown that a general thermodynamic description of amphiphilar aggregation can describe adequately the formation of micelles and of spherical vesicles from amphiphiles with either one or two hydrocarbon tails. For a range of parameters $n_c = 6$ to $n_c = 12$ and $\alpha = 6 \times 10^4$ to $8 \times 10^4$ cal Å$^2$/mole, amphiphiles with single hydrocarbon tails do not form spherical vesicles but only micelles.

Thermodynamically stable spherical vesicles are formed from amphiphiles with two hydrocarbon tails for $n_c$ in the range of 6 to 12 and α in the range $4 \times 10^4$ to $8 \times 10^4$ cal Å$^2$/mole. No stable micellar aggregates form from amphiphiles with two tails at least in the range of parameters considered.

## ACKNOWLEDGEMENT

This work was supported by the National Science Foundation.

## REFERENCES

1. C. Tanford, "The Hydrophobic Effect," Wiley, New York, 1973; J. Phys. Chem. 78, 2469 (1974).
2. P. Mukerjee, Adv. Colloid Interface Sci. 1, 241 (1967).
3. K. Shinoda, T. Nakagawa, B. Tamamushi, and T. Isemura, "Colloidal Surfactants," Academic Press, New York, 1963.
4. E. Ruckenstein and R. Nagarajan, J. Phys. Chem. 79, 2622 (1975).
5. E. Ruckenstein and R. Nagarajan, J. Colloid Interface Sci. (in press).
6. S. Ohki, in "Physical Principles of Biological Membranes," F. M. Snell, Editor, p. 175, Gordon and Breach, New York, 1970.
7. R. Nagarajan and E. Ruckenstein, J. Colloid Interface Sci. (in press).

THERMODYNAMICS OF MICELLE FORMATION

K. S. Birdi

Fysisk-Kemisk Institut, Technical University

Lyngby, Denmark 2800

The free energy of the transfer of a single amphiphile molecule from the aqueous medium to the micelle of aggregation number (N), $\Delta G^o$, can be split into the attractive hydrophobic part and the repulsive part (electrostatic charge repulsion and hydration of the polar groups at the micelle-water surface). In an earlier paper it was concluded that the micellar hydrophobic interactions, which are the main forces responsible for the solubilization of water-insoluble organic compounds, were independent of ionic strength and the aggregation number, N. Application of this finding to extensive data of critical micelle concentration (c.m.c.) and N for alkyl sulfates (of chain lengths with number of carbon atoms 8, 9, 10, 11, and 12) gave a self-consistent theory. These data allowed the calculation of the opposing forces, which was found to be a linear function of the area per polar group at the micelle-water interface. It is concluded from this analysis that the hydrophobic interactions can be estimated accurately from these results, which has not been possible from the current theories found in the literature. This discrepancy arises from the fact that none of the current theories are able to give a satisfactory description of the effect of ionic strength and aggregation number on the hydrophobic interactions.

## INTRODUCTION

Molecules containing both hydrophilic and hydrophobic parts, i.e. amphiphile molecules, under the influence of the hydrophobic interactions form aggregates (micelles) in aqueous systems. In the process of micelle formation the hydrocarbon chains are located inside the micelle core, while the hydrophilic groups maintain contact with water (solvent) at the micelle-water interface. It is now well recognized that the forces which stabilize micelles are similar, in principle, to the forces which are responsible for the formation of bilayers and biological membranes. In other words, the micellar systems have been recognized as very useful model systems for the understanding of the more complex membrane structures.

In the past decade a great many studies have been published on the thermodynamics of micelle formation.[1-8] On the basis of some of these studies, it is now well established that a satisfactory understanding of the thermodynamics of micelle formation is vital in order to be able to describe quantitatively the more complex systems, such as bilayers and membranes. In view of this, it is thus important that the theories on micelle formation be as consistent as possible.

The purpose of this study is to report on the free energy of micellization, which was described by us,[3,10] in accord with other theories,[5] as composed of hydrophobic forces (attractive) and repulsive forces. In general, at this stage in the literature the quantitative estimation of the hydrophobic part of the free energy is possible, and very little deviation between theory and experimental data has been reported. On the other hand, no satisfactory theory is found in literature which can describe the repulsion part. In all current theoretical analyses of micellization, such experimentally observable parameters as critical micelle concentration and micelle aggregation number have been used. In this study we will discuss another parameter, that is, the free energy of solubilization by micellar systems of water-insoluble organic compounds, which was described by us elsewhere.[9] It will then be shown that none of the current theories are able to describe consistently the experimental data; the main reason being that all these theories are not able to satisfactorily describe the effect of added salt and aggregation number on the free energy related to the hydrophobic part.

## THEORETICAL

### General Theory of Micelle Formation

In the following we will recapitulate the thermodynamics of micelle formations, as described previously[2,3] by us and in accordance with literature.[1,4,5,11,12] The process of micellization involves the reversible aggregation of N amphiphile molecules to form a micelle as given in the following:

$$N\, m \rightleftarrows M \tag{1}$$

The equilibrium constant of this process is given by:

$$K_c = C_M/(C_m)^N \tag{2}$$

where $C_m$ and $C_M$ are the concentrations of monomer and micelle, respectively (under the present conditions, concentrations can be used instead of activities, as described elsewhere[2]). At equilibrium we have:

$$N\, \mu_m = \mu_M \tag{3}$$

where $\mu_m$ and $\mu_M$ are the chemical potentials given as:

$$\mu_m = \mu_m^o + RT \ln C_m \quad \text{(monomer)} \tag{4}$$

$$\mu_M = \mu_M^o + RT \ln C_M \quad \text{(micelle)} \tag{5}$$

From these relations the standard free energy of micellization, $\Delta G^o$, per monomer is found to be given as:[1,2,4,5]

$$\Delta G^o = RT \ln C_m - RT/N \ln C_M \tag{6}$$

$$= RT \ln \text{c.m.c.} + RT/N \ln N - RT/N \ln C_M' \tag{7}$$

where $C_M' = N\, C_M$, and in Equation (7) we write that $C_m \simeq$ c.m.c., as described elsewhere.[2]

In order to describe the free energy, $\Delta G^o$, it has been found useful to write it as follows:[2,3,5,11,12]

$$\Delta G^o = \Delta G_\phi^o + \Delta G_e^o + \Delta G_w^o \tag{8}$$

where $\Delta G_\phi^o$ (which will be negative) is the free energy change associated with the transfer of the hydrophobic part of the amphiphile molecule from the aqueous medium to the micelle interior of aggregation N, $\Delta G_e^o$ (which will be positive) is the free energy associated with the electrostatic charge repulsion of the polar head groups and $\Delta G_w^o$ (which will be positive) is the free energy change related to the hydration of the polar groups at the micelle-water interface. Further, in the case of nonionic micelles the term $\Delta G_e^o$ will be absent, while in the case of ionic micelles the opposing forces will be due to $\Delta G_e^o + \Delta G_w^o$. Before discussing the current theories related to the quantitative description of these various forces, Equation (8), it is of interest to consider the micellar solubilization process.

## Micellar Solubilization

It is well known that appreciable amounts of organic compounds which are insoluble in water are solubilized by micellar solutions, i.e. when the amphiphile concentration is greater than c.m.c. In fact, this property has been used in the determination of c.m.c. by various investigators.[13,14] The solubilization phenomena of interest here refers only to those systems where the solute has no effect on $\Delta G^o$, i.e. c.m.c. and N do not change. In other words, we are not going to consider 'solubilization' phenomena where mixed micelles are formed, whereby $\Delta G^o$ is different from that of the pure micelle.

As described elsewhere,[9] in the solubilization experiment the aqueous micellar solution is allowed to attain equilibrium by standing in contact with the excess solute (water-insoluble compound) in the solid state. The aqueous micellar solution can be treated as a pseudo two phase system, and at equilibrium:[9,15]

$$\mu_S^S = \mu_S^{aq} = \mu_S^M \tag{9}$$

where $\mu_S^S$, $\mu_S^{aq}$ and $\mu_S^M$ are the chemical potentials of the solute in the solid state, aqueous phase and micellar phase, respectively. The free energy change involved in the solubilization, $\Delta G_S^\phi$, is given as follows:[9]

$$\Delta G_S^\phi = -RT \ln (c_S^M)/(c_S^{aq}) \tag{10}$$

where $c_S^{aq}$ and $c_S^M$ are the concentrations of the solute in the aqueous phase and in the micellar phase, respectively. Since, in the present study, we will be interested in the variation of $\Delta G_S^\phi$ with the number of carbon atoms in the alkyl chain of the amphiphile molecule, the relation in Equation (10) can be rewritten as:[9]

$$\Delta G_S^o = -RT \ln c_S^M \tag{11}$$

since the term $c_S^{aq}$ remains constant, within the experimental accuracy.

The micellar solubilization in the present case is primarily determined by the hydrophobic interactions, i.e. $\Delta G_\phi^o$. In other words, $\Delta G_S^o$ will be expected to be related to the number of carbon atoms in the alkyl chain. It was shown that plots of $\Delta G_S^o$ versus number of carbon atoms ($n_c$) were linear for a variety of micellar homologous series for different solutes (e.g. DMAB, Orange OT,

# THERMODYNAMICS OF MICELLE FORMATION

Naphthalene, Anthracene). These results are given in Figure 1, since we will analyze these data in a different context here than described earlier.[9]

It is seen that the change in solubilization energy per additional $CH_2$ group, as determined from the slopes in Figure 1, is average magnitude of -837 J (-200 cal)/mole. These data further show that the change in $\Delta G_S^o$ per $CH_2$ is not dependent on the added salt to the solvent. Further, we have calculated the values of $\Delta G_S^o$ for different micellar systems in various salt concentrations, Table I, and these data also show clearly that $\Delta G_S^o$ is not dependent on the ionic strength.

Our results on the solubilization of DMAB in SDS solutions with different ionic strength also showed that $\Delta G_S^o$ is not dependent on the ionic strength.[18] We have further found that this finding also is valid for other solutes, such as naphthalene, anthracene, azobenzene and Sudan II, in the case of SDS and cetyltrimethyl ammonium bromide micellar solutions.[19]

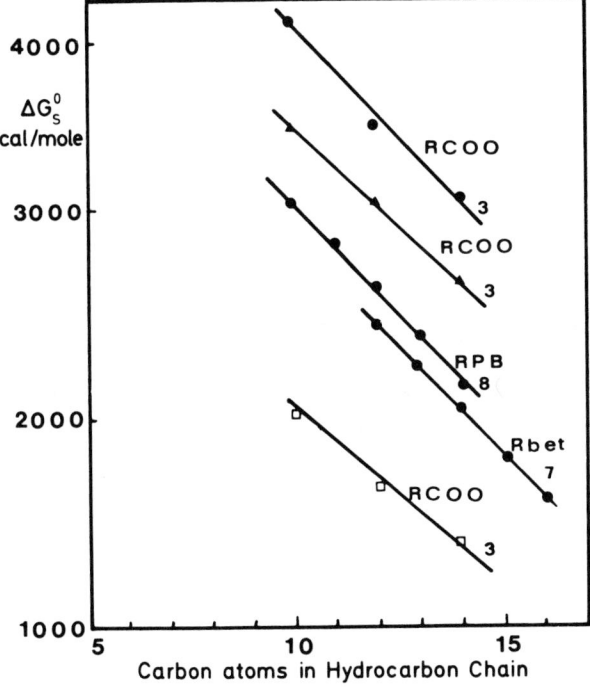

Figure 1. Plots of $\Delta G_S^o$ versus hydrocarbon chain length: RCOO-alkyl-carboxylic acid (water), Rbet-alkylbetaines (water), RPB-alkylpyridinium bromide (0.5 M-NaBr). Solutes: Orange OT ●; DMAB △; Naphthalene □.

Table I. Magnitude of $\Delta G_s^o$ in Various Salt Solutions.

| Solute | | $\Delta G_s^o$ (kJ mol$^{-1}$) | Reference |
|---|---|---|---|
| **SDS** | | | |
| Water | Orange OT | 11.59 | 13 |
| 0.03 mol dm$^{-3}$ NaCl | Orange OT | 11.95 | 13 |
| 0.1 mol dm$^{-3}$ " | Orange OT | 11.77 | 13 |
| Water | Orange OT | 11.63 | 16 |
| 0.03 mol dm$^{-3}$ NaCl | Orange OT | 11.63 | 16 |
| 0.1 mol dm$^{-3}$ " | Orange OT | 11.63 | 16 |
| **DTAB** | | | |
| Water | Orange OT | 11.13 | 17 |
| 0.1 mol dm$^{-3}$ NaBr | Orange OT | 11.02 | 17 |
| 0.51 mol dm$^{-3}$ " | Orange OT | 10.82 | 17 |

(SDS: sodium dodecyl sulfate, DTAB: dodecyltrimethylammonium bromide)

The data given in Figure 1 and in Table I provide very useful information as regards the effect of ionic strength on the various forces which are responsible for the micelle formation, Equation 8. It is of interest to mention that the results in Table I also indicate that $\Delta G_s^o$ is not dependent on the aggregation number, N, since the magnitude of N is known to increase with increasing ionic strength in the case of ionic micelles.[20,21] The above results thus indicate that the hydrophobic forces, $\Delta G_\phi^o$, which are mainly responsible for the solubilization are not dependent on ionic strength or N, and that these forces show a linear dependence on number of carbon atoms, $n_c$, in the alkyl chain of the amphiphile molecule. All the current theories are not able to describe satisfactorily the effect of aggregation number and ionic strength (in the case of ionic micelles) on $\Delta G_\phi^o$ for an amphiphile molecule of given chain length.[4,5,7,22,23] In the following we will discuss these results in detail and present modifications which are consistent with the experimental data.

## DISCUSSION

### Hydrophobic Interactions

In the case of simple amphiphiles consisting of a single hydrocarbon chain with a hydrophilic group at one end, the micelle formation is a cooperative process whereby the alkyl chains are located at a certain distance from each other. It is also well established

# THERMODYNAMICS OF MICELLE FORMATION

that the distance between the hydrocarbon chains is of such magnitude that the core of micelles is assumed to be liquid-like. The opposing forces, $\Delta G_e^o + \Delta G_w^o$, as described below in detail, are responsible for the micelle having a definite size. The purpose of this discussion is to compare the various theories which describe the term $\Delta G_\phi^o$ in relation to the micellar experimental data.

It is well established that the transfer of the alkyl chain of an amphiphile from the aqueous medium to the core of the micelle is, in principle, analogous to the hydrophobic free energy for the transfer of an alkyl chain from the aqueous medium to a nonpolar solvent (liquid hydrocarbon). In order to distinguish the latter process, $\Delta G_\phi^\phi$, from the former, $\Delta G_\phi^o$, we will use the appropriate designations as mentioned. In accord with the earlier theories on the solubilities,[24] it has recently been shown that in the case of systems corresponding to $\Delta G_\phi^\phi$, the hydrophobic free energy is proportional to the surface area of the alkyl chain, rather than the chain length or molar volume.[25,27] However, the hydrophobic energy related to the process of transfer of an alkyl chain from aqueous medium to the bulk hydrocarbon, $\Delta G_\phi^\phi$, is not expected to be of the same order of magnitude as the energy for the transfer of an alkyl chain from the aqueous medium to the interior of a micelle, $\Delta G_\phi^o$. This difference has been generally explained by considering that the molecular motion inside the micelle would be more restricted than in the bulk hydrocarbon.[4,5] In a recent study it was assumed that $\Delta G_\phi^o$ consists of two parts, a constant part independent of micelle size which represents the free energy gained for the complete immersion of the alkyl chain in the interior of the micelle, and a variable part reflecting the positive free energy contribution by residual contacts between the core surface and the solvent.[5]

Considering first $\Delta G_\phi^\phi$, which has been described satisfactorily in the current literature, we find that in the case of linear alkyl chains the magnitude of $\Delta G_\phi^\phi$ is given as follows:[26,27]

$$\Delta G_\phi^\phi = (84.9 + 42.75 + 31.8(n_c-2)) \ 106.7 \ \text{J/mole} \qquad (12)$$

where the surface area of the terminal $CH_3$, alpha $CH_2$, and the rest of the $CH_2$ groups is given by 84.9 $Å^2$, 42.75 $Å^2$, and 31.8 $Å^2$, respectively. This relation thus shows that $\Delta G_\phi^\phi$ changes with each additional $CH_2$ group by -3393 J/mole (-810 cal/mole), in accord with the earlier theories where chain length was used as the parameter for describing the $\Delta G_\phi^\phi$.[4] With this in view, we will in the following describe the various properites of micellar systems which are related to the alkyl chain length, i.e. $\Delta G_\phi^o$, in relation to $\Delta G_\phi^\phi$.

## Variation of c.m.c. with Alkyl Chain Length

It has been shown by many investigators that the c.m.c. decreases logarithmically with increase in the number of carbon atoms, $n_c$, for homologous series of various micellar systems.[1,4,5,7,9,20,22,23,28-30] It has been found that RT d ln cmc/$dn_c$ in the case of nonionic micellar systems is not dependent on the added salt to the solvent, while it is dependent on the salt concentration in the case of ionic micelles.[4,5,7,22,23] Considering the nonionic results, we thus find that the value of RT d ln c.m.c./$dn_c$ is in the range of -2850 J/mole to -2992 J/mole (-715 cal/mole).[1,4,5,31] These results thus show that from relations given in Equations 6 and 8 for defining $\Delta G^o$, the terms $RT/N \ln C_M$ and $\Delta G_w^o$, are not dependent on chain length. In other words, in the case of nonionic systems the opposing forces, i.e. $\Delta G_w^o$, are not dependent on the alkyl chain length. It is also of interest to mention here, that the term $RT/N \ln C_M$ (in Equation 6) is independent of $n_c$. It can be concluded that in the case of nonionic systems the change in $\Delta G_\phi^o$ with the number of carbon atoms is given by RT d ln c.m.c./$dn_c$, in accord with literature.[4,5]

The RT ln c.m.c. versus $n_c$ plots for ionic systems are given in Figure 2, as a function of ionic strength.[20] Even though these data have been known in the literature for a long time, none of the current theories have attempted to analyze these. It is seen that there is a linear relation between RT ln c.m.c. and $n_c$, however, the magnitude of the slopes changes with ionic strength. We are not able to explain why the data points for n = 11 in solutions containing 0.03, 0.1, and 0.3 mol dm$^{-3}$ NaCl do not agree with the plot for the other data points. In order to find a plausible explanation for this discrepancy, work is in progress in this laboratory. In order to analyze these plots it is of interest to express them by the following equations:

Table II. Magnitude of RT d ln c.m.c./$dn_c$ for Sodium Alkyl Sulfates in Various Salt Solutions (21°C)[20]

| | |
|---|---|
| Water | RT ln c.m.c. = 2062 - 408 $n_c$ |
| 0.01 M-NaCl | RT ln c.m.c. = 2011 - 419 $n_c$ |
| 0.03 M-NaCl | RT ln c.m.c. = 3085 - 536 $n_c$ |
| 0.1 M-NaCl | RT ln c.m.c. = 3723 - 627 $n_c$ |
| 0.3 M-NaCl | RT ln c.m.c. = 3876 - 675 $n_c$ |
| | (cal/mole units) |

It is of interest to compare these data with similar relations for a variety of ionic micelles in water,[22] where the value of RT d ln c.m.c./$dn_c$ was reported to be in the range of -1530 J/mole to -1732 J/mole (-414 cal/mole). In other words, in water the change in RT ln c.m.c. per $CH_2$ group is the same for all ionic micellar systems. Unfortunately, no other data is available on other ionic systems than the one given in Table II. Some results on alkyltrimethylammonium bromide[23] indicate that the value of RT d ln c.m.c./$dn_c$ varies with ionic strength as: -1848 J/mole (water), -2194 J/mole (0.025 M-KBr).

These data thus clearly show that in contrast to the nonionic systems, the magnitude of RT d ln c.m.c./$dn_c$ changes from -1707 J/mole (-408 cal/mole) in water to -2824 J/mole (-675 cal/mole) in 0.3 M-NaCl (Table II). Considering the term $\Delta G_e^o$ in Equation 8, we can thus conclude that as the value of this opposing force becomes less and less with increasing ionic strength the slopes in Figure 2 approach the magnitudes given by the nonionic systems. Since added salt has no effect on RT d ln c.m.c./$dn_c$ for the nonionic systems, which means salting-out effect is negligible. From this observation we will argue that the salting-out effect is also negligible in the case of ionic systems. In the current literature the effect of ionic strength on the variation of c.m.c. with $n_c$ has

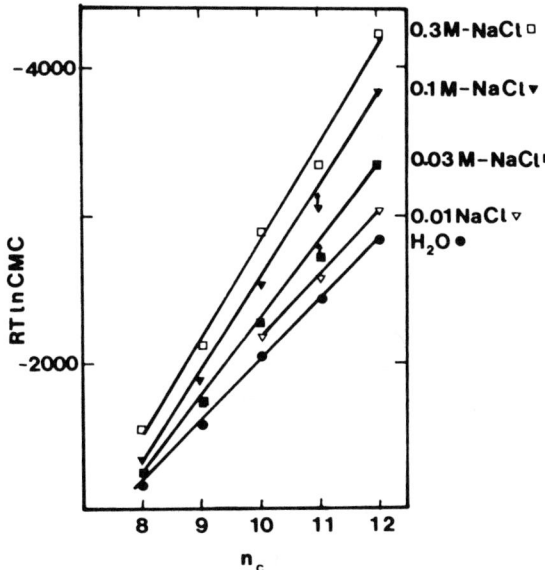

Figure 2. Plot of RT ln c.m.c. of alkyl sulfates as a function of the number of carbon atoms in the chain ($n_c$).

been considered,[4,5,23] however, no satisfactory theory has been given as regards the effect of ionic strength on RT d ln c.m.c./$dn_c$, i.e. $\Delta G_\phi^0$. As discussed earlier, we have shown that in the case of solubilization by micellar systems, no effect of ionic strength on $\Delta G_s^0$ (which is related to $\Delta G_\phi^0$) is observed. With this in view it is of interest to discuss the variation of N with $n_c$ as a function of ionic strength.

## Variation of Aggregation Number with Alkyl Chain Length

The purpose of this study was to describe the various properties of micellar systems which are related to the alkyl chain length of the amphiphiles. In the foregoing we have already discussed the variation of c.m.c. and $\Delta G_s^0$ with $n_c$. As will be apparent further below, any general theory should be able to describe the effect of alkyl chain length on c.m.c., N, and $\Delta G_s^0$. In Figure 3 we give the plots of ln M versus $n_c$, as a function of ionic strength.[20] It is seen that all these plots are linear, as has also been described by some other investigators.[20,23] However, as was recognized by some investigators,[20] these plots have no thermodynamic significance, as such. On the other hand, the plots of ln N versus $n_c$, as given in Figure 4, as a function of ionic strength for alkyl sulfates, are of thermodynamic significance. It is seen that ln N versus $n_c$ plots also show linear relationship. It is not difficult to determine that due to small effects of logarithm of molecular weight of amphiphile molecule in each homologous series gives rise to a negligible effect in the plots in Figure 3.

Considering the data in Figure 4, we find that the slopes are dependent on the ionic strength, analogous to the variation of c.m.c. with $n_c$. These data given in Figure 4 have never been analyzed by any investigator in the literature. The data points for $n_c$ = 11, also, in this case are not in agreement with the plots of other chain lengths, as was observed in the c.m.c. versus $n_c$ plots in Figure 2. These plots can be described by the equations given in Table III below.

Table III. Magnitude of d ln N/$dn_c$ for Sodium Alkyl Sulfates in Various Salt Solutions (21°C)[20]

| | |
|---|---|
| Water | RT ln N = 897 + 122 $n_c$ |
| 0.1 M-NaCl | RT ln N = 464 + 182 $n_c$ |
| 0.3 M-NaCl | RT ln N = 304 + 208 $n_c$ |

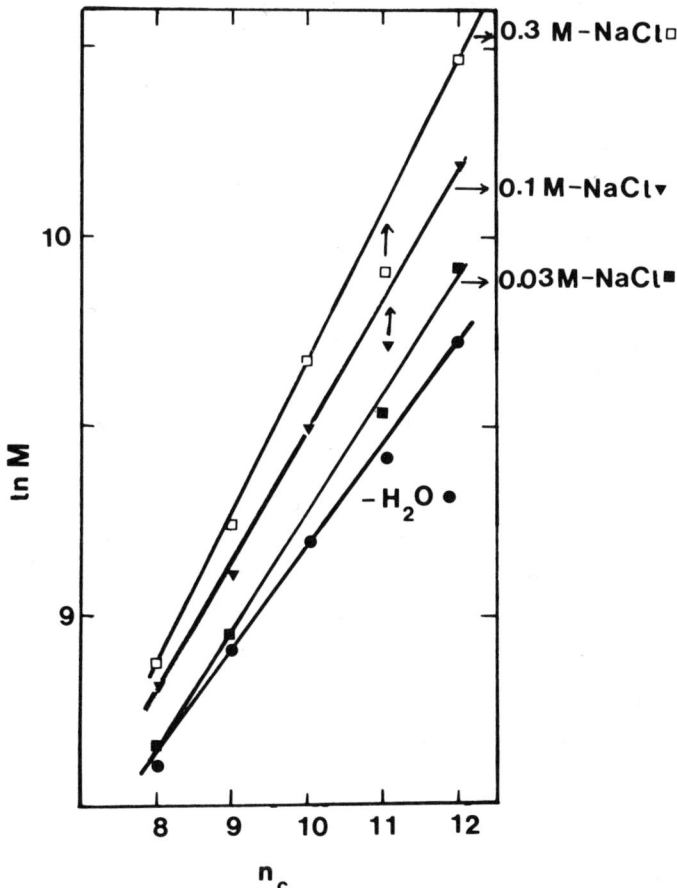

Figure 3. Plot of ln M of alkyl sulfates as a function of number of carbon atoms in the chain ($n_c$). (M—micellar molecular weight).

It is safe to conclude, besides the fact that the data is not as extensive as we would require in order to make any general comment, that there is a linear relation between ln N and $n_c$. It is also seen that the variation of ln N with $n_c$ is dependent on the ionic strength. Since both RT ln c.m.c. and ln N are shown above to be related to $n_c$, we can now proceed in the following to show its thermodynamic significance.

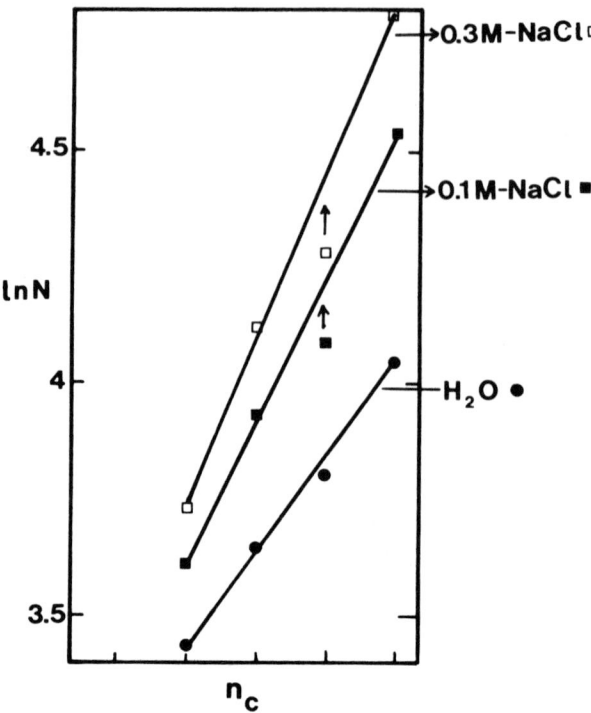

Figure 4. Plot of ln N of alkyl sulfates as a function of number of carbon atoms in the chain ($n_c$).


## Effect of Chain Length on c.m.c., N, and $\Delta G_s^o$

Considering the micellar solubilization results, we found that the hydrophobic interactions which are responsible for the free energy of solubilization, $\Delta G_s^o$, are linear functions of number of carbon atoms, $n_c$. This is in agreement with both the process of transfer of an alkyl group from water medium to a liquid nonpolar solvent, $\Delta G_\phi^\phi$, and as well as for $\Delta G_\phi^o$. The most important finding was that $\Delta G_s^o$ is not dependent on the added salt concentration. This was also the case for the change in $\Delta G_s^o$ with each additional $CH_2$ group.[9] Further, it was also shown that $\Delta G_s^o$ was not dependent on aggregation number, N. It can thus be concluded from these observations that the hydrophobic interactions, $\Delta G_\phi^o$, are independent of ionic strength and N.

The data given here show that the magnitude of RT d ln c.m.c./$dn_c$ is dependent on ionic strength, Table II. The magnitudes given show a significant change already with very low ionic strengths. This observation is difficult to explain, since such small salt additions cannot affect $\Delta G_\phi^o$. In view of the solubilization data discussed above, we will explain these data as follows. In the present study and in the literature,[4,5,7] the phase separation concept has been applied.[1] However, it has been mentioned elsewhere[1] that the mass-action approach is more useful in the case of ionic micelles, especially in order to explain the counter-ion effect.[23] Thus we will argue that the slopes in Figure 2 are different due to the effect of ionic strength on the charge of micelle, as determined by N and the firmly bound counter-ions. As more data is available, we will give a more thorough description of this discrepancy in a later report.

The data on the variation of ln N versus $n_c$, in the case of alkyl sulfates,[20] as a function of ionic strength give analogous plots as for the c.m.c. These data given in Figure 4, and the slopes given in Table III show that the effect of ionic strength is approximately the same as on the slopes of c.m.c., Table II. It was found that the slopes in both Tables II and III vary linearly with log ionic strength. This then indicates that we can consider the micellar c.m.c. and the aggregation number, N, data from Figures 2 and 4, for alkyl sulfates,[20] without taking into consideration the corrections needed for the charge of the micelles.

As noticed by the reader, we have not described in detail the opposing forces $\Delta G_e^o$ or $\Delta G_w^o$, the reason being that none of the current investigators have been able to give a satisfactory theory for these opposing forces.[4,5,7,32] The recent theories have concluded that Debye-Hückel approach has provided a very crude estimation of $\Delta G_e^o$.[4,5] However, in one study where this theory was applied in the case of ionic systems with no added salt, the term $\Delta G_e^o$ was given as[32]

$$\Delta G_e^o = A N e^2/D b \qquad (13)$$

where A is the Avogadro's number, e the protonic charge, D the solvent dielectric constant, and b the micellar radius. These investigators then tried to estimate the effect of neglect of change in N with temperature, which has been pointed out by us earlier to give incorrect enthalpy measurements.[2] The enthalpy correction term described by these investigators: $T(d\Delta G_N^o/dT)_{T,p}(dN/dT)_p$, was found to give such values as 105 kJ/mole (25 kcal).[32] We have analyzed these theories further, and found that the expression for $\Delta G_e^o$ is indeed a crude estimation, since the correction term is found to be in the range of 105 kJ/mole to 630 kJ/mole.[33] It is obvious that since calorimetric micellar enthalpies are in the range of -12.5 kJ/mole to +12.5 kJ/mole, the enthalpy correction magnitudes given by these investigators[32] are not reasonable. In view of these difficulties in determining the magnitudes of both $\Delta G_e^o$ and $\Delta G_w^o$, we will consider the present data by taking the semi-empirical approach suggested by some investigators.[5]

It has been suggested that the magnitude of the opposing forces will depend on the separation between head groups, and that the area per polar head group can be used as a measure of this separation.[5] We will show that the area per head group can be related to N, which is known in the present case, as follows.

Assuming spherical micelles (similar description has been recently given by J. Israelachvili, D. J. Mitchell, B. W. Ninham, J. Chem. Soc. Faraday II, in press) with radius R, we find:

$$N v = 4 \Pi R^3 / 3 \tag{14}$$

$$\alpha = 4 \Pi R^2 / N = (3v)^{2/3} (4\Pi/N)^{1/3} \tag{15}$$

where v is the volume of the alkyl chain of the amphiphile and $\alpha$ is the area per head group at the micelle-water interface. As already evident, we find that c.m.c. and N are related which will satisfy the expression for $\alpha$ in Equation (15) in describing the opposing forces.

Under the present state of experimental data, we will as a first approximation assume that $\Delta G_\phi^o \simeq \Delta G_\phi^o$. Since we will argue that any difference that will be present is due to the factor 106.7 in Equation 12, as the surface area of the alkyl chain will be the same in the two processes under comparison, we will thus calculate $\Delta G_\phi^o$ for the various alkyl chains of the data given in Figures 2 and 4, from Equation 12. Since we showed above that the term RT/N ln $C_M$ does not affect the plots in Figure 2, then after neglecting this term in Equation 6 and combining with Equation 8 we obtain:

$$RT \ln c.m.c. - \Delta G_\phi^o \simeq \Delta G_e^o + \Delta G_w^o = \Delta_{op} \tag{16}$$

where we denote the opposing forces with $\Delta_{op}$. After calculating $\Delta G_\phi^o$ for each alkyl chain (from $n_c = 9$ to $n_c = 12$), we can thus determine the magnitude of the left hand side of Equation 16. It is then interesting to determine how this term varies with the area per polar head group, $\alpha$, which in return is related to N, as given in Equation 15. In Figure 5 we show a plot of $\Delta_{op}$ versus $1/N^{1/3}$ for various alkyl sulfates. It is seen that all the plots show linear relationship; however, the slopes are dependent on the number of carbon atoms, which indicates that the opposing forces are related to the radius of the micelle. These plots can be described by the following linear equations, Table IV.

In view of the various approximations and the accuracy of the data, especially the N, it is safe to conclude that all these plots cross the origin. It is thus possible to conclude that $\Delta_{op}$ as calculated from Equation 16 is related only to $\alpha$. In other words the surface free energy of the micelle, which will be equal to

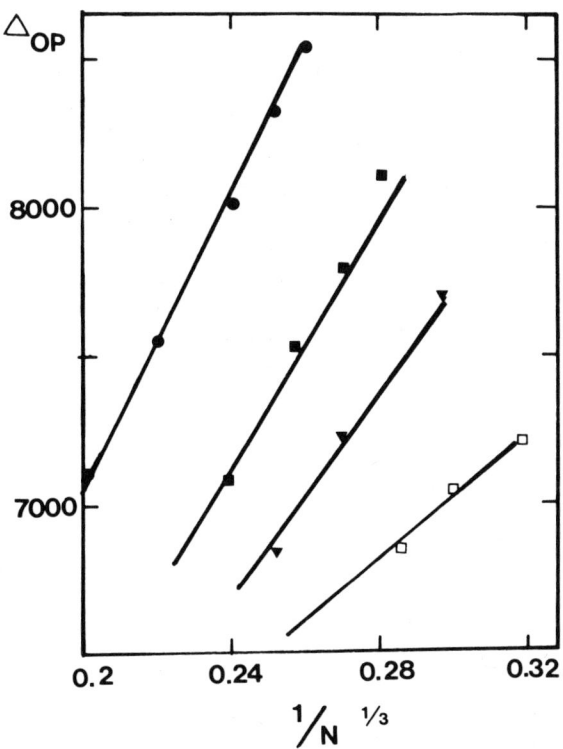

Figure 5. Variation of $\Delta_{op}$ (= RT ln c.m.c. $- \Delta G_\phi^o$, see text) with $1/N^{1/3}$ for alkyl sulfates with number of carbon atoms 12 ●, 11 ■, 10 ▼, 9 □.

Table IV. Variation of $\Delta_{op}$ with Aggregation Number N, as a Function of the Chain Length of Alkyl Sulfates

$n_c = 12$  RT ln c.m.c./55.5 $-\Delta G_\phi^o = -172 + 24480/N^{1/3}$
$\phantom{n_c} = 11$  RT ln c.m.c./55.5 $-\Delta G_\phi^o = -445 + 22045/N^{1/3}$
$\phantom{n_c} = 10$  RT ln c.m.c./55.5 $-\Delta G_\phi^o = -237 + 18881/N^{1/3}$
$\phantom{n_c} = 9$  RT ln c.m.c./55.5 $-\Delta G_\phi^o = 692 + 13238/N^{1/3}$

interfacial tension x α, is proportional to $\Delta_{op}$. Since the various plots are related to the number of carbon atoms, this suggests that the interfacial tension is proporational to the curvature.

In order to give a thorough and general analysis, we need more systematic data analogous to these on alkyl sulfates. In the absence of this information, we will only give a semi-quantitative description along with some data on other micellar systems. Besides the experimental data given on alkyl sulfates,[20] where c.m.c. and N were measured as a function of ionic strength, there is some experimental data in literature on other ionic systems. The c.m.c. and N data as a function of ionic strength have been reported on dodecyl pyridinium salt of bromide (DPyB) and of chloride (DPyC).[34] The magnitude of $\Delta G_\phi^o$ was calculated from Equation 12, as was done above. The values of $\Delta_{op}$, calculated from Equation 16, are given in Figure 6, against $N^{1/3}$. These plots can be described by the following equations, Table V.

Table V. Variation of $\Delta_{op}$ with Aggregation Number N, as a Function of Counter-Ion for Dodecyl Pyridinium Bromide (DPyB) or Chloride (DPyC)[34]

| | |
|---|---|
| DPyB | RT ln c.m.c./55.5 − $\Delta G_\phi^o$ = 2955 + 13311/$N^{1/3}$ |
| DPyC | RT ln c.m.c./55.5 − $\Delta G_\phi^o$ = 5319 + 3259/$N^{1/3}$ |

It is of interest to compare the data in Tables IV and V. In both analyses we have expressed in molar fractions. It is seen that in the case of alkyl sulfates the slopes are related to the alkyl chain length. On the other hand, the cationic systems show the effect of the counter-ion on the slopes, as one might expect. However, we must stress that the plots in Figure 6 show too much scatter which does not allow a quantitative analysis. In any case, it is safe to conclude that the relation between $\Delta_{op}$ and α will provide very useful information, whenever more accurate and systematic data are available for other systems.

## CONCLUSION

The aim of this study was to describe the hydrophobic interactions in micellar systems which are responsible for the solubilization of water-insoluble organic compounds, are not dependent on

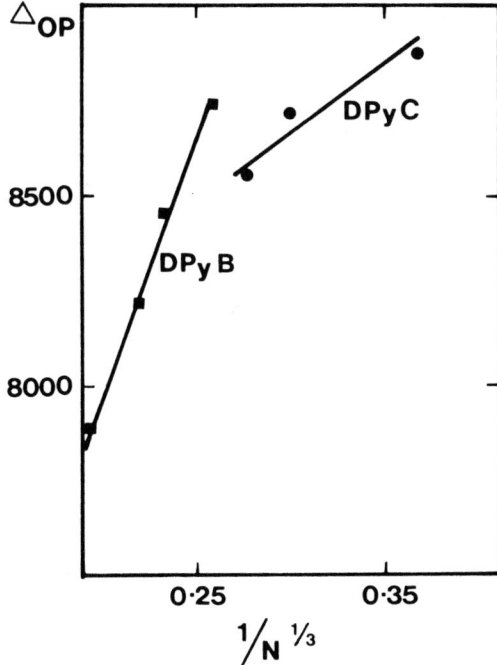

Figure 6. Variation of $\Delta_{op}$ with $1/N^{1/3}$ for DPyB and DPyC in salt solutions.

the added salt as shown by various experimental data.[9] This finding was then applied to the current theories on micellar systems. The hydrophobic interactions in micelles can be satisfactorily described, since it is known that the removal of the alkyl chain from water phase to the hydrocarbon phase is related to the surface area of the molecule in contact with water for a variety of systems.[26,27] Since the surface area of the molecule remains constant, at constant temperature and pressure, then this relation can be applied to micellar systems as well. In this study we have based the analysis on the assumption that $\Delta G_\phi^o$ and $\Delta G_\phi^\phi$ are roughly of the same magnitude, where the former refers to micellar systems while the latter to the process where the alkyl chain is transferred from aqueous medium to the hydrocarbon phase. It is a safe approximation under the present conditions where the magnitude of the opposing forces $\Delta_{op}$, cannot be directly estimated. The magnitude of the opposing forces could thus be estimated based on these assumptions and the knowledge of c.m.c. and aggregation number, N. The analysis of experimental data for alkyl sulfates of chain lengths 9, 10, 11 and 12 carbon atoms, as a function of ionic strength, showed that calculated opposing forces, $\Delta_{op}$, were a linear function of the area per polar group at the micelle-water interface. On the other hand the data for other micellar systems showed that the $\Delta_{op}$ was related to the counterions, that is, in the case of dodecyl pyridinium salts of chloride, DPyC, and bromide, DPyB.

It is shown that a self-consistent analysis is obtained in the description of the extensive data available on the alkyl sulfates,[20] when it is assumed that $\Delta G_\phi^o$ is not dependent on the ionic strength or N (as concluded from the solubilization results),[9] and assuming that $\Delta G_\phi^o \sim \Delta G_\phi^\phi$ (when $\Delta G_\phi^\phi$ can be accurately calculated from Equation 12).

It is obvious that the finding that aggregation number and ionic strength do not affect $\Delta G_\phi^o$ in micellar systems has a strong impact on the bilayer and membrane theories. In order to analyze this subject extensively, more experimental data is being made available in this laboratory, and a more thorough analysis will be given in a future report.

## ACKNOWLEDGMENTS

It is a pleasure to thank Dr. J. Kratohvil and Dr. J. Israelachvili for many helpful suggestions. The author thanks the award of travel grant from Statens Naturvidenskabelige Forskningsfond. The excellent technical assistance of Mrs. J. Klausen is acknowledged.

## REFERENCES

1. D. G. Hall, B. A. Pethica, in "Nonionic Surfactants", M. J. Schick, Editor, pp. 516-577, Marcel Dekker Inc., New York.
2. K. S. Birdi, in "Colloidal Dispersions and Micellar Behavior", K. L. Mittal, Editor, ACS Symposium Series No. 9, pp. 233-238, Amer. Chem. Soc., Washington, D.C., 1975.
3. K. S. Birdi, in "Proc. International Conf. Colloid and Surface Science", E. Wolfram, Editor, pp. 473-479, Budapest, 1975.
4. C. Tanford, in "The Hydrophobic Effect", C. Tanford, Editor, pp. 36-85, John Wiley and Sons, New York, 1973.
5. C. Tanford, J. Phys. Chem., $\underline{78}$, 2469 (1974).
6. D. Stigter, J. Colloid and Interface Sci., $\underline{47}$, 473 (1974).
7. D. Stigter, J. Phys. Chem., $\underline{78}$, 2480 (1974).
8. Y. Zimmels and I. J. Lin, Colloid Polymer Sci., $\underline{252}$, 594 (1974).

10. K. S. Birdi, Colloid Polymer Sci., $\underline{252}$, 551 (1974).
11. D. Attwood and A. T. Florence, Kolloid-Z.u.Z. Polym., $\underline{246}$, 580 (1971).
12. P. Molyneux and C. T. Rhodes, Kolloid-z.u.Z. Polym. $\underline{250}$, 886 (1972).
13. R. J. Williams, J. N. Phillips and K. J. Mysels, Trans. Faraday Soc., $\underline{51}$, 729 (1955).
14. K. S. Birdi, Anal. Biochem., $\underline{74}$, 620 (1976).
15. M. E. L. McBain and E. Hutchinson, "Solubilization and Related Phenomena", p. 75, Academic Press, New York, 1955.
16. H. Schott, J. Phys. Chem., $\underline{70}$, 2966 (1966).
17. H. Schott, J. Phys. Chem., $\underline{71}$, 3611 (1967).
18. J. Steinhardt, N. Stocker, D. Carroll and K. S. Birdi, Biochemistry, $\underline{13}$, 4461 (1974).
19. K. S. Birdi, to be published.
20. H. F. Huisman, Proc. Kon Ned. Akad. Wetensch. B67, 367 (1964), thesis.
21. H. Coll, J. Phys. Chem., $\underline{74}$, 520 (1970).
22. I. J. Lin and P. Somasundaran, J. Colloid Interface Sci., $\underline{37}$, 731 (1971).
23. B. W. Barry and G. F. Russell, J. Colloid Interface Sci., $\underline{40}$, 174 (1972).
24. I. Langmuir, in "Third Colloid Symposium Monograph", p. 3, Chemical Catalog Co., New York, 1925.
25. R. B. Herman, J. Phys. Chem., $\underline{76}$, 2754 (1972).
26. G. L. Amidon, S. H. Yalkowsky and S. Leung, J. Pharm. Sci., $\underline{63}$, 3325 (1974).
27. G. L. Amidon, S. H. Yalkowsky, S. T. Anik and S. C. Valvani, J. Phys. Chem., $\underline{79}$, 2239 (1975).
28. K. A. Wright, A. D. Abbott, V. Sivertz and H. V. Tartar, J. Amer. Chem. Soc., $\underline{61}$, 549 (1939).
29. M. L. Corrin, H. B. Klevens and W. D. Harkins, J. Chem. Phys., $\underline{14}$, 480 (1946).
30. H. B. Klevens, J. Amer. Oil Chem. Soc., $\underline{30}$, 74 (1953).
31. J. M. Corkill, J. F. Goodman and R. H. Ottewill, Trans. Faraday Soc., $\underline{57}$, 1627 (1961).
32. A. Holtzer and M. F. Holtzer, J. Phys. Chem., $\underline{78}$, 1442 (1974).
33. K. S. Birdi and J. Kratohvil, to be published.
34. W. P. J. Ford, R. H. Ottewill and H. C. Parreira, J. Colloid Interface Sci., $\underline{21}$, 522 (1966).

SIZE DISTRIBUTION OF MICELLES: MONOMER-MICELLE EQUILIBRIUM, TREATMENT OF EXPERIMENTAL MOLECULAR WEIGHT DATA, THE SPHERE-TO-ROD TRANSITION AND A GENERAL ASSOCIATION MODEL

Pasupati Mukerjee

School of Pharmacy, University of Wisconsin

Madison, Wisconsin 53706

The forces of interaction causing micelle formation are reviewed briefly. The free energy of micelle formation involves not only hydrophobic interactions and head group self-interactions but the interactions between the head groups and the hydrocarbon core of the micelles as well. Methods of determining different average molecular weights from experimental data on a particular average are presented. It is shown that size-distribution index values can be determined from the concentration-dependence of an average degree of aggregation. Micellar systems fall in two broad classes: 'small' micelles with narrow distributions and large micelles with very wide distributions. An association model for the latter is reviewed. Evidence is presented that both for 'small' and large micelles the stepwise association involves a cooperative region followed by an anticooperative region. A general association model is derived on the basis of a sphere-to-rod transition. The relation of this qualitative association model to the formation of 'small' micelles of narrow distributions and large micelles of wide distributions, as also the transition from 'small' to large micelles, is discussed. Several other qualitative consequences of the general association model are presented.

# INTRODUCTION

The general problems of different patterns of self-association exhibited by hydrophobic solutes of different molecular structures have been reviewed recently.[1,2] The present paper is concerned exclusively with flexible chain surfactants in aqueous media.

## Self-association Equilibria

Assuming ideality, i.e. negligible interaction between species so that the activity coefficient of all individual species is unity, the self-association of monomers, $b_1$, to the dimer, $b_2$, to the trimer, $b_3$, or, in general, to the q-mer, $b_q$, can be expressed either in terms of step-wise self-association reactions

$$2b_1 \overset{K_2}{\rightleftharpoons} b_2; \quad b_2 + b_1 \overset{K_3}{\rightleftharpoons} b_3; \quad \ldots \quad b_{q-1} + b_1 \overset{K_q}{\rightleftharpoons} b_q \tag{1}$$

or overall association reactions for the formation of $b_q$ from $b_1$

$$qb_q \overset{\beta_q}{\rightleftharpoons} b_q \tag{2}$$

Here $K_q$ is the step-wise association constant and $\beta_q$ is the overall association constant in any consistent scale of concentration units. In this article, moles dm$^{-3}$ units will be used although mole fraction units are preferable for thermodynamic calculations.[3] The relation between $K_q$ and $\beta_q$ is given by Equation 3.

$$\ln\beta_q = \sum_2^q \ln K_q \tag{3}$$

The Equations apply to nonelectrolyte systems generally, i.e. without restriction to solvents, or the nature of the interactions responsible for self-association. If the aggregating monomer is ionic, electrical interactions affect the activity coefficients of the various species and counterion participation in the association equilibria as also salting-out (salting-in) effects may be involved.[3,4,5] Depending upon the strengths of the interactions and the aggregation numbers, various approaches are used to take such charge effects into account.[3,4] If the ionic strength and free counterion concentrations are kept constant, however, the activity coefficients of all species and the counterion participation terms in the mass-action equilibria for each aggregation step are kept constant. They can then all be incorporated in $K_q$ or $\beta_q$ and Equations 1-3 remain applicable.[5] For ionic surfactant

# SIZE DISTRIBUTION OF MICELLES

systems in aqueous media, this condition is realized if a neutral electrolyte such as NaCl is present at a high concentration.

The standard free energy changes associated with the overall and stepwise association are given by

$$q\Delta G_q = \Delta G_q^* = -RT \ln \beta_q \qquad (4)$$

and

$$\Delta G_q' = -RT \ln K_q \qquad (5)$$

$\Delta G_q$ is the average free energy change per monomer when the q-mer forms, $\Delta G_q^*$ is the total change, $\Delta G_q'$ is the change associated with the stepwise process of adding a monomer to the (q-1)-mer, R is the molar gas constant, and T is the absolute temperature.

## MONOMER-MICELLE EQUILIBRIUM AND PROBLEMS IN CALCULATING $\Delta G_q$ OR $\Delta G_q'$

For the prediction of size-distributions of micellar systems, it is necessary to calculate absolute and relative values of $\Delta G_q(\beta_q)$ or $\Delta G_q'(K_q)$ to an extremely high degree of accuracy.[1,2] Figure 1, for example, shows a narrow size distribution for a system which exhibits only a shallow minimum in a $\Delta G_q$ vs. q plot (corresponding to the slight hump in the $(\ln \beta_q)/q$ vs. q plot) involving a total variation of less than 1.5% in $\Delta G_q$ as q varies from 70 to 120. Small changes in the $\Delta G_q$ vs. q profile and the position of the minimum can produce major changes in the predicted average aggregation numbers and size distribution, as discussed later.

Accurate predictions of the absolute and relative values of $\Delta G_q$ require similarly accurate calculations of all the attractive and repulsive interactions involved in the formation of micelles from monomers, including their dependence on the size, shape, and structure of the micelles. The interactions to be accounted for are discussed below in a summary fashion.

### Hydrophobic Interactions

The so-called hydrophobic interactions[1-3,6,7-9] are the primary cause for micelle formation.[1-3,9] The interactions themselves are the composite result of several identifiable factors which are not well understood.[1,2] The interesting and important point to be made here is that the expression of hydrophobicity for hydrophobic monomers in water depends very much on the end state of the monomer, i.e. whether the monomer is partitioned into a hydrocarbon, adsorbed

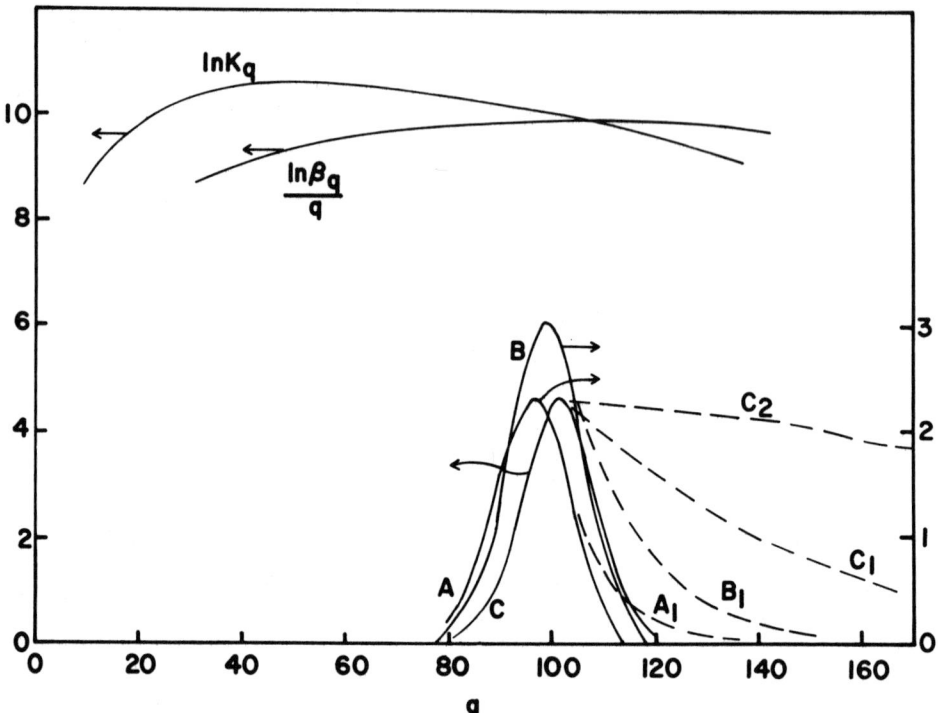

Figure 1. Simulation of micellar size distributions, using the equation $\ln\beta_q = 2(q-1)\ln(q-1) - 0.02(q-1)^2 + 2.7896(q-1)$ for calculating $\ln K_q$ and $(\ln\beta_q)/q$. All micellar concentrations in monomer mol dm$^{-3}$. For curves A and $A_1$, $[b_1] = 4.12 \times 10^{-5}$ mol dm$^{-3}$, ordinate $q[b_q] \times 10^7$, right-hand scale; for curves B and $B_1$, $[b_1] = 4.33 \times 10^{-5}$ mol dm$^{-3}$, ordinate $q[b_q] \times 10^5$, right-hand scale; for curves C, $C_1$ and $C_2$, $[b_1] = 4.55 \times 10^{-5}$ mol dm$^{-3}$, ordinate $q\, b_q \times 10^3$, left-hand scale. Full curves A, B, and C calculated using the $(\ln\beta_q)/q$ curve. Dashed portions $A_1$, $B_1$, $C_1$, and $C_2$ are discussed in the text.

at an interface, or micellized. It has been shown that the unitary free energy of micelle formation of octyl glucoside, after reasonable corrections are applied for the head-group effect, is substantially lower in absolute magnitude than the free energy change expected for the hydrophobic moiety, the octyl group, on partitioning into a hydrocarbon medium.[3] Figure 2 shows a summary of some results for the incremental change in free energy per -CH$_2$- group ($\Delta\Delta G_{CH_2}$) for chains of medium length where the head-group effect may be assumed to be small. Different values are obtained for different processes. If the complete transfer (partition) into a hydrocarbon is assumed to provide the maximal effect of hydrophobicity, it is clear that the other processes involve only partial expressions of hydrophobicity and it is not possible to use any particular process as a model for any other in free energy calculations. The low value of $\Delta\Delta G_{CH2}$ for the transfer process from water to gas suggests the very important role of chain-chain interactions in micelle formation and inter-

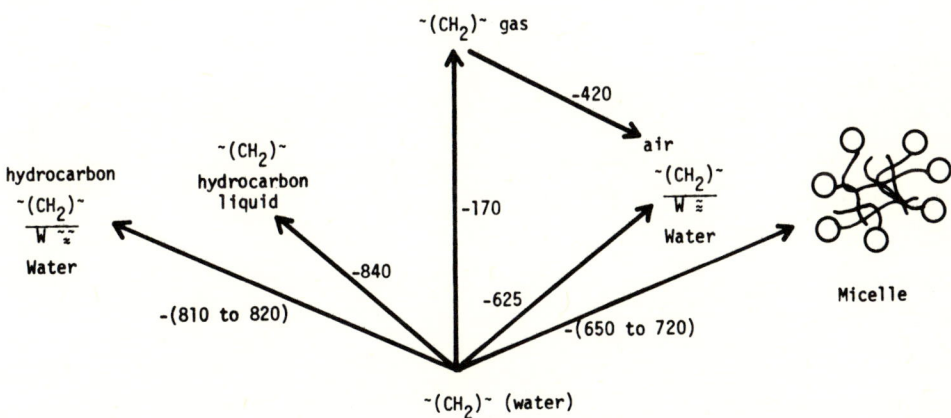

Figure 2. Incremental free energy changes per CH$_2$ group ($\Delta\Delta G_{CH_2}$) obtained for chains of medium length for transfer from dilute aqueous solutions of monomers to the hydrocarbon-water interface, hydrocarbon liquids, the gas phase, the air-water interface and micelles.[3] Also included is the free energy of adsorption of vapors to the air-water interface.[10]

actions of $-CH_2-$ groups at interfaces[3] in the net process. The lower value of $\Delta\Delta G_{CH_2}$ on micellization, when compared to partitioning into hydrocarbons, is indicative of a constrained state and lower entropy of the monomeric chains inside a micelle.[3] These findings are also consistent with the evidence of some solid-like structure in the micelles of sodium alkyl sulfates, as shown by differences between chains of odd and even carbon numbers and the substantial Laplace pressures inside the micelles expected from their curved interfaces.[11]

## Head Group Self-interaction

Head groups of monomers generally oppose micelle formation. Table I shows some critical micellization concentration (c.m.c.)

Table I. Critical Micelle Concentrations of Surfactants Containing Octyl Chains[3,12]

| Compound | Formula | C.m.c. at 25°C (mol dm$^{-3}$) |
|---|---|---|
| Octyl glycol ether | $C_8H_{17}OCH_2CH_2OH$ | $4.9 \times 10^{-3}$ |
| Octyl α-glyceryl ether | $C_8H_{17}OCH_2CH(OH)CH_2OH$ | $5.8 \times 10^{-3}$ |
| β-D-octyl glucoside | $C_8H_{17}OCH_2C_5H_{10}O_5$ | $2.5 \times 10^{-2}$ |
| Octyl trioxyethylene glycol monoether | $C_8H_{17}(OCH_2CH_2)_3OH$ | $7.5 \times 10^{-3}$ |
| Octyl hexaoxyethylene glycol monoether | $C_8H_{17}(OCH_2CH_2)_3OH$ | $9.9 \times 10^{-3}$ |
| Octyl nona-oxyethylene glycol monoether | $C_8H_{17}(OCH_2CH_2)_9OH$ | $1.3 \times 10^{-2}$ |
| Octyl dimethylamine oxide | $C_8H_{17}N(CH_3)_2O$ | 0.16 |
| Octyl N-betaine | $C_8H_{17}N^+(CH_3)_2CH_2COO^-$ | 0.25 (27°) |
| Sodium octyl sulfate | $C_8H_{17}SO_4^-Na^+$ | 0.13 |

data for different surfactants with a common hydrocarbon (octyl) chain. As has been pointed out,[3] these variations can be ascribed to the self-interaction of the head groups at the micellar surface where their concentrations are high. The molecules that comprise the head groups, such as glycerol, glucose, or polyethylene glycols, all show an increase in the activity coefficient on going from a dilute to a concentrated solution.[13,14] This corresponds to a repulsive interaction for the monomer forming a micelle. The zwitterionic head groups involve strong repulsion of dipoles at the micelle surface. As has been pointed out,[3] these repulsive interactions must be evaluated for the local conditions at the micelle surface where the effective dielectric constant may be considerably lower than the value of water.[15]

## Head Group - Core Interaction

The extremely high c.m.c. of octyl-N-betaine, higher than that of octyl glycol ether by a factor of 50, and higher than even the ionic surfactant sodium octyl sulfate, suggests strongly that an additional repulsive interaction should be identified and treated separately. It is well known that electrolytes are negatively adsorbed at air-water and oil-water interfaces and raise interfacial tensions.[10] These interactions have been ascribed to electrostatic image forces: a charge located in a medium of high dielectric constant but close to an interface at the other side of which there is a medium of a low dielectric constant suffers a repulsive interaction. Stigter[16] has recently shown that ionic surfactants with different head groups exhibit different image interactions. Since such an interaction must exist even in the absence of mutual interactions of head groups, it appears that it should be treated separately, although it is likely to be modified by the presence of other head groups in close proximity. Repulsive image forces are expected for dipolar systems also[17] and probably make some contribution to $\Delta G_q$ for non-ionic surfactants. For zwitterionic head groups and the amine oxide (Table I), such interactions are expected to be quite important because of their high dipole moments.

It seems, therefore, that the unitary free energy of micelle formation[3] should be calculated using at least three different contributions, the hydrocarbon chain ($\Delta G_{HC}$), the head group self-interaction ($\Delta G_{HG}$), and the interaction between the head group and the hydrocarbon core ($\Delta G_{HG-HC}$). All of these quantities are likely to be dependent on q. An accurate calculation of $\Delta G_q$ appears to be a formidable problem in the present state of our knowledge.

As discussed before,[3] when the chain contribution to the free energy of micelle formation of sodium dodecyl sulfate is calculated by correcting for the head group self-interaction effect (charge effect) using theories of electrical double layers, the value appears to be lower in absolute magnitude when compared to the estimate from non-ionic surfactants. It appears that the introduction of the core-head group interaction (image interaction) reduces the discrepancy to some extent.

## ANALYSIS OF EXPERIMENTAL MOLECULAR WEIGHT DATA

The complete description of a self-associating system requires a knowledge of the values of $K_q$ or $\beta_q$ for all relevant values of q. This task is extremely difficult experimentally because of propagation of errors: the uncertainty of one $K_q$ value affects those of all subsequent ones.[1] In general, it is difficult to obtain more than two or three parameters characterizing the self-associating system with any degree of precision. For flexible chain surfactants which form the 'classical' type micelles, with a hydrocarbon core and polar groups at the surface, a sharp and well defined c.m.c. can usually be obtained by a variety of methods.[12] The c.m.c. itself can be used to calculate approximate values of $\Delta G$.[3] The additional parameter most frequently measured is $N_w$, or the weight average degree of aggregation of the micelles, from light scattering studies. Only in very few investigations have other averages been measured. The number average, $N_n$, is particularly difficult to obtain directly because vapor pressure osmometry is not of high enough precision when the aggregation number is high, and membrane osmometry is generally impractical because the permeating monomers are in association-dissociation equilibrium with the micelles, and, therefore, allow equilibriation of the micelles.[18]

For ideal micellar systems, it is possible, however, to obtain different average aggregation numbers if either the monomer concentration, $b_1$, or any one average aggregation number is available experimentally over the whole range of concentrations.[5] The previously discussed general approaches are extended and adapted further in the following treatment which deals with some results which are particularly useful for the determination of a size-distribution index for micellar systems.

If we define the following quantities,

$$M = \Sigma[b_q]; \quad C = \Sigma q[b_q]; \quad G = \Sigma q^2[b_q]; \text{ and } H = \Sigma q^3[b_q] \quad \ldots \quad (6)$$

where M is the total molar concentration of all species, C is the usual total concentration in monomer units, and G and H are higher weighted sums, then the average aggregation numbers of the aggregate species are defined as

$$N_n = \frac{C-[b_1]}{M-[b_1]}; \quad N_w = \frac{G-[b_1]}{C-[b_1]}; \quad N_z = \frac{H-[b_1]}{G-[b_1]} \quad (7)$$

# SIZE DISTRIBUTION OF MICELLES

$N_n$, $N_w$, and $N_z$ represent the number, the weight, and the Z-average. From the following derived relations,

$$\frac{d(M-[b_1])}{d[b_1]} = \frac{C-[b_1]}{[b_1]} \tag{8}$$

$$\frac{d(C-b_1)}{d[b_1]} = \frac{G-[b_1]}{[b_1]} \tag{9}$$

$$\frac{d(G-[b_1])}{d[b_1]} = \frac{H-[b_1]}{[b_1]} \tag{10}$$

a number of relations between $[b_1]$, $C$, $N_n$, $N_w$, and $N_z$ can be obtained, of which a few of particular interest are given below.

$$\frac{d \ln (C-[b_1])}{d[b_1]} = N_w \tag{11}$$

$$d(M-[b_1]) = \frac{d(C-[b_1])}{N_w} \tag{12}$$

$$N_n + \left(\frac{N_n}{N_w}\right) \frac{dN_n}{d \ln(C-[b_1])} = N_w \tag{13}$$

$$N_w + \frac{dN_w}{d \ln (C-[b_1])} = N_z \tag{14}$$

$$\frac{d \ln N_n}{d \ln N_w} = \frac{N_w - N_n}{N_z - N_w} \tag{15}$$

All the aggregation numbers used are functions of C and increase with C[1,2] excepting in the special case of a unique (single) multimer.

These equations apply equally well if the average aggregation numbers include the monomer. The number average, $N_n'$, is then given by $C/M$. In this case $[b_1]$ is not subtracted from C, M, G, or H in

Equations 8 to 14. The Equations can, therefore, be used to analyze different sets of data. Thus, for example, C and $[b_1]$ data allow $N_w'$, defined as G/C, to be determined (Equation 11), whereas C and $N_w'$ data allow M and, therefore, $N_n'$ to be determined by integration from infinite dilution to C (Equation 12).

In the case of micellar systems, molecular weights are usually determined by assuming that the micellar concentration in monomer units is C-c.m.c. This is equivalent to replacing $[b_1]$ in the definition of the average aggregation numbers in Equation 7 by the c.m.c., and assuming $[b_1]$ to be effectively constant above the c.m.c. The validity of this approximation improves as the value of $N_w$ increases and as the ration C/c.m.c. increases because of the following relationship.[5]

$$d \ln[b_1] = \frac{d \ln (C-[b_1])}{N_w} \qquad (16)$$

When $N_w$ is large, the approximation is usually valid excepting very close to the c.m.c. where $C-[b_1]$ changes rapidly.

The replacement of $[b_1]$ by the c.m.c. allows the determination of different averages from experimental data on a particular average. In general, the determination of a lower average (for example $N_n$ from $N_w$ data) requires an integration over the whole concentration range (Equation 12). Higher averages can, however, be obtained from data over limited ranges of concentration. Thus Equation 13 can be used to obtain $N_w$ from $N_n$ data, and Equation 14 can be used to determine $N_z$ from $N_w$ data, which are the most numerous in the micellar literature.

The application of these Equations is often hampered by non-ideality effects (intermicellar interactions and virial coefficients). In many cases, however, after reasonable corrections are made for the second virial coefficient, $N_w$ shows very little dependence on concentration.[5] The analysis of such data indicate that the size-distribution indices, $N_w/N_n$ or $N_z/N_w$, from Equations 12 and 14, are close to unity. These indices are unity for strictly monodisperse systems which correspond to $N_w$ values being completely independent of concentration.[5] To take a typical example of special interest, the apparent aggregation number of sodium dodecyl sulfate in o.3M NaCl[19] decreases by about 11% as the micellar concentration C-c.m.c. increases by about a factor of 20 in the experimental range. This is ascribed to a positive second virial coefficient. If a theoretically calculated virial coefficient[19] is now used as a correction factor, $N_w$ increases by about 1 to 2% over the whole concentration range. Considering the uncertainties of the experimental data and of the calculated virial coefficients, $N_w/N_n$ and

$N_z/N_w$ for this system appear to be indistinguishable from unity within an uncertainty of 1 to 2%. Thus, micellar size distributions appear to be quite narrow in many cases.

On the other hand, when very large micelles are investigated, with $N_w$ values considerably higher than what a spherical micelle can accommodate and sometimes exceeding 2,000, $N_w$ has been found to be proportional to $(C-c.m.c.)^{1/2}$ for several surfactants with non-ionic,[5] ionic,[5,20] or zwitterionic[21] head groups. This leads to an $N_w/N_n$ index of 2.0 and an $N_z/N_w$ index of 1.5 for such systems which are, therefore, highly polydisperse.

## Small Micelles and Large Micelles

A majority of the micellar systems appear to fall in one or two categories: systems showing narrow size distributions ($N_w/N_n < 1.05$) and others showing very wide distributions ($N_w/N_n \approx 2$). The former can be described as 'small' micelles although their aggregation numbers may exceed by roughly a factor of 2 the number expected for a spherical micelle with a radius equaling the length of a stretched-out monomer. Thus, sodium dodecyl sulfate in 0.3M NaCl has an aggregation number[19] of 123, roughly twice the number a spherical micelle can accommodate. The narrow size-distribution of the 'small' micelles makes the usual assumption of monodispersity a reasonably valid one in many cases.[1,2] The 'large' micelles exhibit aggregation numbers which go up to some hundred times the value for a spherical micelle.[5,20,21]

A point of caution regarding 'large' micelles involves the misleading conclusions that may be drawn if the c.m.c. is estimated from the turbidity data. Figure 3 shows some turbidity data for a non-ionic surfactant. The use of an apparent c.m.c. value of $8.5 \times 10^{-4}$ g/cm$^3$ from these data results in a Debye plot typical of small micelles, i.e. a linear plot consistent with an aggregation number independent of concentration and a constant positive second virial coefficient as shown in Figure 4. Such a representation may lead to serious overestimates of $N_w$ at low concentrations. The use of the c.m.c. value from surface tension measurements,[22] $5 \times 10^{-6}$ g/cm$^3$, shows how the aggregation number increases rapidly at low concentrations (Figure 4). The expected continued increase in $N_w$ with concentration appears to be overshadowed by the non-ideality effects in this and similar systems at higher concentrations.[5]

<u>Shapes of Micelles</u>. For 'small' micelles, the assumption of sphericity can only be approximately valid in view of the structure of the head groups and the evidence of a rough surface.[23] One important problem is the range of sizes over which 'small' micelles are 'effectively' spherical, at least as far as the hydrocarbon core

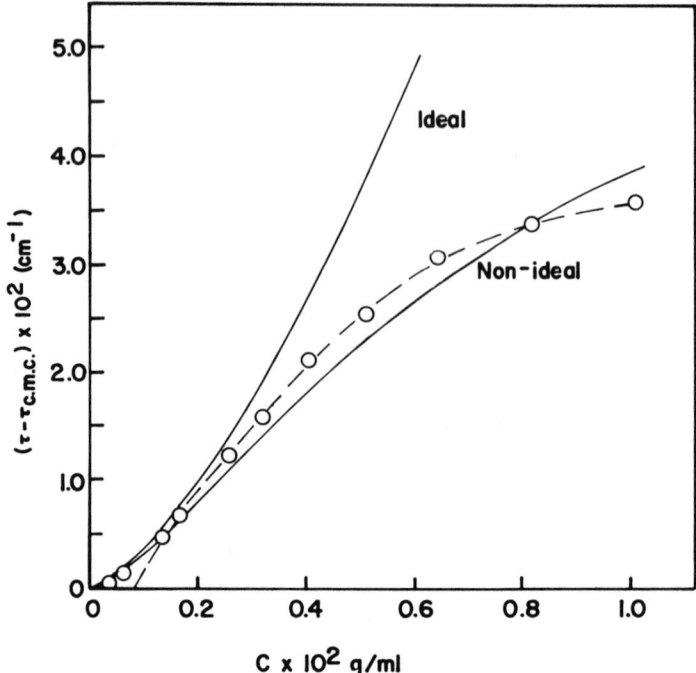

Figure 3. Turbidity data ($\tau$) for $C_{14}H_{29}(OCH_2CH_2)_6OH$ at 30°C, estimated from Balmbra et al.[22] The ideal line represents calculations based on an association model for large micelles (Equation 17) assuming no intermicellar interactions.[5] The non-ideal line is calculated applying corrections for second virial coefficients.[5]

is concerned. Here, the applicability of any strict geometrical constraint for a limiting size of the spherical micelle, such as the radius of the core equaling the length of a stretched-out monomer chain, is not easy to evaluate. It is extremely unlikely that the monomers in a micelle are all in their completely stretched-out configuration. On the other hand, for a given value of the aggregation number, a continuous range of shapes may exist which are compatible with the density of the hydrocarbon core,[24] and the surface is likely to show corrugations because of local fluctuations. To what extent such corrugations of the surface allow a micelle to be 'effectively' spherical and yet have an average radius larger than the length of the stretched-out monomer is not known.

For 'large' micelles, the evidence from several sources indicates a highly asymmetric structure. These micelles appear to be worm-like, i.e. cylindrical (presumably with cupped ends) and flexible to some extent.[20,25,26,27]

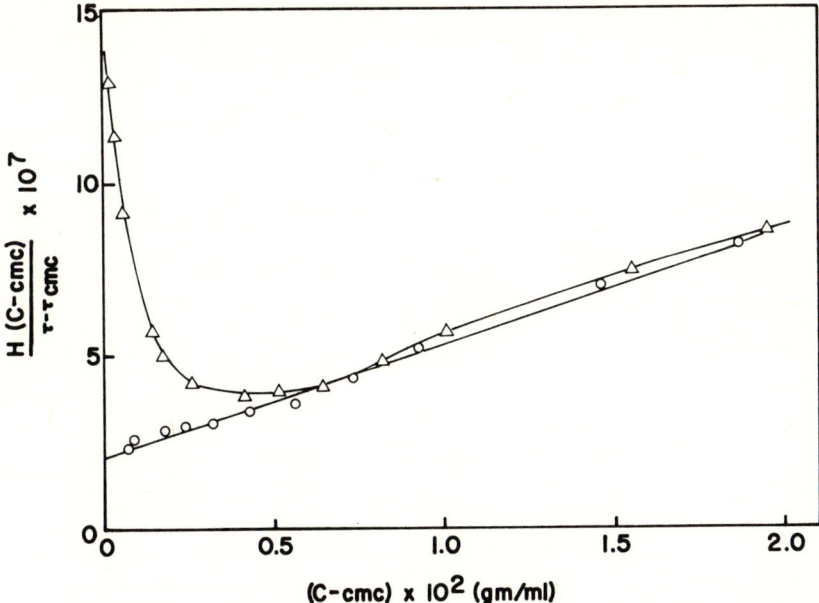

Figure 4. Debye plots based on turbidity data for $C_{14}H_{29}(OCH_2CH_2)_6OH$ at $30°C^{22}$ (Figure 3). H is the optical constant.[22] △—Using c.m.c. from surface tension[22] ($5 \times 10^{-6}$ g/cm$^3$), ○—using c.m.c. ($8.5 \times 10^{-4}$ g/cm$^3$) from turbidity data (Figure 3).

## ASSOCIATION MODEL FOR LARGE MICELLES

When q values are very large, well beyond the range of spherical micelles, some general considerations indicate that for rod-like (or disc-like) micelles of constant radius or thickness, the step-wise aggregation constant $K_q$ becomes independent of q.[5] This constancy is expected to hold when end-effects are negligible, i.e. when the end-to-end distance is comparable to several chain lengths for nonionic surfactants for which all interactions are short range, and, additionally, to several thicknesses of the electrical double layer for ionic surfactants which involve long-range interactions. Considering now a prolate sphero-cylinder of constant radius as a reasonable model for large micelles, when the micelle is large enough such that one end does not 'see' the other end, every monomer added in a step-wise reaction meets the same environment in the micelle, produces the same change in the volume and surface area, and undergoes the same interactions. Therefore ln $K_q$ becomes a constant at high values of q.

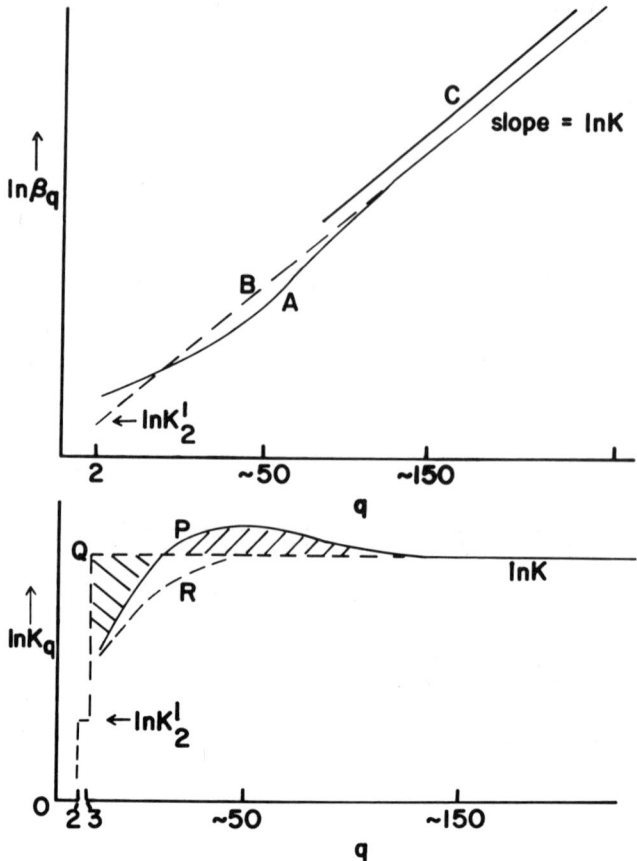

Figure 5. Schematic representation of the variations of $\ln K_q$ and $\ln \beta_q$ with q for a surfactant with a chain length of roughly 12 carbon atoms and their relation to the association model for large micelles[5] (Equation 17).

The considerations above lead to a simple model of aggregation for large micelles. Figure 5 shows schematically the variation in $\ln \beta_q$ with q (curve A) for a surfactant with a chain of about 12 carbons. When values of $N_w$ are very large, in the range of several hundreds or higher, the 'small' micelles coexisting with the large micelles may be assumed to make a negligible contribution to the population. For such a case, curve A is replaced by the straight line B of a constant slope of ln K and a value of $\ln K_2'$ for the dimer. The association model then is represented by an open-ended continuous scheme, i.e. with no upper limit for q, involving two stepwise association constants, a <u>hypothetical</u> $K_2'$ and K.

$$K_2' < K, \quad K_q = K \; (q \geqslant 3) \tag{17}$$

The model leads to a simple formula for $N_w$[5]

$$N_w = 2 \left(\frac{K}{K_2'}\right)^{1/2} \left(\frac{C-c.m.c.}{c.m.c}\right)^{1/2} \tag{18}$$

which is applicable when the c.m.c. is low with respect to C. When $C \gg c.m.c$, $N_w$ is proportional to $C^{1/2}$. The value of K can be determined from the value of $[b_1]$ at the c.m.c. or the c.m.c. itself because, in this model, when $N_w$ is high, $1 - K[b_1]$ is a very small positive number which is usually less than 0.01.[5] This model of stepwise association includes the initial cooperativity of self-association necessary for micelle formation[5] in the condition $K_2' < K$. If $K_2' = K$, the model reduces to a noncooperative open-ended self-association model which leads to the well known most probably distribution.[1,2]

The above association model for large micelles has been found to be applicable to several micellar systems involving ionic, non-ionic and zwitterionic head groups[5,20,21] for which a proportionality of $N_w$ to $C^{1/2}$ has been observed after correcting for non-ideality effects. The model predicts a wide size distribution ($N_w/N_n = 2.0$, $N_z/N_w = 1.5$). The importance of the non-ideality effects is shown in Figure 3, where the turbidities predicted by the above model without non-ideality corrections and with corrections based on rigid rod or the equivalent prolate spherocylinder model are compared with experimental values.

For some short-chain lecithin analogues, Tausk and Overbeek[21] have used a modification of the above model corresponding to curve C in Figure 5. Here the aggregation is assumed to begin at a certain number, n, and continue growth by successive stepwise association involving the same constant K.

The applicability of the model for large micelles to systems where $N_w$ varies from about 500 to 2500[5,21] appears to provide a serious constraint on the allowed variation of lnK over similar wide ranges of q values in real systems. lnK is typically in the range of 10 or 15. Thus as q increased from 500 to 2500, $\ln\beta_q$ typically increases by 20,000 to 30,000. Since a change in $\ln\beta_q$ by 11.5 units increases or decreases the concentration fo $b_q$ by $10^5$, a magnitude likely to alter the size-distribution pattern markedly, it appears that lnK in many systems of large micelles remains constant to within very narrow limits at high q values.

This conclusion has an important consequence regarding permissible shapes of large micelles. It indicates that at high q values, barring highly improbably compensatory changes, $\Delta G'_{HC}$, $\Delta G'_{HG}$ and $\Delta G'_{HG-HC}$ are individually independent of q within very narrow limits over a wide range of q values, thus requiring that the change in surface area per monomer, the volume available per monomer in the micelle and the surface curvature remain effectively the same. These conditions are consistent with the sphero-cylinder model with constant radius but appear to rule out ellipsoids of revolution.

Secondary Aggregation of Small Micelles. The process of aggregations of small (spherical) micelles coalescing to produce large micelles may have some interesting mechanistic and kinetic consequences but it is thermodynamically indistinguishable from any scheme involving monomer-micelle equilibrium as long as all micelles are in equilibrium with monomers. The reversible aggregation of small micelles which maintain their identity has been proposed to explain the very high molecular weights observed for large micelles.[28,29,30] Several arguments indicate that this is rather improbable. For small micelles, the non-ideality effects generally exhibited suggest repulsive interactions. In the case of small ionic micelles, positive virial coefficients have been found at NaCl concentrations of 0.3M or even 1M.[19] Non-ionic micelles containing oxyethylene groups also show repulsive interactions when the micelles are small.[28] It appears then that for spherical micelles to be held together, long-range van der Waals' forces exerted by the cores must be invoked and these do not appear to be strong enough. The model of reversible secondary aggregation of small micelles also requires a barrier which prevents coalescence and a directional character to the secondary aggregation process, i.e. the formation of a roughly linear chain of small micelles, to explain the highly assymmetric character of the large micelles.[25,26]

## COOPERATIVITY AND ANTICOOPERATIVITY

It has been pointed out that the existence of a sharp c.m.c. and the formation of micelles of high aggregation numbers close to the c.m.c. require an initial cooperativity of self-association: $K_q$ increasing with q at the initial stages.[1,2,31] Considering the growth process of very small micelles, structural and energetic considerations suggest that this cooperativity has a long-range character, i.e. $K_q$ increases continuously with q over a substantial range of q values. This cooperativity in real micellar systems is thus different from the one involved in the two constant model for large micelles (Equation 17) where the cooperativity is confined to the change from $K_2'$ to $K_3$. It has been shown that this latter kind of cooperativity can lead to sharp c.m.c. values depending upon the ratio of $K_2'$ to $K^5$ but it is unlikely that any real micellar system behaves according to this model very close to the c.m.c.

Whereas the initial cooperativity must exist for all micellar systems exhibiting a sharp c.m.c., a limit to the growth process implied by the existence of a narrow size distribution for 'small' micelles indicates that the range of cooperativity is followed by a range of anticooperativity,[1,2] i.e. $K_q$ decreasing with q beyond a certain value of q. As shown in Figure 5, this corresponds to the existence of a maximum or a hump in the $\ln K_q$ vs q plot or a shallow minimum in the $\Delta G'_q$ vs q profile.

For the development of a general association model, it is important to note that an analysis of the $K_2'$ and K values needed to fit the molecular weight data of large micelles using Equation 17 indicates that an anticooperative region exists even though the micelles observed experimentally are very large. Some typical values for these constants are given below. For the non-ionic surfactant, $C_{16}H_{33}(OCH_2CH_2)_7OH$, $\ln K$ is 13.22 and $\ln K_2'$ is 7.88, whereas for the ionic surfactant $C_{16}H_{33}N(CH_3)_3Br$ in 0.178M KBr, $\ln K$ is 10.5 and $\ln K_2'$ is 2.17.[5] Figure 5 shows a schematic relation between the $\ln K_q$ vs q curve and the $\ln \beta_q$ vs q curve, $\ln \beta_q$ representing the area under the curve of $\ln K_q$ vs q from 2 to q (Equation 3). Disregarding the uncertainty regarding the relative values of the first two or three constants, $\ln K_2$, $\ln K_3$, etc., which are related to the problem of premicellar aggregation,[3] the true $\ln K_q$ vs q curve must have a slowly rising initial part denoting cooperativity and a linear, horizontal part at very high values of q where $\ln K_q = \ln K$. The replacement of curve A ($\ln \beta_q$ vs q) by curve B to obtain the simple two-parameter model is equivalent to the replacement of the true $\ln K_q$ vs q curve with the two-step curve Q, involving $\ln K_2'$ and a constant value of $\ln K$ for q values of 3 and higher. For large micelles, the area under Q must equal the area under the true $\ln K_q$ vs q curve. Choosing $C_{16}H_{33}(OCH_2CH_2)_7OH$ as an example, the area under Q is 6591 when q = 500, and this must be close to the value of $\ln \beta_{500}$. We are thus dealing with very large numbers. If the ture $\ln K_q$ vs q curve shows no anticooperative region, then it must have a shape corresponding to the curve R, $\ln K_q$ increasing gradually to a value of $\ln K$. In such a case, the area between Q and R within the range of q = 3 to very high values of q must be balanced by the difference between the true $\ln K_2$ and the model parameter $\ln K_2'$. Considering that $\ln K_2'$ is typically only a few units less than $\ln K$, it is extremely unlikely that the difference between $\ln K_2$ and $\ln K_2'$ can be enough to provide the necessary initial cooperativity for large micellar systems. It seems, therefore, that for large micelles also, the shape of $\ln K_q$ vs q curve involves a maximum or a humt (curve P, Figure 5) with a cooperative region preceding it, and an anticooperative region following it. The cross-hatched areas between Q and P cancel to a large extent, allowing a very substantial amount of initial cooperativity to exist. Curve A is a rough representation of the $\ln\beta_q$ vs q curve corresponding to curve P. The anticooperative region of the $\ln K_q$ vs q curve is expected to have some important consequences for the kinetics of formation and growth of large micelles.

## SPHERE-ROD TRANSITION AND A GENERAL ASSOCIATION MODEL

A generally applicable qualitative association model for flexible chain surfactants can now be presented incorporating the three features of the $\ln K_q$ vs q curve derived earlier, an initial cooperativity ($\ln K_q$ increasing with q) up to a maximum, a region of anticooperativity where $\ln K_q$ decreases with q, and finally, at high values of q, $\ln K_q$ becoming independent of q. Figure 6 shows in a rough schematic fashion how such a curve is generated. Very large micelles are assumed to be rod-like although disc-like micelles are not incompatible with the model. As q increases, the micelles go through three regions of growth: (a) a growth of small, roughly spherical micelles, (b) a transition region in which roughly spherical micelles grow into small rod-like micelles with end effects progressively getting less important, and, finally (c) large rod-like micelles where end effects are negligible. For qualitative

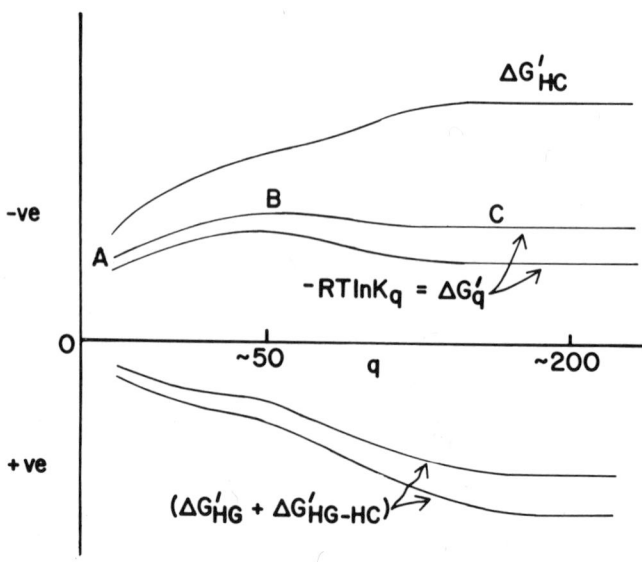

Figure 6. Schematic representation of the general association model for micelle formation of flexible chain surfactants.

reasoning, we assume further that the rods (cylinders) have roughly the same radius as the largest sphere of the above growth region (a). For the hydrophobic contribution, $\Delta G'_{HC}$, two important considerations are the loss in total hydrocarbon-water interface area when a monomer is added to a micelle, as also the amount of volume available to the monomer in the micelle, which is important for the freedom of movement of the chain and its entropy. These considerations indicate that $\Delta G'_{HC}$ becomes more favorable in the growth region (a) as q increases as also between spheres and rods [region (a) to region (c)]. Including the transition region (b), therefore, $\Delta G'_{HC}$ is expected to vary schematically in the manner shown in Figure 6.

The repulsive interactions $\Delta G'_{HG}$ and $\Delta G'_{HG-HC}$ are also dependent on q and the micelle shape. Model calculations show[32] that both for ionic and non-ionic surfactants $\Delta G'_{HG}$ is higher for the rods than for spheres of the same radius as the rods. This is principally due to differences in the density of the head groups at the micelle surface. The surface-to-volume ratio of spheres of radius r, segments of unidorm cylinders of radius r and segments of uniform discs of thickness 2 r are 3/r, 2/r, and 1/r respectively. Thus, for a representative case applicable roughly to hexadecyl chains, if a sphere of radius 2nm (20A°) contains 70 monomers, the surface area per monomer is $0.718 nm^2$ (71.8 sq.A°). The corresponding areas per monomer for a cylinder of 2nm radius and a disc of 4nm thickness are $0.479 nm^2$ (47.9 sq.A°) and $0.239 nm^2$ (23.9 sq.A°) The latter value is indeed so low that it becomes extremely difficult to accommodate moderate sized head groups, and these small surface areas argue strongly against disc-like micelles. The transition from the sphere to the cylinder is also accompanied by a reduction in the curvature which also plays an important role in increasing $\Delta G'_{HG}$ for both ionic and non-ionic surfactants.[32] Since the growth of small, spherical micelles [region (a)] also involves decreasing surface area per monomer, and decreasing curvature, the repulsive interactions are expected to increase in magnitude with q in roughly the manner exhibited in Figure 6, which includes the sphere-to-rod transition region.

The sum of the attractive and repulsive interactions generates the $lnK_q$ curve which now, typically, has three regions shown in Figure 6, a region of cooperativity (AB), a region of anticooperativity (BC), and a horizontal region above C. Several consequences of this general model have been referred to earlier. The absolute values of $ln_q$ [and the corresponding $(ln\beta_q)/q$ or $\Delta G_q$] are reflected in the c.m.c. The sharpness of the c.m.c. is produced by the cooperative region AB. When very large micelles form, the area under the $lnK_q$ vs q curve up to C in comparison with the area obtained by extending the horizontal line beyond C down to q = 2 fixes the value of $lnK_2^1$ (Equation 17 and Figure 5).

Some additional consequences of this general qualitative model are of interest and provide considerable insight into micellar systems generally. One of the most important is that the range and magnitude of the anticooperative region BC arising from the sphere-to-rod transition dictates as to whether the micelles will be small or large. Figure 1 illustrates this with model calculations. With the assumed functions of $\ln\beta_q$ and the corresponding $\ln K_q$, which shows only a cooperative region and an anticooperative region, the size distribution curves of A, B, and C are calculated. These simulate typical small micellar systems. The size distribution curves are fairly narrow. The c.m.c. of this system is about $4.1 \times 10^{-5}$ moles dm$^{-3}$. As the ratio of the total micellar concentration to the c.m.c. changes from roughly 0.1 (curve A) to 11 (curve B) to $1.6 \times 10^3$ (curve C), the curves maintain roughly the same shape, and the average aggregation number increases slightly, as is typical of narrow distributions (Equation 14).

If the $\ln K_q$ curve now is allowed to become independent of q at any value of q somewhat above the maximum (Figure 1), the curve will be similar to the general model proposed in Figure 6. When this is done at q = 116, i.e. if lnK above q = 116 equals $\ln K_{116}$, the micelles containing 116 or more monomers make less than a 2% contribution to the total distribution for curve C and less at lower micellar concentrations. For this case, therefore, up to some $10^3$ times the c.m.c., the size distribution is not affected significantly by the contribution from the large micelles and the distribution remains narrow and typical of small micelles. This is equivalent to stating that the transition from the sphere to the large rod is not favorable enough for the latter to form, and that the hump of the lnq vs q curve around the maximum B in Figure 6, corresponding to the depth and range of the generally shallow minimum in the free energy profile ($\Delta G'_q$ vs q) in Figure 6 'protects' the narrow distribution, i.e. keeps it from getting wide. Thus, if the point C in Figure 6 corresponds to q = 116 or any higher value in Figure 1, the narrow distribution of the micelles is not affected.

On the other hand, if the point C of Figure 6 is moved down to q = 96 in Figure 1, i.e. if lnK above q = 96 is $\ln K_{96}$, then the micelles become large close to the c.m.c. When the micellar concentration is about 2.7 times the c.m.c., $N_w$ has the calculated value of 628. As the micelle concentration increases to 30 times the c.m.c., $N_w$ increases to about 2000. Such a case corresponds to the situation of large micelles where the hump in $\ln K_q$ vs q curve (Figure 5) is not large enough to 'protect' the narrow distribution of small micelles, i.e. prevent the open-ended growth of very large micelles.

The transition between 'small' and large micelles is thus primarily controlled by the anticooperative region BC in Figure 6.

Whether a particular system exhibits 'small' micelles or large micelles, therefore, depends on the interplay of the attractive interactions, $\Delta G'_{HC}$, and the repulsive interactions $\Delta G'_{HG}$ and $\Delta G'_{HG-HC}$ (Figure 6). Model calculations[32] indicate that for a given chain, a reduction in the number of oxyethylene groups for a non-ionic surfactant or an increase in the ionic strength for an ionic surfactant decreases the repulsive interactions relatively more for large rod-like micelles (q values above C in Figure 6) than for spherical micelles (q values corresponding roughly to B in Figure 6). Such changes, therefore, tend to reduce the height of the anticooperative region BC and promote the formation of large micelles, as has been observed experimentally.[20,28,33] Figure 6 shows two such repulsive curves and the resultant $RT\ln K_q$ vs q curves schematically. Transitions between small and large micelles can also be brought about by changing counterions to ones which show stronger specific binding to the micelle surface.[33] Here the effect of specific binding in reducing $\Delta G'_{HG}$ is expected to be more pronounced for large rods than for spheres because of the higher surface charge density for the former.

The extremely sensitive dependence of the transition of small to large micelles on the nature of the anticooperative region BC in Figure 6 is illustrated further by some model calculations in Figure 1. It has been already indicated that for this particular $\ln K_q$ vs q profile, depending upon whether the lnK curve is assumed to become horizontal (independent of q) at q = 96 or q = 116 or higher, very large micelles or small micelles form. Curves C, $C_1$, and $C_2$ are calculated assuming the same monomer concentration. For C, the anticooperative region is assumed to extend to q = 130 according to the $\ln K_q$ vs q curve in Figure 1. For $C_1$ and $C_2$, $\ln K_q$ is assumed to become independent of q at q = 103 and q = 102 respectively. The contribution of the larger micelles changes very markedly for this minor difference in the extent of the anticooperative region. Curves $A_1$, $B_1$, and $C_1$ show the effect of changing the monomer concentration on the relative contribution of the larger micelles, assuming $\ln K_q$ becomes independent of q at q = 103. Systems of this sort can exhibit a transition from small to large micelles over a range of concentration. Such transitions have also been reported.[34]

Several additional points of interest regarding micellar size distirbutions can be derived from the comparison of $\ln K_q$ vs $(\ln \beta_q)/q$ curves in Figure 1. This comparison is equivalent to the comparison of $\Delta G'_q$ for stepwise association (Equation 5) and the average value $\Delta G_q$ (Equation 4). Figure 1 shows a model calculation in which $\ln K_q$ has a maximum at q = 50 whereas $(\ln \beta_q)/q$ has a maximun at q = 100. This separation of the two maxima for small micelles is expected to be a general phenomenon arising from the shape of the $\ln K_q$ curve with its cooperative and anticooperative portions. As shown in curves A, B, and C of Figure 1, the size-distribution curves have average aggregation numbers close to 100, i.e. close to the maximum in $(\ln\beta_q)/q$ curve. This is also expected generally. Thus a sphere-to-rod transition and the commencement of an anticooperative region at q = 50 is not incompatible with a <u>narrow</u> size distribution of larger micelles. Considerations such as these indicate that 'small' micelles need not be spherical to have a narrow distribution and that the average aggregation numbers of small micelles may be considerably higher than the aggregation numbers of the largest possible spherical micelle. A representative case in point here is that of sodium dodecyl sulfate in 0.3M NaCl, as discussed earlier.

On the other hand, many micellar aggregation numbers are consistent with spherical micelles of radius roughly equaling the length of the monomer. Sodium dodecyl sulfate in water, for example, has an $N_w$ value of 57[19] which is consistent with such a spherical micelle. The qualitative relation between the position of the maximum in $\ln K_q$ with that of $(\ln\beta_q)/q$ and the maximum of the size distribution curve indicates that for such systems $\ln K_q$ has a maximum at a q value lower than that of the spherical micelle with a radius equaling the monomer chain length. The relative positions of the maxima of $\ln K_q$ and $(\ln\beta_q)/q$ depend upon the nature of the cancellation of the $\Delta G'_{HC}$ with the repulsive interactions (Figure 6). The frequently observed effect of increasing counter-ion concentration on increasing the $N_w$ values of small micelles is thus readily understandable in terms of a decreasing $\Delta G'_{HG}$ which shifts the maximum in the $(\ln\beta_q)/q$ to higher values of q. Such changes were investigated some years ago by Stigter and Overbeek using a spherical model for ionic micelles.[35] For several ionic surfactant systems with very long chains, e.g. n-octadecyl tributylammonium bromate, McDowell and Kraus[36] found that the equivalent conductance <u>rises</u> sharply at a critical concentration, and then goes through a maximum before decreasing. The rise in conductance indicates the presence of very small micelles. Stigter[37] has analyzed these data to indicate a polydispersity of these very small micelles. In this case, $\Delta G'_{HG}$, because of the extremely low c.m.c. values, i.e. low ionic strength, increases very rapidly with size for small spherical micelles,[32] resulting in a maximum of $(\ln\beta_q)/q$ at very low q values, which is highly sensitive to ionic strength. Systems of this sort may require a range of concentrations before narrow distributions of micelles of moderate size are reached.

An additional important point is that because of the general shape of the $\ln K_q$ curve, $(\ln\beta_q)/q$ is usually only slightly lower than $\ln K_q$ at the peak of the micellar size-distribution curve. The model calculations in Figure 1 show an example where the $\ln K_q$ and the $(\ln\beta_q)/q$ curves cross at a q-value somewhat higher than the maximum of the size distribution curves. If, as discussed earlier, the $\ln K_q$ curve is assumed to become independent of q at some value of q to conform with the general association model of Figure 6, the difference between the $\ln K_q$ and the $(\ln \beta_q)/q$ curves at high values of q is reduced.

The similarity of the $(\ln \beta_q)/q$ and $\ln K_q$ values near the peak of the size distribution curves is implicit in the experimental determination of the corresponding free energy quantities from c.m.c. data. For a nonionic system, from Equations 2 and 4,

$$\frac{\Delta G}{RT} = -\frac{\ln[b_q]}{q} + \ln[b_1]_{c.m.c.} \tag{19}$$

For approximate purposes it is customary to calculate $[b_q]$ by assuming monodispersity, and assuming that the micellar concentration $q[b_q]$ is roughly 2% of the c.m.c. and $[b_1]$ equals the c.m.c.[11] On the other hand, at the peak of the micellar size distribution, i.e. for the 'most probable micelle,'[38] $[b_{q-1}] = [b_q]$ is a reasonable approximation.[39] Using Equations 1 and 5, therefore, we have

$$\frac{\Delta G'}{RT} = \ln[b_1]_{c.m.c.} \tag{20}$$

Since, when q values are large (50-100), the micellar term $(\ln[b_q])/q$ is typically 1 or 2% of $\ln[b_1]_{c.m.c.}$, when mole fraction concentration scales are employed, $\Delta G$ is also typically 1 or 2% lower in absolute magnitude than $\Delta G'$ on the unitary scale. This similarity in the magnitude of the average $\Delta G$ and $\Delta G'$ corresponding to stepwise addition in many micellar systems is a direct consequence of the postulated shape of the lnq vs q curve in the general association model and is likely to be useful in evaluating and formulating detailed theories of micelle formation and models they are based on.

## REFERENCES

1. P. Mukerjee, J. Pharm. Sci., 63, 972 (1974).
2. P. Mukerjee, in "Physical Chemistry: Enriching Topics from Colloid and Surface Science," IUPAC, H. van Olphen and K. J. Mysels, Editors, pp. 135-153, Theorex, La Jolla, CA (1975).
3. P. Mukerjee, Adv. Colloid Interface Sci., 1, 241 (1967).
4. A. K. Ghosh and P. Mukerjee, J. Am. Chem. Soc., 92, 6408 (1970).
5. P. Mukerjee, J. Phys. Chem., 76, 565 (1972).
6. P. Mukerjee and A. K. Ghosh, J. Am. Chem. Soc., 92, 6419 (1970).
7. W. Kauzmann, Advan. Protein Chem., 14, 1 (1959).
8. G. Nemethy, Angew. Chem., Intern Ed., 6, 195 (1967).

9.  C. Tanford, "The Hydrophobic Effect," John Wiley, 1973.
10. J. T. Davies and E. K. Rideal, "Interfacial Phenomena," 2nd ed., Academic Press, NY, 1963.
11. P. Mukerjee, Kolloid-Z.u.Z. Polymere, 236, 76 (1970).
12. P. Mukerjee and K. J. Mysels, "Critical Micelle Concentrations of Aqueous Surfactant Systems," NSRDS-NBS-36, U.S. Govt. Printing Office, Washington, DC (1971).
13. G. Scatchard, W. J. Hamer and S. E. Wood, J. Am. Chem. Soc., 60, 3061 (1938).
14. P. H. Elworthy and A. T. Florence, Kolloid-Z.u.Z. Polymere, 208, 157 (1966).
15. P. Mukerjee and A. Ray, J. Phys. Chem., 70, 2144 (1966).
16. D. Stigter, J. Phys. Chem., 78, 2480 (1974).
17. F. P. Buff and N. S. Goel, J. Chem. Phys., 56, 2405 (1972).
18. K. J. Mysels, P. Mukerjee and M. Abu-Hamdiyyah, J. Phys. Chem., 67, 1943 (1963).
19. F. Huisman, Proc. Kon. Ned. Akad. Wetensch., Ser. B., 67, 388, 407 (1964).
20. N. A. Mazer, G. B. Benedek, and M. Carey, J. Phys. Chem., 80, 1075 (1976).
21. R. J. M. Tausk and J. Th. G. Overbeek, Biophys. Chem., 2, 175 (1974).
22. R. R. Balmbra, J. C. Clunie, J. M. Corkill, and J. F. Goodman, Trans. Faraday Soc., 60, 979 (1964).
23. D. Stigter and K. J. Mysels, J. Phys. Chem., 59, 45 (1955).
24. C. A. J. Hoeve and G. C. Benson, J. Phys. Chem., 61, 1149 (1957).
25. P. Debye and E. W. Anacker, J. Phys. Colloid Chem., 55, 644 (1951).
26. R. J. M. Tausk, C. Oudshoorn, and J. Th. G. Overbeek, Biophys. Chem., 2, 53 (1974).
27. D. Stigter, J. Phys. Chem., 70, 1323 (1966).
28. P. H. Elworthy and C. B. Macfarlane, J. Chem. Soc., 907 (1963).
29. J. M. Corkill, J. F. Goodman, and T. Walker, Trans. Faraday Soc., 63, 759 (1967).
30. C. Tanford, J. Phys. Chem., 78, 2469 (1974).
31. P. Mukerjee and J. R. Cardinal, J. Pharm. Sci., 65, 882 (1976).
32. P. Mukerjee (1973), unpublished work.
33. E. W. Anacker and H. M. Ghose, J. Am. Chem. Soc., 90, 3161 (1968).
34. F. Reiss-Husson and V. Luzzati, J. Phys. Chem., 68, 3504 (1964); J. Colloid Interf. Sci., 21, 534 (1966).
35. D. Stigter and J. Th. G. Overbeek, in "Proc. 2nd Int. Cong. Surface Activity," Volume 2, p. 311, Butterworth, London, 1957.
36. M. J. McDowell and C. A. Krans, J. Am. Chem. Soc., 73, 2173 (1951).
37. D. Stigter, Rec. trav. chim., 73, 611 (1954).
38. R. H. Aranow, J. Phys. Chem., 67, 556 (1963).
39. P. Mukerjee, J. Phys. Chem., 73, 2054 (1969).

IONIC INTERACTIONS IN AMPHIPHILIC SYSTEMS STUDIED BY NMR

Björn Lindman, Göran Lindblom, Håkan Wennerström and
Hans Gustavsson
Division of Physical Chemistry 2

Chemical Center, P.O.B. 740, S-220 07 LUND 7, Sweden

The interaction of small ions with aggregates formed by amphiphile ions may be conveniently followed by magnetic resonance methods which give information on the mode of counterion-amphiphile interaction, the fraction of counterions bound, the mobility of the counterions and on hydration phenomena. Different types of applications of NMR to ion binding studies are reviewed and the results compared with those obtained by other methods. Principally the alkali and halide ions were investigated and both continuous-wave and pulse methods were utilized. Fourier transform techniques were used in certain cases to obtain improved sensitivity. NMR parameters considered include quadrupole relaxation, quadrupole splitting, shielding, water isotope effects and translational diffusion. Both quadrupole relaxation, quadrupole splitting and shielding demonstrate a marked specificity in the alkali ion binding to simple surfactant aggregates, the mechanism of interaction being different for different polar end-groups. On the other hand, for a given polar head the counterion binding is relatively independent of alkyl chain length, amphiphile concentration and phase structure. The water isotope effect in the shielding of $^{133}Cs^+$ and $^{37}Cl^-$ was used to probe into counterion hydration. It is found that the counterions retain their inner hydration sheaths over wide concentration ranges. Lithium ion translational motion, which can be conveniently studied by the NMR spin-echo technique, is shown to give certain information on phase structure.

## INTRODUCTION

The self-association behavior of ionic amphiphiles differs markedly from that of non-ionics and this as well as numerous other types of observations points to the importance of ionic interactions for the properties of ionic amphiphile systems. Illuminating general analyses and discussions concerning the adsorption of counterions at the surface of micelles have been presented by Mukerjee[1-3] and by Stigter[4-7]. It appears to be a rather general result that 50% - 80% of the counterions may be considered to be held inside the shear plane in the Stern layer for micellar solutions[1,2,8]. It is a natural starting-point to analyse counterion binding in terms of the general electric double layer theory of charged interfaces but the specificity in the counterion binding as emphasized, for example, by Mukerjee[1-3] shows the need to consider various specific interactions. The problem is similar to that of counterion binding to polyelectrolytes and in this case recent articles stress the importance of specific interactions[9,10].

The interaction between small counterions and amphiphilic aggregates can depending on the purpose be characterized in different ways but from the point of view of understanding the interactions at a molecular level a description must include the following points, some of which we attempt to represent schematically in Figure 1.

i) The number of counterions bound. For the case where different types of sites at the amphiphile aggregates can be distinguished, the distribution of counterions over these sites is of interest.

ii) The type and strength of interaction between bound counterions and amphiphilie aggregate.

iii) The hydration of counterions and amphiphile ions.

iv) The mobilities of bound and free counterions.

Additionally intermicellar electrostatic interactions greatly influence the solution properties but these interactions will not be considered here.

As regards i) it should be pointed out that it is connected with difficulties to unambiguously divide the counterions into "free" and "bound". Different experimental techniques differ in this distinction and may therefore give different results for the degree of counterion binding, which we denote $\beta$ and define as the ratio of counterions and amphiphile ions in an aggregate. These problems have been discussed by Mukerjee et al.[1] who also review

# IONIC INTERACTIONS IN AMPHIPHILIC SYSTEMS

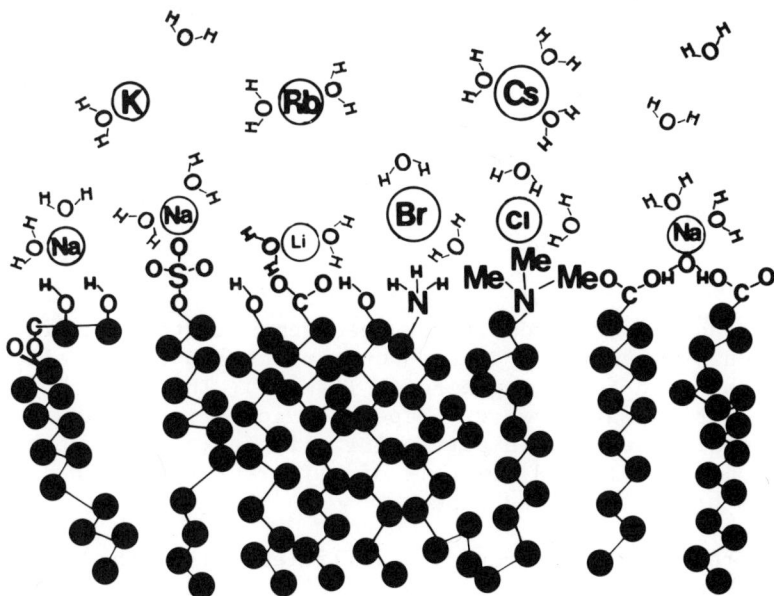

Figure 1. Schematic illustration of the surface of an amphiphile aggregate showing some of the end-groups and counterions considered in the text.

different methods to determine $\beta$. It was found that electric conductivity, light scattering and variation of critical micelle concentration (CMC) with counterion concentration give somewhat different results. A further method which will be considered below involves the study of the translational diffusion of the counterion. Since in the interpretation of the various NMR parameters of the counterions, the degree of counterion binding will be needed the ambiguity mentioned will cause some difficulty. The counterion binding values obtained from translational diffusion, which are based on considering those counterions as bound which move with the micelles, are probably well applicable to counterion quadrupole relaxation rates while their relation to quadrupole splittings and shieldings is unclear. Apparently, for detailed considerations this problem has to be thoroughly examined. On the other hand, as judged from the variation of $\beta$ with the method employed qualitative deduction may certainly be made. Furthermore, for the comparison of a certain NMR parameter between different surfactants, phases etc., the error introduced by the uncertainty in $\beta$ will probably be quite small.

For point ii) we have to consider the electrostatic interaction

of the counterion both with the amphiphile aggregate as a whole and
the local interaction with the particular amphiphile polar head.
For the latter question it is of particular interest to compare
different counterions as a simple electrostatic approach would
give the interaction energy to be directly related to the distance
of closest approach and thus to the effective counterion radius.
As regards overall and local effects the overall effects may through
aggregate charge density and shape be responsible for the approximate counterion distribution while local effects may slightly modify this.

For a number of cases it has been demonstrated that it is the
interaction between the hydrated counterions and the micelle that
is important (see for example Refs. 1 and 4). The degree of counterion hydration may, however, depend on ionic radius and polar
head of amphiphile and, in particular, little is known about hydration phenomena in liquid crystals of different types and in micellar
aggregates of the reversed type.

Concerning iv) different possibilities exist: The bound counterion may remain attached to a particular amphiphile ion for a
certain time or it may diffuse rapidly along an equipotential surface. The mobility of free ions may be influenced by obstruction
effects etc.

We have here given an enumeration of questions (an enumeration
which is by no means exhaustive) which have to be answered before
we have a detailed understanding of the ionic interactions in amphiphile systems. Whereas many previous studies of counterion binding phenomena are of an indirect nature and, therefore, their
bearing on specific molecular interactions in many cases not quite
clear, direct spectroscopic observations should be valuable in this
context. Nuclear magnetic resonance, allowing the direct observation of all the alkali and halide ions, as well as of many other
possible counterions, comes naturally in the foreground because of
its wide range of applicability and its negligible perturbation
of the sample. In our laboratory we have during the last 10 year
period been applying a number of NMR principles for the elucidation
of ion binding phenomena in various systems and the present article
reviews the study of ion binding in surfactant systems by NMR. An
essential part of this work has concerned NMR methodological and
theoretical aspects. These will not be considered here but may be
found in the original work cited and in Ref. 11 reviewing most of
the NMR background. An important incentive for these studies has
been our interest in molecular interactions at interfaces in biological systems and our work has included studies of transport proteins, enzymes, membranes and biological polyelectrolytes. While
the biological aspects will not be described here it should be clear
that many results to be given below may be helpful in discussions

of the mechanisms of certain processes in living systems. We may
mention ion passage through membranes, activation of enzymes, association and transport properties of biological amphiphiles, ion
binding specificity of polyelectrolytes such as nucleic acids and
mucopolysaccharides etc.

## FEATURES OF THE PROGRESSIVE ASSOCIATION OF AMPHIPHILES

Before going into a detailed discussion of the ion binding
phenomena it may be suitable to survey briefly the different phases
and aggregates which have been studied. In aqueous solutions of an
amphiphile there is with increasing concentration a progressively
developing association. As demonstrated by Mukerjee et al.[12] and by
Ekwall[13], micelle formation is generally preceded by the formation
of smaller aggregates and in recent years this aspect of amphiphile
self-association has been characterized in detail by Danielsson and
co-workers[14,15]. Micelle formation is both from experimental findings and from considerations of the mass action law found to start
at a fairly sharp concentration for long-chain amphiphiles while
no well-defined critical micelle concentration (CMC) may be given
for short-chain amphiphiles. The problems of micelle size distributions and the development of micelle shape and size with concentration have been treated by Mukerjee[16] and Tanford[17]. In a number of
cases a transformation from globular to very long rod-like micelles
has been found as the amphiphile concentration is increased and it
has also been observed that solubilization may induce considerable
micelle shape alterations.

In the absence of water it seems that amphiphilic compounds
generally form small complexes in an organic solvent[18,19] while in
the presence of water, aggregates of colloidal dimensions may form[20].
These aggregates are of the reversed type with the polar heads
pointing towards the center where water molecules and counterions
are located while the non-polar ends are directed outwards into the
intermicellar solution.

A characteristic general feature of amphiphile-water systems
is the formation of mesomorphous phases in an intermediate concentration range. Several different liquid crystalline structures have
been documented, the most important of which are the lamellar,
hexagonal and cubic phases. Hexagonal and cubic phases are of at
least two types with either normal or reversed aggregates. For
three-component systems ionic surfactant-water-organic solvent,
phase equilibria may become quite complex with a large number of
different mesophase regions in the phase diagram. The pioneering
work of Ekwall and co-workers in this field has recently been summarized[21]. As an illustrative example we reproduce in Figure 2 the
phase diagram of the classic study of the system sodium octanoate-
-decanol-water and here also the phase structures are outlined.

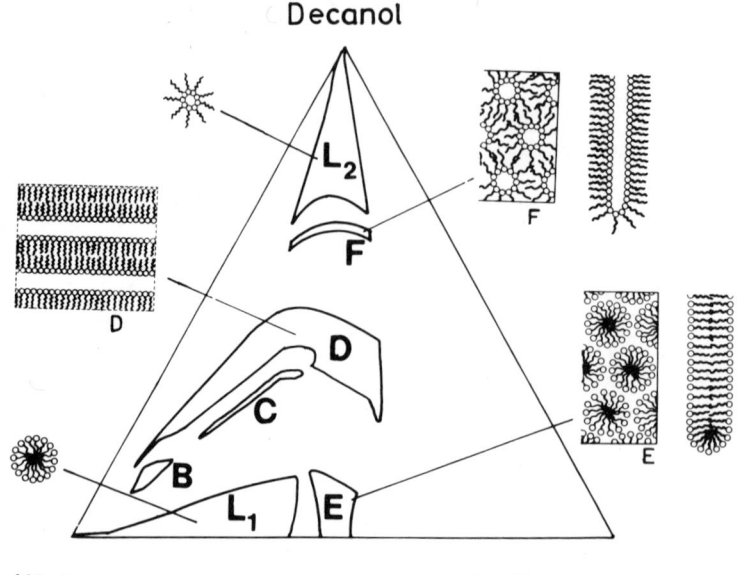

Figure 2. Phase diagram of the ternary system sodium octanoate-decanol-water at 20°C as described by Ekwall and collaborators (cf. Ref. 21). Inserted are schematic drawings of the structures of the aggregates. $L_1$ and $L_2$ denote isotropic solution phases and B, C, D(lamellar), E(hexagonal) and F(reversed hexagonal) regions with liquid crystalline phases.

From the point of view of counterion binding it is significant to note that the presence and extension of the various phases may vary markedly with the counterion. For example, in the systems alkali octanoate-decanol-water the minimal water contents parallel the hydration needs of the counterions suggesting the importance of counterion hydration[22]. As regards counteranions, formation of cubic and hexagonal mesophases may be completely different for chloride and bromide.

## SOME GENERAL ASPECTS ON THE NMR PROPERTIES OF ALKALI AND HALIDE IONS

With the exception of fluorine, the alkali and halogen nuclei form a group with very similar magnetic resonance properties. Therefore, principles of interpretation and fields of application will be essentially the same in all cases. The possibilities in NMR studies may be discussed on the basis of Table I where we have

Table I. Selected Properties of the Alkali and Halogen Nuclei.

| Nucleus | Spin | Frequency at 2.35 T MHz | Relative sensitivity[a] | Natural abundance % | Quadrupole moment $\cdot 10^{28}$ $m^2$ | $T_1^{-1}$ in water[b] $s^{-1}$ | Range of $\sigma$ values[c] ppm | Ionic radius[d] |
|---|---|---|---|---|---|---|---|---|
| $^6$Li   | 1   | 14.71 | 0.0085  | 7.43  | 0.00046 |       |     |      |
| $^7$Li   | 3/2 | 38.87 | 0.294   | 92.57 | -0.03   | 0.055 |     | 0.60 |
| $^{23}$Na | 3/2 | 26.46 | 0.0925  | 100   | 0.14    | 16.2  | 5   | 0.95 |
| $^{39}$K  | 3/2 | 4.67  | 0.00051 | 93.08 | 0.11    | 24    | 15  | 1.33 |
| $^{41}$K  | 3/2 | 2.57  | 0.00008 | 6.91  |         |       |     |      |
| $^{85}$Rb | 5/2 | 9.66  | 0.0105  | 72.8  | 0.27    | 420   |     | 1.48 |
| $^{87}$Rb | 3/2 | 32.72 | 0.175   | 27.2  | 0.13    | 100   | 50  |      |
| $^{133}$Cs | 7/2 | 13.12 | 0.0474 | 100   | 0.003   | 0.075 | 80  | 1.69 |
| $^{35}$Cl | 3/2 | 9.80  | 0.0047  | 75.4  | 0.0789  | 25    | 30  | 1.81 |
| $^{37}$Cl | 3/2 | 8.16  | 0.0027  | 24.6  | 0.0621  | 15    |     |      |
| $^{79}$Br | 3/2 | 25.06 | 0.0786  | 50.57 | 0.33    | 1460  |     | 1.95 |
| $^{81}$Br | 3/2 | 27.01 | 0.0985  | 49.43 | 0.28    | 1050  | 70  |      |
| $^{127}$I | 5/2 | 20.01 | 0.0934  | 100   | -0.69   | 5270  | 100 | 2.16 |

a. Relative to $^1$H at equal number of nuclei and at constant field.
b. At infinite dilution. From Refs. 11 and 23.
c. In aqueous alkali halide solutions. From Refs. 11, 24 and 25.
d. In units of $10^{-10}$ m. From Ref. 26.

listed the general NMR properties and also some pertinent observations of the NMR parameters of the ions. As can be seen all the nuclei have spin quantum numbers $I > 1/2$ and therefore possess electric quadrupole moments. The interaction of these quadrupole moments with the field gradients arising from ionic charges and dipole moments of molecules will in diamagnetic systems have a dominating influence on the appearence of the NMR spectrum. In isotropic phases the fluctuation of the quadrupole interaction constitutes a very effective relaxation mechanism which will determine the NMR line width. In anisotropic phases, the quadrupole interaction also causes a splitting of the NMR signal. The location of the signal (for quadrupole-split signals the central peak) will be determined by the shielding effect exerted by the electrons and this will vary with the molecular interactions.

The possibility of studying the NMR signal will be determined by the intrinsic NMR sensitivity considering also the natural abundance of the isotope and by the line broadening characteristic of the sample. Line broadening, which in the cases considered mainly is due to rapid quadrupole relaxation and static quadrupole effects in an anisotropic phase, will cause the intensity to spread out and the amplitude to decrease and thus limit the concentration range available for study.

As can be inferred from Table I, and from experience, the NMR sensitivity is sufficient for a wide range of applications in the surfactant field for all ions except potassium which therefore has been studied only to a very limited extent. As an example it is possible for isotropic solutions to study $Na^+$ concentrations below $10^{-3}$ M using the Fourier transform technique. For macroscopically oriented mesophases the requirements can be judged by recalling that for $I = 3/2$ an outer peak has 30% of the total intensity. For nonoriented or powder mesophase samples the intensity is spread out over a considerable range, which depends on the magnitude of the quadrupole splitting and therefore considerably higher concentrations are required. However, quadrupole splittings of both $Li^+$, $Na^+$, $Rb^+$ and $Cs^+$ may be conveniently obtained over wide parts of the concentration regions where liquid crystalline phases typically exist. Studies of the quadrupolar halide ions are more difficult to accomplish as a result of low NMR sensitivity in the case of chloride and very rapid relaxation in the case of the bromide and iodide ions. Quadrupole relaxation studies are feasible for $Cl^-$ and $Br^-$ if the water content is not too low while halide ion quadrupole splittings are difficult to obtain for powder samples.

In the subsequent discussion, the division of the material will be according to NMR parameters, i.e. shielding, quadrupole relaxation, quadrupole splitting and translational diffusion whereas at the end some general conclusions and comparisons will be made.

For each parameter, a short mention of the theoretical background and the experimental technique will be given while detailed descriptions are to be found in the original work. If not otherwise stated the results refer to the probe temperature which is 25-30°C depending on the experimental conditions and spectrometer employed.

Before going into the discussion of the results a general remark on the effect of chemical exchange phenomena is appropriate. In the systems studied the counterions are distributed over different sites characterized by different values of the NMR parameters. While in the limit of very slow counterion exchange between the different environments, the NMR spectrum will be a superposition of the spectra characteristic of the different counterion environments, only the weighted average value of a parameter is observable in the limit of rapid exchange. Thus for a general parameter X,

$$X = \sum_i p_i X_i \tag{1}$$

where $p_i$ is the fraction of counterions in site i. Considerations of the exchange rates necessary ($1-10^5$ $s^{-1}$ depending on nucleus and parameter considered), as well as direct investigations of the temperature dependence and the isotope effect (cf. Ref. 11), strongly indicate that the rapid exchange approximation (Equation (1)) may be generally applied for all cases considered below. Assuming a two-site model, considering the counterions either as free (f) or bound (b) gives

$$X = p_b X_b + p_f X_f = p_b(X_b - X_f) + X_f \tag{2}$$

Here $X_f$ is generally known and to obtain $X_b$, which is often the quantity of interest, $p_b$ must be independently determined or estimated (cf. above).

## SHIELDING

As a result of a shielding effect due to the electrons the effective magnetic field at the nuclear position differs from the applied field. The chemical shift, which is the difference in shielding between the sample and some reference, manifests itself as a dependence of the NMR signal position on sample composition. The definition of the chemical shift will be such that a positive shift is upfield and as reference it will be natural to take (by an extrapolation procedure) the ion at infinite dilution in water. The chemical shift was the first NMR parameter finding wide use for chemical problems but its application to ion binding phenomena in surfactant systems is of recent date, the first report[27] being published in 1973. Our studies of counterion chemical shifts in amphiphile systems have been mainly concerned with the $Na^+$, $Cs^+$ and $Cl^-$ ions. While initially we utilized continuous-wave techniques, using a wide-line spectrometer, our recent investigations were performed with Fourier transform techniques thus improving considerably sensitivity and precision.

In studying counterion chemical shifts it is found that at low concentrations the change in shielding is small but that, in the region of the CMC, the signal starts to shift markedly with increasing concentration. While keeping in mind the limitations and the approximative nature of the pseudo-phase separation model of micelle formation, we have found it most useful for a rather good estimation of the properties of the bound counterions. Thus with this model, and assuming the ratio, $\beta$, of counterions to amphiphile ions in the micelles to be independent of concentration, it may easily be shown from Equation (2) for a general parameter X that above the CMC

$$X = X_f + \beta(X_b - X_f) - \frac{\beta m_c}{m_t}(X_b - X_f) \qquad (3)$$

Here $m_c$ denotes the CMC and $m_t$ the total amphiphile concentration. Below the CMC we should have $X = X_f$. Apparently, if $X_b$ and $X_f$ are independent of concentration, a plot of X versus $m_t^{-1}$ should give two straight lines intersecting at the CMC. From the slope or intercept of the high concentration straight line segment the quantity $\beta X_b$ is obtainable and if $\beta$ may be obtained independently we have information on $X_b$. (Of course, if for sensitivity reasons, for example, CMC is not obtainable from the NMR studies, values from the literature may be employed.)

In the case of chemical shifts ($\delta$), taking the free ion as reference, Equation (3) reduces to

$$\delta = \beta\delta_b(1 - \frac{m_c}{m_t}) \qquad (4)$$

As is exemplified in Figure 3, experimental data are in good agreement with the model employed and in Table II we have listed values of CMC and of $\beta\delta_b$ for a number of surfactants[27-30]. We have also calculated values of $\delta_b$ using the $\beta$ values indicated, which were obtained from the studies of counterion translational motion (see below) or estimated (partly from studies by other methods)[31]. The CMC values obtained are in good agreement with literature data. It is clear from these results that the counterion chemical shifts may be conveniently employed for detecting changes in the binding state of the ions. A more important question is if we from the $\delta_b$ values may learn about the binding state of the ion when it is attached to the micellar aggregate. Unfortunately, our knowledge of the mechanism of chemical shifts is far from complete so a quantitative interpretation is out of question. As discussed for example in Ref. 11 the

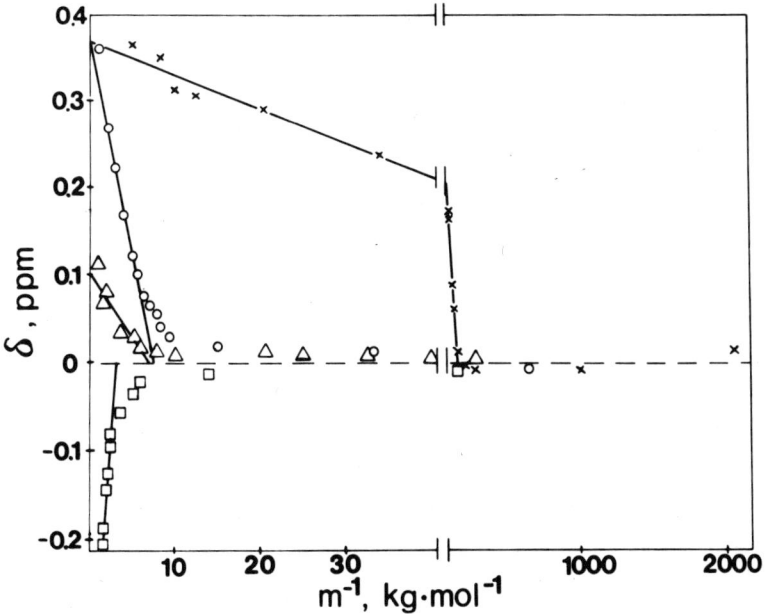

Figure 3. Plots of $^{23}Na^+$ chemical shifts versus the inverse amphiphile concentration (cf. text). Amphiphiles included are

        □ sodium octanoate
        O sodium octyl sulfate
        △ sodium octyl sulfonate
        × sodium dodecyl sulfate

Shifts are given in ppm relative to $Na^+$ ions at infinite dilution in water, with a positive shift being upfield.

Kondo-Yamashita overlap model, attributing the dominating downfield (paramagnetic) shielding term to overlap between the ion's outer p orbitals and the orbitals of other ions or molecules, has had a considerable success in a qualitative rationalization of chemical shift data for simple electrolytes in different solvents. According to this model we can write $\delta_b$ as[11]

$$\delta_b = \sigma_p - \sigma_p^o = -\frac{k}{\Delta E} <r^{-3}>_p (\Sigma A_{M-X} - A_{M-X}^o) \tag{5}$$

Here $\sigma_p$ is the paramagnetic shielding term for micellarly bound counterions and $\sigma_p^o$ that of the free counterion at infinite dilution. k is a constant, $\Delta E$ is an average excitation energy, $<r^{-3}>_p$ the expectation value of $r^{-3}$ for an outer p electron and $A_{M-X}$ appropriate sums of squares of overlap integrals between the outer p orbi-

Table II

Values of CMC, $\beta \cdot \delta_b$ and $\delta_b$ obtained from studies of the concentration dependence of the counterion chemical shift. Shifts are given relative to the ion at infinite dilution, a positive $\delta_b$ value being upfield.

| Amphiphile | Nucleus studied | CMC | $\beta\delta_b$ | $\delta_b$ ($\beta$ used) | Ref. |
|---|---|---|---|---|---|
| $CH_3(CH_2)_6COO\ Na$ | $^{23}Na$ | 0.37 | -0.43 | -0.72 (0.6) | 27 |
| $CH_3(CH_2)_7OSO_3Na$ | -"- | 0.14 | 0.37 | 0.62 (0.6) | 30 |
| $CH_3(CH_2)_8OSO_3Na$ | -"- | 0.077 | 0.37 | 0.62 (0.6) | 30 |
| $CH_3(CH_2)_9OSO_3Na$ | -"- | 0.038 | 0.37 | 0.62 (0.6) | 30 |
| $CH_3(CH_2)_{11}OSO_3Na$ | -"- | 0.0086 | 0.36 | 0.60 (0.6) | 30 |
| $CH_3(CH_2)_7C_6H_4SO_3Na$ | -"- | 0.01 | 0.23 | 0.38 (0.6) | 30 |
| $CH_3(CH_2)_6COO\ Cs$ | $^{133}Cs$ | 0.40 | -8.1 | -13.5 (0.6) | 27 |
| $CH_3(CH_2)_7NH_3Cl$ | $^{35}Cl$ | 0.27 | -10 | -12.5 (0.8) | 29 |
| $CH_3(CH_2)_7N(CH_3)_3Cl$ | -"- | 0.37 | -12.6 | -15.8 (0.8) | 29 |

tals of the ion studied (M) and the orbitals of other species (X). As the chemical shifts are sensitive primarily to direct contact effects, what is determining $\delta_b$ for hydrated ions is the difference in ion-water overlap integrals between the bound counterion and the free counterion.

As can be inferred from Figure 3, as well as from Table II, $\delta_b$ of $Na^+$ depends considerably on the polar end of the amphiphile. This demonstrates that $Na^+$ ions interact distinctly different with different amphiphiles and that there is thus a marked specificity in the counterion binding. As has already been mentioned, and as will be further discussed below, the sodium ion can generally be considered to retain its inner hydration sheath on binding to a micelle. Apparently then, the overlap integrals between bound $Na^+$ and O of $H_2O$ vary markedly with the head group of the amphiphile, the overlap being in the sequence $-CO_2^- > H_2O$ (free ion) $> SO_3^- > SO_4^-$. It seems that in the case of $-CO_2^-$ the polarization of the water molecules in the first hydration layer is greater than for the free ion while it is less for $Na^+$ interacting with $-SO_4^-$. The most plausible interpretation of such polarization effects seems to be in terms of hydrogen bonding, i.e. the hydrogen bonding of $H_2O$ in the first hydration sheath follows the sequence $-CO_2^- > H_2O > SO_3^- > SO_4^-$. The involvement of such hydrogen-bonding is interesting for the understanding of counterion specificity and has already been considered to explain relative counterion affinities in different systems. Apparently the counterion shielding offers a convenient way of studying counterion binding mechanisms.

Of other results obtained from counterion shielding studies the following may be mentioned:

a) There are no marked changes in the $Na^+$ interaction at the transitions from normal to reversed micelles in the systems sodium cholate-decanol-water and sodium octanoate-octanoic acid-water[29].

b) Upfield $^{23}Na^+$ shifts were observed[29] in the isotropic solution phase rich in octanoic acid in the last-mentioned system, suggesting the involvement of the acid in the counterion binding aggregates This correlates well with previous findings[32].

c) According to the $\delta_b$ values given in Table II the interaction of the micellarly bound counterions is essentially independent of the length of the non-polar chain.

d) Increases in the chemical shift above the predictions of the simple model have been noted at higher concentrations in a number of cases[28,30] suggesting changes in counterion binding corresponding to increases in either $|\delta_b|$ or $\beta$.

## WATER ISOTOPE EFFECT IN SHIELDING

It has been known for some time that the shielding of many ions is markedly greater with $D_2O$ as a solvent than with $H_2O$. (For a review see Ref. 11.) For example this water isotope effect in shielding is at low salt concentrations 4.4 ppm for $Cl^-$ and 1.5 ppm for $Cs^+$. The effect can probably be mainly referred to an isotope effect in the mean excitation energy which is supported by UV spectroscopic investigations. Even without a detailed analysis of the effect, which has not yet been achieved, it should be most useful as a sensitive probe of counterion hydration phenomena. Thus if the water isotope effect remains unchanged on attachment of a counterion to an aggregate, an essentially unchanged ion-water contact can be assumed, while in the case of a reduction of the effect a partial dehydration is suggested.

Our investigations of the water isotope effect in shielding are still in an initial stage but preliminary studies for a number of systems have confirmed the utility of the method. According to the $^{133}Cs$ chemical shift data given in Figure 4 for $H_2O$ and $D_2O$ solutions of cesium octanoate there is no significant reduction of the water isotope effect even at quite high amphiphile concentrations. Consequently, no detectable elimination of the counterion hydration on binding of the ion to a micellar aggregate can be

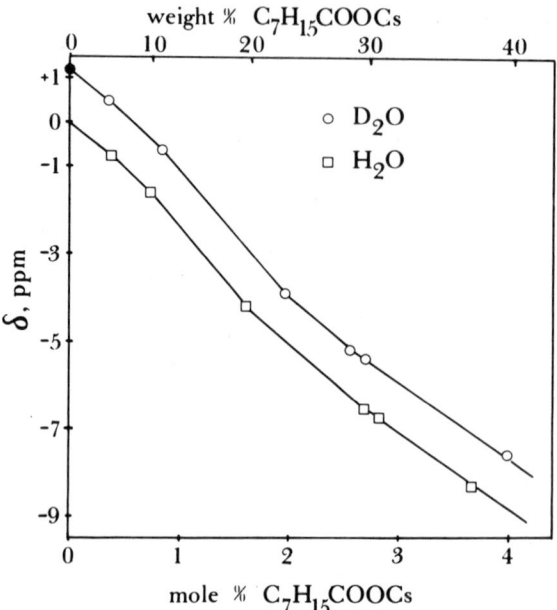

Figure 4. $^{133}$Cs chemical shifts (in ppm) of H₂O (□) and D₂O (○) solutions of cesium octanoate as a function of amphiphile concentration. (From Ref. 29.)

inferred for this system. On the other hand, a small reduction of the isotope effect has been observed at higher concentrations of cesium dodecanoate and dodecyltrimethylammonium chloride. A preliminary study of the isotropic solution phase rich in octanoic acid of the three-component system cesium octanoate-octanoic acid-water has given the results shown in Figure 5. As can be inferred there is a slow elimination of cesium-water contact with decreasing water concentration over a wide concentration range while the final dehydration starts to take place rather abruptly as the number of water molecules per counterion falls below 4.

## QUADRUPOLE RELAXATION

Nuclear magnetic relaxation, i.e. the return of a system of nuclear spins to the equilibrium state after a perturbation, is effected by time-dependent interactions involving the nuclear spins. For the alkali ions, as well as for $Cl^-$, $Br^-$ and $I^-$, the generally dominating interaction is that between the nuclear electric quadrupole moment and electric field gradients sensed by the nucleus. Studies of quadrupole relaxation may be performed in different ways. Our early studies used continuous-wave spectra recorded on a wide-

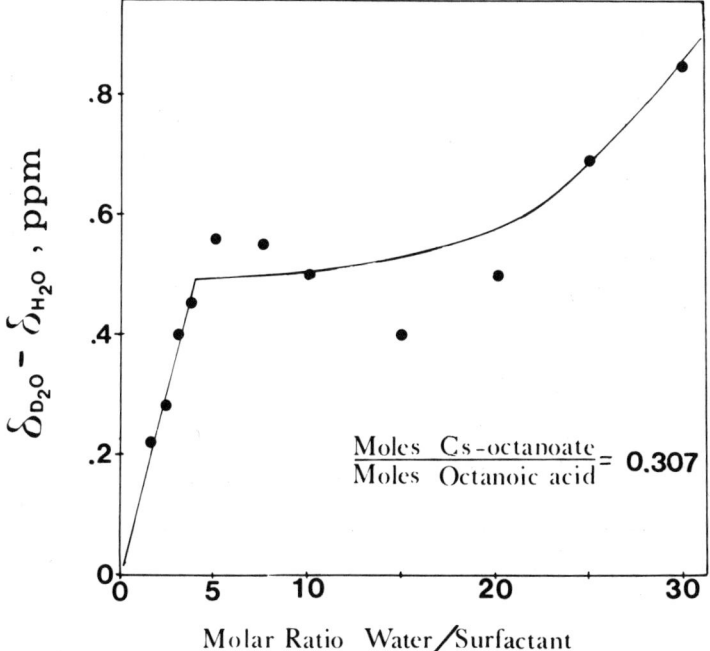

Figure 5. Water isotope effect in shielding of $^{133}Cs^+$ counterions for solutions of cesium octanoate and water in octanoic acid. The ratio of cesium octanoate to octanoic acid is kept constant and the water content is varied.

-line spectrometer making use of the proportionality between line width and transverse relaxation rate, $R_2 = 1/T_2$. For quadrupole-split spectra, $R_2$ may be obtained from the central line provided second-order quadrupole effects may be neglected. In recent years, the relaxation times have been measured directly on a pulse spectrometer using the $180°$-$\tau$-$90°$ pulse sequence for the longitudinal relaxation time, $T_1$, and the Carr-Purcell-Meiboom-Gill method for $T_2$. Transverse relaxation times for low counterion concentrations have been obtained from Fourier transform spectra.

For the rationalization of the concentration dependence of relaxation we may use the same model as discussed above for the shielding. Thus the concentration dependence of the relaxation rate, $R = R_1$ or $R_2$, should be given by

$$R = R_f + \beta(R_b - R_f) - \frac{\beta m_c}{m_t}(R_b - R_f) \tag{5}$$

above the CMC while below this concentration $R = R_f$. Typical plots of $R_1$ of $^{23}Na^+$ versus the inverse surfactant concentration are given

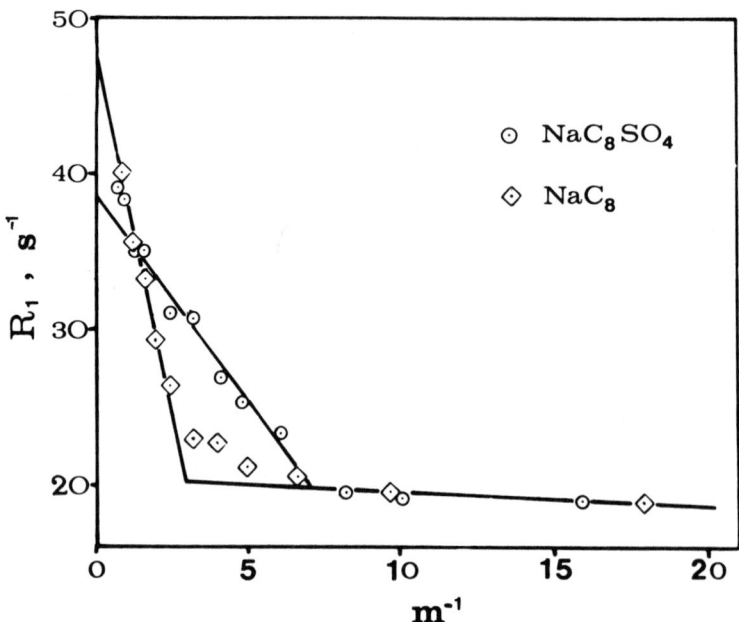

Figure 6. $^{23}$Na$^+$ longitudinal relaxation rate, $R_1$, for aqueous solutions of sodium octyl sulfate and sodium octanoate plotted as a function of the inverse amphiphile concentration. (From Ref. 28.)

in Figure 6 while corresponding plots for $R_2$ may be found in Refs. 33($^{85}$Rb$^+$), 34($^{35}$Cl$^-$ and $^{81}$Br$^-$) and 30($^{23}$Na$^+$). From such plots we have determined values of $\beta \cdot (R_b - R_f)$ and then of $R_b$ using an observed or estimated value of $\beta$. These quantities together with the CMC values obtained from the intersection points are listed in Table III.

The $R_b$ values should contain pertinent information on the mode of interaction and the mobility of bound counterions, but the analysis becomes complicated by the fact that the relaxation rate is determined by the magnitude of the field gradient being modulated as well as by the correlation time ($\tau_c$) which is the time constant of the modulation. For proteins, $\tau_c$ of bound alkali and halide ions is often long enough so that the extreme-narrowing condition is inapplicable and then a separation of $\tau_c$ and the field gradient may be achieved[11]. For counterions in surfactant systems, on the other hand, observation for a large number of cases of an equality of $T_1$ and $T_2$, or their independence of resonance frequency, have shown that the extreme narrowing condition applies[28,34,35]. (It is only for $^{23}$Na relaxation at very low water concentrations in the system

# IONIC INTERACTIONS IN AMPHIPHILIC SYSTEMS

## Table III

Values of CMC, $\beta(R_b-R_f)$ and $R_b$ obtained from studies of the concentration dependence of the counterion quadrupole relaxation. (Data taken from Refs. 28, 30, 33 and 34.)

| Amphiphile | Nucleus | CMC M | $\beta(R_b-R_f) \cdot 10^{-3}$ $s^{-1}$ | $R_b \cdot 10^{-3}$ ($\beta$) $s^{-1}$ |
|---|---|---|---|---|
| $CH_3(CH_2)_8N(CH_3)_3Br$ | $^{81}Br$ | 0.18 | 1.75 | 2.69 (0.8) |
| $CH_3(CH_2)_9N(CH_3)_3Br$ | -"- | 0.079 | 1.69 | 2.44 (0.8) |
| $CH_3(CH_2)_{13}N(CH_3)_3Br$ | -"- | — | 2.88 | 3.76 (0.8) |
| $CH_3(CH_2)_{15}N(CH_3)_3Br$ | -"- | 0.00099 | 3.69 | 4.82 (0.8) |
| $CH_3(CH_2)_8NH_3Br$ | -"- | 0.13 | 5.51 | 6.89 (0.8) |
| $CH_3(CH_2)_9NH_3Br$ | -"- | 0.047 | 5.51 | 6.89 (0.8) |
| $CH_3(CH_2)_9C_4H_4N\ Br$ | -"- | 0.047 | 1.69 | 2.38 (0.8) |
| $CH_3(CH_2)_7N(CH_3)_3Cl$ | $^{35}Cl$ | 0.25 | 0.36 | 0.70 (0.8) |
| $CH_3(CH_2)_7NH_3Cl$ | -"- | 0.27 | 1.07 | 1.54 (0.8) |
| $CH_3(CH_2)_6COO\ Na$ | $^{23}Na$ | 0.37 | 0.028 | 0.066 (0.6) |
| $CH_3(CH_2)_7OSO_3Na$ | -"- | 0.14 | 0.020 | 0.052 (0.6) |
| $CH_3(CH_2)_8OSO_3Na$ | -"- | 0.085 | 0.022 | 0.054 (0.6) |
| $CH_3(CH_2)_9OSO_3Na$ | -"- | 0.038 | 0.023 | 0.057 (0.6) |
| $CH_3(CH_2)_{11}OSO_3Na$ | -"- | 0.0077 | 0.030 | 0.068 (0.6) |
| $CH_3(CH_2)_4COO\ Rb$ | $^{85}Rb$ | 0.53 | 0.627 | 1.47 (0.6) |
| $CH_3(CH_2)_6COO\ Rb$ | -"- | 0.32 | 0.582 | 1.39 (0.6) |

sodium octanoate-octanoic acid-water that a frequency dependence of relaxation has been observed[36].) For this short correlation time case we have for the relaxation rate

$$R_1 = R_2 = \frac{3\pi^2}{10} \cdot \frac{2I+3}{I^2(2I-1)} \nu_Q^2 \tau_c \qquad (6)$$

Here $\nu_Q = \frac{e^2qQ}{h}$ is the quadrupole coupling constant characterizing the modulated interaction. eq is the field gradient (assumed to have cylindrical symmetry) and eQ the quadrupole moment. In addition to the difficulty of separating $\nu_Q$ and $\tau_c$ a further difficulty is to establish which molecular motion is effecting relaxation. These problems have been considered in some detail in Ref. 37 and here we will only make some general comments. As there is not a great variation of $R_b$ with the length of the alkyl chain[30,34] or with micellar shape[34,35], and as continuous varations of the relaxation rate at phase transitions have been observed for a number of systems[38-40], a local motion, and not a motion over the dimensions of the aggregate, must effect relaxation[37]. This local motion can be water molecule rotation or translation or an exchange of the counterion between being bound at the micelle and free in the intermicellar solution. For the latter relaxation mechanism the quadrupole coupling constant may be estimated using a simple electrostatic model[37]. Since we for a number of cases have observed relaxation to proceed ca. 20% more rapidly in $D_2O$ solutions than in $H_2O$ solutions[28,35], the motion causing relaxation must somehow be connected with the motion of water molecules. This is consistent with the view that the counterions retain their hydration sheaths on binding to the micelles. By assuming that the relaxation of the free ion at low concentrations is caused by the rotation of water dipoles in the first hydration layer[23] we may estimate upper limits of the average rotational correlation time of the water molecules in the hydration sheath of micellarly bound counterions. For $Na^+$ bound to octanoate micelles, for example, we obtain $\tau_c \leq 10^{-11}$ s corresponding to a very high moleculear mobility at the surface of micelles.

Irrespective of these interpretation difficulties, counterion quadrupole relaxation is most useful for comparing counterion binding of different systems and for probing into changes is counterion binding with changes in composition, phase structure etc. The markedly different $^{23}Na^+$ $R_b$ values of octyl sulfate and octanoate[28,30], for example, point to a considerable difference in interaction in the two cases. Qualitatively the results are as expected for the model of counterion binding discussed above but alternative interpretations cannot be excluded.

As a rule, deviations from the above simple model of the concentration dependence are observed as much higher concentrations than the CMC are attained[33-35,39]. This corresponds to an enforced counterion binding showing up as an increase in the product $\beta \cdot R_b$, which in turn suggests some structural modifications of the micelles at higher concentrations. For hexadecyltrimethylammonium bromide, for example, the increase in $^{81}Br$ relaxation may be attributed to a change in the micellar form from globular to cylindrical[35]. Since for the corresponding chloride no such effect is observable, a considerable counterion specificity as to the evolution of micellar shape may be noted. For sodium octanoate where a change in counterion binding to the micelles was inferred at ca. 1.3 m it could be argued that changes in both $\beta$ and the interaction of the bound counterions have to occur to explain the data[39].

The relation between counterion binding and solubilization may also be studied conveniently by quadrupole relaxation. For hexadecyltrimethylammonium bromide micelles, some polar solubilisates were found to affect counterion binding both by releasing counterions from the micelles by reducing the micellar surface charge density and by inducing a micellar shape transition[35]. A nonpolar solubilisate showed none of these effects[35]. Solubilisation of long-chain alcohols in octanoate micelles was found to increase $R_b$ of $Na^+$ suggesting the involvement of -OH in the counterion binding process[39].

For a large number of three-component systems, it has been observed that the counterion relaxation remains essentially unchanged at phase transitions[38-40], e.g. micellar → reversed micellar solutions, isotropic solution → mesophase. This independence of counterion binding of phase structure suggests that ionic interactions are not related to the origin of the transitions between different phases[39].

Effects of electrolyte addition on counterion quadrupole relaxation have only been studied to a limited extent but a recent study showed that addition of NaCl to solutions of sodium octanoate causes not only an increase in the number of counterions bound but also a change in the intrinsic relaxation rates of these ions[39].

As the counterion relaxation properties of reversed micellar solutions[38-40] are similar to those of solutions containing normal micelles, support is obtained for the model where the counterions are located in the water cores of the reversed micelles where they can be extensively hydrated. When the water content is reduced to the point where the reversed micelles are ceasing to exist, a rapid increase in counterion relaxation is observed, corresponding to an enforced counterion-amphiphile ion interaction[39-41].

It is a general observation for all types of phases, normal and reversed micellar solutions and hexagonal and lamellar liquid crystals, that the quadrupole relaxation is mainly determined by the water-to-surfactant molar ratio, again emphasizing the significance of counterion hydration[38-40,43]. At the lowest water contents where the phases exist quite dramatic increases in the relaxation rate may occur.

It should finally be mentioned that our quadrupolar relaxation studies have benefited greatly from the cooperation with several experts on surfactant association, in particular P. Ekwall, I. Danielsson and J.B. Rosenholm and furthermore that quadrupole relaxation studies have been reported also by Eriksson et al.[44] and by Robb and Smith[45,46].

## QUADRUPOLE SPLITTINGS

Many of the NMR interactions of interest for the study of surfactant systems are orientation dependent, i.e. they vary with the relative orientation of the magnetic field and some direction on the molecular level. As a result of molecular motion, orientation is time dependent which leads to a reduction of the effective interaction. Of fundamental importance for the general appearance of the NMR spectrum is the relation between the time-scale at which the relevant molecular motion becomes isotropic and the magnitude of the interaction. If molecular motion is not rapid enough, different orientations are characterized by different interaction strengths, and thus of different NMR frequencies, leading to a broad NMR absorption for a so-called powder sample which is not macroscopically aligned. Such effects explain why solids and liquids are characterized by very different NMR spectra.

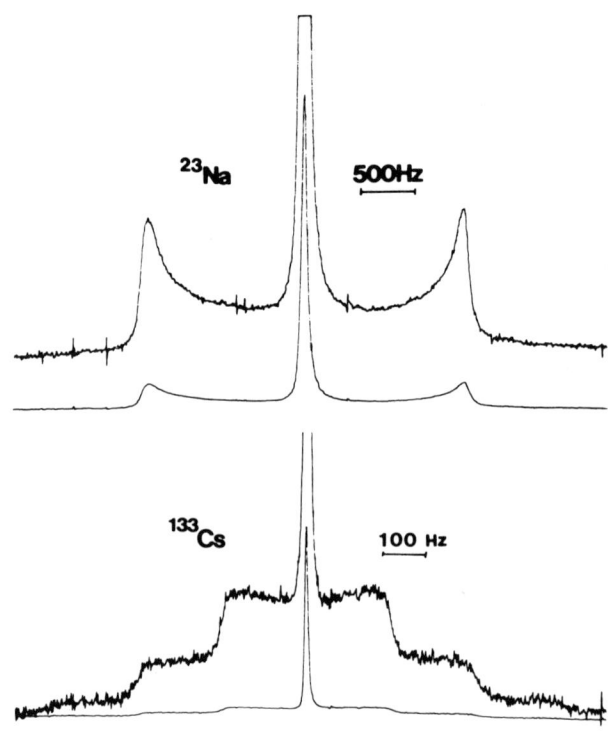

Figure 7. $^{23}$Na$^+$ and $^{133}$Cs$^+$ Fourier-transform NMR spectra of counterions in lamellar mesophases. Sample compositions were (in percent by weight) 32.7% sodium octanoate, 39.6% decanol and 27.7% water, and 42.2% cesium octanoate, 29.6% octanoic acid and 28.2% water.

IONIC INTERACTIONS IN AMPHIPHILIC SYSTEMS 215

For the case of quadrupole interactions of counterions, the relevant interaction strength varies widely with nucleus, sample composition etc. but if we take a typical figure of $10^4$ Hz, molecular motion has to be isotropic in a time of about $10^{-5}$ s or less to completely average out this interaction. This occurs for isotropic solutions and also for cubic liquid crystalline phases but as a rule not for anisotropic liquid crystals. The residual quadrupole interaction not averaged by the molecular motion leads to a splitting of the NMR absorption into (2I+1) component signals. Typical spectra of powder samples may be found in Figure 7. The distance between two adjacent absorption maxima is called the quadrupole splittting, $\Delta$. The magnitude of $\Delta$ is determined by three factors, i.e. the fraction of counterions bound to the anisotropic amphiphile structure, the quadrupole coupling constant of the binding site and by the extent to which the molecular motion reduces the field gradients at the appropriate time-scale. The following expression for $\Delta$ applies for a powder sample[37]

$$\Delta = \frac{3}{4I(2I-1)} \left| \Sigma\, p_i \nu_{Qi} S_i \right| \qquad (7)$$

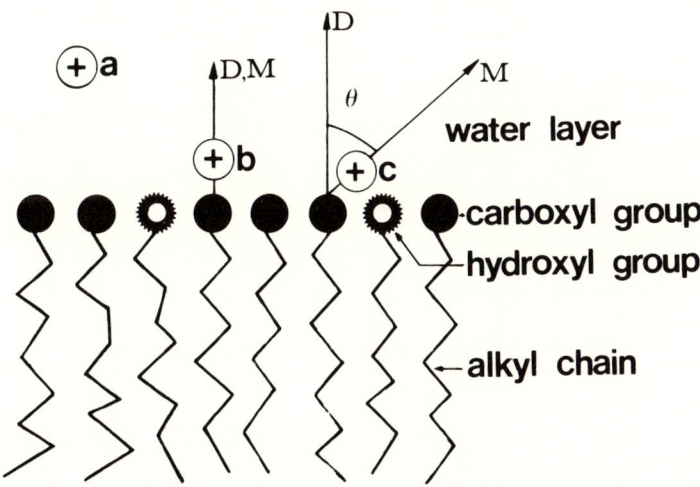

Figure 8. Schematic drawing of the counterion binding at the lamellar surface. The normal to the lamellae is denoted D while M denotes the direction of the field gradient.
    a.  The counterion is moving freely in the water layer.
    b.  The counterion is located symmetrically with respect to the amphiphile polar end-group.
    c.  The counterion penetrates between amphiphile polar head-groups. (From Ref. 47.)

Here $S_i$, the order parameter, describing the degree of orientation of the field gradients, is given by $S_i = \frac{1}{2}\overline{(3\cos^2\theta-1)}$ where $\theta$ is the angle between the symmetry axis (director) of the liquid crystal and the field gradient. (Assumed to have cylindrical symmetry.) The bar denotes an average which may be taken over time or ensemble of molecules. To illustrate the meaning of the order parameter in the case of counterion NMR studies, let us consider the schematic drawing[47] in Figure 8. If $\theta \simeq 0°$ and the range of its fluctuation small then S will be of the order of unity, while S may be small even for a rigid arrangement if $\theta$ is in the region of the magic angle (54°44') where $3\cos^2\theta-1 = 0$. Another reason for a vanishing order parameter is an isotropic fluctuation of the field gradients. This will be the situation for a free counterion not interacting with the amphiphile aggregates. Assuming then that $S_f = 0$ and that there is a single type of counterion binding site at the amphiphile aggregate Equation (7) reduces to

$$\Delta = \frac{3}{4I(2I-1)} P_b |\nu_{Qb} S_b| \qquad (8)$$

It would be of considerable interest to obtain $S_b$ which contains information on the geometry of the molecular arrangement at the amphiphile aggregate surface. To deduce $S_b$ certain estimates and assumptions are necessary. If $P_b$ has the same meaning as in other experimental approaches, for example diffusion, its value may be considered to be known with relatively good precision. It will be 0.6 - 0.9 over the larger parts of the mesophase regions studied. If we additionally estimate $\nu_{Qb}$ by means of a simple electrostatic model[37], taking the amphiphile polar heads as point charges, we obtain $S_b$ to be of the order of unity for sodium ions in some mesomorphous phases where the amphiphile polar head is $-SO_3^-$ or $-OSO_3^-$. This would suggest $\theta$ to be close to zero with a limited range of fluctuation. $\theta = 0°$ corresponds most probably to a symmetrical location of $Na^+$ with respect to the three oxygens pointing out into the aqueous layers. Such a location would be expected for a non-specific electrostatic interaction. The temperature (Figure 9) and concentration dependences of the splitting are small and provide support for the deduced $S_b$ value. An approximately factor of two difference in splitting between the lamellar and hexagonal mesophases agrees with that expected from the differences in phase anisotropy[37].

$^{23}$Na splittings for the case where the surfactant end-group is carboxylate differ totally from those with sulfonate or sulfate end-groups[37,47,48]. Thus the $^{23}$Na splittings are much smaller with the $-CO_2^-$ end-group and the variations of the splitting with both

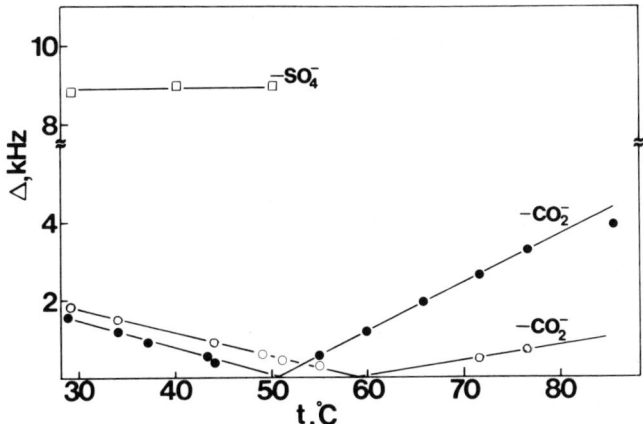

Figure 9. Temperature dependence of the $^{23}$Na quadrupole splitting for a lamellar mesophase sample composed of 24.3% sodium octanoate, 40.4% decanol and 35.3% water (●), a hexagonal mesophase sample composed of 51.0% sodium octanoate and 49.0% water (O) and a hexagonal mesophase sample composed of 58.3% sodium octylsulfate and 41.7% water (□).

temperature and sample composition are complex with vanishing splittings being observed well within the stability ranges of the mesophases. These observations pertain inter alia to the hexagonal and lamellar mesophases of the systems sodium octanoate-decanol-water[47], sodium octanoate-octanoic acid-water[47] and sodium octanoate-pentanol-water[39]. Temperature dependences of the splitting are exemplified in Figure 9. In terms of the two-site model, with the assumption that $S_f = 0$, these observations indicate that $S_b$ is small and that it may change sign with changing composition or temperature. In turn this suggests that the value of $\Theta$ may pass through the magic angle. There would thus seem to be a considerable angle between the normal to the amphiphile aggregate and the direction of the counterion-amphiphile ion vector. An alternative interpretation of these findings would be in terms of a counterion exchange between two anisotropic sites having opposite signs of the quadrupole coupling constant. Irrespectively of which model is the correct one, it appears, however, that one has to consider a partial penetration of the counterions between the polar head groups. Also, a markedly different counterion interaction for $-SO_4^-$ and $-CO_2^-$ end-groups is clearly demonstrated. A hydrogen-bond interaction between water of $Na^+$ hydration and carboxylate (cf. above) seems to explain these observations well.

The quadrupole splitting method is of recent date, the first observations[49,50] being reported in 1971 for $^{35}Cl^-$, $^{37}Cl^-$ and $^{23}Na^+$. More recently we have investigated also $^{81}Br^-$, $^7Li^+$, $^{39}K^+$, $^{85}Rb^+$, $^{87}Rb^+$ and $^{133}Cs^+$ splittings and made detailed studies of certain systems[35,39,47-54]. During the studies of the sodium octanoate--decanol-water and sodium octanoate-pentanol-water systems we have collaborated with G.J.T. Tiddy and J.B. Rosenholm, respectively. Interest in the quadrupole splitting method for studying ion binding in liquid crystals has increased in recent years and also other groups have started investigations[55-58].

It seems that the quadrupole splitting method shows great promise although certain problems of interpretation remain to be solved. It can be noted in conclusion that counterion interactions with dipolar polar heads also may give considerable $^{23}Na$ quadrupole splittings as shown by observations for three-component systems non--ionic amphiphile - sodium chloride-water (unpublished results) and that for model membrane systems built up of phospholipid bilayers, interesting effects of electrolyte addition and of cholesterol have been noted[54,59-61]. For cationic amphiphiles[35,49,52] it seems that the halide counterion order parameter is much larger when the polar head is $\overset{+}{N}H_3$ than when it is $\overset{+}{N}(CH_3)_3$.

## COUNTERION TRANSLATIONAL MOTION

It is clear that counterion binding will reflect itself in the ion's effective translational mobility but the translational motion of counterions may be affected also by structural features of the phase under consideration. These may cause translational diffusion to be restricted and then it is important to consider the time (or alternatively the diffusion distance) during which diffusion is studied by the particular method. Thus different experimental methods respond to diffusion over widely different distances.

Counterion diffusion studies may be employed to different problems but ourselves we have been mainly concerned with obtaining information on the degree of counterion binding ($\beta$) and on structural characteristics. If there are no other restrictions on counterion diffusion than the participation of the counterions in some association process the observable diffusion coefficient can be written

$$D = \sum_i p_i D_i = p_f D_f + p_b D_b \qquad (9)$$

The second relation pertains to the two-site model considered above. The two-site assumption is less critical in the present case as counterions bound in different ways to the same type of aggregate, and diffusing with the aggregate, are characterized by the same

$D_b$ (that given by the translational mobility of the aggregate) and, therefore, need not (and can not) be distinguished. For the case of micellar solutions, $D_b$ may be determined experimentally[62,63] as it should equal the diffusion coefficient of a solubilisate having very low solubility in the intermicellar solution. (Of course, one has to consider the possibility that the solubilisate affects size or shape of the micelles.) Interpretation of the counterion diffusion results is often facilitated by the fact that $D_b \ll D_f$. In many cases, $D_b$ may be neglected with good approximation and then Equation (9) reduces to $D = p_f D_f$. According to our experience, Equation (9) applies well to micellar solutions taking $D_f$ to be the counterion diffusion coefficient at low concentrations since the varation of $D_f$ with concentration due to general interionic interactions is much smaller than that caused by association. Obstruction effects, arising from the prolongment of the diffusion path caused by the presence of large aggregates, likewise seem to be negligible except at very high concentrations.

When the aqueous regions where the counterions are located are somehow limited in size, diffusion will be restricted and, while locally the translational mobility may be high, the diffusion observed over long distances would be very slow. As this type of restriction is connected with structural features, it may give significant information on phase structure which may not be easily obtainable by other methods. The principle has been used previously in connection with amphiphile diffusion[64].

Diffusion coefficients may be obtained by several alternative experimental methods which are differently applicable to different systems. Since 1971 we have performed studies by a radiotracer technique in cooperation with N. Kamenka, B. Brun and H. Fabre at the Faculty of Science in Montpellier to obtain information on association phenomena and phase structures in several surfactant systems. The method used, the so-called open-ended capillary tube method, essentially consists in inserting a capillary with a labelled solution into a large volume of identical solution except for labelling and observing the decay of the capillary's radioactivity content due to diffusion. This technique has the advantage of permitting studies at low concentrations and of permitting very specific labelling. The second experimental method used is the NMR spin-echo mehod which essentially involves determining the influence of an applied magnetic field gradient on the dephasing of the nuclear magnetic moments. As decay of the magnetization due to diffusion must dominate over relaxation effects, nuclei with very small relaxation times are not amenable to study. This excludes, for example $Cl^-$, $Rb^+$ and $Br^-$ counterions while $^7Li^+$ should be well adapted to this type of investigation. The NMR method has the further disadvantage of not allowing studies at low concentrations but has the advantage of being applicable irrespective of sample consistency. For liquid crystalline phases it is a potentially very interesting method.

There seems to be no report of NMR studies of counterion translation in the literature and our own studies have mainly been performed with the radiotracer technique which was used previously by others[65,66]. According to Equation (3) the observable counterion diffusion coefficient may be expected to be

$$D = D_f + \beta(D_b - D_f) - \frac{\beta m_c}{m_t}(D_b - D_f) \qquad (10)$$

above the CMC. As is exemplified in Figure 10 for sodium octanoate the predicted behavior agrees closely with observation[67]. As $D_b$ and $m_c$ may be obtained separately, $\beta$ can be calculated from such data. For sodium octanoate we obtained $\beta$ to be 0.59 in the concentration region just above the CMC (0.38 m) and to start to increase somewhat above 1 m. For sodium dodecyl sulfate, $\beta$ was obtained to be ca. 0.6 in a wide concentration range and for hexadecyltrimethylammonium bromide $\beta$ was found to be ca. 0.7 with a slight increase with concentration[68]. The extent of counterion binding to cholate aggregates is, on the other hand, small except at very

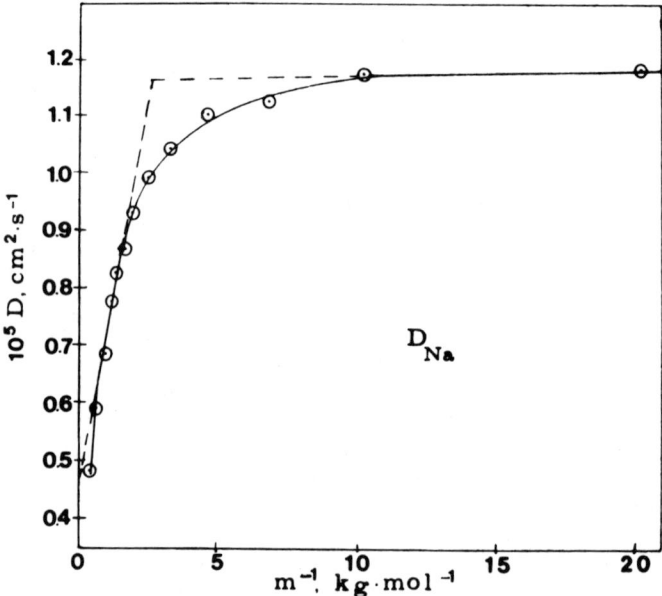

Figure 10. Na$^+$ self-diffusion coefficient in aqueous solutions of sodium octanoate plotted versus the inverse soap molality. (from Ref. 67.)

high concentrations[69]. For pre-micellar aggregates an interesting difference in sodium ion binding between $-CO_2^-$ and $-SO_4^-$ end-groups has been noted[70]. In agreement with previous findings of Mukerjee[12], sodium ions do not appear to participate in the pre-micellar association of dodecyl sulfate while this is the case for octanoate. This difference is also, although less clearly, suggested by the quadrupole relaxation and shielding data. For sodium dodecyl sulfate and sodium octanoate micelles, solubilization of polar compounds has been found to lead to a partial release of counterions while non-polar compounds have small effects on counterion binding[68]. Solubilization effects for hexadecyltrimethylammonium bromide solutions are complex and analogous to the $^{81}Br$ quadrupole relaxation results mentioned above[68].

$^7Li^+$ NMR diffusion studies have been performed for the decanol-rich solution phase of the system lithium octanoate-decanol-water. (The help of J. Andrasko in these measurements is gratefully acknowledged.) $D_{Li^+}$ is found to be very low while we know from quadrupole relaxation studies that the local mobility is considerable. So for example $D_{Li^+}$ is $1.6 \cdot 10^{-11}$ m$^2$/s for a solution composed of (by weight) 9% lithium octanoate, 21% water and 70% decanol. This clearly demonstrates that $Li^+$ diffusion is restricted and correlates well with the accepted view that these solutions in a wide concentration range contain micelles of the reversed type with closed water cores[20]. Diffusion of lithium ions between different reversed micelles apparently is very slow. Previously, radiotracer studies of the diffusion of sodium ions, octanoate ions, decanol and water molecules had given rather detailed information on aggregate composition in decanol-rich solutions of sodium octanoate, decanol and water[71]. $D_{Na^+}$ and $D_{H_2O}$ seem to lie on a low approximately constant level as long as the decanol content is below ca. 90%. Above this concentration they increase rapidly pointing to the dissolution of the reversed micelles. The translational mobility of the reversed micelles could be obtained by studying the diffusion of $Ca^{2+}$ ions[71] as the calcium ions can be assumed to pass only very slowly from one aggregate to another.

## MISCELLANEOUS

Above we have surveyed the most important principles for the direct study of the binding of small ions in amphiphile systems by NMR. This list is by no means exhaustive and, in particular, information on counterion binding may be obtained indirectly in many types of NMR studies.

To study different NMR parameters of the water molecules is a natural way of looking into the problem of counterion hydration. An investigation[72,73] of the variation of the degree of orientation of the water molecules with the alkali ion was performed by deuteron

NMR for a number of amphiphilic mesophases. As $^2$H has I = 1 its NMR signal is split into two components in an anisotropic environment. Since the field gradients are of an intramolecular origin and, therefore, relatively independent of solution composition, the quadrupole coupling constant is known with good precision and average order parameters are directly obtainable. The degree of water orientation depends markedly on counterion and provides further support for the contention that counterions are considerably hydrated. The degree of water orientation was considerably greater with Li$^+$ as counterion than with the other alkali ions. From the dependence of water orientation on sample composition it was found that only a small number (6 or less) of water molecules is oriented with respect to the amphiphilic aggregates. No effect of counterion or ionic group of the surfactant on the orientation of decanol in anionic surfactant-decanol-water mesophases was observed.

$^{14}$N quadrupole splittings have to a little extent been applied to study the binding of NH$_4^+$ in surfactant systems[74]. In this case two origins of the field gradients have to be considered, i.e. either effects as considered above for the monoatomic ions or distortions from tetrahedral symmetry induced by the environment. $^{14}$N quadrupole splittings of the NH$_4^+$ ion in the lamellar mesophase region of the system ammonium octanoate-decanol-water are found, in contrast to the results for the alkali ions, to be only slightly dependent of composition. It is probable that this reflects an essentially concentration independent distortion of the ammonium ion.

Electron spin relaxation of the VO$^{2+}$ ion can be conveniently employed to probe into the counterion mobility at the surface of an amphiphile aggregate[75,76]. A considerable mobility is observed, the rotational correlation time of the hydrated vanadyl ion bound to a micelle of alkylsulfate being ca. $7 \cdot 10^{-11}$ s. A high counterion mobility is also found for lamellar and hexagonal liquid crystals, while the motion of VO$^{2+}$ is more hindered for aggregates of the reversed type. A qualitative correlation between the VO$^{2+}$ correlation time and the area per amphiphile end-group is observed[76].

A very interesting counterion specificity has been observed for hexadecyltrimethylammonium micelles by $^{13}$C, $^{35}$Cl and $^{81}$Br NMR[77]. From the observation of $^{35}$Cl$^-$ and $^{81}$Br$^-$ relaxation in mixtures of hexadecyltrimethylammonium bromide and chloride, the existence of a considerable amount of both globular and rodshaped micelles is indicated. Furthermore, it appears from the quadrupole relaxation data that Cl$^-$ interacts with marked preference to the globular aggregates, while Br$^-$ interacts with the cylindrical micelles. This parallels the above-mentioned observation that the micellar shape transition is observed with Br$^-$ but not with Cl$^-$. Recent $^{13}$C NMR studies are interesting in this connection. Somewhat unexpectedly, two separate $^{13}$C resonance signals of the ω-CH$_3$ group are observed in mixtures of the two amphiphiles, with the relative intensities corresponding to the relative amounts of the two counterions. This provides direct support for a considerable counterion specificity and, furthermore, which is more difficult to understand, the amphiphile exchange between the two types of micelles should be slow on the NMR time-scale, i.e. with average life-times of the order of a second or longer.

## SOME CONCLUDING REMARKS

It should be clear from this description that nuclear magnetic resonance provides us with several different possibilities to directly monitor ionic interactions in amphiphilic systems and also that in many cases we are only in the initial stages of application. Certain rather definite conclusions can be drawn about the general features of counterion hydration and counterion specificity while the more detailed interpretations remain somewhat tentative in many respects.

Counterion specificity has been demonstrated for both anionic and cationic amphiphiles. The different behavior of $Cl^-$ and $Br^-$ ions may have to do with the difference in polarizability which has been shown to lead to a stronger adsorption of $Br^-$ to proteins (Ref. 78 and literature cited therein) and to the air-water interface[79]. The specificity in the alkali ion binding has been known for a long time. Mukerjee et al.[1] demonstrated a variation of the CMC of alkali dodecyl sulfates with the alkali ion and also found a greater degree of dissociation for $Li^+$ than for $Na^+$. In the polyelectrolyte field the opposite alkali ion affinity series of $-SO_4^-$ and $CO_2^-$ groups is well documented in many cases. (See for example Refs. 10 and 80.) Also it has been proposed that hydrogen-bond interactions are important in the case of $-CO_2^-$ end-groups[10,81]. The difference in alkali ion binding of $-SO_4^-$ and $-CO_2^-$ is supported by many of our NMR observations which also permit some insight into the binding mechanism. The partial penetration of sodium ions between amphiphile polar groups in the case of $-CO_2^-$ as suggested by the $^{23}Na$ quadrupole splittings is of interest and will deserve further study. It is interesting in this connection to consider the values of the area per polar group obtained in the X-ray diffraction studies of Ekwall et al.[81,82] for the lamellar phases of the systems sodium octanoate-decanol-water and sodium octyl sulfate-decanol-water. For corresponding molar compositions, the areas per polar group are closely the same although the cross-section of the sulfate group is appreciably larger. This observation might be taken, in accordance with the $^{23}Na$ quadrupole splitting data, to suggest some location of counterions between polar heads[83] but more systematic studies, including studies with the other alkali ions, are required to settle this point.

Hydration of amphiphile aggregates has been a matter of considerable debate in recent years. Muller et al. suggested from their studies[84-86] of $^{19}F$ NMR chemical shifts of partially fluorinated surfactants that there is considerable penetration of water into the interior of the micelles. However, Mukerjee and Mysels[87] have shown the interpretation to be incorrect as the mutual phobicity of hydrocarbon and perfluoro groups will place ω-trifluoro groups at or close to the micellar surface to a considerable extent. Another critique of the water penetration concept has been given by Stigter[7]. J. Ulmius in our laboratory has recently made investigations of $^{19}F$ relaxation of partially fluorinated surfactants in $H_2O$ and $D_2O$ solutions which provide direct support for the views of Mukerjee and Mysels[87]. Thus for micellar solutions of $CH_3(CH_2)_3CF_2(CH_2)_4CO_2^-K^+$ the $^{19}F$ spin-lattic relaxation time is closely the same in $H_2O$ and $D_2O$. Studies[67] of water translational motion, furthermore, show that only a small number of water molecules move with the micelle as a kinetic entity; the hydration number obtained in this way is not much greater than expected for the hydration of the amphiphile polar head and bound counterions. Water penetration beyond the $\alpha$-$CH_2$ group can apparently not be considerable.

Our different types of NMR data suggest that the counterions bound to micellar aggregates retain their inner hydration sheaths over wide concentration ranges and it is also suggested that the hydration water may participate in the counterion binding process. The importance of counterion hydration has been well demonstrated in the work of Mukerjee and Stigter cited above and a further significant observation in this respect is the variation with alkali ion of the minimum amount of water compatible with the existence of certain phases[88].

## REFERENCES

1. P. Mukerjee, K.J. Mysels and P. Kapauan, J. Phys. Chem. 71, 4166 (1967).
2. P. Mukerjee, Adv. Colloid Interface Sci. 1, 241 (1967).
3. P. Mukerjee and A. Ray, J. Phys. Chem. 70, 2150 (1966).
4. D. Stigter, J. Phys. Chem. 68, 3603 (1964).
5. D. Stigter, J. Colloid Interface Sci. 23, 379 (1967).
6. D. Stigter, J. Colloid Interface Sci. 47, 473 (1974).
7. D. Stigter, J. Phys. Chem. 78, 2480 (1974).
8. D. Stigter and K.J. Mysels, J. Phys. Chem. 59, 45 (1955).
9. U.P. Strauss, In "Polyelectrolytes", E. Sélégny, Editor, p. 79, D. Reidel Publishing Company, Dordrecht, 1974.
10. H.P. Gregor, in "Polyelectrolytes", E. Sélégny, Editor, p. 87, D. Reidel Publishing Company, Dordrecht, 1974.
11. B. Lindman and S. Forsén, "Chlorine, Bromine and Iodine NMR, Physico-Chemical and Biological Applications", Springer Verlag, Berlin (1976).

12. P. Mukerjee, K.J. Mysels and C.I. Dulin, J. Phys. Chem. 62, 1390 (1958).
13. P. Ekwall, Finska Kemists. Medd. 72, 59 (1963).
14. I. Danielsson and P. Stenius, J. Colloid Interface Sci. 37, 264 (1971).
15. I. Danielsson, J.B. Rosenholm, P. Stenius and S. Backlund, Proc. Int. Congr. Colloid Surface Sci., Budapest 1975, E. Wolfram, Editor, in press.
16. P. Mukerjee, J. Phys. Chem. 76, 565 (1972).
17. C. Tanford, "The Hydrophobic Effect", John Wiley, New York, 1973.
18. A.S. Kertes, H. Gutmann, O. Levy and G.Y. Markovits, in "Chemie, physikalische Chemie und Anwendungstechnik der grenzflächenaktiven Stoffe", vol. II, p. 1023, Carl Hanser Verlag, München 1973.
19. O.A. El Seoud, E.J. Fendler, J.H. Fendler and R.T. Medary, J. Phys. Chem. 77, 1876 (1973).
20. P. Ekwall, J. Colloid Interface Sci. 29, 16 (1969).
21. P. Ekwall, Adv. Liquid Cryst. 1, 1 (1975).
22. P. Ekwall, Wiss. Z. Friedrich-Schiller-Univ. Jena, Math.-Naturwiss. Reihe 14, 181 (1965).
23. H.G. Hertz, Ber. Bunsenges. Phys. Chem. 77, 531 (1973).
24. C. Deverell and R.E. Richards, Mol. Phys. 10, 551 (1966).
25. E.G. Bloor and R.G. Kidd, Can. J. Chem. 50, 3926 (1972).
26. R.A. Robinson and R.H. Stokes, "Electrolyte Solutions", 2nd Ed., Butterworths, London, 1968.
27. H. Gustavsson and B. Lindman, J.C.S. Chem. Commun. 1973, 93.
28. H. Gustavsson and B. Lindman, J. Amer. Chem. Soc. 97, 3923 (1975).
29. H. Gustavsson and B. Lindman, Proc. Int. Conf. on Colloid and Surface Sci., Budapest 1975, E. Wolfram, Editor, Akadémiai Kiadó, Budapest 1975, p. 625.
30. H. Gustavsson and B. Lindman, presented at the Int. Conference on Colloids and Surfaces, San Juan, Puerto Rico, 1976.
31. P. Mukerjee and K.J. Mysels, Nat. Stand. Ref. Data Ser., Nat. Bur. Stand., No.36 (1971).
32. S. Friberg, L. Mandell and P. Ekwall, Kolloid-Z.Z. Polym. 235, 955 (1969).
33. B. Lindman and I. Danielsson, J. Colloid Interface Sci. 39, 349 (1972).
34. G. Lindblom and B. Lindman, J. Phys. Chem. 77, 2531 (1973).
35. G. Lindblom, B. Lindman and L. Mandell, J. Colloid Interface Sci. 42, 400 (1973).
36. B. Lindman and P. Ekwall, Kolloid-Z.Z. Polym. 234, 1115 (1969).
37. H. Wennerström, G. Lindblom and B. Lindman, Chem. Scr. 6, 97 (1974).
38. G. Lindblom and B. Lindman in "Chemie, physikalische Chemie und Anwendungstechnik der grenzflächenaktiven Stoffe", vol. II, p. 925, Carl Hanser Verlag, München, 1973.
39. J.B. Rosenholm and B. Lindman, J. Colloid Interface Sci. in press.

40. B. Lindman and P. Ekwall, Mol. Cryst. $\underline{5}$, 79 (1968).
41. G. Lindblom, B. Lindman and L. Mandell, J. Colloid Interface Sci. $\underline{34}$, 262 (1970).
42. B. Lindman, "Applications of Nuclear Quadrupole Relaxation to the Study of Ion Binding in Solutions" Thesis, Lund, 1971.
43. G. Lindblom and B. Lindman, Mol. Cryst. Liquid Cryst. $\underline{14}$, 49 (1971).
44. J.C. Eriksson, Å. Johansson and L.-O. Andersson, Acta Chem. Scand. $\underline{20}$, 2301 (1966).
45. I.D. Robb, J. Colloid Interface Sci. $\underline{37}$, 521 (1971).
46. I.D. Robb and R. Smith. J.C.S. Faraday 1, $\underline{70}$, 287 (1974).
47. G. Lindblom, B. Lindman and G.J.T. Tiddy, Acta Chem. Scand. A $\underline{29}$, 876 (1975) and unpublished results.
48. G. Lindblom and B. Lindman, Mol. Cryst. Liquid Cryst. $\underline{22}$, 45 (1973).
49. G. Lindblom, H. Wennerström and B. Lindman, Chem. Phys. Lett. $\underline{8}$, 498 (1971).
50. G. Lindblom, Acta Chem. Scand. $\underline{25}$, 2767 (1971).
51. G. Lindblom, Acta Chem. Scand. $\underline{26}$, 1745 (1972).
52. G. Lindblom, N.-O. Persson and B. Lindman, in "Chemie, physikalische Chemie und Anwendungstechnik der grenzflächenaktiven Stoffe", vol. II, p. 939, Carl Hanser Verlag, München 1973.
53. H. Wennerström, "Theoretical and Experimental Studies in Nuclear Magnetic Relaxation" Thesis, Lund, 1974.
54. G. Lindblom, "Ion Binding in Micellar Solutions, Liquid Crystals and Biological Model Membrane Systems Studied by NMR Techniques" Thesis, Lund, 1974.
55. M. Shporer and M.M. Civan, Biophys. J. $\underline{12}$, 114 (1972).
56. D.M. Chen and L.W. Reeves, J. Amer. Chem. Soc. $\underline{94}$, 4384 (1972).
57. D.M. Chen, K. Radley and L.W. Reeves, J. Amer. Chem. Soc. $\underline{96}$, 5251 (1974).
58. K. Radley, L.W. Reeves and A.S. Tracey, J. Phys. Chem. $\underline{80}$, 174 (1976).
59. N.-O. Persson, G. Lindblom, B. Lindman and G. Arvidson, Chem. Phys. Lipids, $\underline{12}$, 261 (1974).
60. G. Lindblom, N.-O. Persson, B. Lindman and G. Arvidson, Ber. Bunsenges. Phys. Chem. $\underline{78}$, 955 (1974).
61. G. Lindblom, N.-O. Persson and G. Arvidson, Adv. Chem. Ser., $\underline{152}$, 121 (1976).
62. D. Stigter, R.J. Williams and K.J. Mysels, J. Phys. Chem. $\underline{59}$, 330 (1955).
63. J. Clifford and B.A. Pethica, J. Phys. Chem $\underline{70}$, 3345 (1966).
64. T. Bull and B. Lindman, Mol. Cryst. Liquid Cryst. $\underline{28}$, 155 (1975).
65. A.P. Brady and D.J. Salley, J. Amer. Chem. Soc. $\underline{70}$, 914 (1948).
66. J. Clifford and B.A. Pethica, Trans. Faraday Soc. $\underline{60}$, 216 (1964).
67. B. Lindman and B. Brun, J. Colloid Interface Sci. $\underline{42}$, 388 (1973).

68. H. Fabre, "Etude par self-diffusion des phases $L_1$ et $L_2$ des melanges ternaires" Thesis, Montpellier, 1976.
69. B. Lindman, N. Kamenka and B. Brun, J. Colloid Interface Sci. 56, 328 (1976).
70. B. Lindman, N. Kamenka and B. Brun, C.R. Acad. Sci. C278, 393 (1974).
71. N. Kamenka, H. Fabre and B. Lindman, C.R. Acad. Sci, C281, 1045 (1975) and unpublished results.
72. H. Wennerström, N.-O. Persson and B. Lindman, in "Colloidal Dispersions and Micellar Behavior", K.L. Mittal, Editor, ACS Symposium Series No. 9, p. 253, American Chemical Society, Washington, D.C., 1975.
73. N.-O. Persson and B. Lindman, J. Phys. Chem. 79, 1410 (1975).
74. H. Gustavsson, G. Lindblom, B. Lindman, N.-O. Persson and H. Wennerström, in "Liquid Crystals and Ordered Fluids, vol. 2", J.F. Johnson and R.S. Porter, Editors, p. 161, Plenum Press, New York, 1974 and unpublished results.
75. P. Stilbs and B. Lindman, J. Colloid Interface Sci. 46, 177 (1974).
76. P. Stilbs, J. Jermer and B. Lindman, J. Colloid Interface Sci., in press.
77. Refs. 35 and 54 and unpublished results obtained together with S.-O. Eldh.
78. J.-E. Norne, S.-G. Hjalmarsson, B. Lindman and M. Zeppezauer, Biochemistry 14, 3401 (1975).
79. K. Johansson and J.C. Eriksson, J. Colloid Interface Sci. 49, 469 (1974).
80. M. Rinaudo, in "Polyelectrolytes", E. Sélégny, Editor, p. 157, D. Reidel Publishing Company, Dordrecht, 1974.
81. K. Fontell, L. Mandell, H. Lehtinen and P. Ekwall, Acta Polytech. Scand., Chem. Incl. Met. Ser. 74(III) 1 (1968).
82. P. Ekwall, L. Mandell and K. Fontell, Acta Chem. Scand. 22, 1543 (1968).
83. P. Ekwall, in "Liquid Crystals and Ordered Fluids, vol. 2", J.F. Johnson and R.S. Porter, Editors, p. 177, Plenum Press, New York, 1974.
84. N. Muller and R.H. Birkhahn, J. Phys. Chem. 71, 957 (1967).
85. N. Muller and T.W. Johnson, J. Phys. Chem. 73, 2042 (1969).
86. N. Muller and F.E. Platko, J. Phys. Chem. 75, 547 (1971).
87. P. Mukerjee and K.J. Mysels, in "Colloidal Dispersions and Micellar Behavior", K.L. Mittal, Editor, ACS Symposium Series No. 9, p. 239, American Chemical Society, Washington, D.C., 1975.
88. P. Ekwall and L. Mandell, Acta Chem. Scand. 22, 699 (1968).

# ERRORS IN MICELLIZATION ENTHALPIES FROM TEMPERATURE DEPENDENCE OF CRITICAL MICELLE CONCENTRATIONS

Norbert Muller

Department of Chemistry, Purdue University

West Lafayette, Indiana 47907

To test theoretical predictions of micelle stabilities it is desirable to separate the readily available experimental values of free energies of micellization into their enthalpic and entropic contributions. The enthalpy of micellization has often been evaluated from the temperature dependence of the critical micelle concentration (cmc) using the relation

$$- RT^2 \frac{d}{dT} \ln(\text{cmc}) \cong \Delta H_\nu^o \cong \Delta H_\nu^{o\dagger}$$

Here $\Delta H_\nu^o$ is the enthalpy change per monomer for the formation of the most abundant micelle, $A_\nu$, from $\nu$ monomers, while $\Delta H_\nu^{o\dagger}$ is the enthalpy change for the addition of a single monomer to a micelle $A_{\nu-1}$. It is known that this relation is not exact unless the optimal aggregation number, $\nu$, is independent of temperature, and it has been stated that for this reason no reliable micellization enthalpies are obtainable by this procedure. It is shown here that for a Gaussian distribution of micelle sizes with a full width at half height of $2s\nu$, taking $s$ to be temperature independent but $d\nu/dT \neq 0$, $\Delta H_\nu^o$ remains much more nearly equal to $-RT^2 (d/dT)\ln(\text{cmc})$ than $\Delta H_\nu^{o\dagger}$ does. The difference between the two enthalpy changes depends mainly on $\nu$, $s$, and $d\nu/dT$, and may exceed 10 kJ/mol when the distribution is narrow. Semiquantitative calculations based on this model provide a possible rationalization of the recent experimental finding that micellization of sodium alkyl sulfates involves rather large negative values of $\Delta H_\nu^{o\dagger}$ even though $\Delta H_\nu^o$ is positive or nearly zero for several of these surfactants.

## INTRODUCTION

Many published investigations have been aimed at evaluating the changes in thermodynamic variables which accompany micelle formation in aqueous surfactant solutions, both in order to discover what factors are responsible for the stability of micelles and to test theoretical predictions. The most readily measurable thermodynamically interesting property of such solutions is usually the critical micelle concentration (cmc), and it is commonly agreed that the quantity RT ln(cmc) is very nearly equal to the standard free energy of micellization.[1] Somewhat paradoxically, this result may be reached by two dissimilar routes, which involve in principle quite different definitions of this free energy change, essentially as follows.

Since micelles in a given solution define a distribution of aggregation numbers, their formation is more suitably represented by a set of simultaneous equilibria than by a single equilibrium process.[2-5] Most authors prefer to do this using the set of reactions

$$nA \rightleftarrows A_n; \; n = 2,3,4,---, \tag{1}$$

while others have adopted the alternative of writing

$$A_{n-1} + A \rightleftarrows A_n; \; n = 2,3,4,---, \tag{2}$$

The symbol $K_n$ has been used both for the set of equilibrium constants corresponding to Equation (1) and that associated with Equation (2). In order to be able to refer to either scheme as convenience may dictate, one may define

$$K_n = [A_n]/[A]^n; \; \Delta G_n^o = -(RT/n)\ln(K_n), \tag{3}$$

and

$$K_n^\dagger = [A_n]/[A][A_{n-1}]; \; \Delta G_n^{o\dagger} = -RT\ln(K_n^\dagger). \tag{4}$$

The free energy changes $\Delta G_n^o$ and $\Delta G_n^{o\dagger}$ are in general quite different for most values of n, but they become nearly equal when $n = \nu$, where $\nu$ is the aggregation number of the most abundant micelles.

This follows since $[A_\nu]$ and $[A_{\nu-1}]$ must be nearly identical while [A] is essentially equal to cmc, so that to a good approximation $\Delta G^{o\dagger} = RT \ln(\text{cmc})$, while

$$\Delta G_\nu^o = RT \ln(\text{cmc}) - (RT/\nu) \ln[A_\nu], \tag{5}$$

and in a typical surfactant solution the magnitude of the second term of the right member of this Equation is only about 3% as large as that of the first. Thus $RT \ln(\text{cmc})$ can equally well be interpreted as the average free energy change per monomer for the aggregation of $\nu$ monomers to form the species $A_\nu$ or as the free energy change for the addition of a <u>single</u> monomer to a preformed micelle $A_{\nu-1}$.[6]

Unfortunately, micellization does not represent an exception to the general rule[7] that "the free energy is not a sensitive tool for judging the validity of theoretical models and mechanisms." Consequently, much work has been directed to the end of resolving the free energy of micellization into its enthalpic and entropic components. The enthalpy of micellization is accessible by calorimetric measurements,[8-11] but few laboratories are equipped for such determinations, and they become increasingly difficult for systems of progressivley smaller cmc.[9] An alternative and usually far easier approach is to deduce the enthalpy from measurements of the cmc over a range of temperatures.[12] If $\nu$ were strictly constant one could simply combine the result $\Delta G_\nu^o \simeq \Delta G_\nu^{o\dagger} \simeq RT \ln(\text{cmc})$ and the van't Hoff formula $\Delta H^o = -T^2 (d/dT)(\Delta G^o/T)$ to obtain

$$\Delta H_\nu^o \simeq \Delta H_\nu^{o\dagger} \simeq -RT^2 \frac{d}{dT} \ln(\text{cmc}). \tag{6}$$

an expression which has been extensively used to derive micellization enthalpies.

Of the several assumptions involved in deriving Equation (6), the one which introduces the most serious error is the neglect of the temperature dependence of $\nu$, since it has been shown repeatedly[13-15] that micelles may change size dramatically on heating or cooling. Then the symbol $\Delta G_\nu^o$ or $\Delta G_\nu^{o\dagger}$ does not have the same meaning at different temperatures and the straightforward application of the van't Hoff relation is no longer permissible.[16] Nevertheless, experiments have shown that Equation (6) often, though not always, yields results in good agreement with those of calorimetric determinations.[4,10,11] Two recent studies would seem to call for a further analysis of the errors involved in the use of this Equation.

It was shown by Holtzer and Holtzer[16] that the temperature derivative of $\ln(\text{cmc})$ should obey a relation which, with minor changes of notation, is

$$RT^2 \left(\frac{\partial \ln(cmc)}{\partial T}\right)_P = -\Delta H_\nu^{o\dagger} + T\left[\frac{\partial(\overline{G}_{\nu+1}^o - \overline{G}_\nu^o)}{\partial \nu}\right]_{T,P} \left(\frac{\partial \nu}{\partial T}\right)_P, \quad (7)$$

where $\overline{G}_n^o$ is the standard partial molal free energy of a micelle of order n. The last term in Equation (7) is not readily evaluated theoretically, but it was possible at least to estimate the contribution from electrostatic interactions which proved to be far from negligible. It thus appeared that values of $\Delta H_\nu^{o\dagger}$ calculated from Equation (6) could be grossly in error, so that studies of cmc as a function of temperature could give no valid information about micellization enthalpies. The calorimetric results cited above suggest the contrary, which could be taken to imply either that compensating contributions greatly reduce the magnitude of the correction term in Equation (7) or that the quantity $-RT^2(d/dT) \ln(cmc)$ provides a much better approximate value of $\Delta H_\nu^o$, the enthalpy change measured by the calorimetric method, than of $\Delta H_\nu^{o\dagger}$.

Very recently it was reported that heats of incorporation corresponding to the quantity $\Delta H_\nu^{o\dagger}$ can be derived from measurements of the relaxation times associated with micelle formation and dissolution.[17] Values presented for four sodium alkyl sulfates ranged from -6.7 kJ/mol for the octyl sulfate to -31 kJ/mol for the hexadecyl sulfate. Previous work using both calorimetry and the temperature dependence of the cmc had shown that $\Delta H_\nu^o$ is near zero for sodium dodecyl sulfate[10] and positive for the lower homologs.[8] Though there is no obvious physical reason for $\Delta H_\nu^o$ and $\Delta H_\nu^{o\dagger}$ to be equal, they must be so long as Equation (6) is applicable. Again the question arises whether or not Equation (7) allows large differences between these two quantities to be rationalized and, if so, what conditions are conducive to the existence of such differences.

## EVALUATION OF $\Delta H_\nu^{o\dagger}$ AND $\Delta H_\nu^o$ FOR A MODEL SYSTEM

The dependence of the concentrations $[A_n]$ on the aggregation number n has not been experimentally determined. Mathematical convenience suggests taking a Gaussian distribution, and it seems likely that at least for spheroidal micelles this choice is in reasonably good accord with reality. Then

$$[A_n] = Ce^{-a^2(n-\nu)^2} \quad \text{with } a^2 = \ln 2/s^2\nu^2, \quad (8)$$

the expression for $a^2$ being chosen so that the full width at half height is $2s\nu$, that is,

$$[A_{\nu+s\nu}] = [A_{\nu-s\nu}] = \tfrac{1}{2}[A_\nu]. \tag{9}$$

It is further assumed that any changes in $s\nu$ caused by temperature variations parallel changes in $\nu$, so that $s$ (but not $a$) is constant. Since constant-pressure thermodynamic relations are of primary interest, it is also assumed that this condition holds, allowing derivatives such as $(\partial \nu/\partial T)_P$ to be written simply as $d\nu/dT$.

Turning to Equation (7), the derivative $(\partial/\partial\nu)(\overline{G}^o_{\nu+1} - \overline{G}^o_\nu)$ is essentially equal to the finite difference $(\overline{G}^o_{\nu+1} - \overline{G}^o_\nu) - (\overline{G}^o_\nu - \overline{G}^o_{\nu-1})$, which in turn is equal to the standard free energy change of the reaction

$$2A_\nu \rightleftarrows A_{\nu+1} + A_{\nu-1}. \tag{10}$$

Equation (8) requires the equilibrium constant for this process to be $\exp(-2a^2)$ so that $\Delta G^o = 2a^2 RT = 2(\ell n2)RT/s^2\nu^2$. Then Equation (7) may be rewritten as

$$\Delta H^{o\dagger}_\nu = -RT^2 \tfrac{d}{dT} \ln(\text{cmc}) + 1.386(RT^2/s^2\nu^2)(d\nu/dT). \tag{11}$$

When an estimate of the width of the micelle size distribution is available this makes it possible to evaluate at least semiquantitatively the difference between $\Delta H^{o\dagger}_\nu$ and $-RT^2 (d/dT)\ln(\text{cmc})$.

To find an analogous relation for $\Delta H^o_\nu$, it is convenient to begin with the definition

$$\Delta H^o_n = \frac{1}{n} \sum_{i=2}^{n} \Delta H^{o\dagger}_i \tag{12}$$

and seek a procedure for evaluating the required sum. Equations (4) and (8) give, when $[A] = \text{cmc}$,

$$\Delta G^{o\dagger}_n = RT\ln(\text{cmc}) + a^2 RT(2n - 2\nu - 1). \tag{13}$$

This shows that taking a Gaussian distribution is equivalent to assuming that $\Delta G^{o\dagger}_n$ is a linear function of $n$, with

$$\Delta G^{o\dagger}_n - \Delta G^{o\dagger}_{n-1} = 2a^2 RT. \tag{14}$$

Application of the van't Hoff relation now gives

$$\Delta H^{o\dagger}_n - \Delta H^{o\dagger}_{n-1} = -2RT^2(da^2/dT), \tag{15}$$

which has the somewhat unexpected consequence that $\Delta H_n^{o\dagger}$ would be independent of n if the parameter a were invariant with temperature. However, it was assumed above that $da^2/dt \neq 0$, when $dv/dT \neq 0$, and then

$$\Delta H_n^{o\dagger} = \Delta H_v^{o\dagger} - 2(n - v)RT^2(da^2/dT). \tag{16}$$

Formally it is now possible to evaluate the sum in Equation (12), but one obstacle remains. Equation (8) cannot apply for the smallest values of n, certainly not when n = 1, since otherwise one would have at all concentrations $[A_v]/[A] = \exp[a^2(1-v)^2]$, which would be inconsistent with the requirement that there be no micelles when [A] is well below the cmc and that [A] = cmc when the total concentration is above the cmc. Instead, $[A_n]$ must pass through a minimum at some value, say n = r, with $1 < r << v$, and the distribution can be assumed Gaussian only for n > r. Rewriting Equation (12) in the form

$$\Delta H_n^o = \frac{r}{n} \Delta H_r^o + \frac{1}{n} \sum_{i=r+1}^{n} \Delta H_i^{o\dagger}, \tag{17}$$

one may use Equation (16) to obtain

$$\Delta H_n^o = \frac{r}{n} \Delta H_r^o + \frac{(n-r)}{n} \Delta H_v^{o\dagger} - \frac{2RT^2}{n} \frac{da^2}{dT} \left[ \frac{n(n+1)}{2} - \frac{r(r+1)}{2} - v(n-r) \right]. \tag{18}$$

Taking n = v, this yields

$$\Delta H_v^o = \frac{r}{v} \Delta H_r^o + \frac{(v-r)}{v} \Delta H_v^{o\dagger} + \frac{RT^2}{v} \frac{da^2}{dT} (v^2 - v(1+2r) + r(r+1)). \tag{19}$$

An expression for $\Delta H_n^{o\dagger}$ readily obtainable from Equation (13) is

$$\Delta H_n^{o\dagger} = - RT^2 \frac{d}{dT} \ln(cmc) + 2a^2 RT^2 \frac{dv}{dT} - (2n - 2v - 1)RT^2 \frac{da^2}{dT}, \tag{20}$$

or, again taking n = v,

$$\Delta H_v^{o\dagger} = - RT^2 \frac{d}{dT} \ln(cmc) + 2a^2 RT^2 \frac{dv}{dT} + RT^2 \frac{da^2}{dT}. \tag{21}$$

Since s was assumed independent of temperature,

$$da^2/dT = -2(\ln 2)(dv/dT)/s^2 v^3 = -(2a^2/v)dv/dT. \tag{22}$$

Then it is easily shown that Equation (21) reduces to Equation (11) when $v >> 1$, as it should. Further, combining Equations (19), (21), and (22) yields, after simplification,

$$\Delta H_\nu^o = \frac{r}{\nu}\Delta H_r^o - \frac{(\nu-r)}{\nu}RT^2\frac{d}{dT}\ln(cmc) + \frac{2a^2r(\nu-r)RT^2}{\nu^2}\frac{d\nu}{dT}. \qquad (23)$$

Unless the most abundant micelles are very small, this may be further simplified by imposing the condition $\nu \gg r$, so that

$$\Delta H_\nu^o \cong -RT^2\frac{d}{dT}\ln(cmc) + \frac{r}{\nu}\Delta H_r^o + 1.386\frac{RT^2 r}{s^2\nu^3}\frac{d\nu}{dT}. \qquad (24)$$

The last term in this Equation is just $r/\nu$ times the correction term in Equation (11).

## DISCUSSION

Published data make possible some exploration of the numerical implications of Equations (11) and (24). The variation of micelle size with temperature has been more extensively studied for n-alkyl polyoxyethyleneglycol monoethers than for any other class of surfactants.[14,15,18,19] For these materials $d\nu/dT$ is always positive, and two types of behavior are commonly found. At the lowest concentrations and temperatures, relatively small and roughly globular micelles are formed with $(d/dT)\ln(\nu) \simeq 0.025$. Theoretical calculations[20] for globular micelles of ionic detergents produced values of s in the range of 0.2 to 0.25, suggesting that $1/s^2 = 20$ is a reasonable estimate. The correction term in Equation (11) was then evaluated for a typical micelle with $\nu = 50$, and the results are shown in Table I as case A.

Table I. Numerical Values of the Correction Term in Equation (11) for Four Representative Situations.

| Case | $1/s^2$ | $\nu$ | $\frac{d}{dT}\ln(\nu)$ | $\frac{1.386RT^2}{s^2\nu^2}\frac{d\nu}{dT}$ (kJ/mol) |
|---|---|---|---|---|
| A | 20 | 50 | 0.025 | 10.3 |
| B | 2 | 200 | 0.15 | 1.6 |
| C | 20 | 64 | -0.01 | -3.2 |
| D | 100 | 64 | -0.01 | -16. |

Case B represents the situation found at higher concentrations and temperatures, when the micelles are much larger and change size with increasing temperature very rapidly, with $(d/dT)\ln(\nu) \simeq 0.15$. An optimal aggregation number of 200 was taken, and noting that very large micelles are likely to be highly polydisperse[21] the value of $1/s^2$ was reduced by a factor of 10. Under these conditions, the correction term is not much larger than the experimental error in determining $RT^2(d/dT)\ln(cmc)$, which is often[11] of the order of 1 kJ/mol.

Available data for micelles of sodium dodecyl sulfate suggested the parameters used in case C. It appears that for single-chain ionic surfactants $d\nu/dT$ is negative, of the order of -0.4 to -1 in the absence of added electrolyte, but much greater for the large micelles formed by sodium dodecyl sulfate with added sodium chloride.[13,22] For a solution without added salt, $\nu = 64$ and then $(d/dT)\ln(\nu)$ must be near -0.01. The value of $1/s^2$ was taken to be 20 as suggested both by Tanford's calculations[20] and by an estimate of the width of the distribution derived from kinetic data.[17]

The results of case C are not consistent with the finding[17] that for this material $\Delta H_\nu^{o\dagger} = -18$ kJ/mol although both $\Delta H_\nu^o$ and $RT^2(d/dT)\ln(cmc)$ are nearly zero. However, as shown in case D, the inconsistency disappears if one uses a somewhat narrower distribution, with $1/s^2 = 100$. Alternatively, one could reconcile the results with a smaller readjustment of the width combined with a larger value for $(d/dT)\ln(\nu)$. Of course it is also possible that the reported value of $\Delta H_\nu^{o\dagger}$ is somewhat more negative than it should be; the theory of Nemethy and Scheraga[23] as modified by Nelson and de Ligny[24] calls for an enthalpy change of -10 kJ/mol for the transfer of n-dodecane from aqueous solution to the pure liquid, and there is no obvious reason why the enthalpy change for the incorporation of a dodecyl sulfate ion into a micelle should be more favorable than this by nearly a factor of 2.

The correction terms in Equation (24) are more difficult to evaluate. As already noted, the last term is $r/\nu$ times the correction term in Equation (11). Assuming that for each case in Table I $r/\nu$ is of the order 0.1 or less, the respective contributions from this source should not exceed 1.0, 0.2, -0.3, or -1.6 kJ/mol. The only experimental data that give an indication of the probable value of $\Delta H_r^o$ are those of Aniansson et al.,[17] who found that for sodium alkyl sulfates $r\Delta H_r^o \simeq 135$ kJ/mol. Then $(r/\nu)\Delta H_r^o$ is about 2 kJ/mol, and since the last two terms of Equation (24) turn out to have opposite signs their sum should be of the order of 1 kJ/mol. For the nonionics, the two terms probably have like signs, but the actual value of $\Delta H_r^o$ may be appreciably smaller, since it is likely that the large positive value found for the alkyl sulfates reflects a sizeable electrostatic contribution. Again it is quite possible to rationalize the experimental finding that $\Delta H_\nu^o$ is not greatly different from $-RT^2(d/dT)\ln(\text{cmc})$, even when this is far from true of $\Delta H_\nu^{o\dagger}$. It should be noted here that, using a different approach, Corkill and coworkers[4] derived the equation

$$H_{n_{av}}^o = -(1 - 1/n_{av})RT^2\frac{d}{dT}\ln(\text{cmc}) - (RT^2/n_{av}^2)(dn_{av}/dT), \quad (25)$$

where $n_{av}$ is the average aggregation number, not greatly different from $\nu$. Although this result is not identical with Equation (24), it also requires that $\Delta H_\nu^o$ be given fairly accurately by $-RT^2(d/dt)\ln(\text{cmc})$ even when the correction term in Equation (11) is large.

It appears that the model makes it possible to explain a number of seemingly contradictory theoretical or experimental results in a straightforward manner. The calculations support Holtzer and Holtzer's conclusion that reliable values of $\Delta H_\nu^{o\dagger}$ cannot be inferred from the temperature variation of the cmc, but they suggest that the error in using this procedure to evaluate $\Delta H_\nu^o$ will often be of the order of 1 kJ/mol or less. The difference between the two "enthalpies of micellization" is expected to be small only when $d\nu/dT$ is nearly zero or when the micelles are quite large and highly polydisperse, and it may be either positive or negative according to the sign of $d\nu/dT$. For the sake of simplicity, additional corrections that may be necessary in applying Equation (6) to ionic micelles[25] have not been included.

The assumptions defining the model are not, of course, the only possible ones, and the conclusions are somewhat sensitive to certain changes, especially with respect to the temperature dependence of the distribution width. It is clear from Equation (19) that the difference between $\Delta H_\nu^o$ and $\Delta H_\nu^{o\dagger}$ depends crucially on the assumption that $da^2/dT$ is given by Equation (22). With the simplest alternative assumption, i.e. $da^2/dT = 0$, the difference would nearly vanish, while in the more general case where both $s$ and $\nu$ are allowed to be temperature dependent, Equation (22) becomes

$$\frac{da^2}{dT} = -2a^2\frac{d}{dT}\ln(\nu) - 2a^2\frac{d}{dT}\ln(s), \quad (26)$$

and Equation (24) must be modified accordingly. When this is done, one finds that the results are essentially unaffected as long as $(d/dT)\ln(s)$ is small compared with $(d/dT)\ln(\nu)$.

While there is no compelling argument in favor of choosing $ds/dT = 0$, there would seem to be good reason to reject the alternative $da/dT = 0$. It was noted above that when $a$ is independent of temperature $\Delta H_n^{o\dagger}$ becomes independent of $n$, so that the dependence of $\Delta G_n^{o\dagger}$ on $n$ would have to be purely an entropy effect. This seems unlikely in light of the general observation[26] that a variety of factors which affect values of $T\Delta S^o$ for micellization tend to produce roughly compensating changes in $\Delta H^o$. In contrast, returning to the assumption that $ds/dT = 0$ and using case A of Table I as a numerical illustration, it is found that $\Delta G_n^{o\dagger} - \Delta G_{n-1}^{o\dagger} = 0.011RT$, $\Delta H_n^{o\dagger} - \Delta H_{n-1}^{o\dagger} = 0.16RT$, and therefore $T\Delta S_n^{o\dagger} - T\Delta S_{n-1}^{o\dagger} = 0.15\ RT$, providing a quite plausible example of compensation behavior.

## ACKNOWLEDGMENTS

The author is indebted to Professor Alfred Holtzer for providing a very careful review of an earlier version of this article and suggesting important modifications.

## REFERENCES

1. C. Tanford, J. Mol. Biol., 67, 59 (1972).
2. K. H. Meyer and A. van der Wyk, Helv. Chim. Acta, 20, 1321 (1937).
3. M. J. Vold, J. Colloid Sci., 5, 506 (1950).
4. J. M. Corkill, J. F. Goodman, and J. R. Tate, in "Hydrogen Bonded Solvent Systems," A. K. Covington and P. Jones, Editors, p. 181, Taylor and Francis, London, 1968.
5. C. Tanford, J. Phys. Chem., 78, 2469 (1974).
6. M. F. Emerson and A. Holtzer, J. Phys. Chem., 69, 3718 (1965).
7. J. E. Leffler and E. Grunwald, "Rates and Equilibria of Organic Reactions," p. 50, J. Wiley and Sons, New York, 1963.
8. E. D. Goddard, C. A. J. Hoeve, and G. C. Benson, J. Phys. Chem. 61, 593 (1957).
9. J. M. Corkill, J. F. Goodman, and J. R. Tate, Trans. Faraday Soc., 60, 996 (1964).
10. G. Pilcher, M. N. Jones, L. Espada, and H. A. Skinner, J. Chem. Thermodynamics, 1, 381 (1969).
11. J. H. Clint and T. Walker, J. C. S. Faraday I, 71, 946 (1975).
12. G. Stainsby and A. E. Alexander, Trans. Faraday Soc., 46, 587 (1950).
13. P. Debye, Ann. N. Y. Acad. Sci., 51, 575 (1949).
14. R. R. Balmbra, J. S. Clunie, J. M. Corkill, and J. F. Goodman, Trans. Faraday Soc., 58, 1661 (1962).
15. R. R. Balmbra, J. S. Clunie, J. M. Corkill, and J. F. Goodman, Trans. Faraday Soc., 60, 979 (1964).
16. A. Holtzer and M. F. Holtzer, J. Phys. Chem., 78, 1442 (1974).
17. E. A. G. Aniansson, S. N. Wall, M. Almgren, H. Hoffman, I. Kielmann, W. Ulbricht, R. Zana, J. Lang, and C. Tondre, J. Phys. Chem., 80, 905 (1976).
18. P. H. Elworthy and C. McDonald, Kolloid Z., 195, 16 (1964).
19. N. Muller and F. E. Platko, J. Phys. Chem., 75, 547 (1971).

20. C. Tanford, Proc. Nat. Acad. Sci. U.S.A., 71, 1811 (1974).
21. P. Mukerjee, J. Phys. Chem., 76, 565 (1972).
22. N. A. Mazer, G. B. Benedek, and M. C. Carey, J. Phys. Chem., 80, 1075 (1976).
23. G. Nemethy and H. A. Scheraga, J. Chem. Phys., 36, 3401 (1962).
24. H. D. Nelson and C. L. de Ligny, Rec. Trav. Chim. Pays-Bas, 87, 623 (1968).
25. E. Matijevic and B. A. Pethica, Trans. Faraday Soc., 54, 587 (1958).
26. G. C. Kresheck, "Water, A Comprehensive Treatise," F. Franks, Editor, Vol. 4, pp. 145-157, Plenum Press, New York, 1975.

# THE NATURE OF THE LOCAL MICROENVIRONMENTS IN AQUEOUS MICELLAR SYSTEMS

Pasupati Mukerjee, John R. Cardinal, and Narendra R. Desai

School of Pharmacy, University of Wisconsin
Madison, Wisconsin 53706

Characteristic features of the local microenvironments of different parts of aqueous micellar systems are reviewed and some new results are presented. The different parts include the hydrocarbon core, methylene groups close to the polar head groups, the micellar-water interface including the innermost part of the electrical double layer (Stern layer) for ionic micelles, the outer part of the diffuse double layers, and the mantle of polyoxyethylene groups in some non-ionic surfactants. Different factors involved in determining the 'effective' polarity of the micelle-water interface are reviewed and some new results are presented which suggest that the proximity of the hydrocarbon core as also dielectric saturation at the interface are contributory factors for short-range interactions. Some new results on the dissociation constants of indicator dyes solubilized in micelles of non-ionic and zwitterionic surfactants indicate the substantial role of local microenvironments. The microenvironments experienced by solubilized species are discussed. Some results are presented which indicate that the environments of solubilized benzene and naphthalene are quite polar. These observations are rationalized in terms of the predominance of the interfacial region as the locus of solubilization. Thermodynamic arguments indicate that this effect can be ascribed to the expected mild surface activity of benzene in hydrocarbon-water systems which becomes important for micelles because of their extremely high surface-to-volume ratios. The general implications of some of these findings for colloidal, interfacial and membrane systems are briefly reviewed.

## INTRODUCTION

The understanding of the nature of the local microenvironments of micellar systems is important for the understanding of the structure and properties of the micelles, their ability to solubilize additional species, the physical and chemical properties of the solubilized species and the properties of mixed micelles. Because of the simplicity and adaptability of the micellar systems,[1] many questions regarding microenvironments of small colloidal systems or thin interfacial systems can be studied using micellar systems as model systems. Any insight derived for micellar systems is frequently useful for the understanding of related colloidal and interfacial systems such as monolayers, bilayers and membranes.

The present discussion is limited to aqueous solutions of long-chain amphipathic surfactants. The currently accepted picture of the micelles that form in such solutions is basically that of Hartley.[2,3] The micelles have a hydrocarbon core. The polar head groups, which may be ionic, non-ionic, or zwitterionic, attached to the hydrocarbon chain ends are exposed to water at the micelle-water interface. For the purposes of the present paper, we will distinguish between and discuss separately the hydrocarbon core, a small number of $-CH_2-$ groups adjacent to the polar group, the micelle-water interface region where small polar head groups are located, the outer part of the electrical double layer (diffuse double layer) for ionic micelles, and the outer part of the polar sheath or mantle of oxyethylene groups that are present in many non-ionic surfactants for which the oxyethylene groups are often larger and occupy more volume than the core. We emphasize, however, that characterizing the nature of the local microenvironments often requires some arbitrary distinctions or definitions of 'local' regions, i.e. a dissection of the already small micelle into smaller regions. For investigating small-scale systems of this kind, where typical dimensions may only be a fraction of a nanometer (several Ångstroms), theories based on any continuum model of the medium are highly unsatisfactory.[4] Unfortunately, workable theories based on the recognition of discreteness of the solvent and solute species are mostly nonexistent. On the experimental side, spectroscopic methods are, in principle, more useful than thermodynamic studies for characterizing 'local' microenvironments. As discussed later, a serious problem about spectroscopic sensors and added probes often arises from a lack of knowledge about the location and distribution of the sensor and the probe in the micellar or interfacial system.

## EXPERIMENTAL

**Materials.** Benzene (Eastman White Label) and heptane (Aldrich) were purified by standard methods. Naphthalene (Baker and Adamson) was resublimed before use. Sodium dodecyl sulfate from Mann Research and British Drug Houses was used. The latter sample was purified so that surface tension data showed no minimum near the critical micelle concentration. The purified sample gave the same results as the sample from Mann Research. Cetyl trimethyl ammonium chloride (Eastman White Label) was purified before use. Literature methods were used for the synthesis of dodecyl N-betaine[5] and β-D-octyl glucoside[6]. Brij 35 was passed through columns of animal charcoal and aluminium oxide before freeze drying to remove some peroxides and interfering absorbing material.

**Apparatus.** Cary 14 and Cary 16 spectrophotometers were used.

## RESULTS AND DISCUSSION

The absence of any crystal-like organization of the micelle and the primarily liquid-like nature of the core were deduced by Hartley[2] on the basis of the rapid monomer-micelle equilibriation, the minor dependence of the critical micellization concentration (c.m.c.) on counterions or head groups for ionic surfactants[1] and the ability of the micelles to readily dissolve hydrophobic molecules of a variety of structures. Additional evidence derives from the similarity of the heat capacities[7] and compressibilities[8] of the micelles to those of liquid hydrocarbons. To investigate the possibility of any conformational restraints on the head groups, the optical rotatory dispersion of octyl glucoside was studied above and below the c.m.c.[6] Small changes were observed on micelle formation but these could be ascribed to local medium effects. No evidence was found of any conformational restraint on the glucoside head groups at the micelle surface, as would be expected from a liquid-like character of the core.

In contrast to the above studies, a comparison of the free energies of homologous sodium alkyl sulfates containing chains of odd and even number of carbon atoms suggests that the micelle core may have some structure and may have a partial solid-like character.[9] The free energies were calculated using a mass-action model using the extensive data from Huisman,[10] who drew attention to some other differences in the behavior of chains containing odd or even number carbon atoms. Figure 1 shows an example of an irregularity in the variation of the c.m.c. with chain length when the concentration of the counterion, $Na^+$, is kept constant at 0.3 mol $dm^{-3}$. The c.m.c. values were interpolated from Huisman's data,[10] and a deviation plot is exhibited. The c.m.c. values of the three

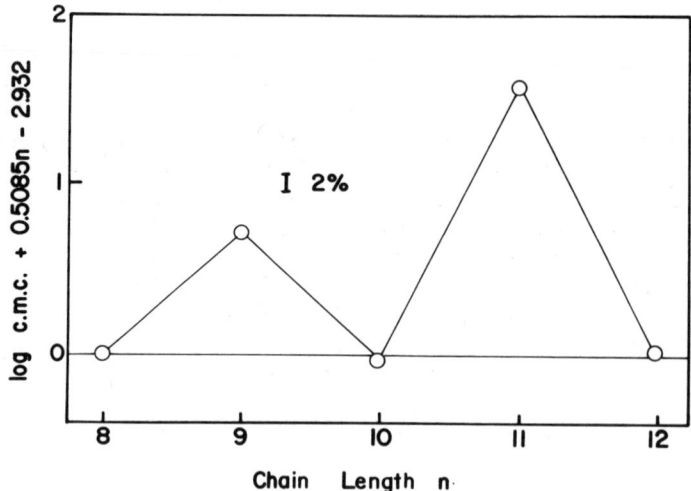

Figure 1. Deviation of the c.m.c. values of sodium alkyl sulfates of different chain lengths at a constant $Na^+$ concentration (0.300 mol $dm^{-3}$) from an equation fitting the even-numbered chains (Equation 1).

compounds with even number of carbon atoms were fitted to an equation

$$\log \text{c.m.c.} = 2.932 - 0.5085n \qquad (1)$$

where n is the number of carbon atoms. The experimental c.m.c. values fit this equation well within experimental error for the even-numbered chains but the odd-numbered chains show appreciable deviation. Paraffin-chain compounds often exhibit such odd-even effects but only when a solid phase is involved.[11]

The deviations observed for the odd-numbered chains are not large enough to demand that the picture of the essentially liquid-like core of the micelle should be given up altogether.[9] It is also not known if the phenomenon is a general one and whether it depends upon the head group and counterions. However, the evidence of some organization in the micelle core and the existence of some order is consistent with the constrained state of the monomeric chains in the core as inferred from free energies of micelle formation and their comparison with transfer free energies of hydrocarbon chain compounds from aqueous solutions to hydrocarbon media.[1]

A factor likely to be of some importance in determining the properties of the hydrocarbon core is the Laplace pressure arising from the presence of the curved interface of micelles.[9] Approximate calculations using fluid-drop models indicate hundreds of atmospheres for small spherical micelles.[9] It is interesting to note that partial specific volumes do not change significantly as added salt concentrations (ionic strengths) vary over a large range for sodium dodecyl sulfate and some cationic surfactants.[12] The aggregation numbers over this range are compatible with roughly spherical micelles. For dodecylammonium chloride, however, when very high aggregation numbers are encountered at high salt concentrations,[13] the partial specific volumes showed significantly higher values.[12] The very large micelles are highly asymmetric and are probably worm-like, i.e. rod-like and with some flexibility.[14-17] The transition from spherical to rod-like micelles is probably accompanied by a considerable reduction in the interfacial curvature, and, because of the higher head group density for the rods, a reduction in the effective interfacial tension as well. These factors should reduce the effective Laplace pressure very markedly, thus providing a qualitative explanation of the increase in partial specific volume and an increase in the effective fluidity of the micellar core indicated by the increase in partial specific volume.[12]

Penetration of Water into the Core. The suggestion that the micellar core contains water came from the work of Muller and Birkhahn[18] who introduced $^{19}F$ nuclear magnetic resonance studies as a valuable approach for micellar systems. Fluorine chemical shifts are much more sensitive to changes in the medium than the corresponding proton shifts. The chemical shifts for some partially fluorinated surfactants ($\omega$-$CF_3$) suggested that the average environment of the terminal $CF_3$ groups was roughly midway between water and hydrocarbons. This gave rise to the suggestion of penetration of water into the hydrocarbon core of the micelles.[18] It has been pointed out recently that the $^{19}F$ chemical shift data permit an alternative explanation based on the pronounced non-ideality of the interactions between fluorocarbons and hydrocarbons.[19] For mixtures of perfluoro and hydrocarbon surfactants, the non-ideality effects are intense enough to lead to a partial demixing of the micelles.[20] In the case of partially fluorinated surfactants, the non-ideality effects are reflected in anomalous critical micellization concentration (c.m.c.) values and lead to a nonrandom sampling of the micelle-water interface by the $\omega$-CF3 groups.[19] The poor solubility of water in hydrocarbons also argues strongly against the presence of much water in the core of the micelle.[19] Podo, Ray, and Nemethy[21] also concluded from their proton NMR studies on the hydrophobic moiety of a non-ionic surfactant that there is little water in the interior of the micelle.

The extent to which the $-CH_2-$ groups attached to the polar group at the micelle surface are wet by water is a matter of some controversy.[19,22-24] The evidence from partial molal volume data[22] and spin lattice relaxation times of micellized hydrocarbon protons[23] suggesting that some methylene groups adjacent to the polar group retain their hydration has been criticized by Stigter,[24] who has argued that the α-methylene group is contained in the hydrocarbon core. In a general sense, considering the fluid nature of the micellar core, it seems that the thermal fluctuations of the monomer at the micelle surface[19] is an important factor as is the possible loss in hydration energy of the polar groups in close contact with the hydrocarbon core. The latter is likely to be more important for ionic and zwitterionic surfactants than for non-ionic surfactants. The hydration energy of ionic head groups is of the order of 30 K.cal/mole. Even a small amount of effective dehydration is equivalent to roughly -0.7 cal/mole, the contribution of a $-CH_2-$ group to the free energy of micelle formation.[1] With regard to the theoretical investigation of such problems, even the finite dimension of the water molecule is likely to be an important factor.

On the experimental side, an important contribution comes from some kinetic studies by Kurz.[25] He showed that the spontaneous hydrolysis of sodium alkyl sulfates, which involves water attack at the α-carbon, is not affected significantly by micelle formation. As stated by Kurz,[25] this implies "that the micelle surface is very 'wet' at least as far as the α-carbon. More precisely stated, the water activity and the solvating power of the medium are essentially unchanged at this depth into the micelle."

## Micelle-Water Interface

This region (layer) contains the head groups. In the case of ionic surfactants, it constitutes the innermost part of the micellar electrical double layer (the Stern layer). The interfacial region is 'rough,[26] as is to be expected from the dimensions of the ionic head groups and thermal fluctuations of the monomers. The roughness of the interface is also related to the penetration of counterions between head groups[26] and their binding.[1,27] The electrostatics of the interface has been investigated extensively. It has been shown from studies on solubilized indicator dyes[28] and estimates of free energies of micelle formation[29] that the Gouy-Chapman theory of electrical double layers leads to an overestimate of the surface potential. This is also indicated by simple calculations of surface counterion concentrations on the basis of the calculated Gouy-Chapman potentials. Thus, for sodium dodecyl sulfate in 0.3M NaCl, the calculated surface potential of 110 millivolts[10] leads to a surface concentration of 22 mol dm$^{-3}$ for sodium ion which is physically implausible. The specificity effects shown by different monovalent counterions[27] also argues against the applicability of the Gouy-Chapman theory. In a series of important papers, Stigter[24,30,31] has introduced many refinements of the Gouy-Chapman model.

Two microenvironmental features of the interfacial region, namely its hydration and its effective polarity, are discussed at somewhat greater length because of their general importance for colloidal and interfacial systems and membranes.

<u>Hydration of Ionic Micelles</u>. Ionic micelles are attractive model systems for hydration studies of charged surfaces because of their simple surface composition. Table I shows some hydration

Table I. Hydration of Ionic Micelles[32]

| Surfactant | Hydration (weight %) | | |
|---|---|---|---|
| | Exp.(a) | Calc.(b) | Monolayer(c) |
| Sodium dodecyl sulfate | 39 ± 4 | 24 | 43 |
| Dodecyl trimethylammonium chloride | 24 ± 8 | 17 | 53 |
| Tetradecyl trimethylammonium chloride | 25 ± 8 | 15 | 45 |

(a) Estimated from viscosity data in 0.1 mol dm$^{-3}$ NaCl assuming spherical geometry and correcting for electroviscous effects.
(b) Calculated assuming 70% counterion binding and primary hydration numbers of 4 for Na$^+$, 2 for Cl$^-$, and 1 for charged head groups.
(c) Calculated for monolayer coverage.

estimates[32] from viscosity data after corrections for the electroviscous effect and assuming a spherical model. Any correction for deviation from sphericity due to roughness of the interface or asymmetry of the core would reduce the estimate of experimental hydration. Even without such corrections, the experimental values are less than the hydration estimated from a uniform monolayer of water, particularly for the cationic surfactants. The hydration for the latter must thus be patchwise. When the experimental hydration values are compared with calculations assuming some primary hydration numbers from ionic solution studies, reasonable agreement is obtained. It appears, therefore, that the hydration of the micelles is consistent with the hydration of the head groups and the counterions in the absence of micellization and there is no need to invoke major contributions to hydration from any ice-like structures at the surface,[33] the postulated increase in the viscosity of water at charged surfaces,[34] or to the cooperative effect of the charges at the micelle surface. It is interesting to note that Hartley presented arguments against the importance of the last factor many years ago.[2]

## Effective Polarity at Interfaces

The characterization of the local 'effective' polarity in an interface region in contact with an aqueous solution is important for understanding the reactivity of molecules or parts of molecules located in such a region. The local dielectric constant of interfacial systems is important also as a parameter for electrical double layers.[35,36]

Some years ago it was shown that the characteristic absorption spectra in the ultraviolet region exhibited by micelles of alkyl pyridinium iodides (as also some other salts) could be ascribed to interionic charge transfer (electron donor-acceptor) interactions.[4,37] These spectra show a high sensitivity to changes in the polarity of the surrounding medium.[38] When comparisons between the micellar spectra and those of ion pairs in suitable hydroxylic solvents and solvent mixtures were made, the 'effective' dielectric constant of the interface (Stern layer) of dodecyl pyridinium iodide micelles was found to be about 36, which is roughly in between the values of 2 for hydrocarbons and 79 for water. The interfacial region, in its polarity, is, therefore, not too far from methanol which has a dielectric constant of 33. Chemical reactivities of groups located at the interface are thus likely to be different when compared to aqueous solutions.

Because of the nature of the short-range interactions between the pyridinium nitrogen and the iodide counterion, the location of the spectroscopic sensor in this case in the innermost part of the electrical double layer or the Stern layer could be assumed with some confidence.[4] Thus, the spectroscopic sensor here may be considered to be 'wet.' At least three identifiable factors, not necessarily independent, can be proposed[4,39] to account for the 'effective' polarity at the interface: (a) the presence of high local concentrations of ions,[35,36,40] (b) partial saturation of the dielectric in the interfacial region because of the intense field present,[40] and (c) the proximity of the hydrocarbon core with its low dielectric constant.[4] Factor (a) was evaluated separately by comparing the spectra of 1-ethyl 4-carbomethoxy pyridinium iodide in solutions of various salts in water and various mixtures of primary alcohols and water.[39,41] It was found that even when bulk dielectric constants were lowered to 30 or less, by adding inorganic salts to water, the microscopic polarity changed very little, in striking contrast to the effect of adding alcohols to water. The low bulk dielectric constants of concentrated salt solutions reflect averages of water molecules 'bound' to the ions and free water molecules. For the short-range interactions between ions in ion pairs and at an interface, the polarity of the immediate environment (the solvation shells) is more important, and it appears to be affected only to a minor extent by the presence of hydrated ions close by.

To investigate the other two factors, studies were conducted on the spectra of dodecyl pyridinium iodide solubilized in different types of micelles.[41] Care was taken to make sure that the results reflected micellar compositions in which the sensor molecule, dodecyl pyridinium iodide, was at a trace concentration (mole fraction less than 0.01). Table II compares the estimated 'effective' dielectric constants of the interfaces of the micelles of two non-ionic surfactants and a zwitterionic surfactant with that of the ionic surfactant dodecyl pyridinium iodide itself. The non-ionic

Table II. 'Effective' Dielectric Constants at the Interfaces of Micelles Containing Different Head Groups and Changes in $pK_a$ Values of a Solubilized Indicator Dye

| Surfactant | Effective Dielectric Constant (a) | $\Delta pK_a$ (b) |
|---|---|---|
| β-D-Octyl glucoside $[C_8H_{17}OCHC_5H_{10}O_5]$ | 46 | 1.32 |
| Brij 35 $[C_{12}H_{25}(OCH_2CH_2)_{23}OH]$ | 36 | 1.81 |
| Dodecyl N-betaine $[C_{12}H_{25}N^+(CH_3)_2CH_2COO^-]$ | 37 | 0.85 |
| Dodecyl pyridinium iodide | 36 | |

(a) From the spectrum of solubilized dodecyl pyridinium iodide using alkanols and alkanol-water mixtures as reference solvents.[4]
(b) Increase in $pK_a$ on solubilization of the sulfonphthalein dye, bromthymol blue.

surfactant octyl glucoside probably comes fairly close to a system where the proximity of the hydrocarbon core is the primary effect. Estimates made from the dielectric constants of concentrated mannitol and sucrose solutions[42] indicate that the structurally similar glucose head groups should not reduce the local dielectric constant by more than 5-10 units from a concentration effect at the micellar interface.

The effect of the proximity of a hydrocarbon core or a region of low dielectric constant is likely to be generally important for many systems such as monolayers, proteins and enzymes, biological membranes and other interfaces. This effect is closely related to the self-potential of ionic groups at such interfaces arising from electrostatic image forces.[41]

The comparison of the 'effective' dielectric constants for octyl glucoside and Brij 35 suggests an additional effect due to the high concentrations of the polyoxyethylene groups,[4] which are mostly ether-like, at the micelle surface of Brij 35.[4] The lower dielectric constants of the zwitterionic and ionic micelles compared to that of octyl glucoside argues strongly for an appreciable contribution from the dielectric saturation effect. The zwitterionic system is likely to have a separation between the average plane of the positive charges attached to the chain and that of the negative charges at the terminus of the head groups leading to the presence of a strong field similar to that of ionic micelles in the innermost part of the electrical double layer.[4]

It seems, therefore, that factors (b) and (c) previously mentioned are both important for the short-range interactions involved in this polarity probe for the interface.

## Indicator Dyes in Micellar Systems

In a previous detailed study referred to earlier,[28] some sulfonphthalein indicator dyes were used to probe the surface potential and surface pH of ionic micelles. The dissociation equilibrium involved for these dyes can be represented as

$$Y^- \rightleftharpoons B^= + H^+ \quad (2)$$

where $Y^-$ represents the yellow, singly-charged form and $B^=$ represents the blue, doubly-charged form. It was inferred in this study that the intrinsic $pK_a$ of these dyes in the solubilized state were higher than in aqueous solution. To examine this effect in a more systematic manner, several indicator dyes have been studied in micellar systems.[41] Figure 2 shows how the indicator ratio $r = [Y^-]/[B^=]$ varies in well buffered solutions of the non-ionic surfactant,

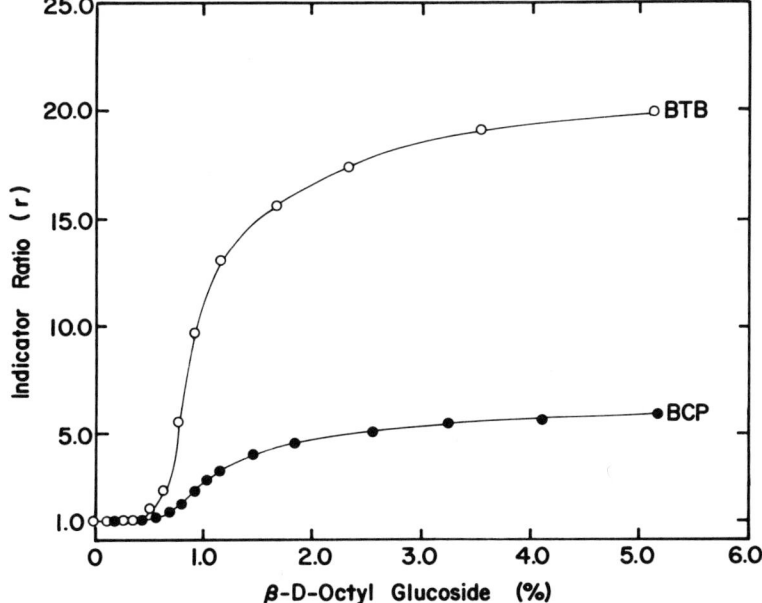

Figure 2. Variation in the indicator ratio r of two sulfonphthalein dyes in aqueous solutions of octyl glucoside in buffers relative to a normalized value of unity for the buffers. BCP: bromcresol purpole, phosphate buffer, pH 6.27, ionic strength 0.1. BTB: bromthymol blue, phosphate buffer, pH 7.00, ionic strength 0.1.

β-D-octyl glucoside. Below the c.m.c. (at about 0.7%), the normalized value of unity in the buffer itself changes very little. Above the c.m.c. the ratio r increases because of progressive solubilization. Two sulfonphthalein dyes show qualitatively similar but quantitatively different results. From the values at higher concentrations, the indicator ratio characteristic of the completely solubilized state can be determined by using a distribution-equilibrium method.[28] When this is done, the apparent dissociation constant of bromthymol blue solubilized in octyl glucoside micelles is found to be reduced by a factor of 21, corresponding to an increase in pK by 1.3 units.

Although different values of $\Delta pK_a$ observed for different sulfonphthalein dyes (see Figure 2) as also many other observations make a quantitative treatment very difficult,[39] qualitatively the rise in $pK_a$ of the anionic dyes and the fall in $pK_a$ of cationic dyes[43] (++ to + type) as also many other observations[41,43] indicate that such changes are marked for colloidal, interfacial, and polymeric systems. Qualitatively the results are consistent with the concept of the lower 'effective' dielectric constant at interfaces previously discussed, and the existence of electrostatic image forces and self-potentials at interfaces.

The complexity of such problems is indicated by some data in Table II which compares the change in $pK_a$ observed for bromthymol blue on solubilization in three different kinds of micelles. While the more pronounced effect for Brij-35 compared to octyl glucoside is consistent with the lower effective dielectric constant at the interface, the less pronounced effect for the zwitterionic surfactant is not. This system is indicative of an additional effect, namely the dissociation-field effect, at the micellar surface of the zwitterionic surfactant.[35,44]

The magnitude of the $\Delta pK_a$ values shows the importance of microenvironmental polarities in affecting chemical reactions.

### Outer Part of the Electrical Double Layer

The diffuse part of the electrical double layer of ionic micelles is of interest with regard to electrokinetic phenomena, which are not discussed here, and intermicellar interactions. The latter involve long-range repulsive interactions and are primarily due to the overlap of the tail or the low potential part of the double layer where the Gouy-Chapman theory is expected to work well. This appears to be the case for the second virial coefficients of sodium alkyl sulfates[10] for which good agreement is found between theory and experiments. However, lithium dodecyl sulfate exhibits a substantially higher virial coefficient compared to the sodium salt,[45] suggesting that specific interactions with counterion near the interfacial region may affect potential distributions far from the interface.[27]

### Microenvironments of Solubilized Species

The location, disposition, and orientation of solubilized species are of fundamental importance in the understanding of the nature of solubilization and the physical and chemical behavior of solubilized species. In general, although the exchange rates of solubilized species between the monomeric state in solution and the solubilized state in the micelles are rapid,[46] the lifetime of the solubilized species in micelles[47] is long enough to allow diffusional and molecular rearrangements to reach equilibrium within the micelle.[19] Thus, depending upon the molecular structure of the solubilized species as also the structure of the micelle, a distribution or rapid interchange between several states of different orientation and energy is to be expected. In some cases, some locations or orientations are greatly preferred. Thus, differences in the location of purely hydrophobic solutes such as heptane and an amphipathic solute (polar-nonpolar) such as heptanol might be expected on structural grounds, the former concentrating in the core, and the latter attaining a disposition similar to amphipathic surfactants themselves with the polar group anchored at the interface. Broad distinctions of this kind and assignments of some relatively static geometrical locations have been discussed at great lengths.[48,49,50]

An interesting point about nonionic micelles containing polyoxyethylene groups is that the voluminous mantle of these micelles containing the head groups can act as an effective locus for solubilization.[51,52] There are indications that in some cases this locus is comparable in importance to the hydrocarbon core.[52] When the solubilizate concentration in the micelle becomes high, non-ideality effects can become important also.[53]

A considerable amount of controversy exists regarding the site of solubilization of aromatic molecules, particularly benzene.[54-58] NMR studies have indicated a surface site[54] whereas comparisons of some features of the ultraviolet (UV) spectrum of solubilized benzene with those in reference solvents have indicated that the site of solubilization is the micellar core.[56,57]

An important problem for such studies concerns the reliability and sensitivity of the calibration procedure used to match the observed behavior in micellar systems to that in reference solvents. In our study of the site of solubilization of aromatic species using UV spectra,[59] several spectroscopic parameters have been chosen which show a graded, nearly monotonic variation in reference solvents of graded polarity including hydrocarbons and water at the extreme ends of polarity and alcohols and water-alcohol mixtures in the intermediate range. We have included here some results of general significance.

Figure 3 shows the long wavelength part of the UV spectra of dilute solutions of benzene in water, heptane, and 0.4 mol dm$^{-3}$ solution of sodium dodecyl sulfate in which the benzene is solubilized to the extent of about 90%. The comparison of the two spectra in the two pure liquids indicates the presence of the first member of a weak band system which is located at 3.6 nm on the long wavelength side of the major peak. The assignment of this band has been a matter of controversy,[60-63] but it is known that it is in the position of the 0-0 band[63] which is forbidden by selection rules and is absent in the vapor phase. The band is quite intense in

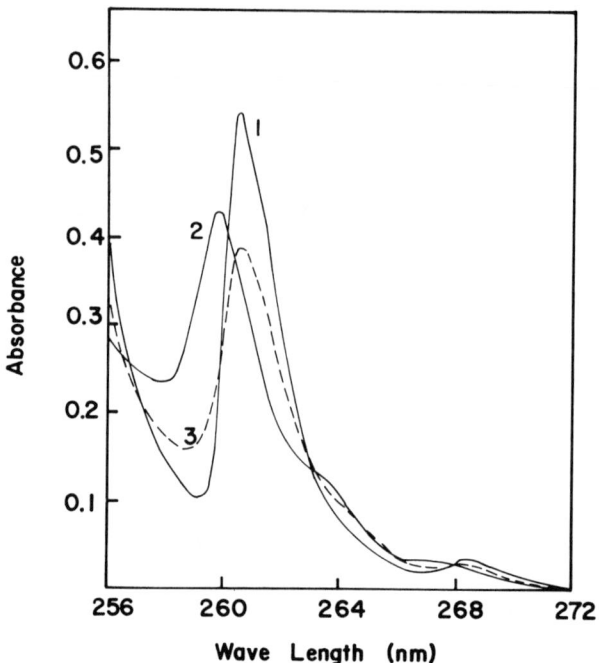

Figure 3. Absorption spectra of dilute solutions of benzene in 1-heptane, 2-water, 3-0.4 mol dm$^{-3}$ sodium dodecyl sulfate in water.

solid benzene[64] and benzene dissolved in some solid organic glasses.[65] In liquid solvents, the apparent intensity of the band, which can be described as a "solvent-induced" band,[66] has been shown to be related to the relative dispersion energy between benzene and the solvent molecules, when the latter are nonpolar.[62] Figure 4 shows how the ratio of the absorbance, $\varepsilon_s$, at the position of the solvent-induced band, i.e. at a wavelength of 3.6nm higher than the nearby peak, and the absorbance at the peak, $\varepsilon_p$, varies in different solvents of graded polarity. The solvents cover the whole range in polarity from heptane to water, and includes several pure alkanols and methanol-water mixtures. The ratio $\varepsilon_s/\varepsilon_p$, i.e. the relative intensity of the solvent-induced band, appears to be a monotonic function of H, which is the molar concentration of -OH groups in these reference solvents relative to that in water (55.5 mol dm$^{-3}$). A substantial but nearly constant part of this ratio is due to the contribution of the tail of the major peak in the region of the solvent-induced band.[66] A correction for this would reduce $\varepsilon_s/\varepsilon_p$ by about 0.1 unit. The correction is, however, difficult to apply and introduces rather serious errors in the ratio.[66] Considering the rough magnitude of the corrections, however, it seems that the corrected $\varepsilon_s/\varepsilon_p$ is quite low in hydrocarbons

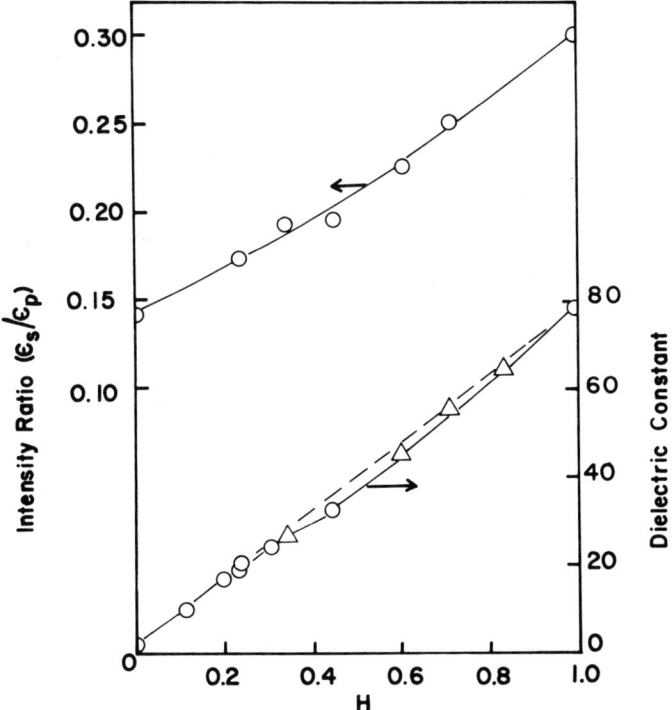

Figure 4. Ratios of absorbances at the solvent-induced band position ($\varepsilon_s$) and the nearby peak ($\varepsilon_p$) in various reference solvents. H is calculated molar concentration of -OH groups in the reference solvents relative to the value of unity in water (concentration 55.5 mol dm$^{-3}$). ○ - Pure solvents, △ - mixtures of alkanols and water. Right-hand ordinate shows dielectric constants of the reference solvents.

where mainly weak dispersive interactions are involved, and increases by nearly a factor of 4 when the medium is changed to water (Figure 3). The nearly linear increase in the ratio can thus be ascribed to a weak, roughly constant, contribution from dispersion forces to which the interactions of benzene with -OH dipoles is added according roughly to the volume concentration of the latter.

It is interesting to note that the bulk dielectric constant of the reference liquids varies roughly linearly with H (Figure 4), and, therefore, an 'effective' dielectric constant value can be assigned to an estimated H-value.

The monotonic variation of the $\varepsilon_s/\varepsilon_p$ ratio with H now provides the calibration curve for solubilized benzene. Figure 3 shows that

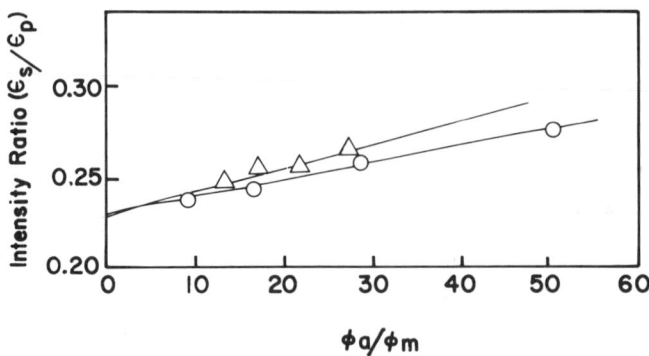

Figure 5. Ratio of the absorbance at the wavelength of the solvent-induced band, $\varepsilon_s$, and the absorbance at the nearby peak, $\varepsilon_p$ (see Figure 3), for benzene at different concentrations of sodium dodecyl sulfate (○) and cetyl trimethylammonium chloride (△). Abscissa shows the ratio of the calculated volume fractions of the aqueous part ($\phi_a$) and the micelles ($\phi_m$).

solubilized benzene has a relative intensity in the position of its solvent-induced band (3.6 nm to the long wavelength side of the peak) which is roughly midway between heptane and water. Figure 5 shows how this intensity varies with micellar concentration for an anionic and a cationic micellar system. Upon extrapolation, the characteristic intensity ratio of the solubilized species is found to be very similar for the two types of micelles. This value, moreover, indicates that the effective average H value of the environment of solubilized benzene is 0.62, corresponding to an 'effective' average dielectric constant of 46. The average microenvironment is thus quite polar.

Figure 6 shows some absorption spectra of naphthalene. In this case also, solvent-induced bands have been observed.[62] The relative intensities of these bands in our reference solvents behave in roughly the same manner as benzene. The effective average H value of the microenvironment of naphthalene is about 0.52. Figure 6 shows that the relative intensity of the solvent-induced band of naphthalene in 0.2 mol dm$^{-3}$ sodium dodecyl sulfate, in which it is nearly completely solubilized, is considerably closer to that of the reference solvent which is a 70-30 mixture by weight of methanol and water than the hydrocarbon heptane. Here also, therefore, the average environment in the solubilized state is quite polar.

Spectroscopic parameters similar to the above, based on the ratios of intensities of peaks and valleys, have been found to be useful for a variety of substituted benzenes.[59] It has been found

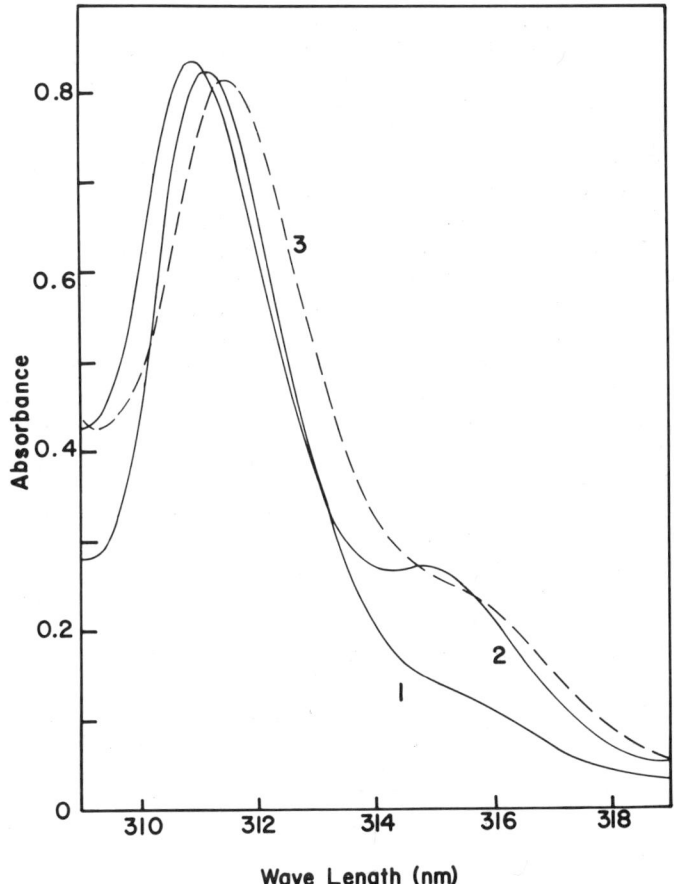

Figure 6. Absorption spectrum of naphthalene. 1- in heptane, 2- in 70-30 mixture of methanol and water (by weight), 3- in 0.2 mol dm$^{-3}$ sodium dodecyl sulfate.

that the surfactant Triton X-100, in which a p-t-octyl phenol group is linked to a polyoxyethylene group by means of an ether linkage, can be used as a solubilized species in trace concentrations in different kinds of micelles. Here the aromatic sensor is attached to the polar group, and must be located close to the interface. The effective H-value for Triton X-100 solubilized in different micelles is similar to that found for benzene.[59]

In view of the previous discussion regarding the lack of water in the micelle core, and the similarity of the polarity of the environment of the aromatic moiety of Triton X-100 and benzene, the best qualitative explanation for the polarity of solubilized benzene is that benzene is located <u>primarily</u> at the micelle-water interface where it is partly exposed to water.

Surface Activity of Solubilized Species. That solubilized benzene is concentrated primarily at the micelle-water interface rather than the hydrocarbon core may appear to be surprising at first sight because of the absence of any obvious polarity in benzene. However, if considerations applicable to ordinary interfaces are extended to micelles, an explanation on thermodynamic grounds becomes readily available. Liquid benzene has a lower interfacial tension against water than liquid aliphatic hydrocarbons. As expected from this, the measured interfacial tensions of mixtures of heptane and benzene against water[59] showed that benzene is slightly surface active as a dilute solution in heptane. Even this mild surface activity, however, predicts on the basis of the calculated surface excess that if the micelles of sodium dodecyl sulfate act like minute drops of a hydrocarbon like heptane, the amount in the surface at low concentrations of benzene exceeds the amount in the bulk by a factor of 4.[59] This very large effect of the surface arises from the extremely large surface areas of micelles per unit volume. It seems, therefore, that if bulk thermodynamic considerations apply even approximately to micellar systems, the interfacial activity of solubilized species becomes an extremely important factor in determining their location and distribution in the micelle. These same considerations apply also to biological membranes and bi-layers for which the surface-to-volume ratios are very high.

Supporting evidence for this picture has been obtained from the study of several alkyl derivatives of benzene.[59] When the alkyl moiety becomes fairly large, as in butyl benzene or p-di-t-butyl benzene, the estimated H-values of the microenvironments of solubilized molecules become appreciably lower, indicating a less polar average environment.

From a molecular point of view, the preference of benzene for the micelle-water interface can be attributed to the ability of π-electron systems to accept hydrogen bonds.[67] The importance of the extremely high surface-to-volume ratio comes from the magnification of small effects. It is likely that considerations such as the above are important for all studies involving solubilized species and spectroscopic probes in micellar and membrane systems when they contain aromatic groups or other groups which are more polar than hydrocarbons. The nitroxide probes for electron spin resonance studies are likely to belong to this list also. It is interesting to note that because of the peculiar properties of benzene solubilized in micelles, benzene itself must be considered a poor 'probe' for studying the properties of the micelles themselves.

## ACKNOWLEDGMENT

Acknowledgment is made to the Donors of the Petroleum Research Fund, administered by the American Chemical Society, for partial support of this research. We are also grateful to the Graduate School, University of Wisconsin, for financial support.

## REFERENCES

1. P. Mukerjee, Adv. Colloid Interface Sci., $\underline{1}$, 241 (1967).
2. G. S. Hartley, "Aqueous Solutions of Paraffin-Chain Salts," Hermann, Paris, 1936.
3. G. S. Hartley, Ann. Rep. Chem. Soc., London, $\underline{45}$, 33 (1948).
4. P. Mukerjee and A. Ray, J. Phys. Chem., $\underline{70}$, 2144 (1966).
5. P. Molyneux, C. T. Rhodes, and J. Swarbrick, Trans. Faraday. Soc., $\underline{61}$, 104 (1965).
6. P. Mukerjee, J. Perrin, and E. Witzke, J. Pharm. Sci., $\underline{59}$, 1513 (1970).
7. E. D. Goddard, C. A. J. Hoeve, and G. C. Benson, J. Phys. Chem., $\underline{61}$, 593 (1957).
8. K. Shigehara, Bull. Chem. Soc. Japan, $\underline{38}$, 1700 (1965).
9. P. Mukerjee, Kolloid. Z.u.Z. Polymere, $\underline{236}$, 76 (1970).
10. F. Huisman, Proc. Kon. Ned. Akad. Wetensch., Ser B, $\underline{67}$, 388, 407 (1964).
11. M. R. Cines, in "Physical Chemistry of Hydrocarbons", A. Farkas, Editor, Vol. 1, Chapter 8, Academic, New York, 1950.
12. P. Mukerjee, J. Phys. Chem., $\underline{66}$, 1733 (1962).
13. L. M. Kushner, W. D. Hubbard, and R. A. Parker, J. Res. Natl. Bur. Std., $\underline{59}$, 113 (1957).
14. P. Debye and E. W. Anacker, J. Phys. Colloid Chem., $\underline{55}$, 644 (1951).
15. F. Reiss-Husson and V. Luzzati, J. Phys. Chem., $\underline{68}$, 3504 (1964).
16. D. Stigter, J. Phys. Chem., $\underline{70}$, 1323 (1966).
17. R. J. M. Tausk, C. Oudshoorn, and J. Th. G. Overbeek, Biophys. Chem., $\underline{2}$, 53 (1974).
18. N. Muller and R. H. Birkhahn, J. Phys. Chem., $\underline{71}$, 957 (1967).
19. P. Mukerjee and K. J. Mysels, in "Colloidal Dispersions and Micellar Behavior," K. L. Mittal, Editor, ACS Symposium Series No. 9, pp. 239-252, American Chemical Society, Washington, DC, 1975.
20. P. Mukerjee and Alex Y. S. Yang, J. Phys. Chem., $\underline{80}$, 1388 (1976).
21. F. Podo, A. Ray, and G. Nemethy, J. Am. Chem. Soc., $\underline{95}$, 6164 (1973).
22. J. M. Corkill, J. F. Goodman, and T. Walker, Trans. Faraday Soc., $\underline{63}$, 768 (1967).
23. J. Clifford, Trans. Faraday Soc., $\underline{61}$, 1276 (1965).
24. D. Stigter, J. Phys. Chem., $\underline{78}$, 2480 (1974).
25. J. L. Kurz, J. Phys. Chem., $\underline{66}$, 2239 (1962).

26. D. Stigter and K. J. Mysels, J. Phys. Chem., $\underline{59}$, 45 (1955).
27. P. Mukerjee, K. J. Mysels, and P. Kapauan, J. Phys. Chem., $\underline{71}$, 4166 (1967).
28. P. Mukerjee and K. Banerjee, J. Phys. Chem., $\underline{68}$, 3567 (1964).
29. P. Mukerjee, J. Phys. Chem., $\underline{73}$, 2054 (1969).
30. D. Stigter, J. Colloid Interface Sci., $\underline{47}$, 473 (1974).
31. D. Stigter, J. Phys. Chem., $\underline{79}$, 1008, 1015 (1975).
32. P. Mukerjee, J. Colloid Sci., $\underline{19}$, 722 (1964).
33. I. M. Klotz, Science, $\underline{128}$, 815 (1958).
34. J. Lyklema and J. Th. G. Overbeek, J. Colloid Sci., $\underline{16}$, 501 (1961).
35. A. Sanfeld, "Introduction to the Thermodynamics of Charged and Polarized Layers," Monographs in Statistical Physics, No. 10, Wiley-Interscience, NY, 1968.
36. M. J. Sparnaay, Rec. Trav. chim., $\underline{77}$, 872 (1958); $\underline{81}$, 395 (1962).
37. A. Ray and P. Mukerjee, J. Phys. Chem., $\underline{70}$, 2138 (1966).
38. E. M. Kosower, J. Amer. Chem. Soc., $\underline{80}$, 3253 (1958).
39. P. Mukerjee and N. R. Desai, Nature, $\underline{223}$, 1056 (1969).
40. J. B. Hasted, D. M. Ritson, and C. H. Collie, J. Chem. Phys., $\underline{16}$, 1 (1948).
41. P. Mukerjee and N. R. Desai, (1971), unpublished work.
42. G. Akerlöf, J. Am. Chem. Soc., $\underline{54}$, 4125 (1932).
43. P. Mukerjee and Alex Y. S. Yang, (1974), unpublished work.
44. L. Onsager, J. Chem. Phys., $\underline{2}$, 599 (1934).
45. K. J. Mysels and L. H. Princen, J. Phys. Chem., $\underline{63}$, 1699 (1959).
46. T. Nakagawa and K. Tori, Kolloid-Z., $\underline{194}$, 143 (1964).
47. J. Oakes, Nature, $\underline{231}$, 38 (1971).
48. P. Mukerjee in "Encyclopaedic Dictionary of Physics," J. Thewlis, Editor, Pergamon, Oxford, 1961.
49. M. E. L. McBain and E. Hutchinson, "Solubilization and Related Phenomena," Academic, New York, 1955.
50. P. H. Elworthy, A. T. Florence, and C. B. Macfarlane, "Solubilization by Surface-Active Agents," Chapman and Hall, London, 1968.
51. T. Nakagawa in "Nonionic Surfactants," M. J. Schick, Editor, Marcel Dekker, New York, 1967.
52. P. Mukerjee, J. Pharm. Sci., $\underline{60}$, 1528 (1971).
53. P. Mukerjee, J. Pharm. Sci., $\underline{60}$, 1531 (1971).
54. J. C. Eriksson and G. Gillberg, Acta Chem. Scand., $\underline{20}$, 2019 (1966).
55. J. E. Gordon, J. C. Robertson, and R. L. Thorne, J. Phys. Chem., $\underline{74}$, 957 (1970).
56. S. J. Rehfeld, J. Phys. Chem., $\underline{74}$, 117 (1970).
57. S. J. Rehfeld, J. Phys. Chem., $\underline{75}$, 3805 (1971).
58. J. H. Fendler and L. K. Patterson, J. Phys. Chem., $\underline{75}$, 3907 (1971).
59. P. Mukerjee and J. R. Cardinal, (1972), unpublished work.
60. J. R. Platt, J. Mol. Spec., $\underline{9}$, 288 (1962).
61. M. Koyanagi and Y. Kanda, Spectrochim. Acta, $\underline{20}$, 993 (1964).

62. M. Koyanagi, J. Mol. Spec., 25, 273 (1968).
63. N. S. Bayliss and N. W. Cant, Spectrochim. Acta, 18, 1287 (1962).
64. H. Sponer, G. Nordheim, A. L. Sklar, and E. Teller, J. Chem. Phys., 7, 207 (1939).
65. J. S. Ham, J. R. Platt, and H. McConnell, J. Chem. Phys., 19, 1301 (1951).
66. J. W. Eastman and S. J. Rehfeld, J. Phys. Chem., 74, 1438 (1970).
67. Z. Yoshida and E. Osawa, J. Amer. Chem. Soc., 87, 1467 (1965); 88, 4019 (1966).

# THE INFLUENCE OF HYDROPHOBIC COUNTERIONS ON THE THERMODYNAMICS AND KINETICS OF IONIC MICELLES

H.Hoffmann, H.Nüsslein and W.Ulbricht

Universität Bayreuth

D-858 Bayreuth, Postfach 3008

Relaxation measurements are reported on surfactant solutions containing Dodecylpyridinium(DPy)- and Tetradecylpyridinium(TPy)-cations and the counterions Benzenesulfonate (BS), Toluenesulfonate (TS), Aminobenzenesulfonate (ABS), Benzoate (B) and the Alkylsulfonates having chain lengths from one to five. The kinetic data will be discussed on the basis of the theory recently given by G.Aniansson and S.Wall.

The residence times of the surfactant molecules in the micelles and the half width of the micellar distribution curves are determined from the fast relaxation process. The residence times are found to increase with the increasing hydrophobicity of the counterions.

The aggregation number of the micellar nucleus, its thermodynamic properties and the number of associated counterions are evaluated from the slow relaxation process. While the small Alkylsulfonates having one to four $CH_2$-groups are not associated with the nucleus that consists of about 8-9 surfactant molecules, the Pentanesulfonate and the aromatic counterions except ABS are highly associated. For some of the studied systems, the micellar aggregation number was determined from the kinetic data.

## INTRODUCTION

Mathematical equations have recently been derived by G. Aniansson et al.[1] to describe the dynamics of micellar equilibria. At the same time experimental evidence was obtained that was in excellent accord with the theoretical model[2]. The theory has since been applied to a series of surfactants and the kinetic parameters that can be obtained from relaxation measurements have been evaluated[3]. The theory predicts and the experiments show two relaxation processes for micellar solutions. The faster of the two relaxation times is due to a shift of the micellar distribution curve and is given by the expression

$$1/\tau_1 = k^-/\sigma^2 + (k^-/n) \cdot (A_{exc}/A_1)$$

where $k^-$ is the dissociation rate constant for the process of a surfactant molecule leaving the micelle, $\sigma$ the variance of the distribution curve, $n$ the aggregation number, $A_{exc}$ the amount of surfactant material in the micelle form and $A_1$ the concentration of the monomer form.

During this process, the number of micelles remains essentially constant.

In a second slower step, the concentration of micelles relaxes to its new equilibrium value. During this process, matter has to pass from the micellar side through the steady state concentration of the distribution minimum to the monomer side or vice versa. As a consequence, the steady state concentration at the distribution minimum enters into this rate equation and hence into $\tau_2$.

The reciprocal relaxation time for this process is given by

$$1/\tau_2 = \frac{1}{R \cdot c_3} \cdot \frac{A_1 + n^2 \cdot c_3}{A_1 + \sigma^2 \cdot c_3}$$

with $R = \sum(1/k_i^- \cdot c_i)$, when $k_i^-$, $c_i$ and $c_3$ are the dissociation rate constant, the concentration at the distribution minimum and the concentration of micelles, respectively. Clearly, the experimental data permit the evaluation of R and hence infromation for particles can be obtained that are present in negligible concentrations under equilibrium conditions. The aggregate at the distribution minimum is called the nucleus of the micelle, because it is the smallest aggregate that has a better chance to grow than to dissociate again.

For a series of surfactants the nucleus was found to have a highly endothermic heat of formation and an aggregation number of about 7-8. It was also shown that the nucleus is still free from

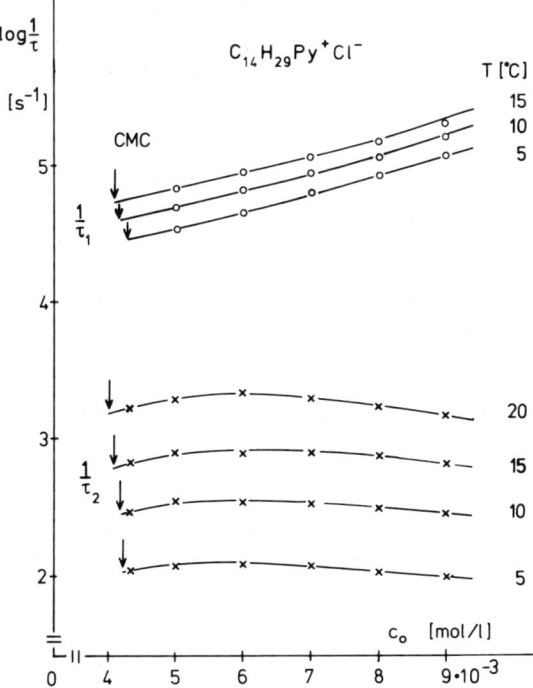

Figure 1. Plot of $\log(1/\tau)$ against the total concentration of TPyCl for different temperatures.

counterions, at least for ionic micelles for which electrostatic interactions only are important for the binding of the counterions. In the case of Alkylpyridiniumiodide, the iodide was found to associate already with the nucleus. In this particular case the bonding of the iodide is enhanced due to charge transfer effects[4].

In the absence of any special interactions, the plots of $1/\tau_2$ against the total concentration were always of the same shape. From a finite value at the cmc, $1/\tau_2$ rises slightly with the total concentration, reaches a maximum at about 1,3cmc and then decreases. At high total concentrations it was found to increase again. A typical plot is shown in Figure 1. The temperature dependence of the $1/\tau_2$-values is very large, but the shape of the curves is not effected by the temperature.

In previous publications, a linear rise of $1/\tau_2$ with the total concentration was often reported[5]. Probably, this type of behavior was caused by impurities in the systems. Measurements on pure substances did not show this type of concentration dependence[6].

The objective of the investigations presented in this article was to find out how the dynamic response of micelles towards a perturbation is changed if hydrophobic counterions are used that interact more strongly with the ionic micelles than ordinary hydrophilic counterions do.

As suitable counterions for the planned measurements proved to be substituted benzene derivates like Benzoate (B), Toluenesulfonate (TS), Benzenesulfonate (BS) and also Alkylsulfonates with small Alkylgroups. As surfactant ions, Dodecylpyridinium (DPy)- and Tetradecylpyridinium (TPy)-ions were used. The halides of these systems had been studied before[7].

## EXPERIMENTAL AND RESULTS

While a number of ionic detergents having hydrophobic counterions have been studied by Lange et al., Mukerjee et al. and Meguro et al.[8a-c], the cmc-values of most of the compounds in the present investigation were not known and had to be determined by conductivity measurements.

All surfactants investigated in this work were prepared from DPy- and TPy-Chloride. These compounds were gifts from Henkel & Cie (Düsseldorf, West Germany). For the preparation of some of the compounds, Sodium-salts of the hydrophobic counterions were added to solutions of TPyCl and DPyCl. The products that precipitated when the solutions were cooled, were purified by successive recrystallizations in water or in mixtures of water and alcohol until no chloride ions could be detected in the solutions.

The same results were obtained when the compounds were prepared from the chlorides via ion exchange columns. Anion exchange columns were charged with those counterions from which the salts were to be prepared. The salts were extracted with solutions of TPyCl and DPyCl, respectively. Quantitative analysis of the solutions thus obtained showed that they could be used directly for the kinetic experiments and conductivity measurements. The ion exchange method was specially used for the preparation of those compounds which did not precipitate at low temperatures.

Typical results are shown for the TPy-Alkylsulfonates in Figure 2 in a plot of the equivalent conductivity $\Lambda$ against $c^{1/2}$. The cmc values for the studied systems are given in Table I. For comparison, the values for some of the Alkylpyridiniumhalides are also included. As can be seen from the Table I and the Figure 2, the cmc-values for the Alkylpyridiniumalkylsulfonates decrease with increasing chain length of the alkyl group in the counterion. Remarkably the cmc-value for $DPyCH_3SO_3$ is even larger than the cmc of DPyCl, for which system only electrostatic interactions between the micellar charge

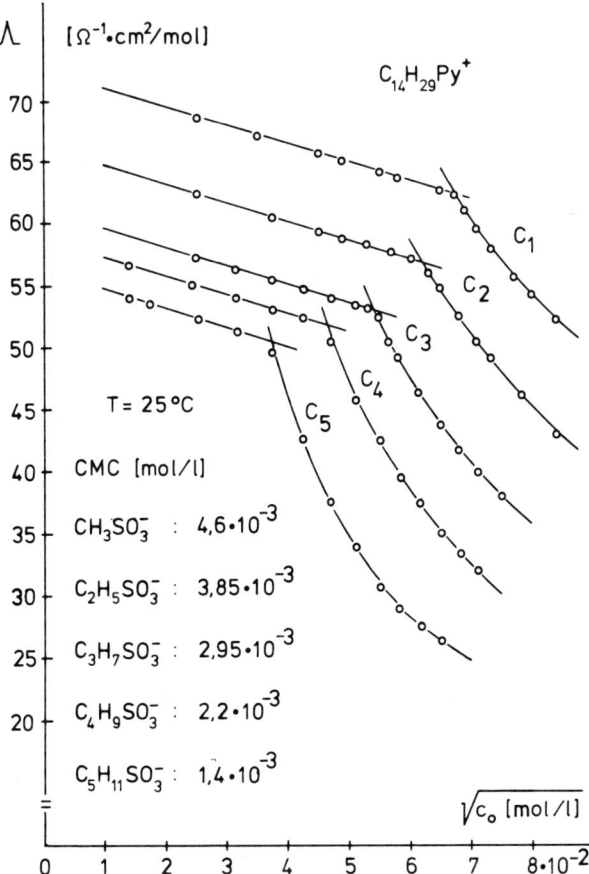

Figure 2. Plot of the equivalent conductivity $\Lambda$ of the TPy-surfactant ion having different counterions against the square root of the total concentration.

and the counterion are usually assumed. For the Alkylsulfonates, it is thus possible to go in small steps from a completely hydrophilic ion to a strongly hydrophobic one. It is interesting to note that the decrease of the cmc with each additional $CH_2$-group in the counterion is dependent on the chain length of the hydrophobic ion. The cmc's for Methane- and Ethanesulfonates vary only by a factor of 1,19, while the cmc's of Butane- and Pentanesulfonates vary by a factor of 1,57. It is also worth mentioning that the fall off of the $\Lambda$-values with increasing $c_o$ after the cmc's in Figure 2 becomes larger with increasing chain length of the hydrophobic counterion. This can be taken as evidence that the dissociation degree of the micelles decreases with increasing chain length of the counterion.

Table I. Values for the cmc and the Ratio between the Equivalent Conductivity $\Lambda$ at the cmc and at a Concentration $c_o = (cmc+10^{-3})$mol/l for DPy- and TPy-Surfactants with Various Counterions at 25°C.

| Counterion | DPy cmc | $\frac{\Lambda(cmc)}{\Lambda(cmc+10^{-3})}$ | TPy cmc | $\frac{\Lambda(cmc)}{\Lambda(cmc+10^{-3})}$ |
|---|---|---|---|---|
| $CH_3SO_3$ | $1,9 \times 10^{-2}$ | 1,05 | $4,6 \times 10^{-3}$ | 1,08 |
| Cl | $1,48 \times 10^{-2}$ | -- | $4,0 \times 10^{-3}$ | 1,10 |
| $C_2H_5SO_3$ | -- | -- | $3,85 \times 10^{-3}$ | 1,10 |
| $C_3H_7SO_3$ | -- | -- | $2,95 \times 10^{-3}$ | 1,17 |
| Br | $1,14 \times 10^{-2}$ | 1,05 | $2,6 \times 10^{-3}$ | 1,20 |
| ABS | $9,8 \times 10^{-3}$ | 1,05 | $2,3 \times 10^{-3}$ | 1,26 |
| $C_4H_9SO_3$ | -- | -- | $2,2 \times 10^{-3}$ | 1,27 |
| B | $5,6 \times 10^{-3}$ | 1,20 | $1,45 \times 10^{-3}$ | 1,50 |
| $C_5H_{11}SO_3$ | -- | -- | $1,4 \times 10^{-3}$ | 1,40 |
| BS | $4,8 \times 10^{-3}$ | 1,20 | $1,1 \times 10^{-3}$ | 1,52 |
| TS | $2,9 \times 10^{-3}$ | 1,25 | $8,5 \times 10^{-4}$ | 1,64 |

Similar conclusions on other systems have been obtained before[8b]. The cmc-values of the salts with aromatic ions compare with those of the Alkylsulfonates having about 5 $CH_2$-groups.

Relaxation measurements using the pressure jump and the shock wave method were carried out on these systems. Some of the results are given in the Tables IIa-IIc. As in previous studies, two relaxation times could be detected in some of the investigated systems. The experimental error on the measured relaxation times is expected to be less than 10%.

For some systems, the slow process was too slow for relaxation studies, while for others the fast process was shorter than 1 μs and could not be resolved by the shock method. In these cases only

Table IIa. Relaxation Times for DPy- and TPy-Surfactants with Various Counterions at Various Temperatures as a Function of the Total Concentration $c_o$.

| Counterion | | DPy | | | | | |
|---|---|---|---|---|---|---|---|
| | | | $\tau_2$-Values (ms) | | | | |
| | T (°C) | | 5 | 10 | 15 | 20 | 25 |
| | $c_o$ (mol/l) | | | | | | |
| $C_5H_{11}SO_3$ | 6,0 x$10^{-3}$ | | --- | 153,5 | 56,3 | 29,7 | 13,5 |
| | 6,5 x$10^{-3}$ | | 295,0 | 121,6 | 48,7 | 25,5 | 11,0 |
| | 7,0 x$10^{-3}$ | | 215,0 | 78,3 | 31,3 | 16,3 | 8,8 |
| | 7,5 x$10^{-3}$ | | 145,0 | 52,3 | 23,0 | 11,9 | 6,3 |
| | 8,0 x$10^{-3}$ | | 77,5 | 32,3 | 15,3 | 8,8 | 5,0 |
| | 9,0 x$10^{-3}$ | | 31,3 | 14,9 | 8,3 | 5,2 | 3,1 |
| | 10,0 x$10^{-3}$ | | 15,4 | 8,5 | 4,9 | 3,2 | 2,2 |
| | 11,0 x$10^{-3}$ | | 10,0 | 6,0 | 3,7 | 2,3 | 1,4 |
| | 12,5 x$10^{-3}$ | | 6,5 | 4,0 | 2,5 | 1,6 | 1,2 |
| BS | 5,12x$10^{-3}$ | | --- | 193,0 | 110,0 | 68,5 | 45,2 |
| | 5,21x$10^{-3}$ | | 280,0 | 162,6 | 98,2 | 58,4 | 37,8 |
| | 5,31x$10^{-3}$ | | 180,0 | 120,6 | 78,8 | 48,0 | 32,2 |
| | 5,69x$10^{-3}$ | | 66,8 | 50,4 | 32,3 | 21,3 | 16,3 |
| | 6,07x$10^{-3}$ | | 34,8 | 24,5 | 17,4 | 14,0 | 11,0 |
| | 6,45x$10^{-3}$ | | 21,6 | 16,1 | 12,3 | 9,3 | 7,7 |
| | 6,83x$10^{-3}$ | | 18,2 | 13,3 | 9,7 | 7,7 | 6,3 |
| | 7,21x$10^{-3}$ | | 11,9 | 9,5 | 7,1 | 5,6 | 4,7 |
| | 7,58x$10^{-3}$ | | 9,6 | 7,8 | 6,0 | 4,4 | 3,4 |
| | 7,96x$10^{-3}$ | | 8,0 | 5,6 | 4,3 | 3,5 | 3,0 |
| | 9,48x$10^{-3}$ | | 3,6 | 2,6 | 2,2 | 1,8 | 1,5 |
| | 11,38x$10^{-3}$ | | 1,8 | 1,5 | 1,2 | 1,0 | 0,85 |
| TS | 3,2 x$10^{-3}$ | | 305,0 | 220,0 | 143,3 | 108,9 | 90,0 |
| | 3,3 x$10^{-3}$ | | 176,0 | 97,5 | 70,0 | 58,7 | 51,0 |
| | 3,6 x$10^{-3}$ | | 27,3 | 16,0 | 13,0 | 8,7 | 8,9 |
| | 4,0 x$10^{-3}$ | | 9,4 | 6,1 | 4,3 | 3,0 | 2,5 |
| | 4,5 x$10^{-3}$ | | 5,4 | 3,5 | 2,5 | 1,6 | 1,25 |
| | 5,0 x$10^{-3}$ | | 3,7 | 2,5 | 1,6 | 1,15 | 0,9 |
| | 5,5 x$10^{-3}$ | | 2,7 | 1,7 | 1,06 | 0,8 | 0,6 |

Table IIb. Relaxation Times for DPy- and TPy-Surfactants with Various Counterions at Various Temperatures as a Function of the Total Concentration $c_o$.

| Counterion | $c_o$ (mol/1) | TPy $\tau_1$-Values ($\mu$s) T(°C) | | | | |
|---|---|---|---|---|---|---|
| | | 5 | 10 | 15 | 20 | 25 |
| Cl | $4{,}6 \times 10^{-3}$ | --- | 25,0 | 15,0 | --- | --- |
| | $5{,}0 \times 10^{-3}$ | 28,0 | 19,0 | 14,0 | --- | --- |
| | $6{,}0 \times 10^{-3}$ | 20,3 | 13,7 | 10,5 | --- | --- |
| | $7{,}0 \times 10^{-3}$ | 13,7 | 10,5 | 8,2 | --- | --- |
| | $8{,}0 \times 10^{-3}$ | 11,0 | 8,3 | 6,3 | --- | --- |
| | $10{,}0 \times 10^{-3}$ | 8,7 | 6,0 | 4,8 | --- | --- |
| $C_3H_7SO_3$ | $3{,}5 \times 10^{-3}$ | 38,0 | 26,0 | 18,5 | 13,5 | --- |
| | $3{,}75 \times 10^{-3}$ | 37,0 | 24,0 | 16,5 | 12,5 | --- |
| | $4{,}0 \times 10^{-3}$ | 35,0 | 20,0 | 14,5 | 11,0 | --- |
| | $4{,}5 \times 10^{-3}$ | 30,0 | 17,0 | 11,5 | 9,0 | --- |
| | $5{,}0 \times 10^{-3}$ | 25,0 | 14,5 | 10,0 | 7,0 | --- |
| | $6{,}0 \times 10^{-3}$ | 18,0 | 10,0 | 7,0 | 5,0 | --- |
| $C_4H_9SO_3$ | $2{,}5 \times 10^{-3}$ | 58,0 | 40,0 | 30,0 | 22,0 | 15,0 |
| | $2{,}8 \times 10^{-3}$ | 50,0 | 35,0 | 26,0 | 19,0 | 13,0 |
| | $3{,}0 \times 10^{-3}$ | 47,5 | 30,0 | 24,0 | 17,5 | 11,5 |
| | $3{,}5 \times 10^{-3}$ | 38,0 | 25,0 | 18,5 | 14,5 | 9,5 |
| | $4{,}0 \times 10^{-3}$ | 35,0 | 21,0 | 15,0 | 11,5 | 8,0 |
| | $4{,}5 \times 10^{-3}$ | 30,0 | 18,5 | 13,0 | 9,0 | 6,0 |
| | $5{,}0 \times 10^{-3}$ | 25,0 | 14,0 | 10,0 | 7,5 | 5,0 |
| | $5{,}5 \times 10^{-3}$ | 22,0 | 12,5 | 8,7 | 6,0 | 4,0 |
| $C_5H_{11}SO_3$ | $1{,}7 \times 10^{-3}$ | 90,0 | 57,5 | 45,0 | 30,0 | 23,0 |
| | $2{,}0 \times 10^{-3}$ | 80,0 | 55,0 | 40,0 | 27,5 | 21,6 |
| | $2{,}5 \times 10^{-3}$ | 62,5 | 45,0 | 34,5 | 25,0 | 18,0 |
| | $3{,}0 \times 10^{-3}$ | 50,5 | 33,0 | 24,8 | 19,0 | 12,7 |
| | $3{,}5 \times 10^{-3}$ | 40,0 | 27,5 | 20,0 | 15,5 | 10,0 |
| | $4{,}0 \times 10^{-3}$ | 37,5 | 22,7 | 17,0 | 12,5 | 8,5 |
| | $4{,}5 \times 10^{-3}$ | 30,0 | 20,0 | 14,0 | 9,5 | 6,5 |
| | $5{,}0 \times 10^{-3}$ | 26,3 | 15,7 | 11,0 | 7,6 | 5,5 |
| ABS | $2{,}6 \times 10^{-3}$ | 60,0 | 40,0 | 28,5 | 20,0 | 15,5 |
| | $2{,}8 \times 10^{-3}$ | 50,0 | 37,5 | 24,0 | 18,0 | 14,5 |
| | $3{,}0 \times 10^{-3}$ | 48,7 | 32,5 | 22,5 | 16,5 | 12,5 |
| | $3{,}5 \times 10^{-3}$ | 40,0 | 25,5 | 19,0 | 14,0 | 9,5 |
| | $4{,}0 \times 10^{-3}$ | 31,0 | 21,7 | 16,0 | 11,5 | 9,0 |
| | $4{,}5 \times 10^{-3}$ | 27,5 | 18,5 | 13,5 | 9,5 | 7,5 |
| | $5{,}0 \times 10^{-3}$ | 23,8 | 15,3 | 12,0 | 8,0 | 6,0 |

# HYDROPHOBIC COUNTERIONS AND IONIC MICELLES

Table IIc. Relaxation Times for DPy- and TPy-Surfactants with Various Counterions at Various Temperatures as a Function of the Total Concentration $c_o$.

| Counterion | $c_o$ (mol/l) | TPy $\tau_2$-Values (ms) | | | | | |
|---|---|---|---|---|---|---|---|
| | T (°C) | 10 | 15 | 20 | 25 | 30 | 35 |
| $C_4H_9SO_3$ | $2,5 \times 10^{-3}$ | --- | 1350,0 | 450,0 | 182,0 | 90,0 | --- |
| | $3,0 \times 10^{-3}$ | 3000,0 | 1300,0 | 445,0 | 180,0 | 88,3 | --- |
| | $3,4 \times 10^{-3}$ | 2950,0 | 1300,0 | 453,0 | 173,0 | 87,5 | --- |
| | $3,8 \times 10^{-3}$ | 3100,0 | 1225,0 | 445,0 | 175,0 | 87,5 | --- |
| | $4,0 \times 10^{-3}$ | 3000,0 | 1300,0 | 450,0 | 183,0 | 87,5 | --- |
| | $4,5 \times 10^{-3}$ | 3000,0 | 1316,0 | 455,0 | 180,0 | 90,0 | --- |
| | $5,0 \times 10^{-3}$ | 3000,0 | 1280,0 | 450,0 | 180,0 | 92,5 | --- |
| ABS | $2,4 \times 10^{-3}$ | --- | 1000,0 | 323,0 | 125,0 | 59,0 | 23,3 |
| | $2,5 \times 10^{-3}$ | --- | 966,0 | 346,0 | 108,0 | 41,6 | 17,5 |
| | $2,6 \times 10^{-3}$ | --- | 823,0 | 285,0 | 98,0 | 35,7 | 14,1 |
| | $3,0 \times 10^{-3}$ | --- | 786,0 | 281,0 | 91,0 | 32,9 | 12,5 |
| | $3,4 \times 10^{-3}$ | --- | 810,0 | 276,0 | 93,0 | 33,6 | 13,4 |
| | $3,8 \times 10^{-3}$ | --- | 777,0 | 270,0 | 93,0 | 36,3 | 13,6 |
| | $4,2 \times 10^{-3}$ | --- | 816,0 | 303,0 | 96,0 | 38,0 | 14,7 |
| | $4,6 \times 10^{-3}$ | --- | 760,0 | 293,0 | 99,0 | 40,0 | 16,0 |

one relaxation time could be observed. But it was always possible to identify the observed process by its concentration dependence as the slow or the fast process.

An attempt was made also to observe the relaxation process that is due to the binding of the hydrophobic counterions. This process should have a concentration dependence that is similar to the fast one of the two relaxation processes but is believed to be several orders of magnitude faster. For this reason sound absorption measurements were carried out on some of the systems, but the process could not be observed[9]. On the basis of these experiments it is therefore likely that the volume change that is associated with the coordination of the counterions is very small and consequently the process is very poorly coupled to the sound waves.

In Figure 3a and 3b the log of the $1/\tau_2$-values for several

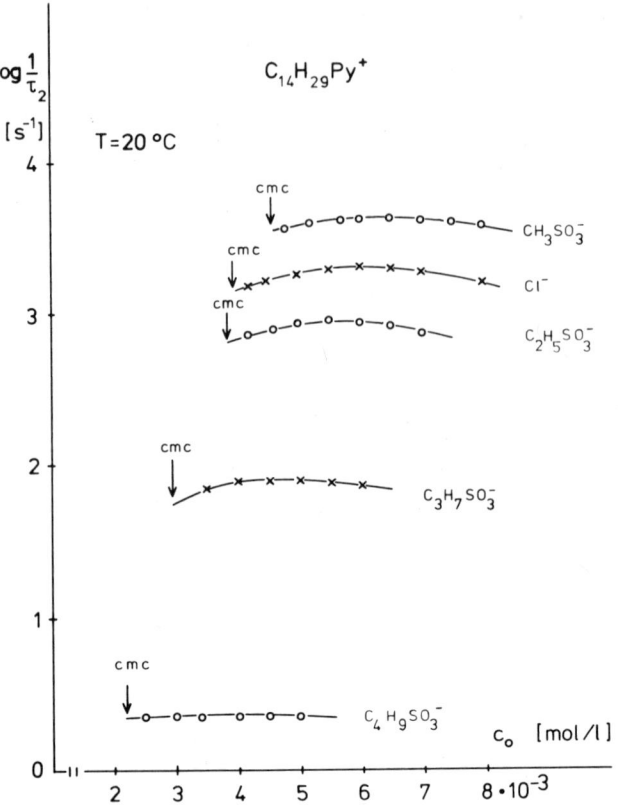

Figure 3a. Plot of the $\log(1/\tau_2)$-values for TPy-surfactants having different counterions against the total concentration.

Alkylsulfonates of the DPy- and TPy-detergents are plotted against the total concentration of the surfactant. With small alkyl chaines the shape of the curves is the same as the one reported for surfactants with hydrophilic counterions. With each additional $CH_2$-group in the counterion, the relaxation processes are shifted to longer times. For the TPy-Pentanesulfonate that has a cmc below $2 \cdot 10^{-3}$ mol/l, the relaxation times $\tau_2$ were too long to be measured. They overlapped with slow conductivity changes that were due to temperature changes caused by the thermostating equipment. The long relaxation time could be measured for the DPy-Pentanesulfonate. For this counterion the concentration dependence of $\tau_2$ is completely different than for the counterions having shorter chain lengths. Judged from the behavior of $\tau_2$ something drastic must have happened regarding the binding of the counterions. The $1/\tau_2$-values are rapidly increasing now with total concentration.

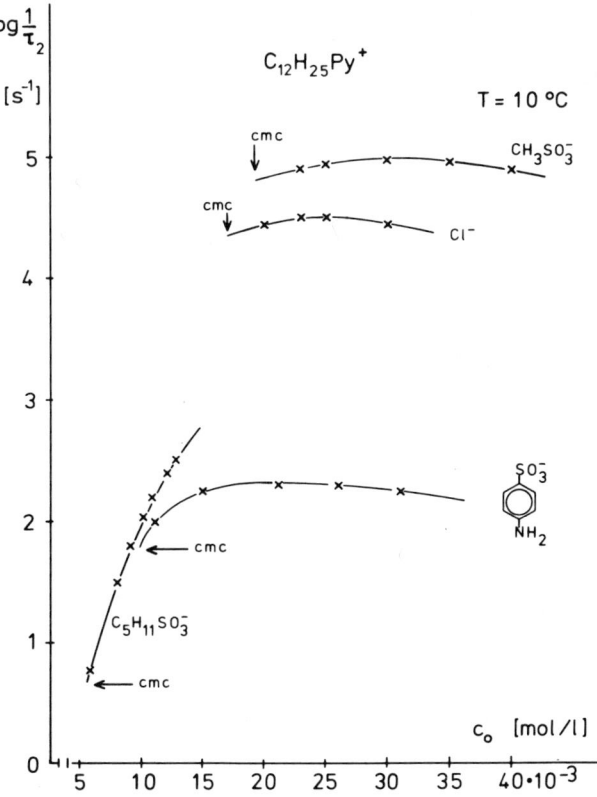

Figure 3b. Plot of the $\log(1/\tau_2)$-values for DPy-surfactants having different counterions against the total concentration

A change in total concentration by a factor of two changes $\tau_2$ by about a factor of 100. A similar behavior is observed for some of the aromatic counterions. In this respect the Pentanesulfonate is like the aromatic ions (see Figure 4).

No drastic changes are observed in the shape of the $1/\tau_1$-c - curves. They are all similar for all the surfactants and similar to the curves observed for micelles with hydrophilic counterions (see Figures 5a and 5b). The difference is only a gradual change.

Due to their lower cmc's the $1/\tau_1$ plots begin at lower concentrations and rise almost linearly with increasing total concentration. For the same concentrations the $1/\tau_1$-values vary no more than a factor of two for the different surfactants. Significant changes become apparent however when the relaxation times for the

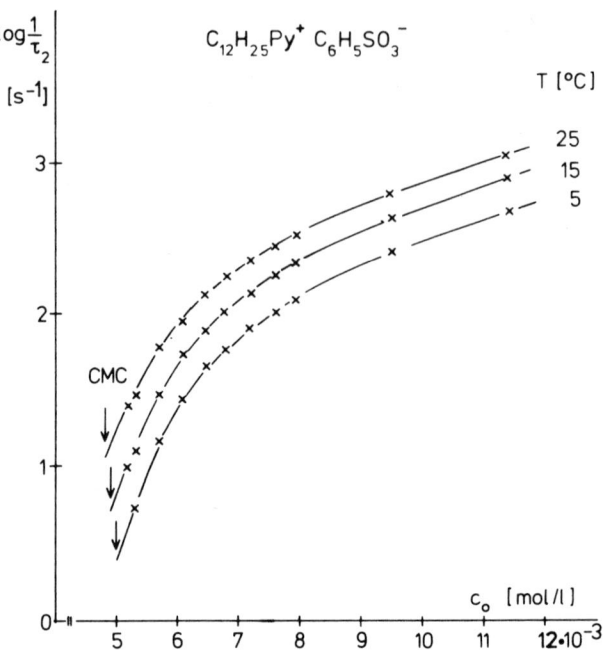

Figure 4. Plot of the $\log(1/\tau_2)$-values for DPyBS against the total concentration for different temperatures.

different systems are normalized in a plot of $(1/\tau_1)/(1/\tau_1)_{cmc} - 1$ against $(c_o-cmc)/cmc$. The slope of such curves gives $\sigma^2/n$ (see Figure 6).

## DISCUSSION

### 1. The cmc-Values

The cmc-values are found to decrease with increasing hydrophobicity of the counterion. Using the mass action model to describe micelles, the cmc is the reciprocal value of the stability constant for the reaction $A_n + A \rightleftharpoons A_{n+1}$ at the distribution maximum. In this case $A_n = A_{n+1}$ and consequently $K_n = 1/A_1$. It is of theoretical interest to ask the question whether the found increased stability of the micelles having hydrophobic counterions is due to an enthalpy or an entropy term.

For many ionic detergents with hydrophilic counterions, the cmc's are known to have a minimum around $20°C$[10]. The detergents

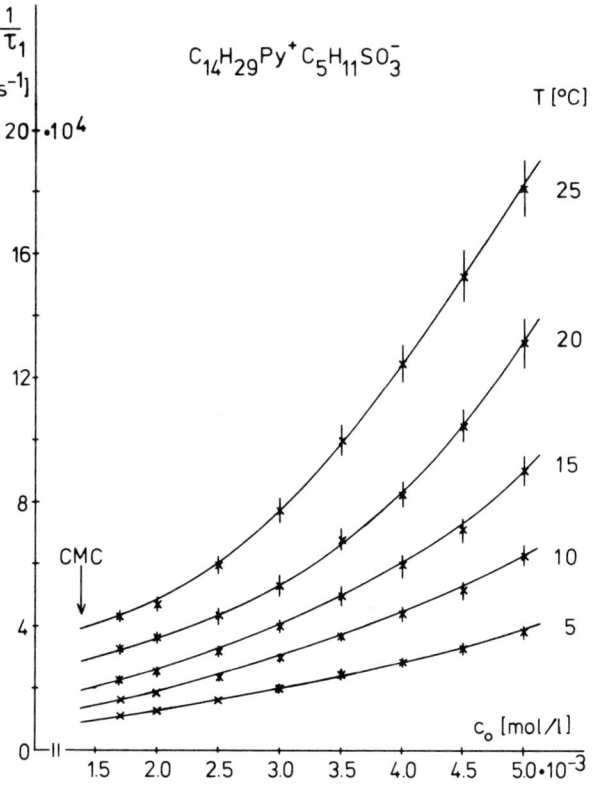

Figure 5a. Plot of the $1/\tau_1$-values for $TPyC_5H_{11}SO_3$ against the total concentration for different temperatures.

with hydrophobic counterions behave in the same way. The minimum is also around 20°C but seems to be somewhat shallower. Around the minimum the cmc is practically independent of the temperature. Very often this is interpreted that the incorporation of a surfactant into the micelle has a negligible heat of reaction[11]. As was shown previously and as will be discussed later on, the situation is much more complicated. At the distribution maximum large negative $\Delta$ H-values are observed, while at the distribution minimum positive $\Delta H$-values are observed.

If the heat of reaction for the transfer of a monomer into the micelle has to be determined, this should be calculated from $K_n \cdot n$ and not from $K_n = 1/cmc$, because $K_n \cdot n$ represents the equilibrium constant for the distribution of a specific detergent molecule between the micellar and the monomer state. In the latter case, sta-

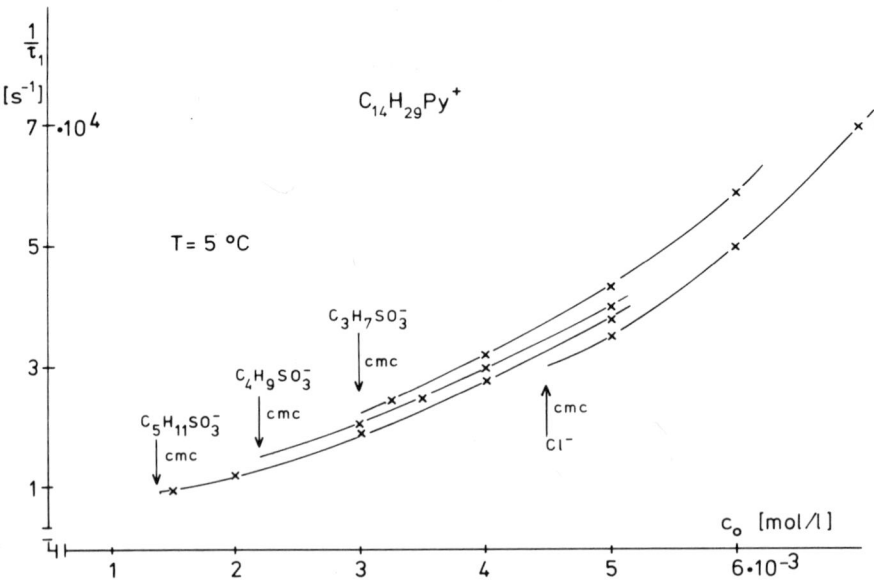

Figure 5b. Plot of the $1/\tau_1$-values for TPy-surfactants having different counterions against the total concentration.

bility constants for different aggregates are compared when the mean aggregation number changes with temperature. From the literature it is known that n really changes with the temperature[12]. It is likely that the mean aggregation number n changes very similar with the temperature for both hydrophilic and hydrophobic counterions. For a few systems where n was measured as a function of the temperature, it was found to decrease with increasing temperature[12]. Since the cmc changes in the same way for both types of systems it is likely that $K_n \cdot n$ changes the same also. For temperatures above the cmc-minimum, the incorporation process could then only be an exothermic one, because both $K_n$ and n decrease. Under the assumption that both $K_n$ and n have the same temperature dependence for hydrophilic and hydrophobic counterions, the increased stability of micelles having hydrophobic counterions would be caused by an entropy term. As will be shown later, the heat of reaction can be determined from the temperature dependence of $k^-/n$. This quantity can be determined experimentally from the slope of the plots of $1/\tau_1$ against the total concentration.

## 2. The Fast Relaxation Process

For the systems where the fast relaxation times could be measured, it is possible to evaluate the parameters $k^-/n$, $k^-/\sigma^2$ and

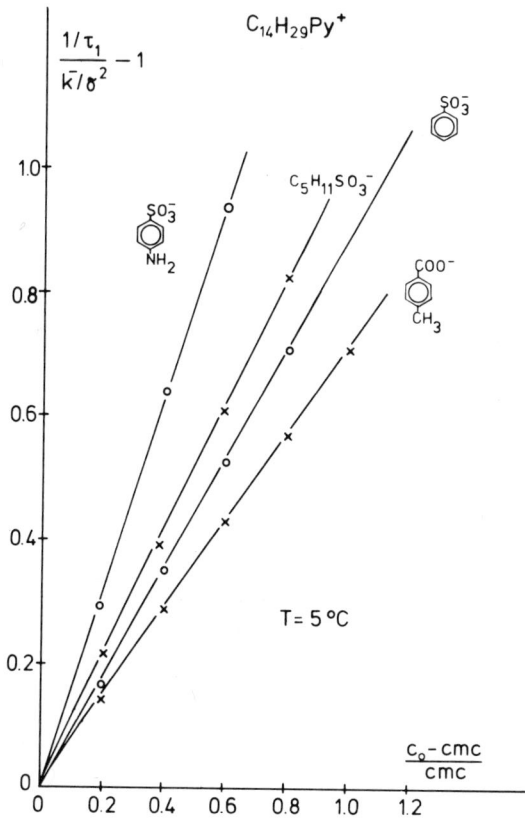

Figure 6. Plot of $((1/\tau_1)/(\bar{k}/\sigma^2)-1)$ for several TPy-surfactants having different counterions against $(c_o-cmc)/cmc$.

$\sigma^2/n$ from the relaxation times. These values are given in Table III. Also given are the values for $k^+/n$ that have been obtained from the $k_+/n_-$ and the corresponding cmc-values according to the equation $k^+/k^- = K = 1/cmc$. The data show that the $\sigma^2/n$-values decrease with increasing hydrophobicity of the counterion. This means that the distribution curves become narrower as compared for micelles with hydrophilic counterions. Both $k^+/n$ and $k^-/n$ decrease with increasing hydrophobicity of the counterion with the change in $k^-/n$ being larger than the change in $k^+/n$. Unfortunately, the values for $k^+$, $k^-$ and $\sigma$ cannot be obtained directly from $\tau_1$, because n-values are not available from the literature for most of the systems. There is some evidence, that n is increasing slightly with the binding of the hydrophobic counterions on the micelle[13]. The reported change in n is considerably smaller than the observed change for the parameters $k^+/n$ and $k^-/\sigma^2$, what makes it likely that the changes in $k^+/n$, $k^-/n$ and $\sigma^2/n$ reflect a change in $k^+$, $k^-$ and also in $\sigma$

Table III. Values of the Kinetic Parameters and the cmc's for the TPy-Surfactant with Various Counterions at Various Temperatures.

| Counterion | T (°C) | cmc (mol/l) | $k^-/n$ (s$^{-1}$) | $k^-/\sigma^2$ (s$^{-1}$) | $\sigma^2/n$ | $k^+/n$ (1/mol·s) |
|---|---|---|---|---|---|---|
| Cl | 5 | $4{,}48 \times 10^{-3}$ | $7{,}0 \times 10^4$ | $3{,}3 \times 10^4$ | $2{,}2$ | $1{,}6 \times 10^7$ |
|  | 10 | $4{,}2 \times 10^{-3}$ | $9{,}3 \times 10^4$ | $4{,}0 \times 10^4$ | $2{,}3$ | $2{,}2 \times 10^7$ |
|  | 15 | $4{,}1 \times 10^{-3}$ | $1{,}1 \times 10^5$ | $4{,}7 \times 10^4$ | $2{,}3$ | $2{,}7 \times 10^7$ |
| $C_4H_9SO_3$ | 5 | --- | $1{,}8 \times 10^4$ | $1{,}4 \times 10^4$ | $1{,}3$ | $9{,}0 \times 10^6$ |
|  | 10 | --- | $3{,}1 \times 10^4$ | $2{,}3 \times 10^4$ | $1{,}3$ | $1{,}4 \times 10^7$ |
|  | 15 | --- | $4{,}4 \times 10^4$ | $3{,}0 \times 10^4$ | $1{,}2$ | $2{,}0 \times 10^7$ |
|  | 20 | --- | $5{,}7 \times 10^4$ | $4{,}1 \times 10^4$ | $1{,}2$ | $2{,}6 \times 10^7$ |
|  | 25 | $2{,}2 \times 10^{-3}$ | $8{,}6 \times 10^4$ | $6{,}0 \times 10^4$ | $1{,}2$ | $3{,}9 \times 10^7$ |
| $C_5H_{11}SO_3$ | 5 | --- | $9{,}8 \times 10^3$ | $9{,}0 \times 10^3$ | $1{,}1$ | $7{,}0 \times 10^6$ |
|  | 10 | --- | $1{,}5 \times 10^4$ | $1{,}4 \times 10^4$ | $1{,}1$ | $1{,}0 \times 10^7$ |
|  | 15 | --- | $2{,}0 \times 10^4$ | $1{,}9 \times 10^4$ | $1{,}1$ | $1{,}4 \times 10^7$ |
|  | 20 | --- | $3{,}4 \times 10^4$ | $2{,}9 \times 10^4$ | $1{,}2$ | $2{,}4 \times 10^7$ |
|  | 25 | $1{,}4 \times 10^{-3}$ | $4{,}6 \times 10^4$ | $3{,}8 \times 10^4$ | $1{,}2$ | $3{,}3 \times 10^7$ |

The change in $k^-/n$ comes probably about by the change of the electrostatic potential in the doulbe layer of the micellar surface. The dissociation of a monomer from the micelle is assisted by the repulsion of a monomer from the micelle due to its like charge. When more counterions are associated into the micelle as inicated by the conductivity plots for the hydrophobic ions from Figure 2 some of the micellar charge is reduced and consequently the repulsion is reduced also what is finally reflected in the reduced rate constant for $k^-/n$. It is more difficult to understand why $k^+/n$ is reduced. Previously it was reported that if $k^+$-values are determind from $k^+/n$ and accepted values for n, values for $k^+$ are obtained that are in the order of diffusion controlled rate constants . For a given ionic strength the rate constant should therefore increase with the reduced micellar charge that is obtained in the case of the counterion association. However, the incorporation leads also to a reduced cmc and hence to a reduced ionic strength which makes the approach of like charges more difficult. This effect then is believed to overcompensate the first one. In the other hand, it seems also possible that the reason for the smaller $k^+/n$-values is brought about by an increase of n.

It is interesting to note that in the case of DPyJ where the charge of the micelles was compensated by the iodide ion forming a charge transfer complex with the DPy-ion, extreme large values for $\sigma^2/n$ had been observed, while in the present case, where the micellar charge is compensated by hydrophobic counterions, extreme small $\sigma^2/n$-values are observed.

On the basis of the previous measurements it was postulated that the compensation of the charge should lead to high values of $\sigma^2/n$, as was observed for DPyJ. This certainly is not found to be the case for the present systems where the charge density of the micellar surface is also lowered. It would be interesting to have data on n from light scattering measurements to decide whether the observed change is due to n or $\sigma$.

### 3. Micellar Aggregation Numbers from Kinetic Data

For the systems for which the two relaxation times could be measured values for n can be calculated from the kinetic data with the help of some assumptions that are probably justified. The calculation is based on the assumption that the micellar distribution curve is of the Gauss type. In this case, the experimentally known value of the concentration of the particles in the distribution minimum $c_i$ can then be expressed as a function of $c_o$, n, $\sigma$ and r (aggregation number of the nucleus that can be obtained independently). This has been done in Equation (3)

$$c_i = c_{max} \cdot \exp(n - r)^2 / 2\sigma^2 \qquad (3)$$

The only two unknowns therefore in this Equation are $\sigma$ and n. Consequently, if $\sigma^2/n$ is known from $\tau_1$, the parameters $\sigma$ and n can be obtained separately, because $c_i$ can be obtained independently from $\tau_2$.

The values that are calculated in this way are given in Table IV. They turn out to be somewhat lower than n-values for similar systems from light scattering data. This could mean that the used assumptions are not strictly valid or that the micelles that are found close to the cmc are really somewhat smaller than micelles at higher concentrations, at which light scattering measurements are usually carried out. Admitting that the values for n thus obtained are associated with a systematic error, they still should indicate the right trend as far as the temperature dependence of n is concerned or for that matter the dependence of n on counterions.

### 4. The Micellar Nucleus

The relaxation times $\tau_2$ carry interesting information on the nucleus of the micelles. There is the unexpected result that $\tau_2$ varies enormously, if for a given surfactant ion the chain length in the counterion is changed. The reason for the large changes in $\tau_2$ can only have its origin in a change of R because all the other parameters that enter into Equation (2) are assumed to vary little if at all. Since the nucleus is free of counterions for all systems that have a concentration dependence of $\tau_2$ that is similar than in the case of hydrophilic counterions, the nucleus has actually to be the same one for these systems and hence the rate constant $k_i$ that enters into R can hardly be different. The large change must therefore have its origin in $c_i$. Actually this is not surprising because the nucleus concentration besides being in equilibrium with the abundant micelles must also be in equilibrium with the monomer concentrations and hence its concentration must vary with $c_1$ because $c_i = K_i \cdot c_1^r$. As a consequence of the different cmc-values, $c_i$ must vary therefore with a power of the monomer concentration that is given by the aggregation number r of the nucleus. The large dependence of $1/\tau_2$ on the counterions can therefore be used to determine the value for r.

In Figure 7 the $\log(1/\tau_2)$-values for different counterions are plotted against the $\log(cmc)$. If systems are used for which the $1/\tau_2$-$c_o$-curves are little dependent on $c_o$, what is indicative that the nucleus is free of counterions, a straight line arises which has a slope of 10. Accordingly the nucleus has an aggregation number of 10. This number is only approximately correct, because

Table IV. Values of the Kinetic Parameters, the Aggregation Numbers and the Distribution Widths for $TPyC_4H_9SO_3$ and TPyABS, Calculated from the Two Relaxation Times, at Various Temperatures.

$TPyC_4H_9SO_3$

| T (°C) | $\sigma^2/n\, k^-/\sigma_1^2$ (s$^{-1}$) | $k^+/n$ (1/mol·s) | cmc (mol/1) | n | $k^+$ (1/mol·s) | $k^-_1$ (s$^{-1}$) | $\sigma$ |
|---|---|---|---|---|---|---|---|
| 15 | 1,21  3,0x10$^4$ | 2,0x10$^7$ | 2,2x10$^{-3}$ | 56 | 1,1x10$^9$ | 2,5x10$^6$ | 9 |
| 25 | 1,16  6,0x10$^4$ | 3,9x10$^4$ | 2,2x10$^{-3}$ | 47 | 1,8x10$^9$ | 4,0x10$^6$ | 8 |

TPyABS

| T (°C) | $\sigma^2/n\, k^-/\sigma_1^2$ (s$^{-1}$) | $k^+/n$ (1/mol·s) | cmc (mol/1) | n | $k^+$ (1/mol·s) | $k^-_1$ (s$^{-1}$) | $\sigma$ |
|---|---|---|---|---|---|---|---|
| 15 | 1,15  3,1x10$^4$ | 1,6x10$^7$ | 2,3x10$^{-3}$ | 53 | 8,2x10$^8$ | 1,9x10$^6$ | 8 |
| 25 | 0,99  5,8x10$^4$ | 2,5x10$^7$ | 2,3x10$^{-3}$ | 42 | 1,1x10$^9$ | 2,4x10$^6$ | 6,5 |

Table V. Thermodynamic Parameters for the Transfer of a TPy-Ion from the Surface of the Micelle into the Micelle and the Reaction Enthalpy for a Micellar Nucleus of DPy Having Various Counterions.

| | TPy | | | DPy | |
|---|---|---|---|---|---|
| Counterion | ΔH (kcal/mol) | ΔG (kcal/mol) | ΔS (cal/mol·°K) | Counterion | ΔH (kcal/mol) |
| Cl | -5,7±1,5 | -7,3±0,3 | +5,9±1,5 | ABS | +34,5±1 |
| $C_4H_9SO_3$ | -9,0±1,5 | -8,0±0,4 | -3,6±1,5 | $C_5H_{11}SO_3$ | +23,0±1 |
| $C_5H_{11}SO_3$ | -9,5±1,5 | -8,4±0,5 | -3,9±1,5 | BS | +15,0±1 |
| | | | | TS | +10,0±1 |

activity coefficients have been neglected in the Equation (2). They should be taken into account because the nuclei for the different systems are formed under different ionic strength conditions.

The correction can be applied using the extended Debye-Hückle Equation and leads to a value for r of 8-9. This shows that the aggregation process has to proceed many steps before a nucleus is reached that has a better chance to grow to a real micelle than to dissociate again. A similar conclusion was reached earlier when the size of the nucleus was estimated from the change of $1/\tau_2$ with the addition of a nonsurface active salt that shifts the cmc of the surfactant.

It is noteworthy that in Figure 7 the straight line for the DPy-surfactants lies below the one for the TPy-surfactants. If surfactants are compared that have about the same cmc like TPyCl and $DPyC_4H_9SO_3$ the relaxation time of the latter system is about a factor 1000 longer than for the first one. While this behavior seems unusual on the first glance, it can be well explained and is consistent with the theory.

From Equation (2) $1/\tau_2$ can be approximated by $1/\tau_2 = (n^2 \cdot c_i \cdot k^-)/A_1$ where $c_i$ is given by $c_i = \prod K_i \cdot c_1^r$ in which $\prod K_i$ stands for the product of the individual stability constants for the oligomers. It is very likely that this term is considerably larger for the TPy-surfactants than for the DPy-surfactants, because the hydrophobic interaction energy is larger in the prior case. The observed behavior that $1/\tau_2$ for the same monomer concentration is larger for $TP_y$ than for $DP_y$, can thus directly be derived on the basic simple equations.

It is also worth mentioning that the points that do not fall on the straight lines are always higher than the lines. This is no accident but a consequence from the fact that the straight line represents the dynamics of micelle formation in the limiting case of the absence of coordinated counterions. This route is always open for the micelle formation. The values for $1/\tau_2$ can therefore never fall below this line. However, if there is a chance to go to micelles in another route that is faster the system is going to use this way. This is the situation for micelles having hydrophobic counterions that interact with the nucleus.

5. The Association of Hydrophobic Counterions with the Micellar Nucleus.

In the presence of interacting counterions we find that $1/\tau_2$ is rapidly increasing with the total surfactant concentration. How can we understand this behavior? It is generally agreed upon that

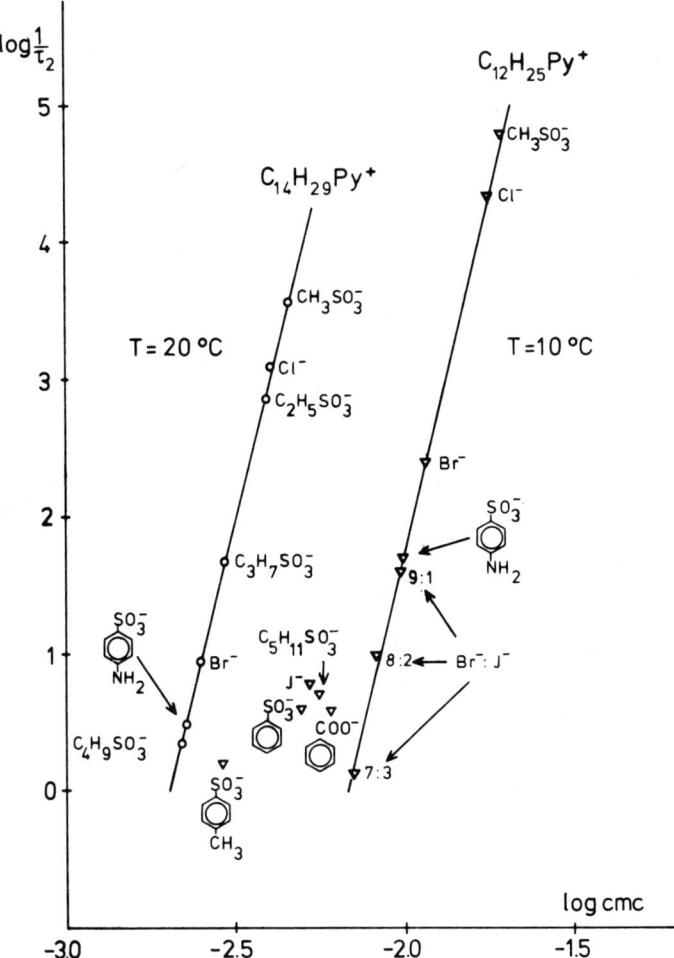

Figure 7. Plot of $\log(1/\tau_2)$ for DPy- and TPy-surfactants having different counterions against their log(cmc)-values.

while the surfactant concentration remains more or less constant once the total concentration reaches the cmc, the counterion concentration keeps growing with $c_0$, because not all counterions are associated on the surface of the micelle. It is probably this fact which makes $1/\tau_2$ grow also because it is likely that the hydrophobic counterions associate with the nucleus and consequently enter into the mass balance of the nucleus and hence into R.

We therefore can formulate the concentration of nuclei in the case of counterion association $c_i = K_i \cdot c_1^r \cdot c_H^m \cdot K_m$, where $c_H$, m and $K_m$ mean the concentration of the hydrophobic counterions, the number of counterions associated with the nucleus and the stability constant for this association process, respectively. If we assume furthermore that for a limited concentration range $K_i \cdot c_1^r$ remains constant and the total change of $1/\tau_2$ is due to $c_H$, it can be shown that

$$(1/\tau_2)/((1/\tau_2)_{cmc}) = c_H^m/(c_H)_{cmc}^m = (((c_H)_{cmc} + \Delta c_H)/(c_H)_{cmc})^m \quad (4)$$

and $\log((1/\tau_2)/(1/\tau_2)_{cmc}) = m \cdot ((c_H)_{cmc} + \Delta c_H)/(c_H)_{cmc}$

Equation (4) can be used to determine the number of counterions that are associated with the nucleus of the micelle. In order to do this we have to know the free ionic concentration of the counterions as a function of the total surfactant concentration. While it is experimentally extremely difficult to obtain exact values for the studied systems, it is nevertheless possible to derive approximate values from the conductivity of the solutions. If we assume that above the cmc the surfactant ion concentration remains constant and furthermore that the contribution of the micelles to the conductivity of the solution is negligible, the change in total conductivity can be related to a change in the counterion concentration. The values for the counterion concentration thus obtained represent an upper limit because as it was shown by Stigter and Mysels[14], the mobility of micelles for many systems is comparable to that of the counterions and the conductivity contribution of the micelles to the total conductivity can usually not be neglected.

For the made assumption, the counterion concentration should vary with $(\varkappa + \Delta \varkappa)/\varkappa$, where $\varkappa$ is the conductivity due to the counterions at the cmc and $\Delta \varkappa$ the total conductivity change for any concentration larger than the cmc from the cmc. In Figure 8a the $\log(1/\tau_2)/(1/\tau_2)_{cmc}$ is plotted against the log of this expression for different systems. In order to avoid ambiguities, the cmc was not used as a reference, but a concentration that was approximately 20% above the cmc. The plot gives indeed a linear relation from which m-values can be obtained. These m-values are minimum values because of the made assumption. A micellar contribution to the total conductivity leads to smaller counterion concentrations and hence to higher m-values.

The number of the associated counterions thus determined increases with the hydrophobicity of the counterions. Their values indicate that the nucleus can become highly saturated with counterions. In the case of TS it is even likely that the number of associated counterions is larger than that of the surfactant molecules themselves and that the nucleus is oppositely charged than the fully developed micelle.

This situation is probably the reason for the lowered reaction enthalpy of the nucleus that can be determined from the temperature dependence of $1/\tau_2$. In the case of no counterion association, large positive $\Delta$ H-values are observed for the formation of the nucleus

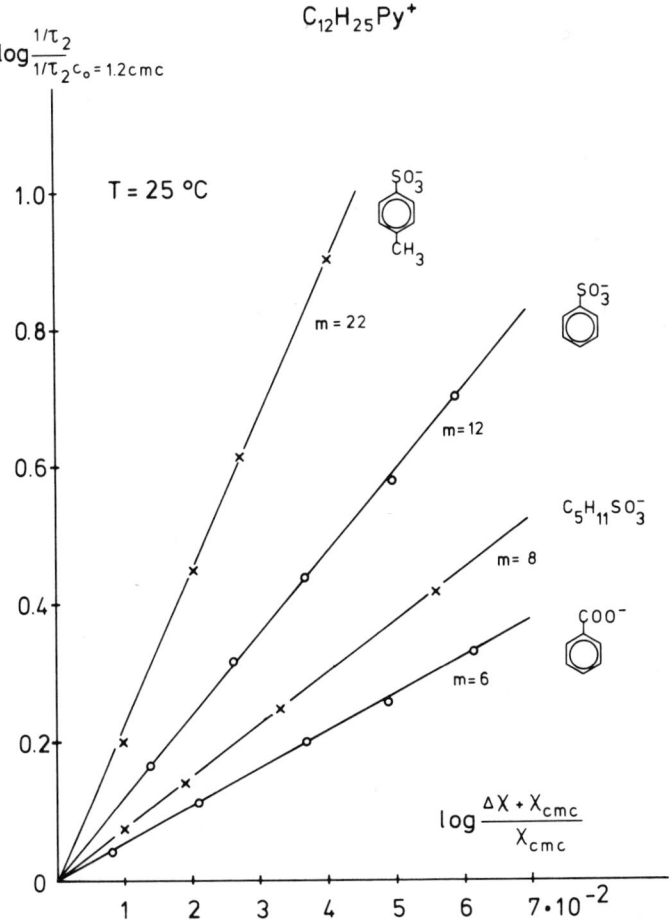

Figure 8a. Plot of log $((1/\tau_2)/(1/\tau_2)_{1.2cmc})$ for DPy-surfactants having different counterions against $\log((x_{cmc} + x_{cmc})/x_{cmc})$ at 25°C.

(see Table V). This endothermic reaction enthalpy can probably be looked at as a heat of reaction that is needed to melt the structured water around a hydrophobic group before two hydrocarbons can get into contact. In the nucleus, some of the hydrocarbon surface of the surfactant is still exposed to water. When the micelles grow, this exposed surface is eliminated and the total alkyl chain of the surfactant is located in the interior of the micelle, while the ionic head group forms the contact layer with the bulk water. In this case, negative reaction enthalpies are observed for the transfer of a monomer into the micelle. Obviously the endothermic heat of reaction that was observed for the aggregation at the nucleus and which was based on purely hydrophobic interaction has now been overcompensated by a second effect. This second effect can be probably looked at as

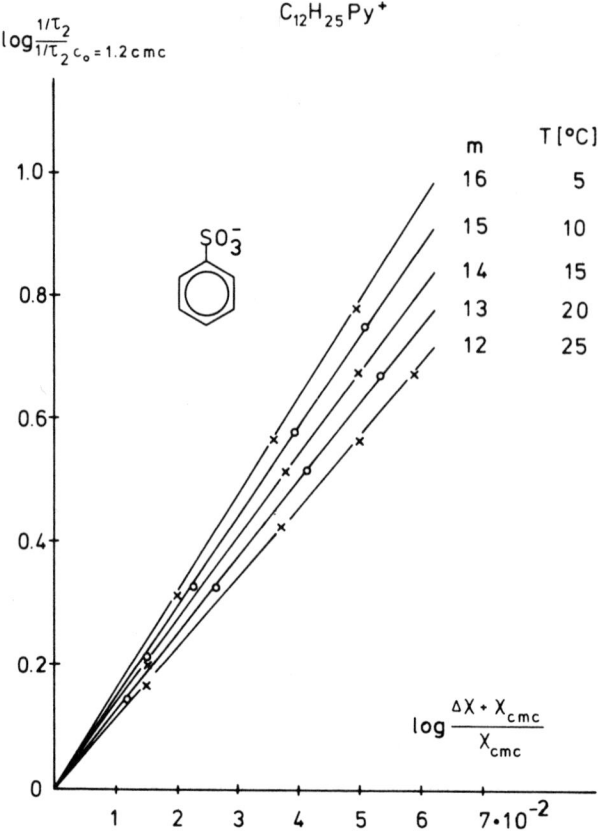

Figure 8b. Plot of $\log((1/\tau_2)/(1/\tau_2)_{1,2cmc})$ for DPyBS at different temperatures against $\log((\varkappa + \varkappa_{cmc})/\varkappa_{cmc})$.

the heat of condensation that is gained when a molecule from the gas phase or from the dispersed form in the solvent water is transferred into the organic phase. The incorporation of the hydrophobic counterions into the nucleus leads to the same effect. Now the hydrophobic endothermic heat of reaction of the nucleus is partially compensated by the heat of condensation of the bounded molecules what finally leads to a reduced endothermic reaction enthalpy for the total nucleus and hence to a smaller temperature dependence of $\tau_2$.

That the incorproation of the hydrophobic counterions into the nucleus is an exothermic process can also be seen from the temperature dependence of the number of the associated counterions for different temperatures. This is shown in Figure 8b, where an analogous plot as in Figure 8a has been drawn for the same system but for different temperatures. The number of associated counterions decreases from 16 to 12 when the temperature is raised from 5 to 25°C.

## 6. The Transfer of a Surfactant Molecule from the Surface of the Micelle into the Micelle

Aniansson et al[1] derived for the rate constant $k^-/n$ the equation

$$k^-/n = (D/l^2) \cdot \exp(-\varepsilon/RT) \tag{5}$$

in which D is the diffusion coefficient of the monomer, l the length of the last remaining $CH_2$-group, over which distance the surfactant ion has to diffuse in order to become free and $\varepsilon$ the total energy for the transfer of a surfactant molecule from the micellar state to the surface of the micelle. The reciprocal value of this rate constant is the residence time of a surfactant molecule inside the micelle. Consequently the reaction enthalpy for this process can be determined from the temperature dependence of $k^-/n$ according to $(d \ln(k^-/n))/(d(1/T)) = \Delta H/RT$.

The $\Delta$ H-values that are obtained from this slope have to be corrected for the temperature dependence of the diffusion coefficient, which leads to a correction term of about 4 kcal. The thus corrected values are given in Table V together with the $\Delta$ G- and $\Delta$ S-values for the same process. The values for $\Delta$ G (= $-\varepsilon$) have been obtained by use of Equation (5) and experimental values for $k^-/n$, D and l. Judged by the $\Delta$ H-values, the transfer of the surfactant from the surface into the micelle becomes more exothermic with increasing hydrophobicity of the coordinated counterion. At the same time, the entropy changes from positive to negative values. It is conceivable that these numbers reflect the packing of counter- and surfactant ions on the micellar surface. With increasing binding of the counterions, the packing and hence the order in the surface is probably increased.

In Figure 9 the log of the $k^-/n$-values are plotted against 1/T. Figure 9 contains also a few plots for the $1/\tau_2$-values from which the reaction enthalpies for the micellar nucleus can be calculated. These values are also given in Table V.

## CONCLUSIONS

Hydrophobic counterions having more than two $CH_2$-groups interact strongly with ionic micelles. They are either incorporated with their hydrophobic part into the micelle or adsorbed at the micellar surface. As a consequence, the surface charge per area is reduced and the cmc-values are shifted to lower concentrations.

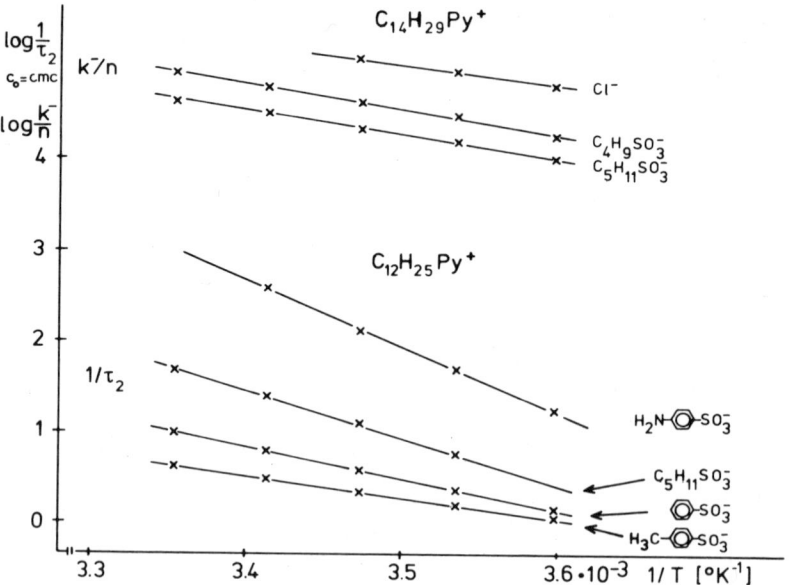

Figure 9. Plot of $\log(1/\tau_2)$ for several DPy-surfactants having different counterions and of $\log(k^-/n)$ for several TPy-surfactants having different counterions against $1/T$.

The residence time of a surfactant ion in a micelle having hydrophobic counterions is increased in comparison to the residence time of the same surfactant ion in an ionic micelle having hydrophilic counterions. The distribution curve of the micelles for a given surfactant narrows with growing hydrophobicity of the counterion. While all these parameters change only gradually with the replacement of the hydrophilic ions by hydrophobic ones, the properties of the micellar nucleus change in principle. For hydrophilic counterions, the nucleus is free of counterions, while in the case of strongly interacting hydrophobic counterions the hydrophobic surface of the aggregate seems to be more or less saturated with the hydrophobic counterions. As a consequence, both the thermodynamics and the kinetics of the formation of the nucleus vary drastically in comparison to the case of hydrophilic counterions.

## ACKNOWLEDGEMENTS

We thank the Deutschen Forschungsgemeinschaft (DFG) for a research grant in support of this work, the Fa. Henkel & Cie GmbH, Düsseldorf, for the donation of the surfactants used in this investigation and the Fonds der chemischen Industrie for general support.

## REFERENCES

1. a. E. A. G. Aniansson and N. S. Wall, J. Phys. Chem. $\underline{78}$, 1024 (1974).
   b. E. A. G. Aniansson and N. S. Wall, J. Phys. Chem. $\underline{79}$, 857 (1975).
2. H. Hoffmann in "Chemical and Biological Applications of Relaxation Spectrometry", E. Wyn-Jones, Editor, p. 181, D. Reidel, Dordrecht, Boston, 1975.
3. E. A. G. Aniansson, M. Almgren, N. S. Wall, H. Hoffmann, I. Kielmann, W. Ulbricht, R. Zana, C. Tondre and J. Lang, J. Phys. Chem. $\underline{80}$, 905 (1976).
4. a. P. Mukerjee and A. Ray, J. Phys. Chem. $\underline{70}$, 2138 (1966).
   b. P. Mukerjee and A. Ray, J. Phys. Chem. $\underline{70}$, 2144 (1966).
5. a. G. C. Kresheck, E. Hamori, G. Davenport and H. A. Scheraga, J. Amer. Chem. Soc. $\underline{88}$, 246 (1966).
   b. G. Bennion, L. Tong, L. Holmes and E. Eyring, J. Phys. Chem. $\underline{73}$, 3288 (1964).
6. R. Folger, H. Hoffmann and W. Ulbricht, Ber. Bunsenges. Physik. Chemie $\underline{78}$, 986 (1974).
7. H. Hoffmann, R. Nagel, G. Platz and W. Ulbricht, Colloid Polymer Sci. $\underline{254}$, 812 (1976).
8. a. H. Lange and M. J. Schwuger, Kolloid Z.u.Z. Polym. $\underline{243}$, 120 (1971).
   b. P. Mukerjee, K. J. Mysels and P. Kapauan, J. Phys. Chem. $\underline{71}$, 4166 (1967).
   c. K. Meguro, T. Kondo, O. Yoda, T. Ino and N. Ohba, Nippon Kagaku Zasshi $\underline{77}$, 1236 (1956).
9. R. Zana, (1976) private communication.
10. P. Mukerjee and K. J. Mysels, "Critical Micelle Concentrations of Aqueous Solutions", NSRDS-BBS No. 36, U.S.A., 1971.
11. C. Tanford, "The Hydrophobic Effect", Wiley-Interscience, New York, 1973.
12. K. Shinoda, T. Nagakawa, B. Tamamushi and T. Isemura, "Colloidal Surfactants", Academic Press, New York, 1973.
13. K. J. Mysels and L. H. Princen, J. Phys. Chem. $\underline{63}$, 1699 (1959).
14. D. Stigter and K. J. Mysels, J. Phys. Chem. $\underline{59}$, 45 (1955).

## ON THE USE OF CHEMICAL RELAXATION METHODS TO DISTINGUISH BETWEEN TRUE MICELLIZATION AND CONTINUOUS ASSOCIATION

R. Zana, J. Lang, S.H. Yiv, A. Djavanbakt and C. Abad

C.N.R.S., Centre de Recherches sur les Macromolécules

6, rue Boussingault, 67083 - Strasbourg Cédex - France

Dodecane-1,12-bis (trimethylammonium bromide and docosane-1,22-bis (trimethylammonium bromide), referred to as $C_{12}Me_6$ and $C_{22}Me_6$, bile acid salts (sodium cholate, deoxycholate and dehydrocholate) and two antihistamines (tripelennamine hydrochloride and chlorpheniramine maleate) have been investigated by means of chemical relaxation methods in order to check some implications of Aniansson and Wall theory for the dynamics of micellar solutions and to gain new informations of the association behavior of these compounds. With the exception of $C_{22}Me_6$, the relaxation behavior of these compounds show important differences with respect to that of classical ionic detergents. The results suggest that $C_{12}Me_6$ form very small and loose aggregates and that bile salts and antihistamines associate in a large range of concentration. In addition to providing kinetic informations, chemical relaxation methods appear very useful to distinguish between "true" micellization and continuous association. In contrast to $C_{12}Me_6$, temperature-jump experiments as well as equilibrium studies indicate that $C_{22}Me_6$ forms micelles. However, this process has been found to be much faster than in a case of a detergent with a comparable cmc. This result is discussed in terms of the conformation of the paraffinic chain of $C_{22}Me_6$.

## INTRODUCTION

Chemical relaxation methods have proved extremely successful in solving a number of problems related with the formation of micelles in solutions of ionic detergents. It must be recalled that measurements of stopped-flow, p-jump, T-jump and ultrasonic absorption have shown[1-4] that dilute micellar solutions are characterized by two relaxation processes, with relaxation times in a ratio between 100 and 1000. The fast relaxation process has been assigned[1] to the association/dissociation of one amphiphilic ion A to/from a micelle $A_n$ (n, aggregation number), according to

$$A_n \underset{k^+}{\overset{k^-}{\rightleftharpoons}} A_{n-1} + A \qquad (1)$$

The slow relaxation process has been attributed[2,3], to the equilibrium (2) where micelles form and dissolve.

$$nA \rightleftharpoons A_n \qquad (2)$$

The correct analytical expressions for the relaxation times $\tau_1$ and $\tau_2$ associated with these two equilibria have been first derived by Aniansson and Wall[5]. The application of these expressions to the results of chemical relaxation measurements on micellar solutions of ionic detergents has permitted Aniansson et al[4] to obtain information thus far not available on these systems : rate constants $k^+$ and $k^-$ of reaction (1), micellar polydispersity, etc..

The model used by Aniansson and Wall[5] to obtain the expressions of $\tau_1$ and $\tau_2$ suggests two extensions of the work described in Reference 4. The first one is a study of bolaform detergents which are compounds whose both ends are electrically charged. Indeed, the expression of the rate constant $k^-$ was derived[5] assuming a diffusive motion of a detergent ion out of a micelle, at right angle to the micellar surface. Such a model is clearly restricted to detergents with only one charged end. In the case of micelle forming bolaform detergents such a motion becomes very unlikely because it implies the crossing of the micelle hydrophobic core by a charged end group. The dissociation of bolaform ions from micelles should therefore proceed by means of a different process. As a result micelle forming bolaform detergents should show a kinetic behavior differing from that of singly charged detergents. This prediction led us to undertake the study of dodecane-1,12-bis (trimethylammonium bromide) and of docosane-1,22-bis (trimethylammonium bromide), referred to as $C_{12}Me_6$ and $C_{22}Me_6$, respectively. Micelle formation has been recently reported[6] in $C_{12}Me_6$ solutions on the basis of dye spectral change and surface tension measurements. To our knowledge $C_{22}Me_6$ has never been investigated thus far. For this reason the conductivities and emf of $C_{22}Me_6$ solutions were determined in order to obtain information on the formation of micelle in this system. Similar experiments as well as density measurements were performed on $C_{12}Me_6$ solutions because micelle formation in this system was not well documented.

The second type of experiments that Aniansson and Wall[5] theory suggested to us concerns compounds whose self association behavior differs from a true micellization. In the Aniansson and Wall treatment, the slow relaxation process results from the presence in the aggregate size distribution curve of a minimum which corresponds to the most unstable aggregates, i.e., those which can hardly be detected in any other way than by chemical relaxation methods. In the case of compounds which undergo a continuous association resulting in a size distribution curve showing no minimum the theory would predict only one relaxation time or a fairly narrow distribution of relaxation times. Thus, chemical relaxation methods should permit a distinction between true micellization and continuous association. Note that the results obtained in studies of compounds such as purine derivatives and dyes, which show a continuous self association through vertical stacking of aromatic rings, confirm this prediction[7-10]. In this work we have undertaken the study of two series of compounds whose association behavior has been already investigated in aqueous solutions but is not well understood. The first series is constituted by bile acid (cholic, deoxycholic and dehydrocholic acids) salts whose association has been extensively investigated in view of their biological importance[11]. The association behavior of these salts appears complex and the existence of several critical micelle concentrations (cmc) has been reported[12,13]. However, a close examination of the results suggests that in fact a progressive association may be taking place in a large concentration range. Antihistamine drugs constitute the second series of compounds where occurs a continuous association rather than a true micellization. In a recent study Attwood and Udeala[14] concluded that the association of some of these drugs is not micellar in nature but no definite conclusion could be reached about the association behavior of other antihistamines. In no instance, however, was a cmc detected by means of conductivity and light scattering measurements.

## INVESTIGATIONS OF BOLAFORM DETERGENTS

### Materials and Equipments

$C_{12}Me_6$ and $C_{22}Me_6$ were synthesized and purified as described by Menger and Wrenn[6]. For the sake of comparison, the parent compounds dodecane and tetradecane trimethylammonium bromide ($C_{12}Me_3$ and

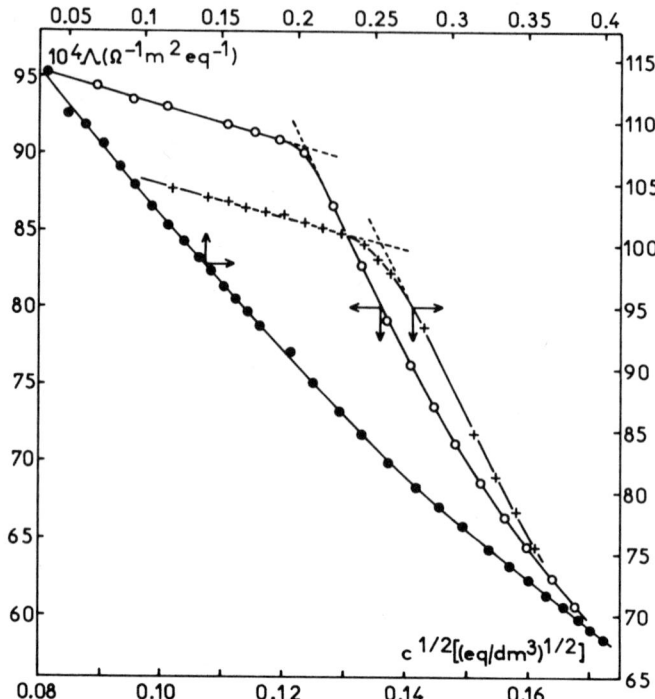

Figure 1. Variation of the equivalent conductivity with the square root of the concentration for $C_{12}Me_6$ (●), $C_{22}Me_6$ (+) and $C_{12}Me_3$ (o) solutions. The values of $c^{1/2}$ for $C_{22}Me_6$ have been shifted by + 0.06 units. Temperature : 25°C.

and $C_{14}Me_3$) were also synthesized and investigated. The purity of these compounds was checked by elementary chemical analysis.

The equipments used in this study were the same as those described in previous work[1-4,15].

## Results

The conductivity and emf of $C_{12}Me_3$ solutions show a sharp change at a concentration of about 0.015M, i.e., close to the reported cmc[16] for this detergent (see Figures 1 and 2). Also, a volume increase of 8 cm³/mol. has been measured upon micelle formation in $C_{12}Me_3$ solutions[17a]. On the contrary the measurements of density, conductivity and emf of $C_{12}Me_6$ solutions revealed no rapid change in the concentration range between 0.02 and 0.05M, where micellization has been reported to take place[6]. The calculated apparent molal

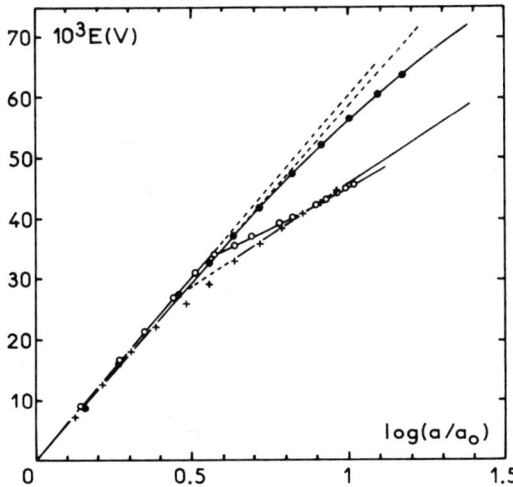

Figure 2. Variation of the emf at 25° with log $(a/a_o)$ for $C_{12}Me_6$ (●), $C_{22}Me_6$ (+) and $C_{12}Me_3$ (o) solutions. The initial concentrations corresponding to the activity $a_o$ were 0.012, 0.0008 and 0.004 mole/dm$^3$ for $C_{12}Me_6$, $C_{22}Me_6$ and $C_{12}Me_3$, respectively. The activity a was calculated from the concentration using Davies equation.

volume shows a continuous decrease of about 3 cm$^3$/mol. as the concentration c is increased from 0.01 to $1\underline{M}$[15], whereas a volume increase usually occurs upon micelle formation[17]. The equivalent conductivity $\Lambda$ also shows a continuous decrease (see Figure 1). Finally the emf E increases linearly up to c ~ $0.1\underline{M}$, again well above the reported[6] micellization range. At c > $0.1\underline{M}$, the emf plot shows a small curvature (see Figure 2) which may indicate some association. For $C_{22}Me_6$ both $\Lambda$ and E show a rapid change at concentrations of about $0.0029$ and $0.0025\underline{M}$, respectively (see Figures 1 and 2), which reveals micelle formation[16,18]. Note that if the cmc of $C_{22}Me_6$ is taken as $0.0027\underline{M}$, the cmc of $C_{12}Me_6$ should be about $2.5\underline{M}$, on the assumption of a doubling of the cmc for each methylene group substracted from the $C_{22}Me_6$ chain[19]. All these results may be summarized by stating that micelle formation appears to be well established for $C_{12}Me_3$ and $C_{22}Me_6$ but that density, conductivity and emf investigations of $C_{12}Me_6$ show no evidence for this process. The chemical relaxation investigations reported below lead to the same conclusion.

The excess absorption of $C_{12}Me_3$ solutions with respect to water showed the usual behavior found for many ionic detergents.

The excess absorption is pratically negligible at c < cmc, and then increases fairly steeply and linearly with c, above the cmc. At higher concentration the absorption goes through a maximum. The cmc value found in these measurements (0.014$\underline{M}$) is in good agreement with that reported by other workers[16]. On the contrary, the excess absorption of $C_{12}Me_6$ solutions remained very small up to concentrations of about 0.15$\underline{M}$, i.e., well above the range where micellization has been reported to occur[6]. At c > 0.3$\underline{M}$, the absorption of $C_{12}Me_6$ solutions increases and a change of slope in the absorption vs c curve is observed at about 0.6$\underline{M}$. However at such a high concentration the distance between two bolaform ions is comparable to the length of the bolaform. Strong solute-solute interactions are therefore to be expected, even though they may not result in association, and may explain the increased excess absorption. Thus ultrasonic absorption measurements do not provide evidence of micelle formation in solutions of $C_{12}Me_6$.

A relaxation signal of fairly large amplitude was found with $C_{12}Me_3$ with the p-jump and shock tube equipments. The corresponding relaxation time $\tau_2$ was in good agreement with that determined by T-jump in the same conditions, and showed the usual increase upon increasing concentration[2,3]. On the contrary no relaxation signal could be detected with $C_{12}Me_6$ in the range where micellization was reported to take place[6].

$C_{12}Me_6$ was also investigated by means of T-jump with a spectrophotometric detection, in the presence of the anionic dye eosine. Indeed, $C_{12}Me_6$ does not absorb light in the range of wavelength available on our equipment. However we have shown[20] that a dye may be used to probe the kinetics of micellar equilibria, provided that the (dye)/(detergent) ratio is kept below 0.01. Relaxation signals of very small amplitude were found above and below the reported range of cmc. In addition the associated relaxation time appeared to be dependent on the dye concentration. These findings indicate that the observed relaxation process is probably due to some eosine-$C_{12}Me_6$ interaction. Thus, as in the case of ultrasonic absorption, T-jump, p-jump and shock tube investigations provide no evidence of micelle formation in solutions of $C_{12}Me_6$.

On the contrary, T-jump investigations of $C_{22}Me_6$ solutions revealed a large relaxation signal. The associated relaxation time ($\tau_2$) is independent of the eosine concentration, if the ratio (eosine)/($C_{22}Me_6$) is kept below 0.01. $\tau_2^{-1}$ decreases as the detergent concentration is increased (see Figure 3), as in the case of $C_{12}Me_3$ and other detergents[2-4]. In the absence of added salt (KBr), the variation of $\log(1/\tau_2 T)$ with $1/T$ yields an apparent activation energy of 117 kJ/mol., i.e., close to the values obtained for sodium alkylsulfates[4] (about 125 kJ/mol. and only slightly dependent on the alkyl chain length). Finally, $1/\tau_2$ and the activation energy show

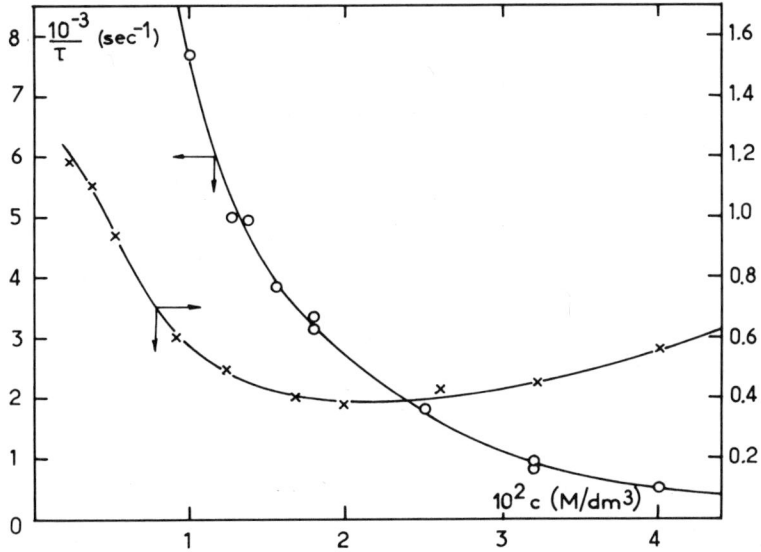

Figure 3. Variation of $1/\tau_2$ for $C_{22}Me_6$ at 25°, in the absence of added KBr (o) and in the presence of 0.02M KBr. (X).

upon increasing ionic strength a decrease similar to that found for most ionic detergents[2-4].

## Discussion

All the results obtained in this study point against micelle formation in solutions of $C_{12}Me_6$. This conclusion appears to contradict that of Menger and Wrenn[6] who reported that micellization takes place in $C_{12}Me_6$ solutions, on the basis of changes of surface tension and of absorption spectrum of a dissolved dye with concentration. The two sets of results can be reconciled, however, by assuming that the aggregates of $C_{12}Me_6$ are quite small (possibly only dimers or trimers) and very loose. Our results in fact clearly indicate that the $C_{12}Me_6$ aggregates are extremely loose. Indeed the density measurements have shown that the volume change $\Delta V$ upon association in the reported micellization range is negligible. This result explains that the excess ultrasonic absorption of $C_{12}Me_6$ solutions is very small and that no relaxation signal can be detected by means of p-jump and shock-tube. Indeed, the relaxation amplitude is proportional to $\Delta V^2$ for the first method and to $\Delta V$ for the other two. Moreover the enthalpy change $\Delta H$ for the aggregation process must also be very small because the amplitude of the T-jump relaxation signal which is proportional to $\Delta H$ was negligibly small. $\Delta H$ and $\Delta V$ being essentially due to changes of hydrophobic hydration upon association, it is concluded that the associated $C_{12}Me_6$ ions probably retain most of their hydration water and, therefore, that the aggregates are very loose. On the other hand, the shape of the surface tension vs c curve of Reference 6 as well as the $\Lambda$ vs $c^{1/2}$ and E vs $\log a/a_0$ curves of Figures 1 and 2 indicate that the aggregation number of the associated species must be very small. On the basis of these conclusions the words micelle and cmc hardly refer to the aggregates found in $C_{12}Me_6$ solutions.

Although the fast relaxation process has not been investigated yet, some qualitative informations can be gained from the T-jump results relative to $C_{22}Me_6$. For this purpose we have compared $C_{22}Me_6$ to $C_{14}Me_3$ because the cmc of these two detergents are fairly close : 0.0027M and 0.0035M[16], respectively, and because the cmc is an important parameter in the treatment of Aniansson and Wall[5]. At a given temperature and value of the ratio a = (c - cmc)/cmc, the value of $\tau_2^{-1}$ for $C_{22}Me_6$ is more than one order of magnitude larger than for $C_{14}Me_6$. In an attempt to give a qualitative explanation of this difference we have used the approximative Equation 3 where n is the average aggregation number of the proper micelles and $\sigma$ the width of the micelle size distribution curve[5]. R is given[4,5] by Equation 4 where $[A_r]$ refers to the molar concentration of the aggregate $A_r$ at the minimum of the micelle size distribution curve and $k_r^-$ is the rate constant for the dissociation of one detergent ion from $A_r$. The emf results show that the ionization degree of $C_{22}Me_6$ is larger than that of $C_{12}Me_3$ and therefore of $C_{14}Me_3$.

$$\frac{1}{\tau_2} \simeq \frac{n^2}{R \cdot cmc} \left[ 1 + \frac{\sigma^2}{n} a \right]^{-1} \qquad (3)$$

$$R \simeq \frac{1}{k_r^- \cdot [A_r]} \qquad (4)$$

It is therefore likely that the value of n of $C_{22}Me_6$ is smaller than that of $C_{14}Me_3$. On the other hand we have shown[6] that $\sigma/n$ increases and $\sigma$ decreases or levels off as n decreases. These results indicate that at a given value of a the quantity $n^2/[1+(\sigma^2/n)a]$ should decrease in going from $C_{14}Me_3$ to $C_{22}Me_6$. The difference between the $\tau_2^{-1}$ values for these two detergents must therefore arise from $R^{-1}$, that is, from $k_r^-$ and/or $[A_r]$. Differences in $k_r^-$ may be interpreted by assuming a folded conformation of the $C_{22}Me_6$ ion, similar to that found for its shorter homologs at the air-water interface[6]. On this assumption the dissociation of one bolaform from a micelle would require less energy and would occur faster than if one of the bolaform charged groups crossed the micelle hydrophobic core or if the bolaform diffused out of the micelle following a direction at right angle to the bolaform axis. We are presently attempting to investigate the fast relaxation process in order to check this tentative explanation.

## INVESTIGATIONS OF BILE SALTS AND OF ANTIHISTAMINES

### Materials

Cholic (C), deoxycholic (DOC) and dehydrocholic (DHC) acids were purchased from Fluka. Aqueous solutions of the sodium salts (NaC, NaDOC and NaDHC) were prepared by stoichiometric neutralization of the acids by NaOH 1N. The solutions were used shortly after complete dissolution of the acid. The stock solutions were stored overnight at below 5°C.

The antihistamines used in this work were generous gifts from Ciba-Geigy (Switzerland) and Allen and Hamburys Researd Ltd. (England) and were used as received. The measurements were performed on about 10 different drugs but we shall only report the results of two typical examples : tripelennamine hydrochloride (TPHCl) and chlorpheniramine maleate (ClPM). Attwood and Udeala[14] have shown that the light scattering data for the first one may be accounted for using a micellar or a non-micellar association model, but that the association behavior of the second one is not of the micellar type.

### Results

The ultrasonic absorption measurements performed at 25°, pH = 8.4 and 9.06 MHz showed that the excess absorption of NaDHC solutions with respect to water is very small even at fairly large concentrations. On the contrary NaDOC and NaC show an excess absorption which increases with concentration to very large values. In the whole concentration range the excess absorption increases in the sequence NaDHC << NaC < NaDOC. The apparent molal volumes $\phi_v$ of these salts have been determined as a function of concentration[21]. In the dilute range, $\phi_v$ increases with c, owing to the association of the bile ions. This increase is larger with NaDOC than with NaC, in agreement with the ultrasonic absorption data. The excess absorption of NaDOC and NaC is largely increased by the presence of 0.2M NaCl in the solution. This effect is relatively more important in the dilute range. For NaDOC a decrease of pH from 8.4 to 7.4 resulted in a large increase of excess absorption. Finally, the absorption of NaDOC in methanol was found to be much smaller than in water. All these results (see Figure 4) parallel the known association

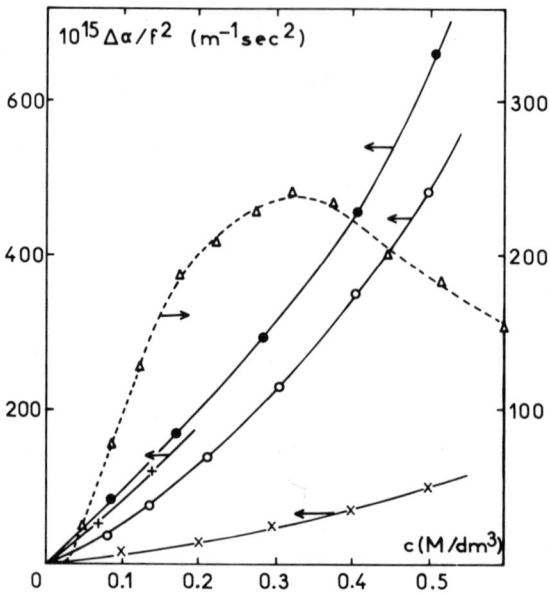

Figure 4. Variation of the excess ultrasonic absorption $\Delta\alpha/f^2$ of NaDOC with concentration at 9.06 MHz, 25° and pH = 8.4 in water (o); $H_2O$ - 0.2$\underline{M}$ NaCl (●) and methanol (X). The crosses (+) refer to results in pure water at pH = 7.46. The curve in dotted line is relative to potassium dodecanoate at 25° and 3.37 MHz from Reference 1. $\Delta\alpha$ = difference between the ultrasonic absorption coefficients of the bile salt solution and the solvent ($H_2O$, $H_2O$ - 0.2$\underline{M}$ NaCl or methanol) ; f = ultrasonic frequency. For the sake of clarity the experimental points at concentrations below 0.05-0.1$\underline{M}$ have been deleted. In all instances the measurements extended down to 0.01$\underline{M}$.

behavior of bile salts and clearly indicate that the observed excess absorption of NaC and NaDOC is due to the association of the bile ions. They also establish clear differences between the ultrasonic absorption behavior of bile salts and that of classical ionic detergents. For the purpose of illustrating these differences the absorption vs c curve for potassium dodecanoate[1] (cmc ≃ 0.026$\underline{M}$) at 3.37MHz and 25°C has been plotted on Figure 4 in dotted line. The main difference is that bile salt solutions present an excess absorption with respect to water, even at the lowest concentration investigated (about 0.01$\underline{M}$) while at c < cmc the excess absorption of ionic detergents remains negligible. On the other hand at fairly high concentration the absorption of ionic detergents goes through a maximum. Such was not found with bile salts in the concentration range investigated. In fact all of the absorption vs c curves showed a faster than linear but monotonous increase of absorption. It must be recalled that three cmc ranges have been reported[12,13] for NaC, at around 0.015, 0.045 and 0.11M. Our results show no evidence of the two higher cmc's. Moreover, although our measurements were not sensitive enough to reach a clear cut conclusion about the lower cmc of NaDOC, the extrapolations of the absorption vs c curves clearly go through the origin for both NaC and NaDOC. In addition a small but definite excess absorption was measured with NaC at c ≃ 0.01M, i.e., below the reported cmc. These results point against the existence of the lower cmc in NaC. The same remarks hold for NaDOC.

The absorption vs c curves for the two antihistamines are plotted on Figure 5. They show no evidence of a cmc, thereby confirming the findings of Attwood and Udeala[14]. As for NaC and NaDOC the excess absorption becomes quite large at high concentration and appears to increase with the extent of aggregation (the limiting aggregation numbers have been found to be 7 for TPHCl and 14 for ClPM[14]). In contradistinction to bile salts, additions of NaCl up to 0.2 M resulted in a small increase of about 10% of the excess absorption of ClPM and left unchanged the excess absorption of TPHCl.

The ultrasonic relaxation spectra of fairly concentrated solutions of NaC and NaDOC were determined in the frequency range between 1 and 250 MHz. The observed decrease of absorption upon increased frequency was found to be much slower than for a single relaxation process. The spectra are likely to be characterized by a distribution of relaxation frequencies centered at about 10 to 20 MHz, the distribution being more narrow for NaDOC than for NaC. In this respect also bile salt solutions behave differently than ionic detergents which are usually characterized by relaxation spectra with one relaxation frequency for dilute solutions, and two relaxation frequencies for more concentrated solutions[1]. The ultrasonic spectra of solutions of TPHCl and ClPM have also been determined. They were found to be close to that for a single relaxation process. The relative decrease of ultrasonic absorption upon increasing frequency was steeper than for bile salts.

Finally, T-jump investigations with the spectrophotometric detection revealed no relaxation signals which could be attributed to the association of NaC, NaDOC, TPHCl and ClPM. The measurements were performed in the presence of acridine orange in the case of bile salt solutions because they do not absorb light. For the antihistamines the investigations were performed on solutions of the pure drug in the ultraviolet range at wavelengths close to the absorption band of the drug.

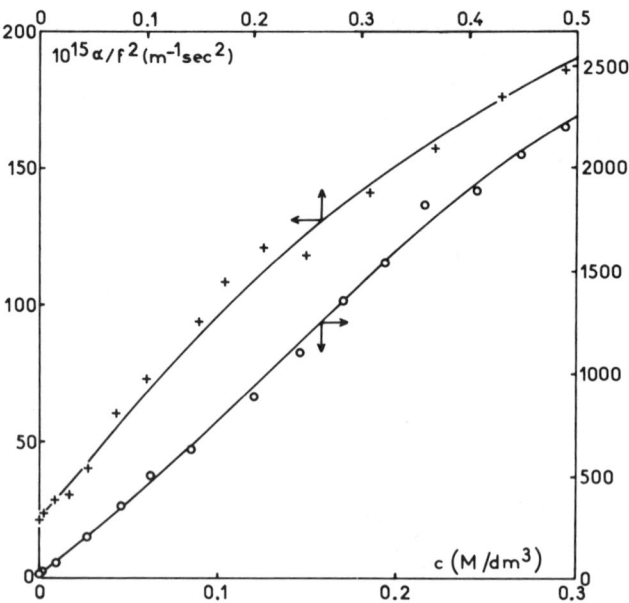

Figure 5. Variation of the ultrasonic absorption of TPHCl(+) and ClPM(O) with concentration at 2.82 MHz and 25°. $\alpha$ is the absorption coefficient of the solution.

## Discussion

The chemical relaxation methods used in this work have shown clear differences between the association behavior of ionic detergents and that of bile acids, ClPM and TPHCl. For all of these systems the absence of the slow relaxation process which is attributed to the micelle formation-dissolution equilibrium (2) coincides with the absence of a cmc. On the basis of the Aniansson and Wall[5] treatment we are then led to conclude that the absence of a maximum and a minimum on the aggregate size distribution curve correlates with the absence of a cmc and thus of a process of true micellization. This conclusion reached on purely kinetic grounds is in agreement with the mathematical definition of the cmc recently given by Ruckenstein and Nagarajan[22]. These authors have used the results of Tanford's[23] semiempirical theory of the thermodynamics of micelle formation. In their treatment the cmc is defined as that concentration where the aggregate size distribution curve shows an inflection point with a horizontal slope. Above the cmc the size distribution function goes through a minimum and a maximum. Below the cmc it only shows a decrease. Numerical calculations indicate that at the defined cmc the amount of aggregated material is negligible with respect to the total concentration and that it increases very rapidly with c above the cmc. It can be expected that for compounds which undergo a continuous association the mathematical conditions which define the cmc (first and second derivatives of the size distribution function equal to zero at the same value of the concentration) are not fulfilled.

In recent reviews, Mukerjee[24] showed that considerable caution should be exercised when using the words micelle and cmc to refer to the association behavior of many compounds. It appears, however, that both theoretical tools (References 5, 22 and 23), as well as experimental methods are now available which permit a clear distinction between "true" micellization (two well separated relaxation times, and a micelle size distribution curve going through a maximum and a minimum), and continuous association (one relaxation time or a distribution of relaxation times, and a monotonous size distribution curve). It is therefore hoped that in the future, chemical relaxation methods will be extensively used in order to reassess the association behavior of the many systems where clear cut conclusions could not be reached thus far.

## REFERENCES

1. E. Graber, J. Lang and R. Zana, Kolloid Z.Z. Polym., 238, 470, (1970) ; E. Graber and R. Zana, ibid., 238, 479 (1970).
2. R. Folger, H. Hoffmann and W. Ulbricht, Ber. Bunsenges. Phys. Chem., 78, 986 (1974).
3. J. Lang, C. Tondre, R. Zana, R. Bauer, H. Hoffmann and W. Ulbricht, J. Phys. Chem., 79, 276 (1975).
4. E.A.G. Aniansson, S. Wall, M. Almgren, H. Hoffmann, I. Kielmann, W. Ulbricht, R. Zana, J. Lang and C. Tondre, J. Phys. Chem., 80, 905 (1976).
5. E.A.G. Aniansson and S. Wall, J. Phys. Chem., 78, 1024 (1974) and ibid., 79, 857 (1975).
6. F. Menger and S. Wrenn, J. Phys. Chem., 78, 1387 (1974).
7. D. Pörschke and F. Eggers, Eur. J. Biochem., 26, 490 (1970).
8. F. Garland and R.C. Patel, J. Phys. Chem., 78, 848 (1974) ; F. Garland and S. Christian, ibid., 79, 1247 (1975).
9. B.H. Robinson, A. Seelig-Löffler and G. Schwarz, J.C.S. Faraday Trans. I, 71, 815 (1975).
10. A.D. James and B.H. Robinson, Adv. Molec. Relax. Proc., in press.
11. D.M. Small, in "The Bile Acids", P. Nair and D. Kritschewsky, Editors, Plenum Press, Vol. I, Chapt. 8, pp. 249-356, New York 1971 ; M.C. Carey and D.M. Small, Amer. J. Medecine, 49, 590, (1970) ; M.C. Carey and D.M. Small, Arch. Intern. Med., 130, 506 (1972).
12. K. Fontell, Kolloid Z.Z.Polym., 244, 246 (1972) ; ibid., 244, 253 (1972) ; ibid., 246, 614 (1972) ; ibid., 246, 710 (1972).
13. P. Ekwall et al., Acta Chim. Scand., 10, 327 and 681 (1956) ;

ibid., 11, 190, 568 and 590 (1957) ; ibid., 12, 1622 (1958).
14. D. Attwood and O.K. Udeala, J. Phys. Chem., 79, 889 (1975).
15. S. Yiv, K. Kale, J. Lang and R. Zana, J. Phys. Chem., in press.
16. P. Mukerjee and K. Mysels, "Critical Micelle Concentrations of Aqueous Surfactant Systems", NSRDS-NBS 36, US Govt Printing Office, 1970 ; E. Anacker, R. Rush and S. Johnson, J. Phys. Chem., 68, 81, (1964).
17. (a) J.M. Corkill, J. Goodman and T. Walker, Trans. Farad. Soc., 63, 768 (1967) ; (b) P. Leduc and J. Desnoyers, Can. J. Chem., 51, 2993 (1973) ; P. Leduc, J.L. Fortier and J. Desnoyers, J. Phys. Chem., 78, 1217 (1974) ; K. Shinoda and T. Soda, J. Phys. Chem., 67, 2072 (1963) ; F. Franks, M. Quickenden, J. Ravenhill and H.T. Smith, J. Phys. Chem., 72, 2668 (1968).
18. E. Keh, C. Gavach and J. Guastella, C.R. Acad. Sci. (Paris), 263C, 1488 (1966) ; K. Shirahama, Bull. Chem. Soc. Jap., 47, 3165 (1965) and references therein.
19. K. Shinoda, T. Nakagawa, B. Tamamushi, and T. Isemura, "Colloidal Surfactants", Ch. I, p. 42, Academic Press, New York, 1963.
20. C. Tondre, J. Lang and R. Zana, J. Colloid Interface Sci., 52, 372 (1975).
21. A. Djavanbakht, K. Kale and R. Zana, J. Colloid Interface Sci., in press.
22. E. Ruckenstein and R. Nagarajan, J. Phys. Chem., 79, 2622 (1975).
23. C. Tanford, Proc. Nat. Acad. Sci. USA, 71, 1811 (1974) and J. Phys. Chem., 78, 2469 (1974).
24. P. Mukerjee, J. Pharm. Sci., 63, 972 (1974) and in "Physical Chemistry : Enriching Topics from Colloid and Surface Science", H. van Olpen and K. Mysels, Editors, Chapter 9, p. 135, Theorex, La Jolla, California, 1975.

KINETIC STUDY ON MICELLIZATION OF IONIC SURFACTANTS BY MEANS OF

RELAXATION METHODS

>Kunio Takeda, Tatsuya Yasunaga
>and Hiromoto Uehara
>Department of Chemistry, Faculty of Science
>Hiroshima University
>Hiroshima 730, Japan

Kinetics of micellization have been studied on the typical ionic surfactants by means of ultrasonic absorption and pressure-jump techniques. The relaxation phenomena were observed only above the critical micelle concentration (CMC). The reciprocal fast relaxation time, $\tau_1^{-1}$, linearly increases with an increase of the surfactant concentration over a wide range between the first and the second CMC. On the other hand, the reciprocal slow relaxation times, $\tau_2^{-1}$, were found to be independent of the total concentration in aqueous solutions of surfactants. In addition, it has become apparent that the appearance of the concentration dependence of $\tau_2^{-1}$ is probably due to the presence of a trace amount of the surfactant containing different carbon numbers in the hydrophobic groups.

A series of results were interpreted by the following two-step mechanism consisting of both the micelle formation reaction and the reaction of micellar conformation change with the increase and decrease of aggregation number,

$$nA \underset{\underset{\text{fast}}{k_{21}}}{\overset{k_{12}}{\rightleftarrows}} A_n \underset{\underset{\text{slow}}{k_{32}}}{\overset{k_{23}, mA}{\rightleftarrows}} A_{n+m}$$

where A is a surfactant monomer, n and n+m are aggregation numbers of the micelle and $k_{ij}$ is the overall rate constant. This mechanism could successfully explain the concentration dependence of not only the relaxation times for both processes but also the relaxation strength for the fast process. The standard volume change of micelle formation reaction, $\Delta V^0$, calculated kinetically, were in fairly good agreement with those obtained from the results by the static measurements and were found to be expressed by the CMC and the mean aggregation number, $\bar{n}$,

and
$$\Delta V^0 = (\frac{1}{CMC})^{\alpha_1} \beta_1$$

$$\Delta V^0 = \bar{n}^{\alpha_2} \beta_2$$

where $\alpha_1 \simeq 1/2$, $\alpha_2 \simeq 3/2$ and $\beta_i$ is constant.

## INTRODUCTION

The dynamic studies of the micelle have attracted special interest from the viewpoint that the investigation of its physicochemical properties is important and unavoidable in order to make clear the representative characteristics of micelles such as solubilization or catalysis. Further, the micelle has also been noted as a model of the biomembrane or the globular protein, which is considered to have a structure similar to that of the micelle.

Although the equilibrium properties of the micelle have been studied to some extent, no kinetic studies have been performed until the development of the fast reaction techniques which have enabled kinetic investigations of fast reaction with half time down to about $10^{-9}$ sec.[1-3] The kinetic studies of micellization have been made on many surfactant systems by means of the various relaxation techniques for the last ten years.[4,27]

It has been verified that two relaxation processes do exist in aqueous solutions of sodium dodecyl sulfate (SDS) and sodium decyl sulfate (SDeS) by means of the pressure-jump, temperature-jump techniques[7,9] and the ultrasonic absorption technique,[18,23] and more recently the pressure-shock technique.[12] For the slow relaxation process, most of the papers including ours have reported that $\tau_2^{-1}$ increases linearly with the total surfactant concentration above the CMC. Furthermore, it goes through its maximum or minimum at a certain concentration, and decreases or increases monotonously with increasing concentration in the more recent paper.[12] These experimental results lead to a confused situation about the concentration dependence of $\tau_2^{-1}$.

On the other hand, the various models[4,18,20,21,27] have been proposed by many researchers but are deficient in considering both the existence of two relaxation processes and the reliable results for the dependences of $\tau_2^{-1}$ on the total surfactant concentrations. The first purpose of this investigation is to make clear experimentally the concentration dependences of $\tau_2^{-1}$. The second is to propose the reaction mechanism of micellization on the basis of these results.

## EXPERIMENTAL

**Materials.** Octylammonium chloride (OAC) was prepared from octylamine by the neutralization with hydrochloric acid. Octyltrimethylammonium bromide (OTAB) and potassium n-caprylate ($KC_8$) were donated by Professor Tanaka of Fukuoka University[28] and Professor Kimizuka of Kyushu University, respectively. Sodium n-caprylate ($NaC_8$) and sodium n-caproate ($NaC_6$) were purchased from Tokyo Kasei Co. The CMC values of these surfactants determined by the electric conductivity method are listed in Table I. SDS, lithium dodecyl sulfate (LDS) and SDeS were prepared from dodecyl and decyl alcohols mostly according to the method of Dreger et al.[29] The purities of these starting materials were determined to be above 99.5 and 99.8%, respectively, by means of gas chromatography. The values of the CMC determined by the electric conductivity method at 25°C were 8.3, 10 and 31.5 x $10^{-3}$ M, respectively. Dodecyltrimethylammonium bromide (DTAB) was donated by Professor Tanaka of Fukuoka University and its CMC was determined to be 1.55 x $10^{-2}$ M at 25°C.[30] Sodium tetradecyl sulfate (STS), used as an additive, was donated by Dr. Uehara of Kinki University and its CMC was 2.21 x $10^{-3}$ M at 35°C. The purities of the various alcohols used as additives are above 99.5%.

**Kinetic Measurements.** The measurements of ultrasonic absorption were performed by the pulse technique over the frequency range from 5 - 95 MHz. The details of the apparatus have been described elsewhere.[31] The sound velocity was measured by the sing-around method at 1.92 MHz. The pressure-jump apparatus described elsewhere[7] was partly modified. Increasing pressure is applied to the sample cell arrangement until a brass diaphragm bursts at pressures of approximately 20 - 100 atm. After the sudden decrease of pressure, the concentration change of the species in solution was followed by means of the electric conductivity method.

## RESULTS

**Fast Process.** The ultrasonic relaxation absorptions were observed above the CMC in surfactant solutions. The relaxation spectra obtained were represented by the single relaxation equation,

$$\alpha/f^2 = \frac{A}{1 + (f/f_r)^2} + B$$

where $\alpha$ is the ultrasonic absorption coefficient, f is the frequency, $f_r$ is the relaxation frequency, A and B are the relaxing and non-relaxing absorption, respectively. The typical absorption spectra are shown in Figure 1 and these experimental parameters are listed in Table I. Figure 2 shows the representative concentration dependences of the relaxation frequencies, $f_r$, in the case of OAC. As clearly shown in this figure, $f_r$ linearly increases with the surfactant concentration between the first and the second CMC.

On the other hand, the existence of the second CMC has been verified as one of the characteristics in the aqueous surfactant solution.[32-36] The beginning point of the deviation of $f_r$ in Figure 2 approximately agrees with the second CMC. Accordingly, the cause of the deviation of $f_r$ should be connected with the second CMC.

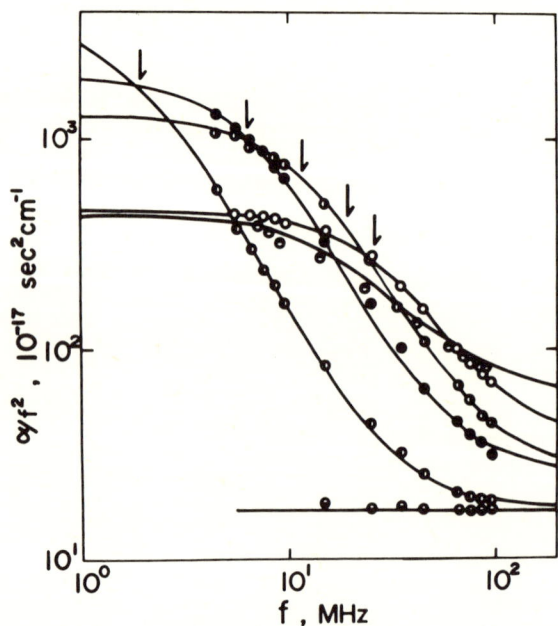

Figure 1. Ultrasonic absorption spectra of octylammonium chloride solutions at various concentrations at 30°C. ●, 0.17 M; ●, 0.38 M; ◐, 0.60 M; o, 1.0 M; ◒, 1.5 M; ◔, 0.17 M in 3.0 M $NaNO_3$.

Table 1. Ultrasonic Parameters for Ionic Surfactant Solutions at 30°C.

| | T °C | $C_0$ M | $f_r$ MHz | $\mu_{max}$ $10^{-3}$ | CMC M |
|---|---|---|---|---|---|
| OAC | 30 | 0.26 | 4.0 | 4.6 | 0.22 |
| | | 0.38 | 6.5 | 11 | |
| | | 0.48 | 9.6 | 13 | |
| | | 0.60 | 12 | 13 | |
| | | 0.80 | 21 | 12 | |
| | | 1.0 | 26 | 12 | |
| | | 1.25 | 28 | 7.6 | |
| | | 1.5 | 21 | 7.0 | |
| OTAB | 30 | 0.33 | 10 | 3.5 | 0.23 |
| | | 0.42 | 13 | 5.6 | |
| | | 0.52 | 17 | 6.1 | |
| | | 0.65 | 25 | 7.1 | |
| | | 0.72 | 27 | 7.1 | |
| | | 0.80 | 33 | 7.4 | |
| $NaC_8$ | 30 | 0.49 | 6.0 | 10 | 0.31 |
| | | 0.56 | 7.0 | 13 | |
| | | 0.65 | 9.0 | 14 | |
| | | 0.69 | 13 | 14 | |
| | | 0.86 | 16 | 13 | |
| | | 1.1 | 19 | | |
| | | 1.2 | 19 | | |
| $KC_8$ | 30 | 0.42 | < 7.0 | | 0.38 |
| | | 0.65 | 8.0 | 9.7 | |
| | | 0.82 | 13 | 11 | |
| | | 1.0 | 18 | 9.7 | |
| $NaC_6$ | 30 | 0.84 | 25 | 1.1 | 0.74 |
| | | 0.94 | 25 | 1.8 | |
| | | 1.0 | 23 | 3.9 | |
| | | 1.15 | 27 | 6.7 | |
| | | 1.3 | 32 | 10 | |
| | | 1.6 | 44 | 15 | |
| | | 2.0 | 51 | 13 | |

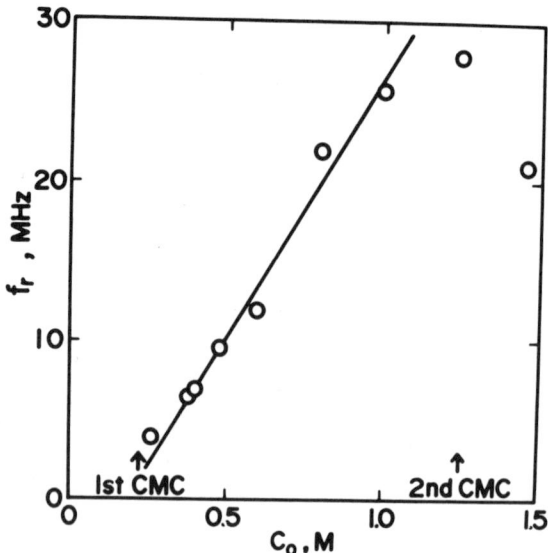

Figure 2. Plot of $f_r$ vs. concentrations of octylammonium chloride at 30°C.

This phenomenon also appears in other surfactant solutions as seen in Table I. This table also shows that the maximum excess absorption per wavelength, $\mu_{max}$, tends to indicate the maximum value at a certain concentration.

Slow Process. The relaxation effects were observed only above the CMC. The values of $\tau_2^{-1}$ obtained on SDS at various temperatures are plotted against the total surfactant concentration in Figure 3 and the results of $\tau_2^{-1}$ clearly differ from those by Hoffmann et al.[10-12]

The concentration dependences of $\tau_2^{-1}$ do hardly appear regardless of the charge of surfactant. Unfortunately, however, we had already reported that $\tau_2^{-1}$ increases linearly with the total surfactant concentration in SDS[7] and SDeS[9] systems using the pressure-jump and the temperature-jump techniques. A marked distinction between the results in this work and those in the previous ones is considered to be due to the difference of the purity of surfactants as indicated by Folger et al.[12] Then the following examinations were performed in SDS solution at 15°C where the experimental precision was greater above the Kraft point, 10 - 11°C[37,38] as seen in Figure 3. The results of microanalysis of the previous SDS: $C_{10}$, 0.9%; $C_{12}$, 96.8%; $C_{14}$, 2.0%; water, 0.56%; inorganic salt,

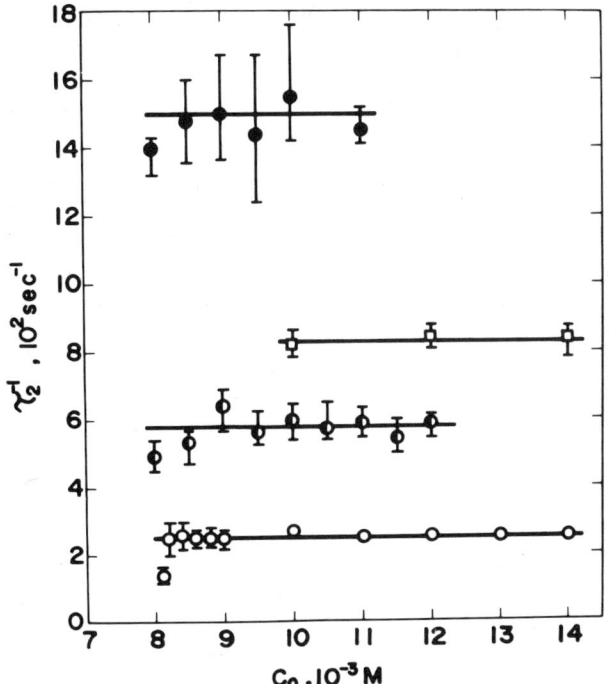

Figure 3. Plot of $\tau_2^{-1}$ vs. concentration of sodium and lithium dodecyl sulfates. Sodium dodecyl sulfate: o, at 15°C; ◐, at 20°C; ●, at 25°C. Lithium dodecyl sulfate: □, at 15°C. The vertical lines were drawn through the experimental points.

trace; lead to the anticipation that the sulfates of alcohols with 10 and 14 carbon numbers primarily act as impurities. Therefore, SDeS and STS were added to SDS used in this work to become the same composition as the previous SDS. The values of $\tau_2^{-1}$ in this mixture system are shown together with those in the mixture system of SDS and SDeS and in the previous SDS system in Figure 4. This figure leads to the conclusion that the concentration dependences of $\tau_2^{-1}$ in the vicinity of the CMC as seen in Figure 3 are due to the impurities below 0.5% or less. Then, the above facts probably lead to the experimental conclusion that $\tau_2^{-1}$ is completely independent of concentration, at least in the typical surfactant systems such as SDS.

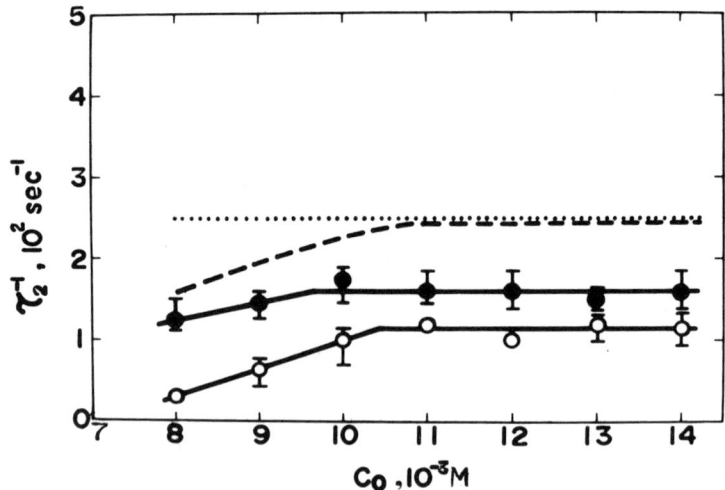

Figure 4. Plot of $\tau_2^{-1}$ vs. concentration of sodium dodecyl sulfate under the various conditions at 15°C. ●, sodium dodecyl sulfate – sodium decyl sulfate – sodium tetradecyl sulfate (96.8 : 0.9 : 2.0) mixture system; o, sodium dodecyl sulfate used in the previous work. A dotted line indicates the values of $\tau_2^{-1}$ in pure system of sodium dodecyl sulfate used in this work and a dashed line does those under the addition of only sodium decyl sulfate (0.4 Wt.%/ sodium dodecyl sulfate).

## DISCUSSION

Reaction Mechanism. Although several reaction models of micellization have been proposed in order to explain only one step[4,18,20,21,27], such models cannot be directly applied to the results in the present work. This is mainly due to the fact that these models have not been constructed on the basis of the new data that two relaxation processes do exist in a surfactant system and $\tau_2^{-1}$ is independent of the total surfactant concentration.

In order to interpret these facts and a series of results, the micellization mechanism is imagined as illustrated in Figure 5, in which only the stable micelles are in the spherical state and have the electric double layers, and the unstable ones are namely ionic pre-micelles with less aggregation numbers without counterions on their surfaces in the simply encountered lamellar state. In the micelle formation process (process A), the hydrophobic groups of surfactant can freely bind up one another and the intermediate aggregates can dissociate readily because the charged double layers do not exist on the surfaces of them and water molecules which break the hydrophobic interaction, penetrate more freely into the unstable intermediates than into the stable micelles with the charged double layers. On the other hand, it is assumed in connection with the process of the increase and decrease of aggregation number (process B) that surfactant monomers must pass through the charged double layers on the micellar surface in entering into a micelle and must cut the stronger hydrophobic interaction in leaving from a micelle.

Under these considerations, the following two-step mechanism is assumed for micellization process as follows,

$$nA \underset{\underset{\text{fast}}{k_{21}}}{\overset{k_{12}}{\rightleftarrows}} A_n \underset{\underset{\text{slow}}{k_{32}}}{\overset{mA \quad k_{23}}{\rightleftarrows}} A_{n+m} \qquad (1)$$

where A is a surfactant monomer, n and n+m are aggregation numbers of the micelle, $k_{ij}$ is the overall rate constant. The first step is the micelle formation and the second one is the micellar conformation change with the increase and decrease of aggregation number. The rate equations of these two processes and the condition of small perturbation give relations such as

$$\frac{d\delta[A_n]}{dt} = -a_{11}\delta[A_n] - a_{12}\delta[A_{n+m}] \qquad (2)$$

$$\frac{d\delta[A_{n+m}]}{dt} = -a_{21}\delta[A_n] - a_{22}\delta[A_{n+m}] \qquad (3)$$

where $a_{ij}$ is given by

$$a_{11} = k_{12}n^2[A]_e^{n-1} + k_{21} + k_{23}[A]_e^m - k_{23}nm[A]_e^{m-1}[A_n]_e$$

$$a_{12} = k_{12}n(n+m)[A]_e^{n-1} - k_{23}m(n+m)[A]_e^{m-1}[A_n]_e - k_{32} \qquad (4)$$

$$a_{21} = k_{23}nm[A]_e^{m-1}[A_n]_e - k_{23}[A]_e^m$$

$$a_{22} = k_{23}m(n+m)[A]_e^{m-1}[A_n]_e + k_{32}$$

On the other hand, an assumption that the distribution of aggregation number of micelle is normal and

$$m \ll n \qquad (5)$$

yields

$$[A_n] \simeq [A_{n+m}] \qquad (6)$$

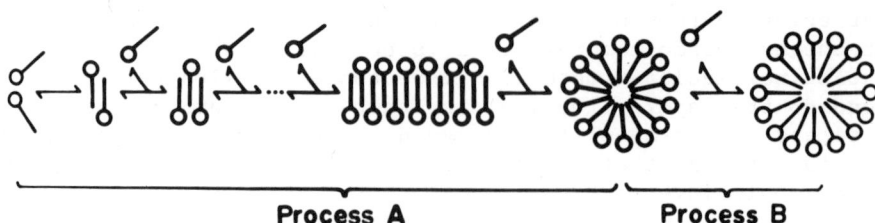

Figure 5. Schematic drawing of micelle formation (process A) and micellar conformation change with increase and decrease of aggregation number (process B).

where the bracket represents concentration. Then the following relations are obtained for $\tau_1^{-1}$ and $\tau_2^{-1}$ which differ by a factor of two or three.

$$\tau_1^{-1} = \frac{nk_{21}}{2[A]_e} C_0 - (\frac{n}{2} - 1)k_{21} \tag{7}$$

and

$$\tau_2^{-1} = k_{32} + k_{23}[A]_e^m \frac{1 + \frac{2m}{n} + \frac{m^2}{n^2} + \frac{m}{n}\frac{k_{32}}{k_{21}} - \frac{k_{32}}{k_{21}}\frac{[A]_e}{[A_n]_e}\frac{1}{n^2}}{1 - \frac{m}{n}\frac{k_{32}}{k_{21}} + \frac{k_{32}}{k_{21}}\frac{[A]_e}{[A_n]_e}\frac{1}{n^2} + \frac{[A]_e}{[A_n]_e}\frac{1}{n^2}} \tag{8}$$

where $C_0$ is the total surfactant concentration. Combining the assumption expressed by Equation (5) and the experimental conditions satisfying $[A_n]_e n^2 > [A]_e$ makes it possible to simplify Equation (8) to the relation

$$k_{32} = \frac{1}{2}\tau_2^{-1} \tag{9}$$

Hoffmann et al.[12,17] have proposed the model which has the opposite situation to the one described above. They have postulated both the micelle formation process and the exchange process of a monomer to and from a micelle, and have assumed that the latter is faster than the former. There are two main problems for their assignment of the two relaxation times. One is the assumption, first introduced by Muller,[26] that the forward and backward rate constants of the elementary steps in the micelle formation process are all the same except the dimerization process. Since the cleavage of any given monomer must take place n times faster than that of a specific monomer, this is related to the conclusion that the fast relaxation time can only be connected with the leaving of a monomer from a micelle, that is, the exchange process of a monomer to and from a micelle. According to Muller's assumption, one can obtain the next relation,

$$[A]_e K = \frac{[A_3]_e}{[A_2]_e} = \frac{[A_4]_e}{[A_3]_e} = \cdots = \frac{[A_n]_e}{[A_{n-1}]_e} = \frac{[A_{n+1}]_e}{[A_n]_e} = \cdots \quad (10)$$

where K is the equilibrium constant of each elementary step. Then, if $[A]_e K \leq 1$,

$$[A_2]_e \geq [A_3]_e \geq \cdots \geq [A_n]_e \geq [A_{n+1}]_e \geq \cdots \quad (11)$$

and on the contrary, if $[A]_e K > 1$,

$$[A_2]_e < [A_3]_e < \cdots [A_n]_e < [A_{n+1}]_e < \cdots \quad (12)$$

These relationships cannot be anticipated from the characteristics of the aqueous solution of ordinary surfactant. The value of $[A_i]_e$ may increase up to n = n̄ and decrease thereafter, as suggested by the reviewer. The above consideration also suggests the presence of the rate determining step rather than that of elementary steps having a comparable rate constant. The other is the results of the magnetic resonance measurements[39-50] that the lifetime of a monomer in the micelle is shorter than one msec. However, the exchange reaction of monomers between the micellar and the inter-micellar states is also considered to take place via the micelle formation and dissociation in the magnetic resonance studies.[39-44] The question as to whether the monomer state changes through the micelle formation reaction or the exchange reaction in the meaning described here, cannot be solved by the present magnetic resonance studies. Our model as described above is, of course, not supported by these studies.

The Rate Constant and the Activation Parameters. The rate constants, $k_{21}$ and $k_{32}$, obtained from Equations (7) and (9) are listed in Tables II and III and are plotted against the CMC in Figure 6. It is worth noting that the rate constant, $k_{21}$, is correlated with the CMC by a single linear relationship independently of the charge of the surfactant, while the rate constant, $k_{32}$, is not followed by a single straight line regardless of the charge. This fact affords a direct assistance in justification of the assumption that the electric double layer plays a considerable role only in the reaction of conformation change with the increase and decrease of aggregation number. Further, Figure 6 gives

$$k_{21} = (CMC)^{\alpha_1} \beta_1 \quad (13)$$

Table II. Rate Constant and Thermodynamic Parameters for Micelle Formation Reaction.

| | T °C | $n^a$ | $k_{21}$ $10^6 sec^{-1}$ | $\Delta V^0$ ml/mole | $\Delta H_{21}^{\ddagger}$ kcal/mole | $\Delta S_{21}^{\ddagger}$ e.u. at 25°C |
|---|---|---|---|---|---|---|
| $SDeS^b$ | 40 | 46 | 0.22 | 180 | | |
| | 17.5 | 31 | 1.9 | | | |
| | 20 | 31 | 2.1 | | 6.6 | -12 |
| | 25 | 31 | 2.6 | | (below 25°C) | |
| $SOS^c$ | 30 | 31 | 3.5 | 140 | | |
| | 35 | 31 | 4.2 | | 5.1 | -16 |
| | 40 | 31 | 4.5 | | (above 30°C) | |
| | 50 | 30 | 6.2 | | | |
| OAC | 30 | 28 | 3.4 | 170 | | |
| OTAB | 30 | 26 | 3.9 | 150 | | |
| $NaC_8$ | 30 | 24 | 4.7 | 130 | | |
| $KC_8$ | 30 | 22 | 6.2 | 93 | | |
| $NaC_6$ | 30 | 18 | 10 | 63 | | |

a) calculated through Equation (15),
b) calculated with the results by Rassing et al.[24],
c) calculated with the results by Rassing et al.[23].

where $\alpha_1 = 1$ and $\beta_i$ is constant. On the other hand, the value of $k_{32}$ clearly increases with increase of the CMC in a series of the same charged surfactants as shown in Figure 6, which gives

$$k_{32} = \alpha_2 \log(CMC) + \beta_2 \qquad (14)$$

where $\alpha_2 \simeq 5 \times 10^3$. This tendency shows that monomers are apt to leave from a micelle in the order of more value of the CMC and in the order of less carbon numbers in the hydrophobic groups. While a mean aggregation number of micelle, $\bar{n}$, was found to correlate with the CMC as follows;[51]

$$\bar{n} = (\frac{1}{CMC})^{\alpha_3} \beta_3 \qquad (15)$$

where $\alpha_3 \simeq 0.29$ and $\beta_3 \simeq 17$. Then, one obtains

$$\log k_{21} = \alpha_4 \log \bar{n} + \log \beta_1 \beta_3^{-\alpha_4} \qquad (16)$$

# KINETIC STUDY ON MICELLIZATION OF IONIC SURFACTANTS

Table III. Activation Parameters for Conformation Change of Micelle with Exchange of Monomers to and from a Micelle.

|      | T °C | $k_{32}$ sec$^{-1}$ | $\Delta H_{32}^{\ddagger}$ kcal/mole | $\Delta S_{32}^{\ddagger}$ e.u. at 25°C |
|------|------|---------------------|--------------------------------------|------------------------------------------|
| SDeS | 5    | $0.90 \times 10^3$  | 23                                   | 39                                       |
|      | 7    | 1.4                 |                                      |                                          |
|      | 10   | 2.1                 |                                      |                                          |
|      | 11   | 2.3                 |                                      |                                          |
|      | 12   | 2.7                 |                                      |                                          |
|      | 15   |                     |                                      |                                          |
| SDS  | 15   | $1.3 \times 10^2$   | 29                                   | 53                                       |
|      | 20   | 2.9                 |                                      |                                          |
|      | 25   | 7.5                 |                                      |                                          |
| LDS  | 15   | $4.2 \times 10^2$   |                                      |                                          |
| DTAB | 15   | $1.1 \times 10^2$   |                                      |                                          |
|      | 30   | 8.0                 |                                      |                                          |

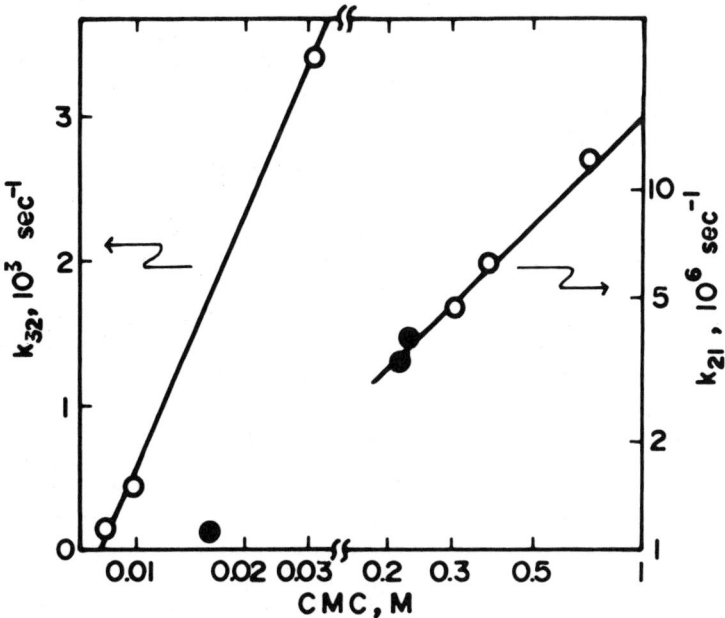

Figure 6. Plot of $k_{21}$ (30°C) and $k_{32}$ (15°C) vs. CMC. o, anionic surfactants; •, cationic surfactants.

and

$$k_{32} = \alpha_5 \log \bar{n} + \alpha_2 \log \beta_3^{-\alpha_5} + \beta_2 \tag{17}$$

where $\alpha_4 \simeq -3$ and $\alpha_5 \simeq -2 \times 10^4$.

The temperature dependence of $k_{21}$ was examined using the results of sodium octyl sulfate (SOS) by Rassing et al.[23] The temperature dependence of $k_{21}$ does not display the normal behavior and is divided into two regions. This result is coupled with the fact that the CMC of this surfactant[52] also reaches a minimum in the temperature region where $k_{21}$ does not show the normal behavior. In addition, taking into account that the concentration of the free surfactant monomer is the important factor in the reaction rate of the fast process as clearly indicated in Equation (7), the above facts lead to the anticipation that the fast relaxation process observed by the ultrasonic absorption technique is attributed to the micelle formation reaction.

The thermodynamic parameters listed in Tables II and III were calculated according to the equation of the absolute reaction rate theory:[53]

$$k = \frac{k_B T}{h} e^{-\Delta H^{\ddagger}/RT} e^{\Delta S^{\ddagger}/R} \tag{18}$$

It is difficult to assess these values definitely because the aggregation number of the micelle itself changes with temperature. However, it is noteworthy that $\Delta H_{21}^{\ddagger}$ is a small value compared with $\Delta H_{32}^{\ddagger}$ and $\Delta S_{21}^{\ddagger}$ is negative sign against the positive and large $\Delta S_{32}^{\ddagger}$. The small values of $\Delta H_{21}^{\ddagger}$ could be originated in the absence of the electric double layers which participate only in the slow process as discussed above.

The negative values of $\Delta S_{21}^{\ddagger}$ indicate that the transition state involves the exposure of the hydrophobic groups to the solvent. This process is necessarily negative due to the solvent ordering process around the hydrophobic groups. This contribution to $\Delta S_{21}^{\ddagger}$ must be superior to that from the disordered monomers produced by the micelle dissociation. The positive and large values of $\Delta S_{32}^{\ddagger}$ indicate the characteristics of the activation state in the slow process. The release of m monomers seems to cause both the increase of entropy in a micelle and the decrease of entropy in bulk by accelerating to make the water structure around a monomer. However, the condition (5) indicates that the enthalpy and the entropy for species $A_{n+m}$ are almost similar to those for $A_n$. Accordingly, only the activation state of the release process of monomers would be less stable than two states of $A_n$ and $A_{n+m}$. The $\Delta S_{32}^{\ddagger}$ decreases in magnitude as the carbon numbers in the hydrophobic groups decrease, suggesting the transition state becomes more ordered. Of course, $\Delta S_{32}^{\ddagger}$ may include a large conformational term corresponding to micellar conformation change containing the rearrangement of the electric double layers.

*The Relaxation Strength for the Fast Process.* The condition that very small periodic perturbations are applied to the system gives the next relations about the concentrations and the rate constants:

$$[A] = [A]_e + \delta[A]\exp(i\omega t)$$
$$[A_n] = [A_n]_e + \delta[A_n]\exp(i\omega t) \qquad (19)$$
$$k'_{12} = k_{12} + \delta k_{12}\exp(i\omega t)$$
$$k'_{21} = k_{21} + \delta k_{21}\exp(i\omega t)$$

where [ ] and [ ]$_e$ are the actual and the equilibrium concentration, respectively, and $k_{ij}$ is the actual rate constant and $\delta$ means the amplitude of the small variations associated with the sound wave. Then, the following relation for $A_n$ is obtained from the rate equation with the condition of the small perturbation:

$$\delta A_n = \frac{[A]_e^n \delta k_{12} - [A_n]_e \delta k_{21}}{i\omega + n^2 k_{12}[A]_e^{n-1} + k_{21}} \qquad (20)$$

With the small change and the pressure dependence of the equilibrium constant, this is further converted into

$$\frac{\delta A_n}{\delta P} = -\frac{(\Delta V^0) k_{21}[A_n]_e}{RT(i\omega + n^2 k_{12}[A]_e^{n-1} + k_{21})} \qquad (21)$$

where $\Delta V^0$ is the standard volume change in the micelle formation reaction. The adiabatic compressibility, $\beta$, is related to the relaxational compressibility, $\beta_r$, namely

$$\beta = \frac{\beta_r}{(i\omega\tau + 1)} \qquad (22)$$

with

$$\beta_r = \frac{2\mu_{max}}{\pi\rho v^2} \qquad (23)$$

where $\rho$ is the density. Finally, one obtains

$$\mu_{max} = \frac{\pi\rho v^2}{2} \cdot \frac{(\Delta V^0)^2}{RT} k_{21}\tau_1[A_n]_e \qquad (24)$$

This equation can successfully explain the concentration dependence of $\mu_{max}$ except two points in the high concentration range as shown in Figure 7. Here the aggregation number, n, was determined by the

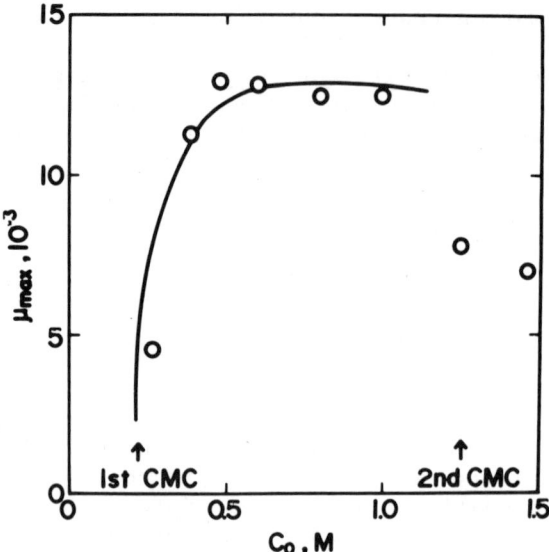

Figure 7. Maximum excess absorption per wavelength, $\mu_{max}$, of octylammonium chloride solutions at various concentrations at 30°C. A solid line indicates the theoretical values of $\mu_{max}$.

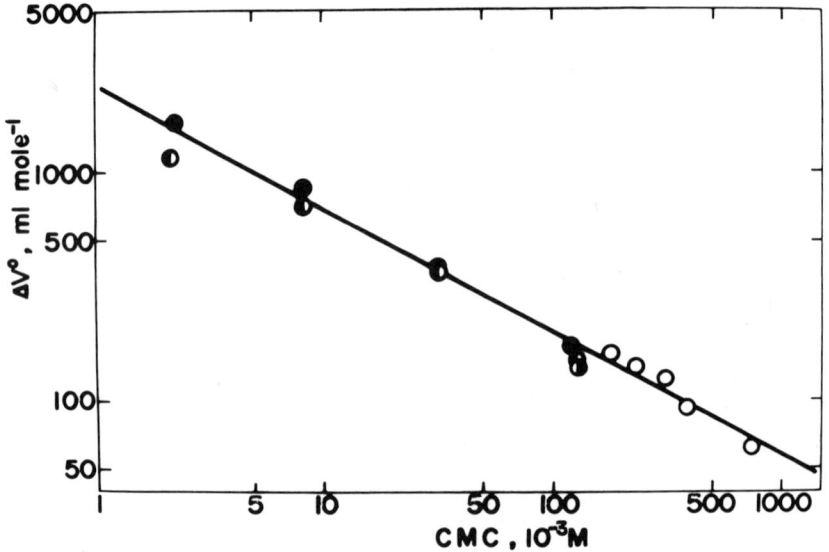

Figure 8. Plot of $\Delta V^0$ vs. CMC. o, calculated from the kinetic results at 30°C in this work; ◐, calculated from the kinetic results of ultrasonic absorption measurements at 30°C.[23] ◉, obtained from the pressure dependencies of the CMC at 30°C;[30,38,54,55] ●, obtained from densities at 25°C.[54,56,57]

relation (15). The standard volume change, $\Delta V^0$, was evaluated by the consideration of $\mu_{max}$ using Equation (24) derived for the fast process of the proposed two-step reaction mechanism. As seen in Table II, the values of $\Delta V^0$ themselves are larger than those for the ordinary reactions but well agree with those calculated through the following equation using the results of the static measurement by Tanaka et al.[30,38,54,57]

$$\Delta V^0 = \bar{n} \Delta V^0_{mo} \tag{25}$$

where $\Delta V^0_{mo}$ is the partial molar volume change per monomer. The rationality of the mechanism proposed in this paper is also demonstrated by the agreements between the experimental and calculated values of $\mu_{max}$, and between the values of $\Delta V^0$ which were separately determined on the basis of the kinetic and static measurements.

The relation between the CMC and these $\Delta V^0$ is shown in Figure 8. These two kinds of $\Delta V^0$ remarkably follow a straight line which is substantially expressed by

$$\Delta V^0 = \left(\frac{1}{CMC}\right)^{\alpha_6} \beta_6 \tag{26}$$

where $\alpha_6 \simeq 0.5$. The $\Delta V^0$ increases with a decrease in the CMC and approximately with an increase of the carbon numbers of the hydrophobic group. On the other hand, the relation (15) has been found to be especially satisfactory in the case where the counterion is univalent independently of the sign of charge of surfactant.[51] In the case of $\Delta V^0$, Equation (26) is also valid regardless of the sign of charge. Then, the next relation is obtained between $\Delta V^0$ and $\bar{n}$ by the combination of Equations (15) and (26):

$$\Delta V^0 = \bar{n}^{\gamma_1} \frac{\beta_6}{\beta_3^{\gamma_1}} \tag{27}$$

where
$$\gamma_1 = \alpha_6/\alpha_3 \simeq \frac{3}{2}$$

As the aggregation number, $\bar{n}$, is considered to be one of the parameters to indicate the surface area of the micelle, Equation (27) means that the square of volume is proportional to the cube of the area. Furthermore, the propriety of this simple relation is demonstrated by the consideration stated below. The relation of $\bar{n}$ and the carbon numbers in the hydrophobic group, N, in the case of sodium alkyl sulfates, gives

$$\bar{n} = N^{\alpha_7} \beta_7 \tag{28}$$

where $\alpha_7 \simeq 2$. In addition, if one can ignore the contribution of ionic head group to the volume change, the following relation can be obtained

$$\Delta V^0_{mo} = N\Delta V^0_{me} \qquad (29)$$

where $\Delta V^0_{me}$ is the partial molar volume change per methylene group. Then, one obtains the same conclusion as expressed by Equation (27):

$$\Delta V^0 = \bar{n}^{\gamma_2} \frac{\Delta V^0_{me}}{\beta_7^{1/\alpha_7}} \qquad (30)$$

where $\gamma_2 = 3/2$.

Furthermore, the combination of Equation (13) with Equation (26) gives

$$\log k_{21} = \alpha_8 \log \Delta V^0 + \log \beta_1 \beta_6^{-\alpha_8} \qquad (31)$$

where $\alpha_8 \simeq -2$. The relation indicated by Equation (31) suggests that water molecules play an important role in this fast process, because these changes of the reaction volume are considered to be caused primarily by the behavior of water molecules.

The Elementary Reaction. Although the elementary steps in the fast process are anticipated to be complicated, the micelle formation consists of the following successive bimolecular reactions as proposed by Alexander et al.:[58]

$$A + A \underset{k_{-1}}{\overset{k_1}{\rightleftharpoons}} A_2 \qquad \tau_1^*$$

$$A_2 + A \underset{k_{-2}}{\overset{k_2}{\rightleftharpoons}} A_3 \qquad \tau_2^*$$

$$\vdots$$

$$A_{n-1} + A \underset{k_{-(n-1)}}{\overset{k_{n-1}}{\rightleftharpoons}} A_n \qquad \tau_{n-1}^*$$

where $k_i$ and $\tau_i^*$ are the rate constant and the relaxation time in each step, respectively.

If the relation

$$\tau_1^* \ll \tau_2^* \ll \cdots \ll \tau_{n-1}^* \qquad (33)$$

is satisfied, one obtains

$$\tau_1^{*-1} = k_1([A]_e + [A]_e) + k_{-1} \tag{34}$$

$$\tau_2^{*-1} = \frac{K_1 k_2 ([A]_e + [A]_e)([A_2]_e + [A]_e)}{K_1([A]_e + [A]_e) + 1} + k_{-2} \tag{35}$$

$$\vdots$$

$$\tau_{n-1}^{*-1} = \frac{2 \prod_{i=1}^{n-2} K_i \prod_{i=1}^{n-2}(K_i[A_i]_e + 1)[A]_e^{n-1} k_{n-1}}{\sum_{i=0}^{n-2} \prod_{j=0}^{i} K_j[A_j]_e} + k_{-(n-1)} \tag{36}$$

where $K_i$ is the equilibrium constant of the i-th elementary step and

$$K_0 A_0 = 1 \tag{37}$$

Then, the last step in the reactions (32) turns out to be the rate determining one of the fast processes and so the overall rate constants, $k_{12}$ and $k_{21}$, can be given by

$$k_{12} = \frac{2 \prod_{i=1}^{n-2} K_i \prod_{i=1}^{n-2}(K_i[A_i]_e + 1)}{n^2 \sum_{i=0}^{n-2} \prod_{j=0}^{i} K_j[A_j]_e} k_{n-1} \tag{38}$$

and

$$k_{21} = k_{-(n-1)} . \tag{39}$$

The assumption (33) is more appropriate for the micelle with very small n, which is not the case here. The relations such as Equations (34) - (36) also become valid in the case of small n, because all aggregates with the individual aggregation number, 2-n, are usually anticipated to exist at the equilibrium state.

On the other hand, since the aggregates with only a certain aggregation number, n, are formed above the certain concentration, namely, CMC in the ordinary surfactant system, it is naturally expected that the aggregates with aggregation number above n can hardly exist and those with aggregation number below n are in very low concentrations and in stationary state. Under these conditions, the next rate equation is obtained for the aggregate with aggregation number, n, in the elementary steps (32);

$$\frac{d[A_n]}{dt} = \frac{k_1 k_2 \cdots k_{n-1}[A]^n - k_{-1}k_{-2}\cdots k_{-(n-1)}[A_n]}{(k_{-1}+k_2[A])(k_{-2}+k_3[A])\cdots(k_{-(n-2)}+k_{n-1}[A])} \quad (40)$$

The above condition can also permit the dissociation rate of $A_i$ to $A_{i-1}$ to be compared with the association of $A_i$ to $A_{i+1}$, that is,

$$k_i \simeq k_{i+1}[A]_e \quad (41)$$

Therefore, one obtains a relation such as

$$\tau_{n-1}^{-1} \simeq \frac{1}{2} n^2 \prod_{i=1}^{n-1} K_i k_{n-1}[A]_e^{n-1} + \frac{1}{2} k_{-(n-1)} \quad (42)$$

which gives

$$k_{12} \simeq \frac{1}{2} \prod_{i=1}^{n-1} K_i k_{n-1} \quad (43)$$

and

$$k_{21} \simeq \frac{1}{2} k_{-(n-1)} \quad (44)$$

The incorporation of monomers leading to the micellar conformation change consists of the elementary steps such as,

$$\begin{aligned} A_n + A &\underset{k_{-n}}{\overset{k_n}{\rightleftarrows}} A_{n+1} \\ A_{n+1} + A &\underset{k_{-(n+1)}}{\overset{k_{n+1}}{\rightleftarrows}} A_{n+2} \\ &\vdots \\ A_{n+m-1} + A &\underset{k_{-(n+m-1)}}{\overset{k_{n+m-1}}{\rightleftarrows}} A_{n+m} \end{aligned} \quad (45)$$

On the other hand, Mukerjee has reported that the micelle size-distributions are indeed narrow and the size-distribution index, $N_w/N_n$, is close to unity in many ordinary surfactant systems.[59] Here, $N_n$ and $N_w$ are the number and weight average degree of association, respectively. The value of m would be expected to be very small from these results.

If m = 1

$$k_{23} = k_n \tag{46}$$

and

$$k_{32} = k_{-n} \tag{47}$$

and if m > 2, the assumption that the activation energy is the same in each process of a monomer release and incorporation, enables to give

$$k_n \simeq k_{n+1} \simeq \ldots \simeq k_{n+m-1} \tag{48}$$

and

$$k_{-n} \simeq k_{-n(n+1)} \simeq \ldots \simeq k_{-(n+m-1)} \tag{49}$$

Then, one obtains finally

$$k_n \simeq \ldots \simeq k_{n+m-1} \simeq \sqrt[m]{k_{23}} \tag{50}$$

and

$$k_{-n} \simeq \ldots \simeq k_{-(n+m-1)} \simeq \sqrt[m]{k_{32}} \tag{51}$$

In conclusion of this part, the relations between the overall rate constants and the rate constants in the rate determining step were obtained for two processes. In the micelle formation process, the rate determining step is considered to exist and involve not only the capture and release of a monomer but also the conversion from the lamellar structure to the spherical one as described in Figure 5. However, the rate determining step would not exist in the process of the conformation change with the increase and decrease of aggregation number. In addition, the above consideration also means that the micelle systems are approximately divided into two groups according to the presence and the absence of the intermediate aggregates at the equilibrium state.

The concentration independence of $\tau_2$ is especially demonstrated on SDS, DTAB and so on and the micellization mechanism is discussed in this paper. As it is, there is no experimental confidence that micelle formation is the fast process or the fast process is exchange process of a monomer to and from a micelle. What is more important at present is to verify experimentally that two relaxation processes observed till now, are undoubtedly attributed to these two reactions. Therefore, the consideration on these problems is no better than the guesswork, but the various experimental results could be interpreted successfully to some extent by the model described above. In addition, it gave the interesting relations among some parameters.

## REFERENCES

1.  I. Amdur and G. G. Hammes, "Chemical Kinetics: Principles and Selected Topics," pp. 131-162, McGraw-Hill, Inc., 1966.
2.  G. H. Czerlinski, "Chemical Relaxation", Marcel Dekker Inc., New York, 1966.
3.  M. Eigen and L. DeMayer, "Technique of Organic Chemistry," Vol. VIII, 2nd ed., pp. 895-1054, S. L. Friess, E. S. Lewis and A. Weissberger, Editors, Interscience Publishers, Inc., New York, 1963.
4.  G. C. Kresheck, E. Hamori, G. Davenport and E. M. Scheraga, J. Amer. Chem. Soc., $\underline{88}$, 246 (1966).
5.  B. C. Bennion, L. K. J. Tong, L. P. Holmes and E. M. Eyring, J. Phys. Chem., $\underline{73}$, 3288 (1969).
6.  B. C. Bennion and E. M. Eyring, J. Colloid Interface Sci., $\underline{32}$, 286 (1970).
7.  K. Takeda and T. Yasunaga, ibid., $\underline{40}$, 127 (1972).
8.  J. Lang and E. M Eyring, J. Polym. Sci., Part A-2, $\underline{10}$, 89 (1972).
9.  K. Takeda and T. Yasunaga, J. Colloid Interface Sci., $\underline{45}$, 406 (1973).
10. U. Herrmann and M. Kahlweit, Ber. Bunsenges. Phys. Chem., $\underline{77}$, 1119 (1973).
11. T. Janjic and H. Hoffmann, Z. Phys. Chem., (Neue Folge) $\underline{86}$, 322 (1973).
12. R. Folger, H. Hoffman and W. Ulbricht, Ber. Bunsenges. Phys. Chem., $\underline{78}$, 986 (1974).
13. T. Yasunaga, K. Takeda and S. Harada, J. Colloid Interface Sci., $\underline{42}$, 457 (1973).
14. K. Takeda, N. Tatsumoto and T. Yasunaga, ibid., $\underline{47}$, 128 (1974).
15. N. Tatsumoto, K. Takeda, S. Isshiki and T. Yasunaga, Bull. Chem. Soc. Jap., $\underline{47}$, 289 (1974).
16. T. Yasunaga, K. Takeda, N. Tatsumoto and H. Uehara, "Chemical and Biological Applications of Relaxation Spectroscopy," (Proceeding of the NATO Advanced Study Institute) E. Wyn-Jones, Editor, pp. 143-157, Reidel Pub., Holland, 1975.
17. J. Lang, C. Tondre, R. Zana, R. Bauer, H. Hoffmann and W. Ulbricht, J. Phys. Chem., $\underline{79}$, 276 (1975).
18. T. Yasunaga, H. Oguri and M. Miura, J. Colloid Interface Sci., $\underline{23}$, 352 (1967).
19. T. Yasunaga, S. Fujii and M. Miura, ibid., $\underline{30}$, 399 (1969).
20. E. Graber and R. Zana, Kolloid-Z. Z. Polym., $\underline{238}$, 470, 479 (1970).
21. P. J. Sams, E. Wyn-Jones and J. Rassing, Chem. Phys. Letter, $\underline{13}$, 233 (1972).
22. J. Rassing and E. Wyn-Jones, ibid., $\underline{21}$, 93 (1973).
23. J. Rassing. P. J. Sams and E. Wyn-Jones, J. Chem. Soc. Faraday Trans., II $\underline{69}$, 180 (1973).
24. J. Rassing, P. J. Sams and E. Wyn-Jones, ibid., $\underline{69}$, 180 (1973).

25. D. A. W. Adair, V. C. Reinsborough, N. Plavac and J. P. Valleau, Can J. Chem., 52, 429 (1974).
26. N. Muller, J. Phys. Chem., 76, 3017 (1972).
27. T. Nakagawa, Colloid Polym. Sci., 252, 56 (1974).
28. M. Tanaka, S. Kaneshina, W. Nishimoto and H. Takabatake, Bull. Chem. Soc. Jap., 46, 364 (1973).
29. E. E. Dreger, G. I. Keim and G. D. Miles, Ind. Eng. Chem., 36, 610 (1944).
30. M. Tanaka, S. Kaneshina, S. Kuramoto and R. Matsuura, Bull. Chem. Soc. Jap., 48, 432 (1975).
31. N. Tatsumoto, J. Chem. Phys., 47, 4561 (1967).
32. M. Miura and M. Kodama, Bull. Chem. Soc., Jap., 45, 428 (1972).
33. M. Kodama and M. Miura, ibid., 45, 2265 (1972).
34. M. Kodama, Y. Kubota and M. Miura, ibid., 45, 2953 (1972).
35. Y. Kubota, M. Kodama and M. Miura, ibid., 46, 100 (1973).
36. M. Kodama, J. Sci. Hiroshima Univ. Ser A, 37, 53 (1973).
37. E. Gotte, Fette U. Seifen, 56, 583 (1954).
38. M. Tanaka, S. Kaneshina, T. Tomida, K. Noda and D. Aoki, J. Colloid Interface Sci., 44, 525 (1973).
39. N. Muller and R. H. Birkhaln, J. Phys. Chem., 71, 957 (1967).
40. N. Muller and R. H. Birkhaln, ibid., 72, 583 (1968).
41. N. Muller and T. W. Johnson, ibid., 73, 2042 (1969).
42. N. Muller and F. Platko, ibid., 75, 547 (1971).
43. T. Florence and R. T. Parfitt, ibid., 75, 3554 (1971).
44. E. J. Fendler, J. H. Fendler, R. T. Medary and O. A. El Seoud, ibid., 77, 1432 (1973).
45. T. Nakagawa and K. Tori, Kolloid-Z. A. Polym., 194, 143 (1964).
46. H. Inoue and T. Nakagawa, J. Phys. Chem., 70, 1108 (1966).
47. R. Haque, ibid., 72, 3056 (1968).
48. T. Nakagawa, H. Inoue, H. Jizomoto and K. Horiuchi, Kolloid-Z. Z. Polym., 229, 159 (1969).
49. F. Tokiwa and K. Tsujii, J. Phys. Chem., 75, 3560 (1971).
50. K. Fox, Trans. Faraday Soc., 67, 2802 (1971).
51. K. Tyuzyo, Bull. Chem. Soc. Jap., 31, 117 (1958).
52. E. D. Goddard and G. C. Benson, Can. J. Chem., 35, 986 (1957).
53. S. Glansstone, K. J. Laidler and H. Eyring, "The Theory of Rate Processes," McGraw-Hill, New York 1941.
54. M. Tanaka, S. Kaneshina, K. Shinno, T. Okajima and T. Tomida, J. Colloid Interface Sc., 46, 132 (1974).
55. S. Kaneshina, M. Tanaka, T. Tomida and R. Matsuura, ibid., 48, 450 (1974).
56. K. Shinoda and T. Soda, J. Phys. Chem., 67, 2072 (1963).
57. J. M. Corkill, J. F. Goodmann and T. Walker, Trans. Faraday Soc., 63, 768 (1967).
58. A. E. Alexander and P. Johnson, Colloid Sci., Vol. II, Clarendon Press, Oxford, (1949).
59. P. Mukerjee, J. Phys. Chem., 76, 565 (1972).

ON THE KINETICS OF REDISTRIBUTION OF MICELLAR SIZES

M. Almgren, E. A. G. Aniansson,
S. N. Wall and K. Holmåker
Chalmers University of Technology
and University of Gothenburg
Gothenberg, Sweden

The kinetic equations for step-wise micelle association and dissociation can be cast into a form of differential-difference equations for relative excess variables. For the process that involves solely a rearrangement of the micellar size distribution solutions have previously been obtained for Gaussian distributions with size-independent dissociation rate constants, by approximating differences by differentials.

This treatment has now been extended in the following ways for small deviations from equilibrium:
1. A general formal solution of the system of difference-differential equations for the fast process has been obtained.
2. Complete solutions for those special cases in which the relaxation modes are given by orthogonal polynomials have been obtained.
3. A necessary and sufficient condition for the relaxation of the monomer concentration to be strictly exponential has been found.
4. A discussion is given of the concentration dependence of the monomer relaxation time constant as obtained above.
5. Numerical and theoretical results are presented for systems that do not fulfill the condition mentioned in 3.

## INTRODUCTION

Relaxation kinetic studies of micellar surfactant solutions seem to indicate that the approach to equilibrium proceeds in two distinct stages:[1-4] First there occurs a redistribution of micellar sizes to a temporary equilibrium distribution. Thereafter the number of micelles adjusts to final equilibrium through net dissolution or formation of micelles. In the case of ionic surfactants these two processes are most probably preceded by a relaxation of the counterion distribution that is expected to be much faster than the first mentioned one.[5] In the following it is assumed that this process is so fast that the counterions can be considered to be in instantaneous equilibrium with the micelles in the rearrangement process.

On the basis that the processes are step-wise ones:

$$A_1 + A_1 \underset{k_2^-}{\overset{k_2^+}{\rightleftarrows}} A_2, \quad A_1 + A_2 \underset{k_3^-}{\overset{k_3^+}{\rightleftarrows}} A_3, \quad \ldots\ldots, \quad A_1 + A_{s-1} \underset{k_s^-}{\overset{k_s^+}{\rightleftarrows}} A_s, \quad (1)$$

where $A_s$ denotes an aggregate with aggregation number s, the kinetic equations were given a form closely resembling equations for flow in "aggregation space".[2,6,7] Letting $A_s$ also denote the concentration of the species, $\bar{A}_s$ its equilibrium value, and introducing the relative deviation from equilibrium, $\xi_s = (A_s - \bar{A}_s)/\bar{A}_s$, the net flow of particles from s-1 to s will be

$$J_s = -k_s^- \bar{A}_s (\xi_s - \xi_{s-1} - \xi_1 - \xi_1 \xi_{s-1}) \qquad (2)$$

At small deviations from equilibrium the last term within the brackets is negligible. The expression is now reminescent of an equation for diffusion in a tube of varying cross section if $k_s^-$ is taken as the analogue of the diffusion constant, $\bar{A}_s$ of the tube cross sectional area, $\xi_s - \xi_{s-1}$ of the concentration gradient. $\xi_1$ plays the less familiar role of a space-independent driving force.

The ordinary kinetic equations take the form

$$\bar{A}_s \frac{d\xi_s}{dt} = -(J_{s+1} - J_s) \qquad (3)$$

which clearly exhibits the analogy with the continuity equation.

Since $\bar{A}_s$ is large at the monomer end and in the micellar region the tube analogue is one with two thick ends connected by a very thin section. The flow through this section which corresponds to a net dissolution or formation of micelles will therefore ordinarily be a much slower process than the attainment of a temporary equilibrium within the "thick ends". The latter process can therefore be treated as if the former one did not exist.

Within this framework "the fast process", i.e. the redistribution over micellar sizes at a constant number of micelles, the Equations (2) and (3) were solved in a particular case by actually replacing differences by derivatives.[6,7,8] $A_s$ was taken to be Gaussian, which corresponds to the assumption that the minimum in the free energy of the micellar species as a function of s be nearly of second degree in s. $k_s^-$ was supposed to be sufficiently slowly varying with s in the micellar region so that it could be taken to be constant.

The method of solution consisted of expanding $\xi_s$ in a suitable set of orthogonal functions of a reduced variable of s taking the coefficients in the expansion to be time dependent. The first order ordinary differential equations of these coefficients could be solved very simply. It was found that the monomer concentration was governed by a single relaxation time which depended on the constant $k_s^- = k$, the average aggregation number n, its variance $\sigma^2$, and the ratio a of surfactant material in micellar and monomer form through the relation

$$1/\tau = k/\sigma^2 + ka/n \tag{4}$$

Assuming $\sigma^2$ and n to be essentially invariant with total surfactant concentration, the slope of $1/\tau$ versus a would give k if n is known, and the intercept would give $\sigma^2$, a quantity previously unknown for low-disperse micelles.

Ultrasonic experiments[2] conformed to (4) if $\sigma^2$, n and $\bar{A}_1$ were assumed to be constant above c.m.c. whereas pressure-jump experiments[2,7] displayed somewhat curved plots of $1/\tau$ versus concentration.

The further work has been done along several lines in order to get a more comprehensive view of the dependence of the relaxation time on the form of the micellar distribution and on the s-dependence of $k_s^-$.

The linear difference-differential equation resulting from (2) and (3) by dropping the last term in (2) has been solved along lines closely analogous to those followed in the above-mentioned solution of the approximate differential equation.[9]

A simple necessary and sufficient condition has been found for the monomer relaxation to be governed by a single relaxation time. When the condition is fulfilled the relaxation time is again governed by Equation (4).[9]

Numerical solutions of the equations have been obtained for a variety of cases, some close to observing the above condition and some more strongly deviating from it.[10,11]

Finally a general expression for a so-called initial relaxation time has been derived and discussed in relation to probable "experimental" relaxation times.[10]

## EXACT SOLUTIONS OF THE DIFFERENCE-DIFFERENTIAL EQUATION

When the nonlinear term of Equation (2) is dropped and $J_S$ then inserted into Equation (3) the resulting equation can be written

$$\bar{A}(x) \frac{\delta \psi(x;t)}{\delta t} = \Delta \{k(x) \bar{A}(x) [\bar{\Delta} \psi(x;t) - \xi_1(t)]\} \quad (5)$$

The variable x has been introduced to emphasize that this variable now pertains to the micellar region only. For the same reason $\bar{A}_s$ has been replaced by $\bar{A}(x)$, $k_s^-$ by $k(x)$ and $\xi_s$ by $\psi(x;t)$. $\Delta$ denotes the forward difference $\Delta f(x) = f(x+1) - f(x)$ and $\bar{\Delta}$ the backward difference $\bar{\Delta} f(x) = f(x) - f(x-1)$. The monomer excess $\xi_1(t)$ is determined by the $\psi$:s through the conservation of surfactant material

$$\bar{A}_1 \xi_1(t) = - \Sigma \, x \, \bar{A}(x) \, \psi(x;t) \quad (6)$$

The sum extends over the micellar region and it is assumed that dimers, trimers, etc. occur in so small quantities that their contribution can be neglected in this connection.

If solutions can be found to the eigenvalue problem

$$\Delta [k(x) \bar{A}(x) \bar{\Delta} u(x)] = -\lambda \, \bar{A}(x) \, u(x) \quad (7)$$

Equations (5) with (6) can be solved in the same way as used for the differential approximation mentioned above. $\psi(x;t)$ is expanded in the orthonormal eigenfunctions $u_m(x)$, $m=0,1,2,\ldots$ with time dependent coefficients $T_m(t)$:

$$\psi(x;t) = \sum_m T_m(t) \, u_m(x) \quad (8)$$

When the $u_m(x)$ are polynomials in x insertion of (8) into Equation (6) will yield $\xi_1(t)$ as a function only of $T_0(t)$ and $T_1(t)$. When further this expression for $\xi_1(t)$ and the expansion (8) are introduced into (5) the use of (7) and the orthogonality of the $u_m(x)$ will result in easily solved differential equations for the $T_m(t)$. $T_0(t)$ will be a constant and $T_1(t)$ will be determined by a single relaxation time, that of Equation (4) with k defined as the number average of $k(x)$

$$k = \sum_x k(x) \bar{A}(x) / \sum_x \bar{A}(x) \quad (9)$$

# KINETICS OF REDISTRIBUTION OF MICELLAR SIZES

The same, single relaxation time will also govern the monomer relaxation since $\xi_1$ only depends on $T_0$ and $T_1$.

The four polynomials fulfilling the requirements above are given in Table I together with the respective micelle distributions and functions $X(x)$, related to the rate constants (see Table II).

## A NECESSARY AND SUFFICIENT CONDITION FOR THE MONOMER RELAXATION TO BE GOVERNED BY EXACTLY ONE RELAXATION TIME

The condition can be formulated in the following way

$$\Delta [k(x) \bar{A}(x)] = -k \bar{A}(x) (x-n)/\sigma^2 \tag{10}$$

To prove its sufficiency we differentiate Equation (6) with respect to the time and insert Equation (5). It will yield

$$\bar{A}_1 \frac{d\xi_1}{dt} = -\xi_1 \sum_x k(x) \bar{A}(x) + \sum_x k(x) \bar{A}(x) \bar{\Delta} \psi(x;t) \tag{11}$$

if the fact that

$$\sum_x x \, \Delta f(x) = -\sum_x f(x) \bar{\Delta}(x) = -\sum_x f(x) \tag{12}$$

is used.

The last sum in (11) can be turned around in a way similar to (12) and (10) thereafter inserted so that there results

$$\bar{A}_1 \frac{d\xi_1}{dt} = -\xi_1 \sum_x k(x) \bar{A}(x) + \frac{k}{\sigma^2} \sum_x (x-n) \bar{A}(x) \psi(x;t) \tag{13}$$

The sum $\sum_x x\bar{A}(x) \psi(x;t)$ can be expressed in $\xi_1$ by Equation (6). The sum

$$\sum_x \bar{A}(x) \psi(x;t) = \delta c_3, \tag{14}$$

is the excess number of micelles which is taken to be constant during the process considered. We further use (9) and the fact that, from the definition of n and a,

$$\sum_x \bar{A}(x) = \frac{1}{n} \sum x \bar{A}(x) = \frac{1}{n} a \bar{A}_1 \tag{15}$$

Table I. The Fundamental Cases Yielding Exponential Monomer Relaxation.

| Orthogonal polynomials | Distribution function | $X(x)$ | Interval and parameters |
|---|---|---|---|
| 1. Hahn polynomials $$\frac{(-1)^m \binom{m+a}{m}\binom{m+b}{m} m!}{\binom{x+a}{x}\binom{N-x+b}{N-x}} \Delta^m \left\{ \binom{x+a}{x-m}\binom{N-x+m+b}{N-x} \right\}$$ | $$\frac{\binom{x+a}{x}\binom{N-x+b}{N-x}}{\binom{N+b}{N}}$$ | $(x+a)(x-N-1)$ | $0 \leq x \leq N$ $n = N(1+a)/(a+b+2)$ $a > -1, b > -1$ |
| 2. Krawtchouk polynomials $$\frac{(-1)^m x!}{m! p^x q^{N-x}} (N-x)! \, \Delta^m \left[ \frac{p^x q^{N-x+m}}{(x-m)!(N-x)!} \right]$$ | binomial $$\binom{N}{x} p^x q^{N-x}$$ | $N+1-x$ | $0 \leq x \leq N$ $n = pN, \; \sigma^2 = nq$ $p > 0, \; q > 0, \; p+q = 1$ |
| 3. Meixner polynomials $$\frac{x!}{\Gamma(\beta+x)} c^{-x-m} \Delta^m \left[ \frac{c^x \Gamma(\beta+x)}{(x-m)!} \right]$$ | negative binomial $$\frac{c^x \Gamma(x+\beta)}{x! \, \Gamma(\beta)}$$ | $x+\beta-1$ | $x \geq 0$ $c = 1-n/\sigma^2; \; \beta = n^2/(\sigma^2-n)$ $\beta > 0; \; 0 > c > 1$ |
| 4. Charlier polynomials $$\frac{x!}{a^x} \Delta^m \left[ \frac{a^{x-m}}{(x-m)!} \right]$$ | Poisson $$e^{-a} \frac{a^x}{x!}$$ | 1 | $x \geq 0$ $a = n = \sigma^2$ |

Table II. Distinguishing Characteristics of the Five Main Types of Models with Exponential Monomer Relaxation, and Their Relationship to the Fundamental Cases in Table I. M and N are the lower and upper limits of the micelle distribution: $A(x) > 0$ for $M < x < N$.

|   | $dk^-(x)/dx$ | $dk^+(x)/dx$ | $M_3/\sigma^2$ | M-n | N-n | Case in Table I |
|---|---|---|---|---|---|---|
| a | $> 0$ | $> 0$ | $> 1$ | $\overbrace{\phantom{xxxxxx}}^{-2\sigma^2}$ | $\infty$ | 3 $\Big\}$ $x \to x-M$ |
| b | $> 0$ | $= 0$ | $= 1$ | $M_3/\sigma^2+1$ | $\infty$ | 4 $\quad k^+(x)\overline{A_1} \propto X(x-M)$ |
| c | $> 0$ | $< 0$ | $-1 < M_3/\sigma^2 < 1$ | | | 2 |
| d | $= 0$ | $< 0$ | $= -1$ | $-\infty$ | $\overbrace{\phantom{xxxxxx}}^{-2\sigma^2}$ | 4 $\quad x \to N-x$ |
| e | $< 0$ | $< 0$ | $< -1$ | $-\infty$ | $M_3/\sigma^2-1$ | 3 $\Big\}$ $k^-(x) \propto X(N-x+1)$ |

so that finally

$$\frac{d\xi_1}{dt} = -\frac{k}{\sigma^2}\left(1+\frac{\sigma^2}{n}a\right)\xi_1 - \frac{kn\delta c_3}{\sigma^2 \bar{A}_1} \cdot \quad (16)$$

The solution of this equation shows that $\xi_1$ is governed by a single relaxation time as proposed and that furthermore its value is once again that of Equation (4).

The necessity of the condition can also be proved (see Reference 9) but requires a larger number of steps.

When the equilibrium condition (for clarity a minus sign suffix is attached to the former $k(x)$)

$$k^+(x+1)\,\bar{A}(x)\,\bar{A}_1 - k^-(x+1)\,\bar{A}(x+1) = 0 \quad (17)$$

is used the condition (10) can be written

$$k^+(x+1)\,\bar{A}_1 - k^-(x) = -\frac{k}{\sigma^2}(x-n) \quad (18)$$

Since $\bar{A}_1$ in general varies with total concentration Equation (18) can be fulfilled for a given set of rate constants if and only if both are polynomials of x of the first degree. Of the exactly soluble cases in Table I the last three fulfill this more stringent condition.

## SYSTEMS WITH EXPONENTIAL MONOMER RELAXATION

Five main types may be distinguished among models with rate constants linearly dependent on x. Representative examples of rate constants and micelle distributions in these five cases are presented in Figure 1. The distinguishing characteristics of the five cases are given in Table II, and also their relationship to the fundamental solutions in Table I.

Using the equilibrium condition (17) as a recurrence relation for the micelle distribution we may express the coefficients in the linear forms of $k^-(x)$ and $k^+(x)\bar{A}_1$ in $k$, $n$, $\sigma^2$ and $M_3$ (the third central moment of the distribution). We obtain

$$k^-(x)/k = 1 + (M_3/\sigma^2 + 1)(x-n)/(2\sigma^2) \quad (19)$$

$$k^+(x)\bar{A}_1/k = 1 + (M_3/\sigma^2 - 1)(x-n-1)/(2\sigma^2) \quad (20)$$

# KINETICS OF REDISTRIBUTION OF MICELLAR SIZES

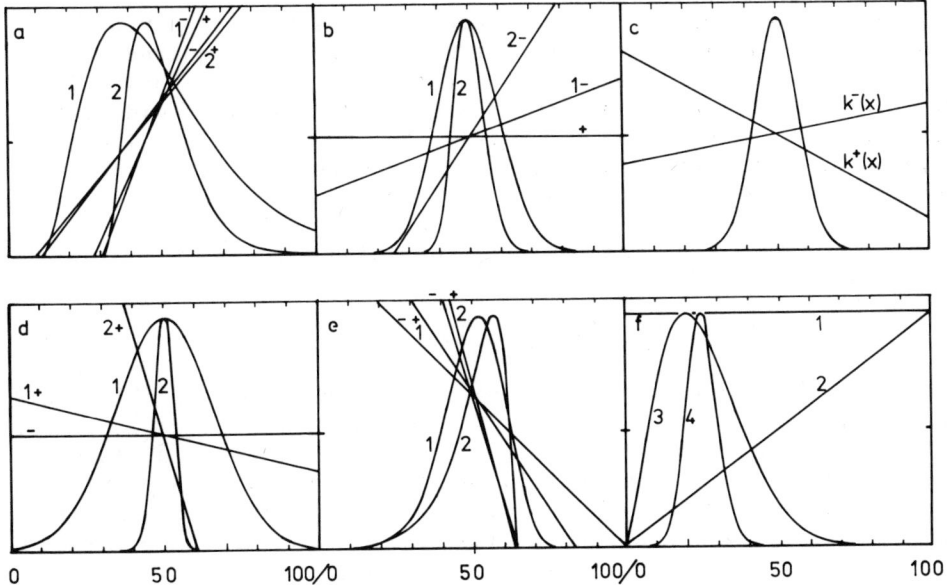

Figure 1. Representative examples of micelle distributions and rate constants for models with strictly exponential monomer relaxation. a-e are of types a-e in Table II, f contains distributions belonging to case 1 in Table I. The straight lines marked + and - represent $k^+(x)\ \bar{A}_1$ and $k^-(x)$, respectively.

a. 1: $n=50$, $\sigma^2=500$, $M_3=25\sigma^2$    2: $n=50$, $\sigma^2=100$, $M_3=10\sigma^2$
b. 1: $n=50$, $\sigma^2=100$, $M_3=\sigma^2$    2: $n=50$, $\sigma^2=100$, $M_3=\sigma^2$
c.    $n=50$, $\sigma^2=50$, $M_3=-0.50\sigma^2$
d. 1: $n=50$, $\sigma^2=250$, $M_3=-\sigma^2$    2: $n=50$, $\sigma^2=10$, $M_3=-\sigma^2$
e. 1: $n=50$, $\sigma^2=100$, $M_3=-5\sigma^2$    2: $n=50$, $\sigma^2=100$, $M_3=-15\sigma^2$
f. 1: $a=b=0$, 2: $a=1$, $b=0$, $n=66,67$, 3: $a=2$, $b=8$; $n=25$
4: $a=49$, $b=159$, $n=25$. By interchanging the values of parameters a and b distributuions are obtained that are the mirror images in $x=50$ to those shown.

All the models shown may be translated along the x-axis and still belong to the same types of Tables I and II. However, on a change of total concentration the models of f will change in such a way that condition (10) is violated. The deviation from this condition will normally be very small, and the monomer relaxation practically exponential with a relaxation time given to a good approximation by Equation (4).

All choices of $M_3$, $\sigma^2$ and n are not possible for a representation of real micellar systems. It must always be required that the rate constants and the micelle concentration be non-negative in the micellar region. As seen from Figure 1 and Table II a wide variety of micelle distributions is possible with these models. The most asymmetric ones are, however, always accompanied by very rapidly varying rate constants.

Equation (4) for the monomer relaxation time is valid in all these cases:

$$\tau^{-1} = k/\sigma^2 + (A_{tot}/\overline{A}_1 - 1)\, k/n$$

Normally all parameters are concentration dependent and $\tau^{-1}$ should not be strictly linear in $A_{tot}$. For the case that $k^-(x)$ is proportional to x k/n is constant. If further n = $\sigma^2$ also k/$\sigma^2$ is constant and only the slight concentration dependence of $\overline{A}_1$ on $A_{tot}$ remains. If activity factors may be regarded as constant this concentration dependence may be taken into account iteratively and true values of a calculated. A strictly linear variation of $\tau^{-1}$ with a is thus predicted in this case ($M_3 = \sigma^2 = n$; a Poisson distribution). Since both $k^-(x)/x$ and $k^+(x)$ are independent of x, this represents the simplest case of micellar equilibria with both entrance and exit rates of a monomer independent of the size of the micelle. The other models represent various types of departures from this case.

With $k^-(x)$ proportional to x the deviations from linearity are entirely due to the term $k/\sigma^2$. If $\sigma^2$ is large this term is small, and a small curvature close to a = 0 is to be expected. With $\sigma^2$ small, this term is large. Since $dn/dA_{tot} = \sigma^2/(1 + na + \sigma^2 a/n)$, the variation of n and therewith k with $A_{tot}$ is small. Since $d\sigma^2/dA_{tot} = M_3/(1 + na + \sigma^2 a/n)$, large variations of $\sigma^2$ may be expected only for very asymmetric distributions and for small a-values. Thus, we again expect deviations from linearity to appear only close to a = 0.

When $k^-(x)/x$ is x-dependent the situation is more complex. Larger deviations from linearity may be expected. However, at high concentrations n and $\sigma^2$ tend to constant values (more rapidly for systems with large n) and we should always expect a linear dependence of $\tau^{-1}$ on a at sufficiently large a-values. Some examples of calculated $\tau^{-1}$,a plots are shown in Figure 2.

## SYSTEMS WITH NON-EXPONENTIAL RELAXATION.
## THE INITIAL RELAXATION TIME.

As soon as condition (10) is substantially violates more than one relaxation mode will be of importance for the monomer relaxation. For systems exhibiting small deviations from condition (10)

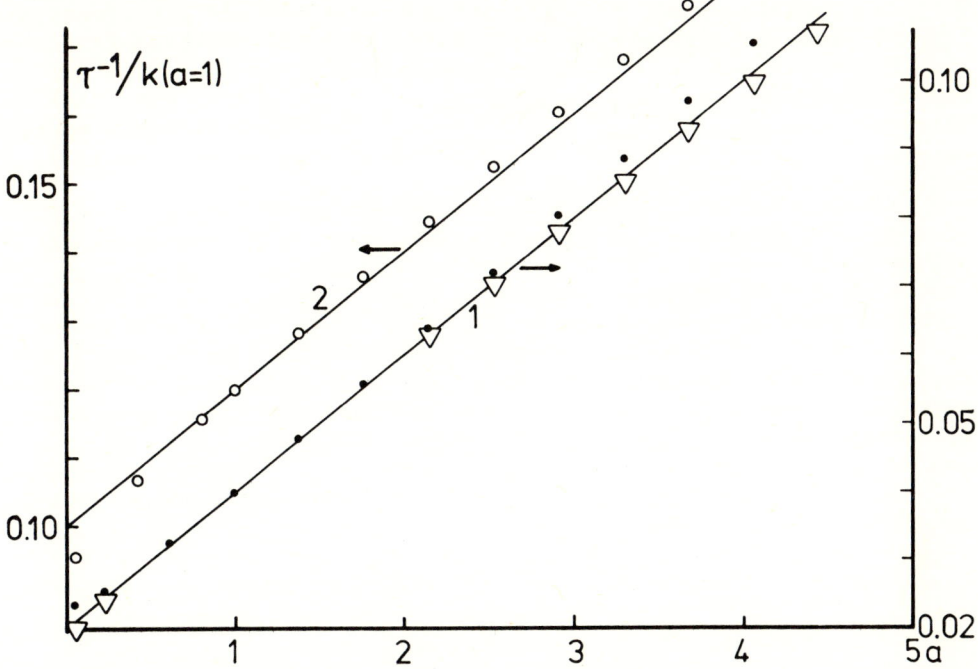

Figure 2. $\tau^{-1}$, a plots for models with exponential monomer relaxation. The straight lines represent $\tau^{-1}$ from Equation (4) with $\sigma^2 = n = 50$ (1, right hand scale) and $\sigma^2 = 10$, $n = 50$ (2, left hand scale). A Poisson distribution model with $n = \sigma^2 = 50$ follows line 1. o represents a model of type a, Table II, with $\sigma^2 = n = 50$ and $M_3 = 5.67\ \sigma^2$ at $a = 1$. ▽: type c, $\sigma^2 = n = 50$, $M_3 = -0.50\ \sigma^2$ at $a = 1$. 0: type d, $n = 50$, $\sigma^2 = 10$, $M_3 = -\sigma^2$ at $a = 1$.

one intuitively expects that the monomer relaxation should remain nearly exponential and that Equation (4) should remain valid for the experimental relaxation time. The first part of this expectation is met if the effect of a small deviation from the condition is to introduce a number of small contributions from relaxation modes besides a dominating one, or a few dominating ones with nearly equal relaxation times. We will see that the second part of the expectation then follows.

The time-dependence of the relative excess concentration of monomers most generally follows

$$\xi_1(t) = \xi_1^* + \sum_{j=1}^{r} \alpha_j \exp(-t/\tau_j) \qquad (21)$$

where $\xi_1^*$ denotes the relative excess concentration at the end of the fast process. We define the initial relaxation time by

$$\frac{1}{\tau_i} = - \lim_{t \to 0} \frac{d \ln[\xi_1(t) - \xi_1^*]}{dt} = \sum_{j=1}^{r} \alpha_j \tau_j^{-1} \Big/ \sum_{j=1}^{r} \alpha_j, \qquad (22)$$

If the experimental monomer relaxation curve appears exponential then $\tau_i$ should be close to the experimental relaxation time, unless there are present relaxation modes with $\alpha_j$ small but $\alpha_j \tau_j^{-1}$ appreciable. By physical reasons this seems unlikely. Such relaxation mode must have a very short relaxation time and therefore closely spaced nodes. It will appear only if the micelle concentration or the rate constants change abruptly with x somewhere, and it will be important only if the initial conditions show a corresponding jump.

Let us assume that $f(x)$ measures the deviation from condition (10), i.e.

$$\Delta(k^-(x) A(x)) / A(x) = - k (x-n)/\sigma^2 + f(x) \qquad (23)$$

We then obtain for $\tau_i$

$$\tau_i^{-1} = \frac{k}{\sigma^2} \left(1 + \frac{\sigma^2 a}{n}\right) + \frac{\Sigma f(x) \, \xi_x(0) \, A(x) \, (1 + \sigma^2 a/n)}{\bar{A}_1 \, a \, (\sigma^2 \xi_1(0)/n - \delta \ln n)} \qquad (24)$$

where $\xi_x(0)$ is the initial value of $\xi_x$ and $\delta n = n \delta \ln n$ is the difference in mean aggregation number between the initial distribution and the equilibrium distribution. We note that if $f(x)$ is small Equation (24) tends towards the simple Equation (4). This equation is always obtained for the case that $\xi_x(0)$ is independent of x, since $\Sigma f(x) A(x) \equiv 0$. Unfortunately those initial conditions are not easily realized experimentally.

Assuming that $f(x)$ is of the second order in x

$$f(x) = \alpha + \beta \, k(x-n)/\sigma^2 + \gamma \, k(x-n)^2/\sigma^2$$

we obtain an expression for $\tau_i$ containing $\gamma$ and in principle measurable quantities

$$\tau_i^{-1} = \frac{k}{\sigma^2} \left(1 + \frac{\sigma^2 a}{n}\right) \left(1 + \gamma \, \frac{\sigma^2/n \, \delta \ln \sigma^2 - M_3/\sigma^2 \, \delta \ln n}{\sigma^2/n \, \delta \ln \bar{A}_1/n - \delta \ln n}\right) \qquad (25)$$

We see that $\gamma k/\sigma^2$ is the difference between the coefficients for $x^2$ in the second order expressions for $k^+(x)\overline{A}_1$ and $k^-(x)$, or $\gamma = (\overline{A}_1 d^2k^+(x)/dx^2 - d^2k^-(x)/dx^2)\sigma^2/2k$. $\gamma$ may be large if there are substantial deviations from linearity over a length $\sigma$ in the variation of $k^+(x)\overline{A}_1$ and $k^-(x)$ with x, and especially if these deviations have significant signs for the two rate constants. The coefficient of $\gamma$ is complicated. The $\delta$-quantities represent the change in $\sigma^2$, n, and $\overline{A}_1$ (approximately cmc) with the perturbation ($\Delta P$, $\Delta T$, or $\Delta E$). Little is known about these changes — most about the temperature change of cmc and n, and, from kinetic measurement, of $\sigma^2$. All the $\delta$-quantities may very well be of comparable magnitudes. The $\gamma$-coefficient will be large if the denominator is small. The latter is proportional to $\xi_1(0) - \xi_1^*$, the monomer amplitude in the first fast process, so it will not be close to zero in interesting cases. The numerator may be large for broad and/or asymmetric distributions.

The concentration dependence of $\gamma$ and the $\delta$-quantities in Equation (25) is about as mild as that of $\overline{A}_1$, n, $\sigma^2$, etc. We would therefore expect the same types of deviations from linearity as for systems with exponential monomer relaxation, but probably more pronounced.

## NUMERICAL STUDY OF MODEL SYSTEMS

Computer calculations have been performed by two methods: a matrix method for solving the linearized Equations (15) and (6), yielding all relaxation modes and their contributions to the monomer relaxation, and a numerical integration method for solving the nonlinear equation system (2,6) which yields the time evolution of the concentrations. Here the models characterized in Table III will be discussed. Fuller accounts for the calculations and the results in these and other cases are given elsewhere.[10,11]

In all cases the results have verified the fundamental separation of the relaxation processes into a rapid and a slow stage.

Models 1-r have (truncated) Gaussian micelle distributions centered around x = 25 with $\sigma \simeq 3,5$. As seen from the two representative examples in Figure 3a the deviations from condition (10) are small in this region. Typically two — for some models at some concentrations up to four — relaxation modes contribute appreciably (5% or more) to the monomer relaxation. The relaxation times of these important modes are always rather close, and the resulting monomer relaxation curves appear exponential, Figure 4. The experimental relaxation time is then close to $\tau_i$. Plots of $\tau_i$ versus a are shown in Figure 5a for the four Gaussian models. The estimates of $k/n$ and $k/\sigma^2$ obtained from the regression lines agree

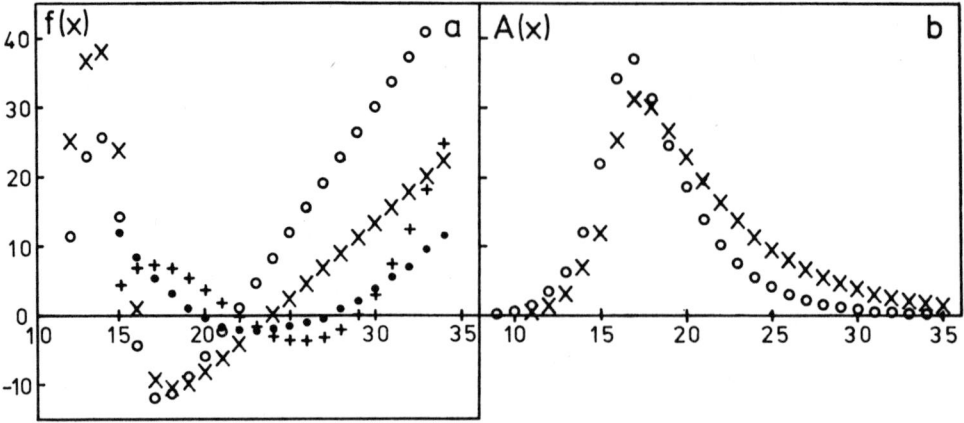

Figure 3. (a) The deviations from condition (10) for Model 1 at a = 0,39 (O), Model 3, a = 2,37 (+), Model 5, a=0,316 (●) and a= 3,50 (x). (b) Micelle distribution Model 5 at a = 3.316 (O) and a = 3.50 (x).

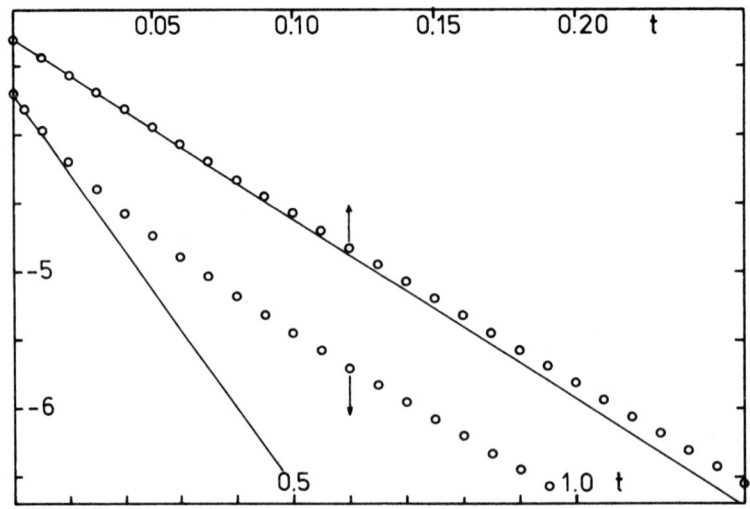

Figure 4. $\ln |\xi_1(t) - \xi_1^*|$ vs. t for (1) Model 4b at a = 2.37 (top time scale) and (2) Model 5b at a = 0,31 (bottom time scale). The straight lines have slopes $-\tau_i^{-1}$.

# KINETICS OF REDISTRIBUTION OF MICELLAR SIZES

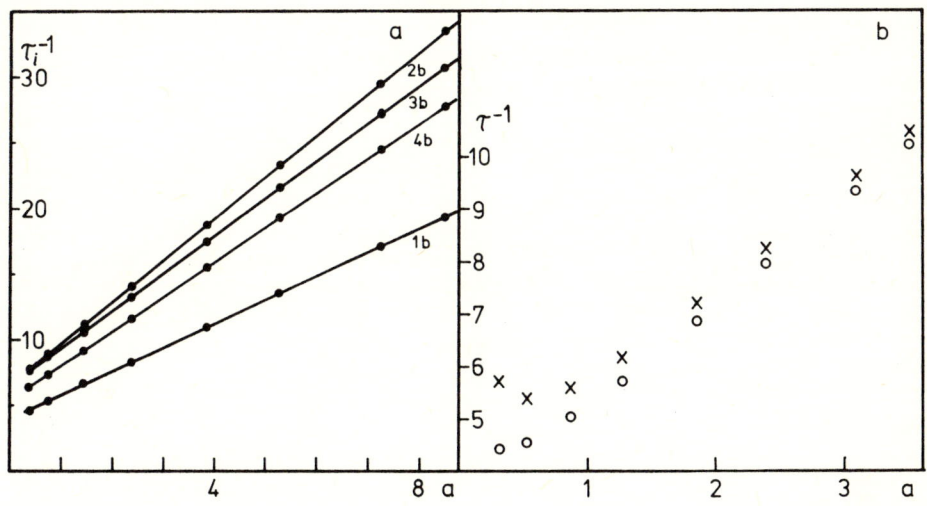

Figure 5. (a) $\tau_i^{-1}$ vs. a for the four Gaussian models. In these cases the experimental relaxation time is close to $\tau_i$. (b) $\tau_i^{-1}$ vs. a for model 5 (x) and $\tau^{-1}$ calculated from Equation (4) with the true values of n, $\sigma^2$, and k at each concentration (o).

Table III. Models Treated Numerically.

| Model no | Micelle distribution | $\bar{k}(x)$ | Initial conditions |
|---|---|---|---|
| 1 | truncated Gaussian $15 \leq x \leq 35$ $n = 25, \sigma = 3,5$ | constant | a. $\xi_x(0)$ constant |
| 2 | | $\propto x$ | b. $\xi_x(0)$ from an initial distribution slightly different from the equilibrium distribution. |
| 3 | | $\propto(-\dfrac{x^2}{150}+\dfrac{x}{3}-\dfrac{5}{2})$ | |
| 4 | | $\propto(\dfrac{x^2}{300}-\dfrac{x}{10}+\dfrac{7}{4})$ | |
| 5 | $\exp[(15-x)/4]-\exp(15-x)$ | constant | |

within a few percent with the model values of these parameters in the middle of the concentration interval.

Model 5 has a skew micelle distribution, Figure 3b, and shows large deviations from condition (10) in the micelle region, Figure 3a. The relaxation spectrum is complicated with a large number of relaxation modes contributing to the monomer relaxation. The latter is clearly non-exponential, Figure 4. The concentration dependence of $\tau_i^{-1}$ is shown in Figure 5b, together with $\tau^{-1}$ calculated from Equation (4) with the (concentration dependent) model values of k, n and $\sigma$. At high concentrations both plots are nearly linear. The estimates of n, $\sigma$ (assuming k known) from the four last points deviate about 25% from the model values. However, with present experimental methods it is probably difficult to obtain a good estimate of $\tau_i$ in such a case - the relaxation might even be mistaken as an exponential relaxation mixed up with a "physical" relaxation process in the equipment.

## CONCLUDING REMARKS

We have here discussed the size redistribution stage of the relaxation of a micellar system. We have found the condition, Equation (10), under which the relaxation of the monomer is strictly exponential, and shown that the relaxation time is then given by Equation (4). We have seen that the monomer relaxation appears exponential with an effective relaxation time approximately given by Equation (4), for systems that show small deviations from condition (10). Most types of micellar systems are probably covered by models that fulfill condition (10), or nearly do so. An important exception is systems in which both nearly spherical and long, rod-like micelles are present. Such systems should have broad micelle distributions with a mean aggregation number that increases rapidly with the total concentration. Model 5 of the previous section is an attempt to represent such a system (within the frame of 35 components given by our computer program). In this case the monomer relaxation is clearly non-exponential. A well-defined measurable quantity is then the initial relaxation time which is given by Equation (24), a counterpart to Equation (4) with a complicated correction term. It is possible to obtain approximate information about rate constants and micelle size distributions even in this case.

## ACKNOWLEDGMENT

This work has been financially supported by grants from the Swedish Natural Science Research Council.

## REFERENCES

1. J. Lang, C. Tondre, R. Zana, R. Bauer, H. Hoffmann and W. Ulbricht, J. Phys. Chem., $\underline{79}$, 276 (1975).
2. E. Wyn-Jones, Editor, "Chemical and Biological Applications of Relaxation Spectrometry", pp. 133-264, D. Reidel Publ. Co., 1975.
3. N. Muller, J. Phys. Chem., $\underline{76}$, 3017 (1972).
4. T. Nakagawa, Colloid. Polym. Sci., $\underline{252}$, 56 (1974).
5. H. H. Grünhagen, J. Colloid Interface Sci., $\underline{53}$, 282 (1975).
6. E. A. G. Aniansson and S. N. Wall, J. Phys. Chem., $\underline{78}$, 1024 (1974).
7. E. A. G. Aniansson, S. N. Wall, M. Almgren, H. Hoffmann, J. Kielmann, W. Ulbricht, R. Zana, J. Lang, C. Tondre, J. Phys. Chem., $\underline{80}$, 905 (1976).
8. E. A. G. Aniansson and S. N. Wall, J. Phys. Chem., $\underline{79}$, 857 (1975).
9. M. Almgren, E. A. G. Aniansson and K. Holmåker, Chem. Phys., in press.
10. M. Almgren, E. A. G. Aniansson and S. N. Wall, to be published.
11. S. N. Wall, forthcoming dissertation.

ULTRASONIC RELAXATION STUDIES ASSOCIATED WITH MONOMER-MICELLE EXCHANGE PROCESSES

W. J. Gettins, J. E. Rassing and E. Wyn-Jones

Department of Chemistry and Applied Chemistry

University of Salford, Salford M5 4WT, U.K.

The ultrasonic relaxation method has been used as a probe to monitor the dynamic interchange between the surface active agent ions and the micellar units. For both single and mixed micelle systems the kinetics associated with this exchange process are treated in terms of the Langmuir adsorption theory and in all cases there is good correspondence between theory and experiment. When the studies were extended to antihistamine type drugs exhibiting colloidal behaviour, more than one association phenomena appeared to be occurring.

The kinetics of micelle formation have been extensively studied by a wide variety of fast reaction techniques including temperature-jump, pressure-jump, shock tube, ultrasonics, stopped flow and magnetic resonance methods[1]. Most of the studies have been carried out using the chemical relaxation techniques, and it is now generally accepted[2,3] that the relaxation spectra of aqueous solutions of pure surfactants above their critical micelle concentration (c.m.c.) are characterized by two relaxation times separated by a factor of c.a. $10^2-10^3$. The fast relaxation process has been extensively studied in this laboratory[4-7] using the ultrasonic method, where it was shown that the sound wave can be used as a probe to monitor the dynamic interchange between the surface active agent ions and the micellar unit. This exchange process can be envisaged as a collision between a small object, the free monomer, and the large spherical micelle. The rate at which the monomers associate to the micelle was assumed to be proportional to the concentration of free monomers in solution, $A_1$, the concentration of micelles, $A_n$, and to the cross-sectional area of the micelles. Using the Hartley spherical micelle model, in which the monomers are located on the micelle surface, the cross-sectional area now becomes proportional to the actual number of monomers in the micelle. The rate of association $r_f$, is expressed as:

$$r_f = k_f (A_1)(A_n) n \tag{1}$$

where $k_f$ is the proportionality constant. If the micelles are not monodisperse:

$$r_f = k_f (A_1) \sum_{i=n'}^{n''} (A_i) i \qquad n' < n < n'' \tag{2}$$

Since the overall concentration, C, is given by:

$$C = (A_1) + \Sigma i (A_i) \tag{3}$$

we find that:

$$r_f = k_f (A_1)(C - (A_1)) \tag{4}$$

The rate at which the monomers dissociate from the micelles, $r_b$, is assumed to be proportional to the concentration of micelles and to the number of monomers in the micelle. This can be compared to a monomer dissociating from a polymer in which all the monomers are equally bonded and where the probability of dissociation is proportional to the number of monomers present.

Thus, $r_b = k_b (A_n) n$

$$= k_b (C - (A_1)) \tag{5}$$

The overall rate equation is:

$$-\frac{d}{dt}(A_1) = k_f(A_1)(C-(A_1)) - k_b(C-(A_1)) \tag{6}$$

from which it can be shown that the relaxation time, $\tau$, associated with this exchange process is given by:

$$1/\tau = k_f C - k_b \tag{7}$$

A typical plot of relaxation data is given in Figure 1. A consequence of the above treatment is that at equilibrium:

$$A_1 = k_b/k_f \approx (c.m.c.) \tag{8}$$

This is in good agreement with most of the relaxation behaviour[2] and also with equilibrium observations that the free monomer concentration is approximately constant above the c.m.c.

In the above model it is obvious that association will only occur when the free monomer collides with that part of the micelle surface not covered by hydrophilic head groups of the monomers. This "steric hinderance" can be taken into account[2] by a consideration of the packing of the monomers on the surface, and the adsorping surface can be defined as a function of overall concentration of the micelle forming agent. On this basis the total micelle surface area, S, present in solution is given by the

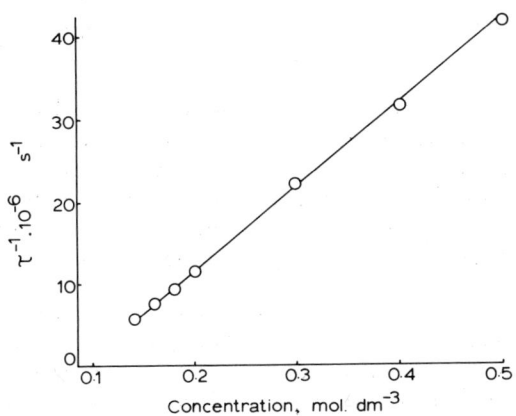

Figure 1. $1/\tau$ against concentration for sodium octyl sulphate at 298K.

equation:

$$S = a_1(C-(A_1))\tag{9}$$

where $a_1$ is the area occupied on the micelle surface by one mole of monomer. This equation is based on the assumption that each monomer in the micellar state contributes an equal area to the total micelle surface. The fraction of this total surface area covered by monomers, $\alpha$, is defined as:

$$\alpha = a_o(C-(A_1))/S = a_o/a_1\tag{10}$$

where $a_o$ is the area of one mole of the monomer head group on the micelle surface. The kinetics associated with the exchange of free monomers with the micellar unit can now be considered in terms of the principles of Langmuir adsorption kinetics, and this leads to the rate equation:

$$\frac{-d}{dt}(A_1) = k_1(A_1)(1-\alpha)S - k_{-1}\alpha S\tag{11}$$

which in turn gives a single relaxation time, $\tau$, given by:

$$1/\tau = k_1 a_o((1-\alpha)/\alpha)C - k_{-1} a_o\tag{12}$$

This is a more detailed relaxation equation than Equation (7). It is also of interest to note that this extended treatment is not limited to spherical micelles.

The amplitude, $\mu_m$, of the ultrasonic relaxation becomes:

$$\mu_m = \text{const.}\ V_A^2 (1 - \frac{V_{AM}}{V_A})^2 (A_1) (\frac{C/(A_1)-1}{C/(A_1)+1})\tag{13}$$

where $V_A$ and $V_{AM}$ are the molar volumes of the free monomer and the monomer in the micelle respectively. For a variety of cationic and anionic surfactants it has been shown[2] that the concentration dependence of the relaxation times and amplitudes can be quantitatively understood in terms of monomer/micelle exchange in which the kinetics are governed by the Langmuir principles.

During the course of this work Aniansson and Wall[8,9] proposed an alternative model for micelle kinetics. This model is based on the analogy between micelle formation and one-dimensional heat conduction and predicts two relaxation times, in accordance with experimental findings. The fast relaxation time of this model is associated with redistribution amongst the micelles, in general agreement with the exchange phenomena discussed in this paper,

The relaxation equation for this fast relaxation time is[10]:

$$1/\tau = \frac{\bar{k}}{\sigma^2} + \frac{\bar{k} A_{exc.}}{n(A_1)} \qquad (14)$$

where $\bar{k}$ is the dissociation rate constant in the region of proper micelles, n is the average aggregation number, $\sigma$ is the width of the Gaussian micellar distribution curve and $A_{exc.} = C - (A_1)$. This equation can be rearranged as follows:

$$1/\tau = \frac{\bar{k}}{n} \left[ \frac{C}{(A_1)} - \left(1 - \frac{n}{\sigma^2}\right) \right] \qquad (15)$$

In all cases $n/\sigma^2 < 1$, which means that the relaxation equation for the fast process of the Aniansson and Wall model predicts exactly the same concentration dependence of $1/\tau$ as Equation (12) in the present model. i.e. $1/\tau$ is linear with respect to overall surfactant concentration and has a negative intercept. This is in general agreement with the majority of available ultrasonic data. It also means that experimental results alone cannot be used to judge the correctness of these models.

The fast relaxation time in surfactants has been measured by ultrasonic[2] and shock tube[10,11] methods. At concentrations in excess of the c.m.c. the majority of the ultrasonic data show a linear dependence of $1/\tau$ with concentration. On the other hand the shock tube measurements were carried out very close to the c.m.c. and the results indicate that $1/\tau$ increases more rapidly with increasing concentration than predicted by Equations (12) and (14). Aniansson and Wall claim that their model applies at concentrations very close to the c.m.c. and consequently, using the shock tube measurements, they have used the slope in this region when analysing the results. It is difficult to understand how they justify deriving $\sigma$, the width of the micellar distribution curve, with results so close to the c.m.c. when in fact at these concentrations micelles are only starting to be formed. On the other hand when this theory is applied to measurements at concentrations well above the c.m.c., where $\sigma$, according to this theory, is expected to play a dominant role, it turns out that for most surfactants $n/\sigma^2 \approx 0$. Finally, we are unable to understand the consequences of this theory for monodisperse systems where $\sigma \to 0$.

The Aniansson and Wall model also predicts the slow relaxation time, whereas our present model is concerned only with the faster process. However, since these two relaxation times are separated by an order of magnitude of $10^2 - 10^3$ the fast time can be treated independently since the slower process will be practically frozen out.

In this laboratory the main emphasis of our research is concerned with obtaining a better understanding of the aqueous solution properties of systems containing components of biological and industrial significance. The kinetic studies on these systems have been exclusively concerned with the fast process and for such complex systems it is preferable to adhere to the simplest available model which at the present time is the collision model based on Langmuir adsorption kinetics. As an example we here show how this model has been further extended to consider the relaxation spectrum associated with the exchange process in aqueous solutions of mixed micelles containing two surfactants. The total surface area, S, of such a system of mixed micelles is given by:

$$S = a_1(C_1 - (A_1)) + a_2(C_2 - (A_1'))$$

$$= a_1(C_1 + \kappa C_2 - (A_1) - \kappa(A_1')) \tag{16}$$

where the a's are the areas taken up by the monomers on the micelle surface, $\kappa = a_2/a_1$ and the subscripts 1 and 2 refer to the individual surfactants. On the basis of the Langmuir kinetic treatment the overall rates at which the monomers $A_1$, $A_1'$ exchange with the mixed micelle are given by the following equations:

$$\frac{-d}{dt}(A_1) = k_1(A_1)(1-\alpha_1-\alpha_2) \, S - k_{-1}\alpha_1 S \tag{17}$$

$$\frac{-d}{dt}(A_1') = k_2(A_1')(1-\alpha_1-\alpha_2) S - k_{-2}\alpha_2 S \tag{18}$$

where the α's are the fractions of the total micelle area covered by the monomers. This model is characterized by a spectrum of two relaxation times, $\tau_1$ and $\tau_2$, given by:

$$1/\tau_1 = a_1 k_1 (1 - \alpha_1 - \alpha_2)(C_1 + \kappa C_2) - (a_1\alpha_1 k_{-1} + a_2 k_1 k_{-2}\alpha_2/k_2) \tag{19}$$

and

$$1/\tau_2 = a_1 k_2 (1 - \alpha_1 - \alpha_2)(C_1 + \kappa C_2) - (a_2\alpha_2 k_{-2} + a_1\alpha_1 k_2 k_{-1}/k_1) \tag{20}$$

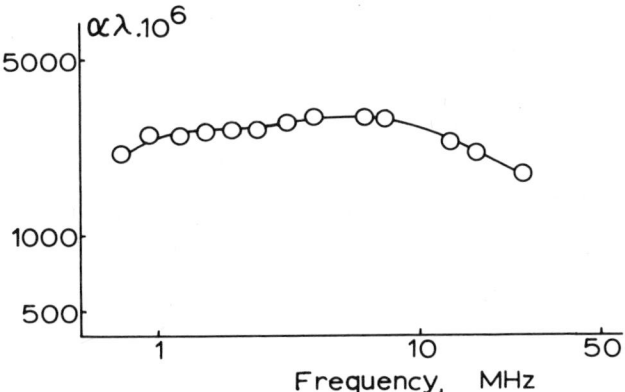

Figure 2. A plot of $\alpha\lambda$ against frequency for a mixture of 0.3M sodium hexyl sulphate and 0.1M sodium nonyl sulphate.

The ultrasonic spectrum of a mixture of sodium hexyl sulphate/ sodium nonyl sulphate is shown in Figure 2, and confirms the existence of two relaxation times. However, the two relaxation frequencies are so close that it is difficult to carry out a quantitative analysis of the data which is precise enough to test the relaxation equations. For this purpose we explored the possibility of choosing a more definitive system. One of the consequences of our earlier work[2] on micellization in solutions of pure surfactants is the knowledge that the exchange process between the monomer and micelle becomes progressively slower as the length of the hydrocarbon tail of the detergent increases. It was found that in the frequency range of our present ultrasonic apparatus these monomer-micelle exchange processes in surfactants with chain lengths greater than $C_{12}$ were too slow to show any relaxation effects. Thus in the mixed micelle system decyltrimethylammonium bromide (DTAB)/cetyltrimethylammonium bromide (CTAB), it is only the exchange of the DTAB monomers with the mixed micelles that can be monitered by the ultrasonic method. The relaxation spectra of this system is always characterized by a single relaxation time as shown in Figure 3. It is clear from Equation (20) that $1/\tau$ should be linear with respect to $C_{CTAB}$ if $C_{DTAB}$ is kept constant and vice versa. These plots are shown in Figure 4 and the lines are parallel indicating that $\kappa = 1$. This is not surprising since $\kappa$ is the ratio of the areas occupied on the micelle surface by the hydrophilic head groups of the monomers, which for both surfactants in this mixed micelle system is the trimethylammonium cation.

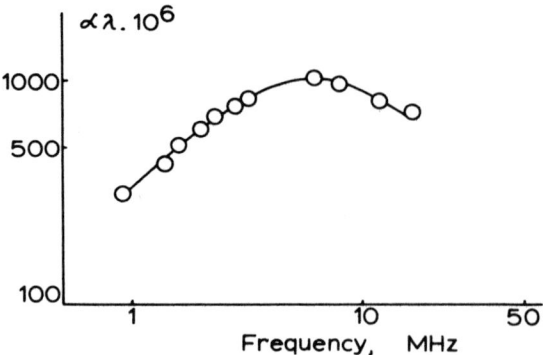

Figure 3. A plot of αλ against frequency for a mixture 0.2MCTAB and 0.3MDTAB.

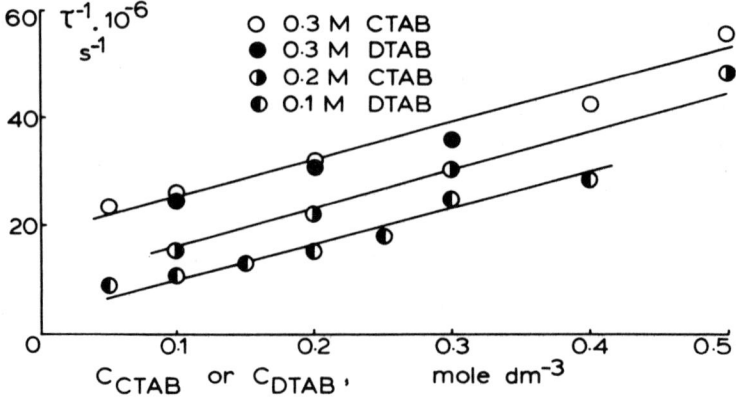

Figure 4. Plots of 1/τ against $C_{CTAB}$ and against $C_{DTAB}$.

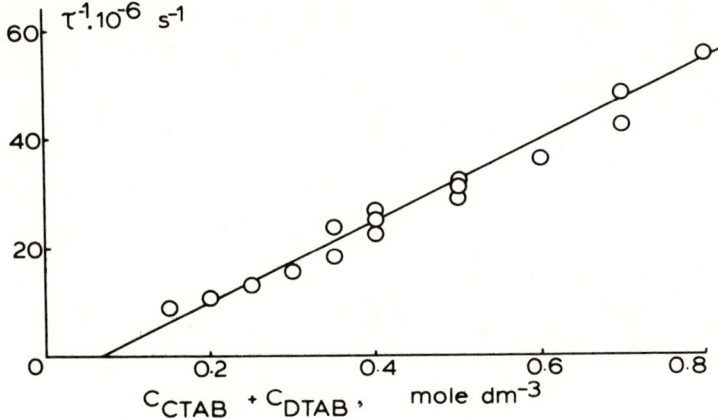

Figure 5. A graph of $1/\tau$ against $(C_{CTAB} + C_{DTAB})$.

With $\kappa = 1$, $1/\tau$ was plotted as a function of $(C_1 + C_2)$ as shown in Figure 5, and with equation (20) yields the following kinetic results for DTAB in the mixed CTAB/DTAB micelle system;

$$k_2 a_1 (1 - \alpha_1 - \alpha_2) = 71.10^6 \text{ dm.}^3\text{mole.}^{-1}\text{s}^{-1} \quad (136.10^6)$$

$$a_2 \alpha_2 k_{-2} + a_1 \alpha_1 k_2 k_{-1}/k_1 = 4.10^6 \text{ s}^{-1} \quad (7.10^6)$$

The rate constants for pure micellization of DTAB are indicated in parentheses. This clearly demonstrates that our model of the exchange process between monomers and mixed micelles described by Langmuir adsorption kinetics is consistent with the ultrasonic measurements.

The work has also been extended to consider several drugs which have colloidal properties[12]. The purpose of these studies is to obtain a better understanding of the solution properties of these drugs and to determine how these affect their biological activity. The association, and in particular micellar, behaviour of several antihistamines reported recently[13-16] prompted us to carry out ultrasonic relaxation measurements on several of these compounds. Those studied included, diphenhydramine hydrochloride, thenyldiamine hydrochloride, mepyramine maleate and chlorpheniramine maleate. It was found however, that several molecular phenomena contributed to their ultrasonic spectra. In addition to aggregation phenomena a large proton transfer effect was observed in those antihistamines based on a pyridine nucleus and containing a maleate counterion. This is rather an unusual interionic effect which can be mimicked in the model system pyridine-sodium maleate and eleminated at pH∼1.8. After removing these proton transfer effects by pH adjustment, the aggregation in all of the compounds can only be understood in terms of two competing association phenomena, namely, base stacking, promoted by the π-electrons in the aromatic ring system, and micellization, which occurs due to their surface active properties. When both these phenomena occur simultaneously it becomes necessary to distinguish the system below the c.m.c. and the system above the c.m.c. For example, as the concentration of antihistamine increases, base stacking occurs which results in the monomer concentration increasing with the overall concentration. Eventually the monomer concentration reaches the c.m.c. and micelles are formed. The quantitative theoretical interpretation of the ultrasonic relaxation data for these antihistamines is currently being considered in terms of these two association mechanisms. This work involves using several model systems to obtain kinetic parameters which can be used in simulation calculations.

## ACKNOWLEDGEMENTS

We thank the S.R.C. for research grants and the following companies for kindly donating samples of antihistamines: Parke, Davis and Co., Burroughs Wellcome and Co. (India) Private Ltd., Smith, Kline and French Lab. Ltd., Winthrop Laboratories, May and Baker Ltd., Hoechst U.K. Ltd., Allen and Hanburys Research Ltd., and A. H. Robins Co. Ltd.

## REFERENCES

1. E. Wyn-Jones, Editor, "Chemical and Biological Applications of Relaxation Spectrometry", p. 133, D. Reidel Publishing Co., Dordrecht (1975).
2. P. J. Sams, J. E. Rassing and E. Wyn-Jones, J. Chem. Soc. Faraday II, 70, 1247 (1974).
3. J. Lang, C. Tondre, R. Zana, R. Bauer, H. Hoffmann and W. Ulbricht, J. Phys. Chem. 79, 276 (1975).
4. P. J. Sams, J. E. Rassing and E. Wyn-Jones, Chem. Phys. Letters 13, 233 (1972).
5. P. J. Sams, J. E. Rassing and E. Wyn-Jones, J. Chem. Soc. Faraday II, 69, 180 (1973).
6. J. E. Rassing and E. Wyn-Jones, Chem. Phys. Letters 21, 93 (1973).

7. P. J. Sams, J. E. Rassing and E. Wyn-Jones, Adv. Mol. Relax. Processes 6, 255 (1975).
8. E. A. G. Aniansson and S. Wall, Reference 1, p. 223.
9. E. A. G. Aniansson and S. Wall, J. Phys. Chem. 78, 1024 (1974).
10. E. A. C. Aniansson, S. N. Wall, M. Almgren, H. Hoffmann, I. Kielmann, W. Ulbricht, R. Zana, J. Lang and C. Tondre, J. Phys. Chem. 80, 905 (1976).
11. H. Hoffmann, Reference 1, p. 181.
12. W. J. Gettins, R. C. Greenwood, J. E. Rassing and E. Wyn-Jones, in press.
13. D. Attwood, J. Pharm. Pharmac. 24, 751 (1972).
14. D. Attwood and O. K. Udeala, ibid. 26, 854 (1974).
15. D. Attwood and O. K. Udeala, J. Phys. Chem. 79, 889 (1975).
16. D. Attwood and O. K. Udeala, J. Pharm. Pharmac. 27, 395 (1975).

THE SIZE, SHAPE AND THERMODYNAMICS OF SODIUM DODECYL SULFATE (SDS) MICELLES USING QUASIELASTIC LIGHT SCATTERING SPECTROSCOPY

N.A. Mazer*, M. C. Carey and G. B. Benedek*
Department of Physics, M.I.T.
Cambridge, Massachusetts  02139*
Department of Medicine, Peter Bent Brigham Hospital
Boston, Massachusetts  02115

From measurements of the autocorrelation function and average scattered light intensity of sodium dodecyl sulfate (SDS) solutions, we have determined the mean size, shape, aggregation number ($\bar{n}$) and polydispersity of SDS micelles as a function of temperature (10-85°C) and NaCl concentration (0.15-0.6 M) for detergent concentrations which appreciably exceed the CMC. In addition we measured the NaCl dependence of the critical micellar temperature (CMT) and have characterized super-cooled micellar solutions over a 7-10°C temperature range beneath the CMT. At high temperatures (>50°C) a minimum micellar size corresponding to a sphere with a hydrated radius of ~25 Å is asymptotically approached in all NaCl and SDS concentrations. As the temperature is lowered, micellar growth occurs and the enlarged micelles are rod-like. The extent of micellar growth increases markedly with increases in NaCl concentration and SDS concentration. The values of $\bar{n}$ for $6.9 \times 10^{-2}$ M SDS in 0.6 M NaCl increase dramatically from ~60 at 85°C to ~1600 at 18°C and the micellar polydispersity is appreciable at these low temperatures. Micellar size and shape have the same dependence on temperature, NaCl concentration and SDS concentration below the CMT as above the CMT. From these results a quantitative theory of micelle formation was developed and we have deduced the values of the thermodynamic parameters governing SDS micellar growth. The effects of added NaCl are found to be consistent with theoretical calculations based on the

Gouy-Chapman theory of the electric double layer. Finally, we have shown that the dependence of the CMT on NaCl concentration can be quantitatively understood by an extension of the Murray-Hartley theory of detergent solubility.

## INTRODUCTION

Surprisingly little is known about the sizes, shapes and aggregation numbers of micelles which exist at detergent concentrations much greater than the critical micellar concentration (CMC). This lack of knowledge results from the fact that the experimental techniques which have been employed to study micellar systems typically require extrapolation to the CMC in order to eliminate the often sizable effects of micellar interaction. Nonetheless, data obtained at high detergent concentrations is vital to the theoretical understanding of micelle formation.

For these reasons we have employed[1] the technique of quasielastic light scattering spectroscopy[2] (QLS) to investigate micellar size, shape, and polydispersity in aqueous SDS solution at detergent concentrations which appreciably exceed the CMC. We studied these micellar characteristics in the presence of various NaCl concentrations and over a wide range of temperatures including the supercooled region below the critical micellar temperature (CMT) phase boundary.[1]

In this paper, we will first describe the QLS technique and explain why it is well suited for studying micellar systems at concentrations well above the CMC. We will then present our results for the SDS micellar system together with an analysis and interpretation of the data according to a quantitative thermodynamic model of micelle formation. This model will be seen to provide a simple physical explanation for the influence of temperature and NaCl concentration on the size and shape of micelles which exist at detergent concentrations well above the CMC.

## THE QLS TECHNIQUE

Quasielastic light scattering spectroscopy is a technique which utilizes the temporal fluctuations in the intensity of the scattered light for measuring the translational diffusion coefficients of macromolecules in solution. These temporal fluctuations are produced by the Brownian movement of the macromolecules and can be detected by measuring the autocorrelation function[2] of the scattered light intensity. By analyzing the time decay of the autocorrelation function, the <u>mean</u> diffusion coefficient, $\overline{D}$, of the macromolecules

# SIZE, SHAPE AND THERMODYNAMICS OF SDS MICELLES

and the variance, V, an index of solution polydispersity are obtained.[1] The precise definition of $\bar{D}$ is given by:

$$\bar{D} \equiv \Sigma\, G_i D_i \qquad (1)$$

where $D_i$ is the diffusion coefficient of the $i^{th}$ macromolecular species and $G_i$ is the fraction of the total light intensity scattered by the $i^{th}$ species. The V value in equation 2 provides a measure of the normalized width of the distribution of diffusion coefficients.

$$V \equiv 100 \times (\overline{D^2} - \bar{D}^2)^{\frac{1}{2}}/\bar{D}\%$$
$$\text{where}\quad \overline{D^2} \equiv \Sigma G_i D_i^2 \qquad (2)$$

From our measurements of $\bar{D}$ we have calculated the mean hydrodynamic radius, $\bar{R}$, of the SDS micelles, using an equation analogous to the simple Stokes-Einstein relation.[1]

$$\bar{D} = \frac{kT}{6\pi\eta\bar{R}} \qquad (3)$$

where k is Boltzmann's constant, T is the absolute temperature and η is the viscosity of the solvent. In using the above equation, we are assuming that the values of $\bar{D}$ measured at moderately high SDS concentrations (i.e. 2 g/dl) will not be substantially affected by micellar interactions. This assumption is consistent with a variety of experimental and theoretical results concerning the effect of macromolecular interactions on the diffusion coefficients measured by the QLS technique.[3,4] A brief development of these theoretical considerations follows.

For a monodisperse solution of macromolecules at a concentration c(w/v), the diffusion coefficient measured by QLS, D(c), has been shown[5] to obey the generalized Stokes-Einstein relation:[4]

$$D(c) = \frac{\frac{\partial \pi}{\partial c}(1-\bar{v}c)}{f(c)} \qquad (4)$$

where $\partial\pi/\partial c$ is the osmotic compressibility, $(1-\bar{v}c)$ is a reference frame correction factor[4] ($\bar{v}$ is the partial specific volume of the macromolecules) and f(c) is the hydrodynamic frictional factor of the macromolecules which is a function of their concentration. As a first order approximation, the concentration dependence of $\partial\pi/\partial c$ will be given by:

$$\frac{\partial \pi}{\partial c} = kT(1 + Bc) \qquad (5)$$

where B is the first osmotic virial coefficient. Similarly, the dependence of f(c) on concentration will be given by:

$$f(c) = 6\pi\eta R(1 + B_f c) \tag{6}$$

where R is the hydrodynamic radius of the macromolecules and $B_f$ is the first virial coefficient of the frictional factor. Substituting these expressions into equation (4) we obtain:

$$D(c) = \frac{kT}{6\pi\eta R}\left[\frac{(1 + Bc)(1-\bar{v}c)}{1 + B_f c}\right] \tag{7}$$

The quantity in brackets accounts for the effects of macromolecular interactions on D(c). In the case of uncharged macromolecules which interact as hard spheres, it has been shown[3,4] that an almost complete cancellation occurs between the numerator and denominator of the bracketed quantity. For such macromolecular systems D(c) is thus to a good approximation given by the simple Stokes-Einstein relation:

$$D(c) \approx \frac{kT}{6\pi\eta R} \tag{8}$$

In the case of charged macromolecules, equation (8) is expected to be valid provided that the ionic strength is adequate to screen out the electrostatic interactions which tend to increase the value of D(c) above the "hard sphere" limit.[1,3]

From the above reasoning we can expect that at concentrations well above the CMC the interactions between the charged SDS micelles will have a much weaker effect on our $\bar{D}$ measurements than on the parameters obtained in conventional techniques such as classical light scattering sedimentation equilibrium or osmometry. For the SDS and NaCl concentrations employed in this study we estimate[1] that the effect of micellar interactions (hard sphere and electrostatic) on $\bar{D}$ will be small* and hence we may reliably deduce $\bar{R}$ using equation (3).

## EXPERIMENTAL RESULTS

As the experimental results have recently been published[1] together with a detailed account of the materials, methods and calculations employed, only a brief summary will be given here.

---

*At high NaCl concentration (>0.45 M) the effect of interactions (primarily hard sphere) should be negligible[1] (<2%) whereas at low NaCl concentration (<0.45 M) interactions (primarily electrostatic) may increase $\bar{D}$ by at most 12%.[1]

SIZE, SHAPE AND THERMODYNAMICS OF SDS MICELLES  363

The temperature dependence of the mean hydrodynamic radius, $\bar{R}$, for a fixed SDS concentration, $6.9 \times 10^{-2}$ M (2 g/dl) in various NaCl concentrations (0.15-0.6 M) is plotted in Figure 1 and the temperature dependence of $\bar{R}$ for a fixed NaCl concentration (0.6 M) in three SDS concentrations ($1.7 \times 10^{-2}$, $3.5 \times 10^{-2}$ and $6.9 \times 10^{-2}$ M) is plotted in Figure 2. Taken together both figures show that R is markedly dependent on temperature, NaCl and SDS concentration. At high temperature, $\bar{R}$ asymptotically approaches a minimum value (~25 Å), which is essentially independent of NaCl and SDS concentration. This value is quite close to the length

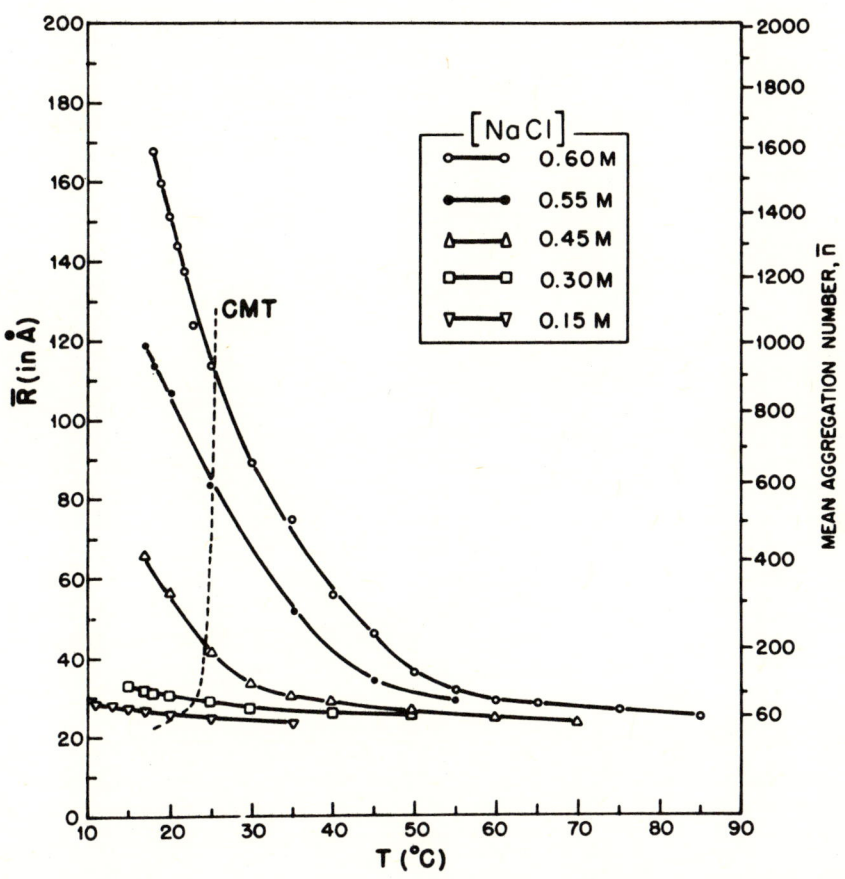

Figure 1. Temperature dependence of mean hydrodynamic radius ($\bar{R}$) for $6.9 \times 10^{-2}$ M SDS in various NaCl concentrations (0.15 M-0.6 M). Vertical axis on right provides a scale of mean aggregation numbers ($\bar{n}$).

of an extended dodecyl sulfate anion (23 Å), as measured on a Stuart-Briegleb molecular model[1], a finding which is consistent with the classical view that the minimum size of a detergent micelle corresponds to a spherical aggregate whose radius is approximately the length of the extended detergent chain[6]. For this reason we believe that at high temperatures SDS micelles possess a minimum spherical structure with an $\bar{R}$ of ~25 Å. As the

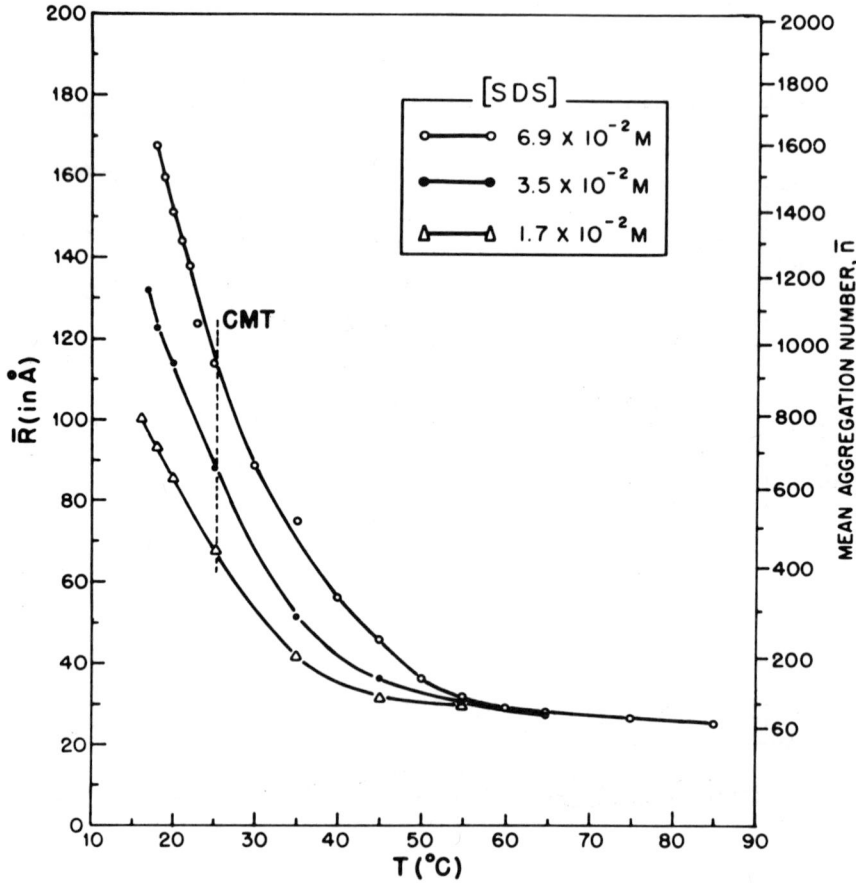

Figure 2. Temperature dependence of $\bar{R}$ in 0.6 M NaCl for three SDS concentrations. Vertical axis on right provides a scale of mean aggregation numbers ($\bar{n}$).

# SIZE, SHAPE AND THERMODYNAMICS OF SDS MICELLES

temperature is lowered (Fig. 1) $\bar{R}$ increases dramatically in NaCl concentrations greater than 0.3 M, whereas with smaller NaCl concentration, $\bar{R}$ increases slightly. The largest $\bar{R}$ value (167 Å) is attained in 0.6 M NaCl at 18°C. As shown in Figure 2 (0.6 M NaCl), the $\bar{R}$ values at low temperatures increase appreciably with successive doubling of the SDS concentration, indicating that detergent concentration itself can greatly affect micellar size. Finally, it is important to recognize that in solutions which have been supercooled below the CMT (regions to the left of the dashed lines in Figures 1 and 2), $\bar{R}$ has the same dependence on temperature, NaCl and SDS concentration as above the CMT. We shall return later to the theoretical significance of the CMT and its relation to micellar size.

On the right hand sides of Figures 1 and 2 the vertical axes provide a scale of mean aggregation numbers ($\bar{n}$) that correspond to the respective $\bar{R}$ values assuming that the micellar shape is represented by a prolate ellipsoid with a semiminor axis of 25 Å. In order to deduce micellar shape, measurements of the temperature dependence of the mean intensity of the scattered light, $\bar{I}(T)$, were obtained for the solutions containing $6.9 \times 10^{-2}$ M SDS in 0.6 M NaCl. This intensity data is shown in Figure 3 as an intensity ratio $\bar{I}(T)/\bar{I}_{min}$ where $\bar{I}_{min}$ represents the intensity scattered by the minimum spherical micelles, measured at 85°C. The ratio $\bar{I}(T)/\bar{I}_{min}$ increases from an asymptotic value of 1 at 85°C to about 19 at 18°C.

The intensity ratio provides information on the temperature dependence of the micellar mass. If the micelles are treated as monodisperse and noninteracting, and the scattering by small molecules neglected, then the intensity ratio will be given by[1]:

$$\frac{\bar{I}(T)}{\bar{I}_{min}} = \frac{M(T)P(T)}{M_{min}} \quad (9)$$

where $M(T)$ is the micellar mass and $P(T)$ is the scattering form factor appropriate to the micelle formed at temperature T, and $M_{min}$ is the micellar mass of the minimum spherical micelle. Even when the complications of micellar interactions are taken into account, we can expect the above equation to be a good approximation since the effect of interactions on the absolute intensities $\bar{I}(T)$ and $\bar{I}_{min}$ will tend to cancel in the intensity ratio.[1]

By plotting $\bar{I}(T)/\bar{I}_{min}$ vs $\bar{R}$ measured at the corresponding temperature, we obtain a curve which offers a sensitive experimental test of micellar shape. Model calculations of the functional dependence of $\bar{I}/\bar{I}_{min}$ or $\bar{R}$ were carried out[1] for three possible shapes: prolate ellipsoid (I), the oblate ellipsoid (II), and sphere (III). In the case of the ellipsoids, it is assumed that the semiminor axis is fixed at 25 Å, i.e., the radius of the minimum spherical micelle. The spherical shape is chosen to represent a spherical "grapelike" aggregate of minimum spherical micelles. The dependence of $\bar{I}/\bar{I}_{min}$ on $\bar{R}$ for each of the three models was calculated using equation (9) (see reference 1 for details) and the curves are plotted with the experimentally derived curve in Figure 4. The experimental intensity ratios clearly increase with $\bar{R}$ in a manner close to that predicted by the prolate model. The agreement between theory and experiment is not exact but in view of the simplifying assumption made in the model calculation, the result provides compelling evidence that SDS micelles become elongated rod-like structures under the experimental conditions of low temperature, high detergent and high NaCl concentration.

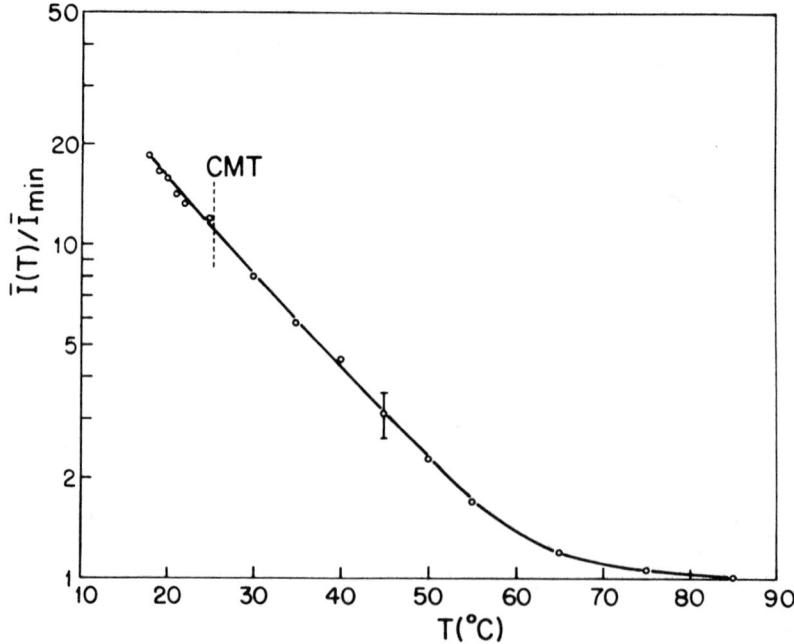

Figure 3. Temperature dependence of the intensity ratio $\bar{I}(T)/\bar{I}_{min}$ for $6.9 \times 10^{-2}$ M SDS in 0.6 M NaCl. Dashed vertical line represents the critical micellar temperature (CMT). Brackets indicate the maximum uncertainty of ±15% in all points.

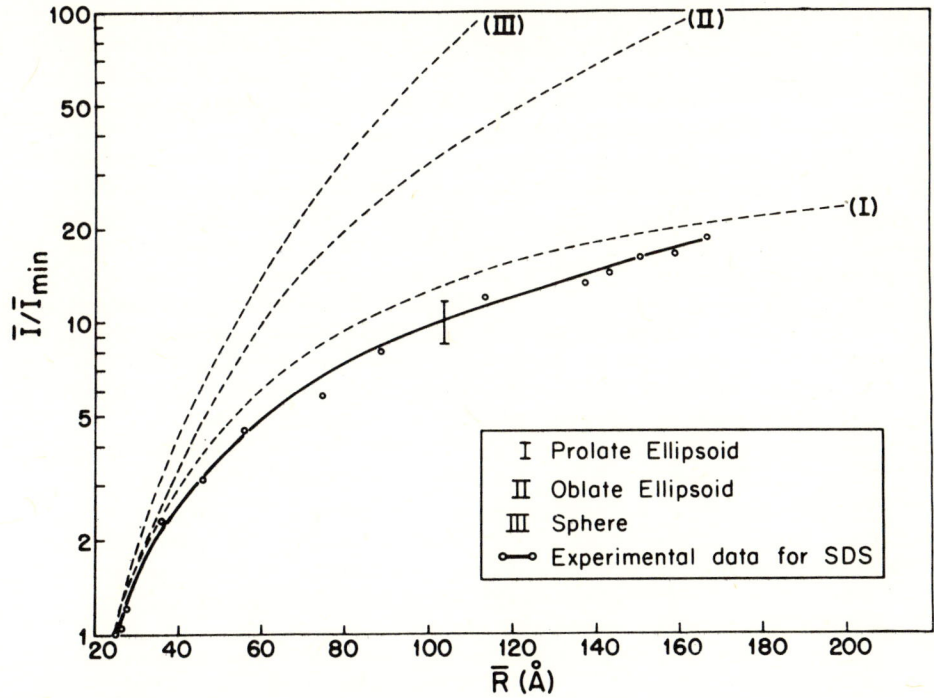

Figure 4. Intensity ratio $\bar{I}/\bar{I}$ min as a function of $\bar{R}$. Dashed curves represent model calculations. Circles represent experimental dependence obtained from Figure 2 and 3. Bracket indicates the maximum uncertainty (±15%) in all data points.

The deduction of micellar shape enables us to estimate the mean aggregation numbers $\bar{n}$, corresponding to different values of $\bar{R}$. Making the reasonable assumption that the partial specific volume of the SDS micelle is independent of micellar size, it can be shown that for the prolate shape (model I), $\bar{n}$ will be given by[1]:

$$\bar{n} = n_{min} \left( \frac{a_I(\bar{R})}{25} \right) \quad (10)$$

where $n_{min}$ is the aggregation number of the minimum spherical micelle, and $a_I(\bar{R})$ is the semimajor axis (in Å) of a prolate micelle with a hydrodynamic radius $\bar{R}$ and a semiminor axis of 25 Å. We estimate the value of $n_{min}$ to be 60 which corresponds closely to the aggregation number measured for SDS micelles in water at the CMC. Using equation 10, we have constructed the "$\bar{n}$ - axis" on the right hand sides of Figures 1 and 2. These axes show that for $6.9 \times 10^{-2}$ M SDS in 0.6 M NaCl, $\bar{n}$ undergoes an extraordinary increase from 60 to 1600 as the temperature decreases from 85° to 18°C. In figure 5 we compare the NaCl dependence of $\bar{n}$ at 25°C with the aggregation numbers of Mysels and Princen[7] obtained at the CMC by conventional light scattering. The figure shows that for NaCl concentrations $\leq 0.3$ M the $\bar{n}$ value at $6.9 \times 10^{-2}$ M is comparable to the value measured at the CMC. However, above 0.3 M NaCl the value of $\bar{n}$ undergoes an increase with NaCl concentration that is far more dramatic than the increase seen at the CMC. Thus in high NaCl concentrations $\bar{n}$ is markedly dependent on the SDS concentration. In this regard our three $\bar{n}$ values at 0.6 M NaCl show that with successive doubling of the SDS concentration $\bar{n}$ increases by a factor of about 1.5.

With regard to the polydispersity observed with these SDS solutions, the typical value of V, (see equation 2), was 35% and did not vary appreciably with temperature, NaCl or SDS concentration. This V value probably overestimates the true micellar polydisperisty for solutions containing the very small micelles (i.e., at high temperatures, low NaCl), as the V values measured in such weakly scattering solutions are sensitive to the slightest amount of microscopic impurity[1]. On the other hand, for solutions containing the large rod-shaped micelles (i.e., at low temperature and high NaCl), the V value of 35% is meaningful and does imply a significant degree of micellar polydispersity. We have estimated[1] from the V value that the distribution of aggregation numbers extends 70% above and below the weight averaged aggregation number.

## THERMODYNAMIC ANALYSIS

We begin by outlining a thermodynamic model of micelle formation which offers a simple physical interpretation of the results described in the previous section. Let us assume that SDS molecules can form aggregates no smaller than the minimum spherical micelle (aggregation number $n_o^*$) and are capable of forming larger micelles with aggregation numbers $n_o + 1$, $n_o + 2$, ... etc. The larger aggregates are modeled here as prolate spherocylinders, a shape consistent with our previous experimental deductions which will be seen to provide a simple conceptual basis for understanding the thermodynamics of micellar growth. For a micelle with aggregation number $n_o + k$, the micellar structure (depicted in Figure 6) consists of k molecules in a cylindrical portion which is capped by hemispheres each containing $n_o/2$ molecules. If we assume that

---

*$n_o$ is taken to be 60 for the SDS micelle.

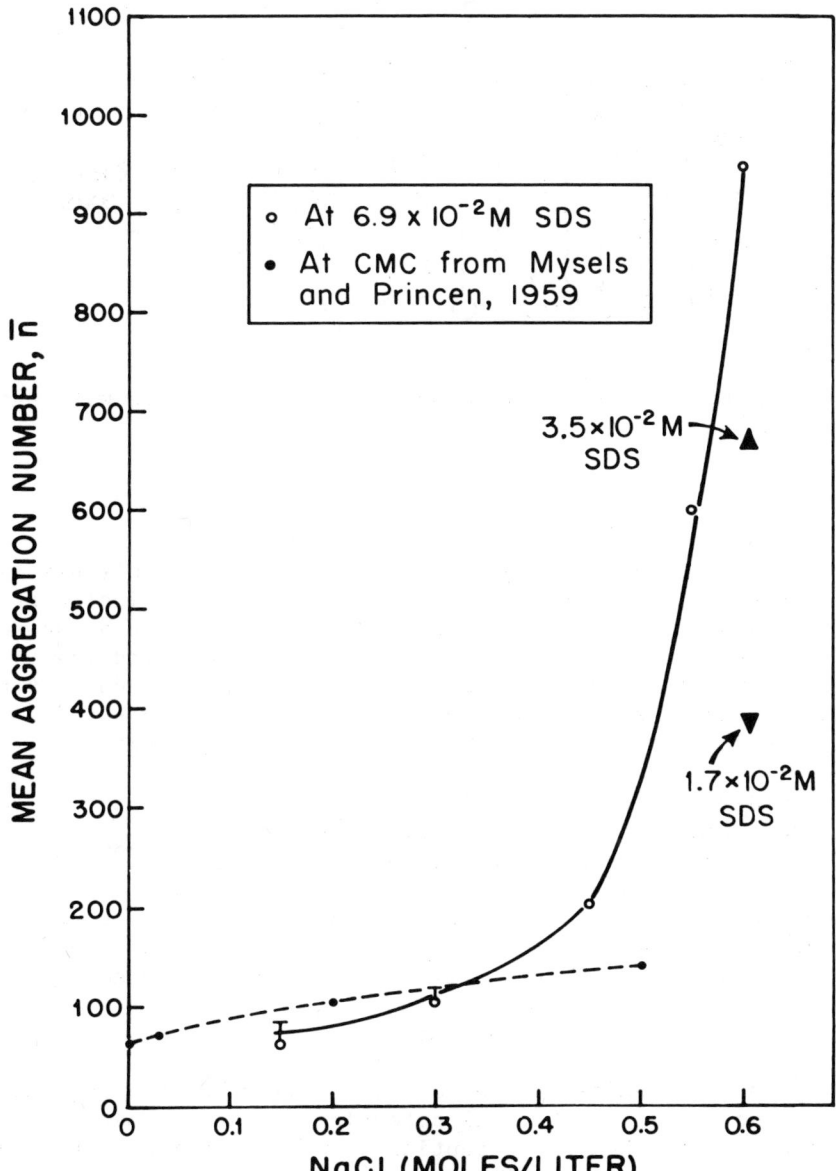

Figure 5. Mean aggregation number ($\bar{n}$) as a function of NaCl concentration at 25°C. Continuous curve represents the results from this study. The range of values represented by $\bar{o}$ at low NaCl concentrations indicate uncertainties due to electrostatic interactions.

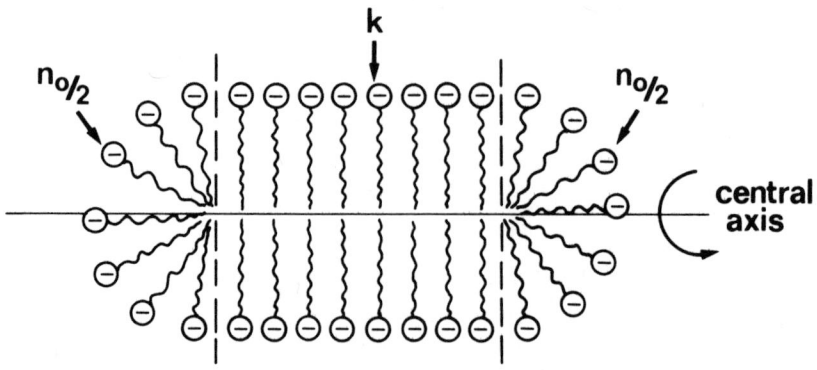

Figure 6. Structure of prolate spherocylinder micelle with aggregation number $n_o + k$.

the volume occupied by the hydrocarbon tail of each molecule is the same whether the molecules are located in the cylindrical portion or in the spherical portion of the micelle then it can be shown that the area per molecule at the surface of the cylindrical portion will be less than or equal to 2/3 of the area per molecule at the surface of the spherical portion. This difference in surface area plays a central role in the following thermodynamic analysis.

Consider an aqueous solution consisting of detergent molecules with a mole fraction X and with NaCl concentration, $C_{NaCl}$, in chemical equilibrium at temperature T, where T is assumed to be greater than the CMT. We wish to determine the mole fraction of monomers $X_1$, and the mole fraction of micelles in the different states of aggregation $\{X_n | n \geq n_o\}$ which exist at equilibrium. Stigter[8] has shown that monomer-micellar equilibrium is achieved when:

$$\mu_n = n \mu_1 \quad (n \geq n_o) \tag{11}$$

where $\mu_n$ is the chemical potential of the micelle with aggregation number n and $\mu_1$ is the chemical potential of the monomers. Using the dependence of chemical potential on mole fraction appropriate for dilute systems (i.e. mole fraction << 1) we obtain

$$\mu_n^o + kT \ln X_n = n(\mu_1^o + kT \ln X_1) \tag{12}$$
$$(n \geq n_o)$$

where the superscript (o) indicates the standard chemical potential.

# SIZE, SHAPE AND THERMODYNAMICS OF SDS MICELLES

Conservation of matter provides the following constraint on the mole fractions.

$$X = X_1 + \Sigma \, n X_n \qquad (13)$$

$$n \geq n_o$$

To complete the model it is essential to specify the dependence of $\mu_n^o$ on n. We assume this dependence is given by:

$$\mu_n^o = \mu_{n_o}^o + (n-n_o)\,\mu^o \quad (n \geq n_o) \qquad (14)$$

where $\mu_{n_o}^o$ is standard chemical potential associated with the two hemispherical portions of the micelle (equivalently the standard chemical potential of the minimum spherical micelle) and $\mu^o$ is the standard chemical potential associated with each detergent molecule in the cylindrical portion of the micelle which contains $(n-n_o)$ molecules. The linear dependence of $\mu_n^o$ on $(n-n_o)$ presumes that the detergent molecules within the cylinder are experiencing only local interactions among themselves and with the surrounding solvent molecules. Murkerjee[9] has previously developed a model of micelle formation in which $\mu_n^o$ increases linearly with n; however, his model assumes that micelle formation begins with dimerization. In view of our experimental observation of a minimum spherical micelle, we feel that the present model*
is more appropriate for quantitatively analyzing our data.

The model is analyzed as follows. From equations (12) and (14) we can relate $X_n$ to $X_1$:

$$X_n = (BX_1)^n/K \qquad n \geq n_o \qquad (15)$$

where B is given by:

$$B = \exp\{(\mu_1^o - \mu^o)/kT\} \qquad (16)$$

and K is given by:

$$K = \exp\{(\mu_{n_o}^o - n_o \mu^o)/kT\} \qquad (17)$$

Substituting equation (15) into (13) and performing the infinite sum we obtain:

---

* A similar model has been used by Tausk and Overbeek[10] to analyze the behavior of Zwitterionic detergents; di C6, C7, C8 lecithins.

$$X = X_1 + K^{-1}(BX_1)^{n_o}\left[\frac{n_o}{1-BK_1} + \frac{BX_1}{(1-BX_1)^2}\right] \quad (18)$$

Taken together equations (15–18) permit the determination of $X_1$, and the set $\{X_n \mid n \geq n_o\}$ in terms of the concentration X and the standard chemical potentials $\mu_{n_o}^o$, $\mu^o$ and $\mu_1^o$, which will be functions of T and $C_{NaCl}$. From the set of mole fractions, it is then possible to calculate using equations (1) and (3) the mean hydrodynamic radius, $\bar{R}$, that would be obtained in an actual QLS measurement*. Likewise, it is also possible to calculate the number and weight averaged aggregation numbers ($\bar{n}_N$ and $\bar{n}_W$) as well the degree of micellar polydispersity, V. When these calculations are performed, a major simplification results when the total concentration X greatly exceeds the monomer concentration $X_1$ ($\approx$ CMC), the precise conditions of our experiment. Under these conditions, $\bar{R}$ (and similarly $\bar{n}_N$ and $\bar{n}_W$) are independent of the parameter $\mu_1^o$ and are found to depend only on the product KX, where K is given by equation (17).

This important results comes about in the following way. If $X \gg X_1$ then we can neglect the $X_1$ term in equation (18) and by multiplying through by K, we obtain:

$$KX = (BX_1)^{n_o}\left[\frac{n_o}{1-BX_1} + \frac{BX_1}{(1-BX_1)^2}\right] \quad (19)$$

This equation shows that the value of $BX_1$ depends only on the quantity KX. From equation (15) and (19) we can also see that the set $\{X_n \mid n \geq n_o\}$ and consequently $\bar{R}$ will also be functions of KX. The dependence of $\bar{R}$ on KX is given in Figure 7. For values of KX between $10^{-3}$ and $10^3$, $\bar{R}$ remains close to 25 Å; however, as KX exceeds $10^4$, $\bar{R}$ increases dramatically. In general one can show that micellar growth will be appreciable only when $KX > n_o^2$. In this limit $\bar{n}_w \approx 2\sqrt{KX}$, a result which is similar to one that Mukerjee has obtained [9].

Our model of micelle formation thus predicts that the extent of micellar growth critically depends on the magnitudes of the thermodynamic parameter K and the detergent concentration X in terms of their product. From equation (17) we see that K can be given a simple physical interpretation. It is a measure of the difference in free energy between forming a minimum spherical micelle from $n_o$ molecules ($\mu_{n_o}$) and assembling the same number of molecules into the cylindrical portion of a micelle ($n_o\mu^o$). We

---

*This calculation requires specifying the diffusion coefficients, $D_i$, of each micellar species. These were determined from Perrin's equation for prolate ellipsoids[11]. The $G_i$ values that were used to calculate $\bar{R}$, incorporated the scattering form factors appropriate for rodlike particles.[12]

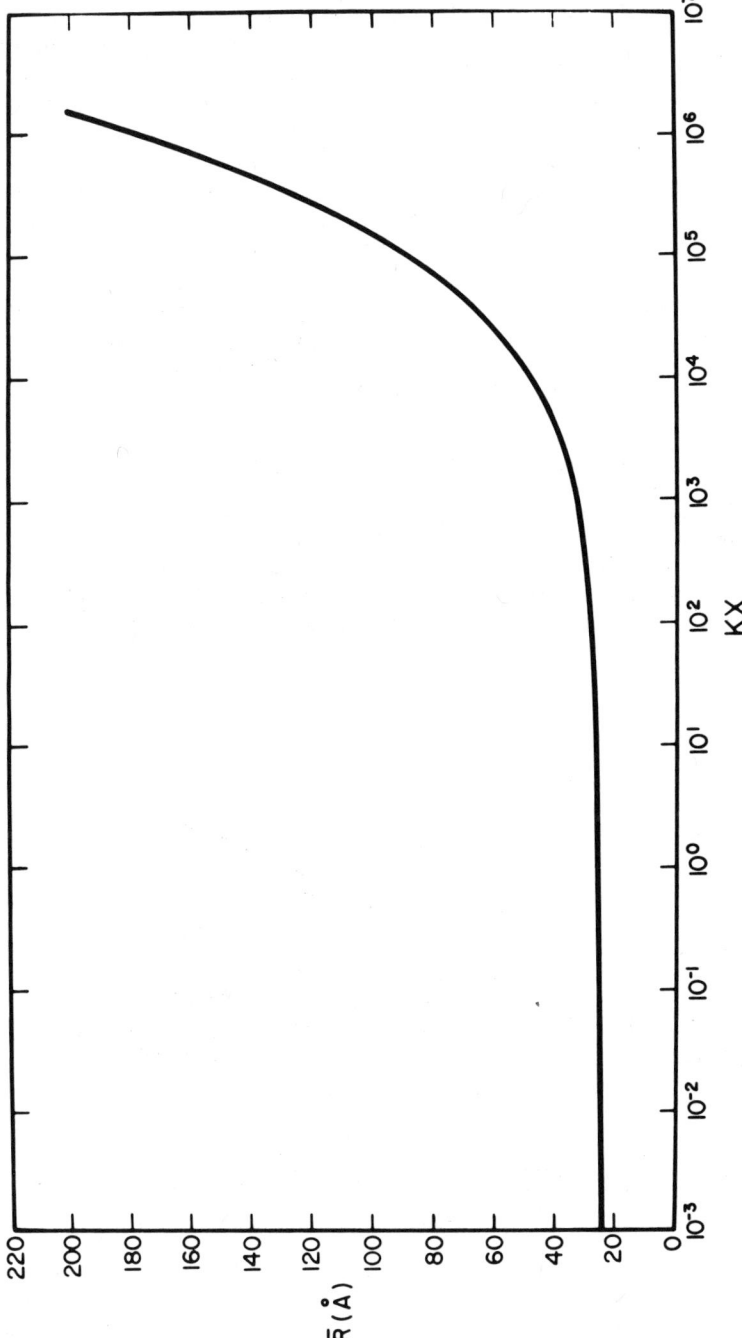

Figure 7. The theoretical dependence of $\bar{R}$ on $KX$ (see text for derivation).

shall henceforth refer to this cylindrical structure as the "equivalent micellar cylinder." If the free energy to form the "cylinder" is much less than the free energy to form the sphere then K will be large and micellar growth will be favored. Conversely if the "equivalent cylinder" requires more free energy than the sphere then K will be small and micellar growth will be negligible. The observed influences of temperature and NaCl concentration on the $\bar{R}$ values of SDS micelles are the result of the changes in K produced by these variables. We shall now assess the numerical magnitudes of these changes.

Using the theoretical curve in Figure 7 and the experimental data of Figure 1 we have deduced the temperature and NaCl dependence of the parameter K for the SDS system. We find that ln K can be represented by the following functional form:

$$\ln K(T, C_{NaCl}) = \alpha/T + \beta(C_{NaCl}) \qquad (20)$$

where $\alpha = 17110°K$ and $\beta(C_{NaCl})$ is plotted in Figure 8. The functional form given in equation (20) implies that only 6 parameters ($\alpha$ and the 5 values of $\beta$) should be needed to adequately fit the 49 $\bar{R}$ values of Figure 1. The agreement between theory and experiment is quite satisfactory as seen in Figure 9.

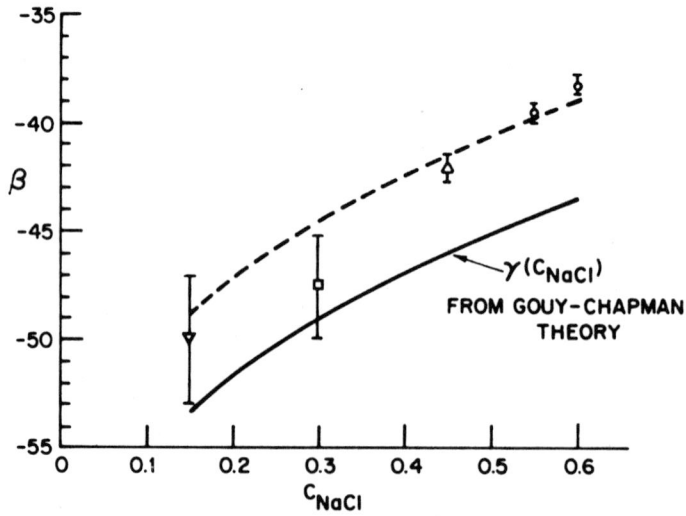

Figure 8. The dependence of $\beta$ (see equation 20) and $\gamma$ (see equation 23) on NaCl concentration [M]. Brackets represent uncertainty in the deduced values of $\beta$. Dashed curve shows that an upward displacement of the $\gamma$ curve by 4.5 units provides a reasonable fit to the $\beta$ values.

# SIZE, SHAPE AND THERMODYNAMICS OF SDS MICELLES

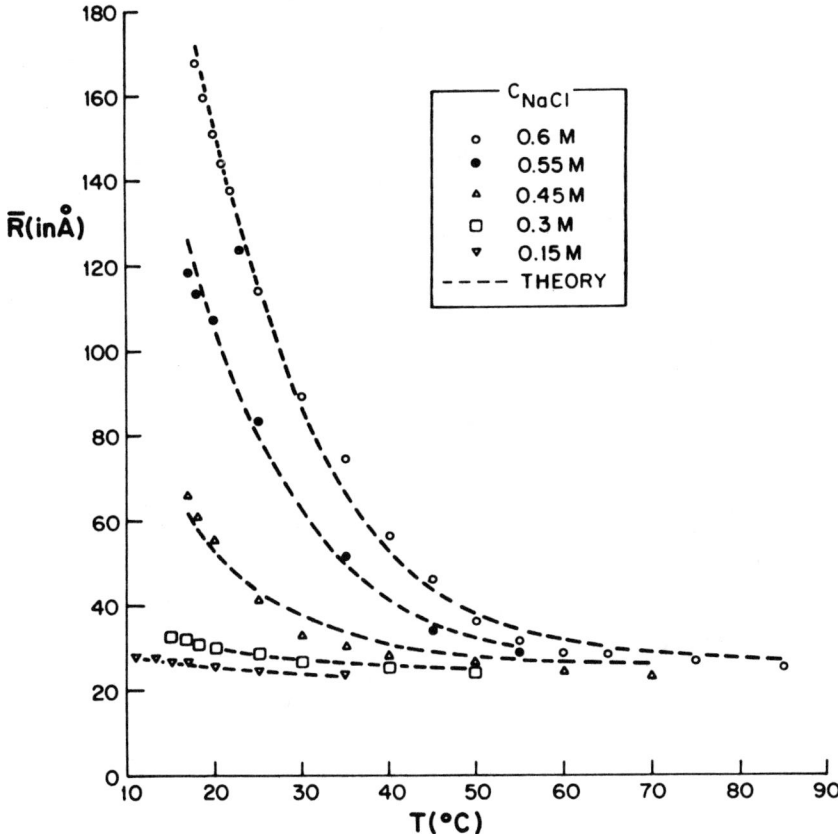

Figure 9: Fit of theory to experimental data of Figure 1 to give $\bar{R}$ vs $T(^{\circ}C)$.

The thermodynamic significance of the parameters $\alpha$ and $\beta$ can be interpreted using a simple theory for the dependence of $\ln K$ on $C_{NaCl}$ and temperature. We decompose $\mu_{n_0}^o - n_o \mu^o$ into two terms:

$$\mu_{n_0}^o - n_o \mu^o = (\mu_{n_0}^{el} - n_o \mu_o^{el}) + (\bar{\mu}_{n_0} - n_o \bar{\mu}_o) \quad (21)$$

The first term, $\mu_{n_0}^{el} - n_o \mu_o^{el}$, represents the difference in the chemical potentials of the electric double layers which surround the minimum spherical micelle and the equivalent micellar cylinder, and is a function of both $C_{NaCl}$ and temperature. The second term, $\bar{\mu}_{n_0} - n_o \bar{\mu}$, represents the difference in chemical potential associated with the non-electrical (i.e., hydrophobic, van der Waals, etc.) interactions of the minimum sphere and equivalent cylinder, and is expected to depend on temperature but not on

$C_{NaCl}$. We have used the Guoy-Chapman (G-C) theory[13] of the flat electric double layer to calculate $\mu_{n_0}^{el} - n_0\mu^{el}$ as a function of $C_{NaCl}$ and T. According to the G-C theory, the free energy per unit area of double layer in an aqueous NaCl system, $F^{el}$, is given by:

$$F^{el} = \frac{2kT\sigma}{e} \left[ \ln(S + \sqrt{S^2+1}) + \frac{1}{S} - \sqrt{1 + \frac{S^2}{S}} \right] \quad (22)$$

where

$$S = \sigma \sqrt{\frac{\pi}{2C_{NaCl}\varepsilon(T)kT}}$$

$\sigma$ represents the surface charge per unit area, $C_{NaCl}$ is the NaCl concentration in the bulk solution, $\varepsilon(T)$ is the dielectric constant of water as a function of temperature and e is the electron charge. As the parameter S is nearly independent of T, over the temperature range of our experiments it follows from equation (22) that $F^{el}$ will be proportional to kT. This implies that the dependence of $\mu_{n_0}^{el} - n_0\mu^{el}$ on $C_{NaCl}$ and T, will have the following form:

$$\mu_{n_0}^{el} - n_0\mu^{el} = kT\gamma(C_{NaCl}) \quad (23)$$

Using equation (22), we have calculated $\gamma(C_{NaCl})$ assuming that the detergent molecules are fully ionized and choosing reasonable estimates of the area per detergent head group on the surface of the minimum spherical micelle (100 $\text{Å}^2$) and the equivalent micellar cylinder (57.6 $\text{Å}^2$). The $\gamma$ values plotted in Figure 8 are negative but decrease in magnitude as $C_{NaCl}$ increases.

For the non-electric term, $\bar{\mu}_{n_0} - n_0\bar{\mu}$, we assume the following form:

$$\bar{\mu}_{n_0} - n_0\bar{\mu} = (\bar{H}_{n_0} - n_0\bar{H}) - T(\bar{S}_{n_0} - n_0\bar{S}) \quad (24)$$

where $\bar{H}_{n_0} - n_0\bar{H}$ and $\bar{S}_{n_0} - n_0\bar{S}$ represent respectively the differences in enthalpy and entropy associated with the non-electrical interactions of the minimum sphere and cylinder. Combining equations (17, 21, 23 and 24) we obtain the functional dependence of lnK on $C_{NaCl}$ and temperature which is predicted by our theoretical analysis.

$$\ln K = (\bar{H}_{n_0} - n_0\bar{H})/kT + \gamma(C_{NaCl}) - (\bar{S}_{n_0} - n_0\bar{S})/k \quad (25)$$

From equation (25) we see that only three parameters — an enthalpy difference, an entropy difference and the electrical double layer parameter ($\gamma$) are needed to account for the effect of temperature and $C_{NaCl}$ on K. As $\gamma$ has been explicitly calculated

SIZE, SHAPE AND THERMODYNAMICS OF SDS MICELLES

from the Gouy-Chapman theory, only 2 unknown parameters remain to characterize the SDS system and these can be readily deduced. Notice that equation (25) has the identical form as the empirical result given in equation (20). By comparing the two equations we find that $\alpha$ and $\beta$ correspond to:

$$\alpha = (\bar{H}_{n_o} - n_o\bar{H})/k \qquad (26)$$

$$\beta (C_{NaCl}) = \gamma (C_{NaCl}) - (\bar{S}_{n_o} - n_o\bar{S})/k \qquad (27)$$

Thus from our value of $\alpha$ we deduce that $H_n - n_o\bar{H}$ is equal to 34 Kcal/mole. In Figure 8 we see that the dependence of $\beta$ on $C_{NaCl}$ nearly parallels the dependence of $\gamma$ on $C_{NaCl}$, a result which is consistent with equation (27). From the displacement of the two curves in Figure 8, the entropy difference $(\bar{S}_{n_o} - n_o\bar{S})$ can be deduced and is found to be $-8.9$ cal/mole·deg. Let us now investigate whether these deduced enthalpy and entropy differences are reasonable by attributing them solely to the hydrophobic interactions experienced by the minimum spherical micelle and equivalent micellar cylinder. Empirically, it has been shown[14] that the free energy of hydrophobic interaction between a hydrocarbon structure and water molecules is proportional to the hydrocarbon-water interfacial area. On this basis the quality $\bar{\mu}_{n_o} - n_o\bar{\mu}$ reflects the difference in the hydrocarbon-water interfacial area between the minimum spherical micelle and the equivalent micellar cylinder. We estimate that the interfacial area associated with the minimum spherical micelle (measured at the surface of the hydrophobic core of the micelle, a radius of $\sim 19$ Å from the center) will be about $1.5 \times 10^3$ Å$^2$ larger than the interfacial area associated with the micellar cylinder. Dividing this difference in area into enthalpy and entropy differences given above we deduce that the enthalpy ($\Delta H$) and entropy ($\Delta S$) of hydrophobic interaction per Å$^2$ of hydrocarbon-water interface should equal:

$$\Delta H = 22.4 \text{ Cal/mole/Å}^2 \qquad (28)$$

$$\Delta S = -5.9 \times 10^{-3} \text{ Cal/mole·deg/Å}^2 \qquad (29)$$

Although $\Delta H$ and $\Delta S$ represent fundamental parameters in the theory of hydrophobic interactions, we have been unable to find any previous estimates in the literature for them. However, the total free energy of hydrophobic interaction, $\Delta G = \Delta H - T\Delta S$, has been estimated by Reynolds, et al[14] to be between 20-25 cal/mole/Å$^2$ at T=298°K. Using equations (28) and (29) we estimate $\Delta G(298°)$ to be 24.2 cal/mole/Å$^2$, in excellent agreement with the literature value. The consistency obtained here suggests that we have obtained reasonable values of $\bar{H}_{n_o} - n_o\bar{H}$ and $\bar{S}_{n_o} - n_o\bar{S}$ from our analysis and also implies that hydrophobic interactions may be the most important of the non-electrical interaction subsumed in $\bar{\mu}_{n_o} - n_o\bar{\mu}$.

Interestingly, we are forced to conclude that the enthalpy of hydrophobic interaction is quantitatively more significant than the negative entropy term.

The above analysis has been an attempt to understand the thermodynamic basis of micellar growth and to interpret the dramatic effects of temperature and NaCl concentration observed for the SDS system. We have seen that micellar growth is unfavorable from the standpoint of the electrical free energy required to assemble the micelle, whereas micellar growth is favorable from the standpoint of hydrophobic interactions.

Having completed our analysis of micellar growth let us briefly consider a final question, alluded to earlier. What is the physical significance of the CMT, its dependence on NaCl concentration and its relation to micellar size? We have previously[1] shown that the phenomenon of the CMT (i.e., the existence of a temperature above which almost any concentration of precipitated hydrated solid detergent will dissolve into a micellar solution), and its dependence on NaCl concentration can be qualitatively explained by the Murray-Hartley theory [15] of detergent solubility. According to this theory the CMT phenomenon results from the coupling of the monomer-solid equilibrium to the monomer-micelle equilibrium. The Murray-Hartley theory predicts that the CMT value will be slightly greater than the temperature at which the aqueous solubility of the detergent monomer is equal to the CMC. If $S_1(T)$ represents the monomeric solubility as a function of temperature, the CMT will be given by:

$$CMT = T_k + \delta T \qquad (30)$$

where $S_1(T_k)$ = CMC and $\delta T$ is a small temperature increment. For the SDS system both $S_1(T)$[16] and the CMC[17] have been determined as functions of the NaCl concentration. Using these functions to calculate the dependence of $T_k$ on NaCl concentration and adding to it a small temperature increment of 3.5°C ($\delta T$) we have obtained a prediction of the NaCl dependence of the CMT (plotted in Figure 10). The prediction agrees quite well with the experimental dependence and provides strong support for the Murray-Hartley theory. In addition, the Murray-Hartley theory is consistent with our observation of the continued growth of SDS micelles below the CMT. In the supercooled solutions, the concentration of detergent monomers ($\approx$ CMC) exceeds $S_1(T)$. Nevertheless as no monomers have precipitated, micelle formation will occur in exactly the same manner as above the CMT.

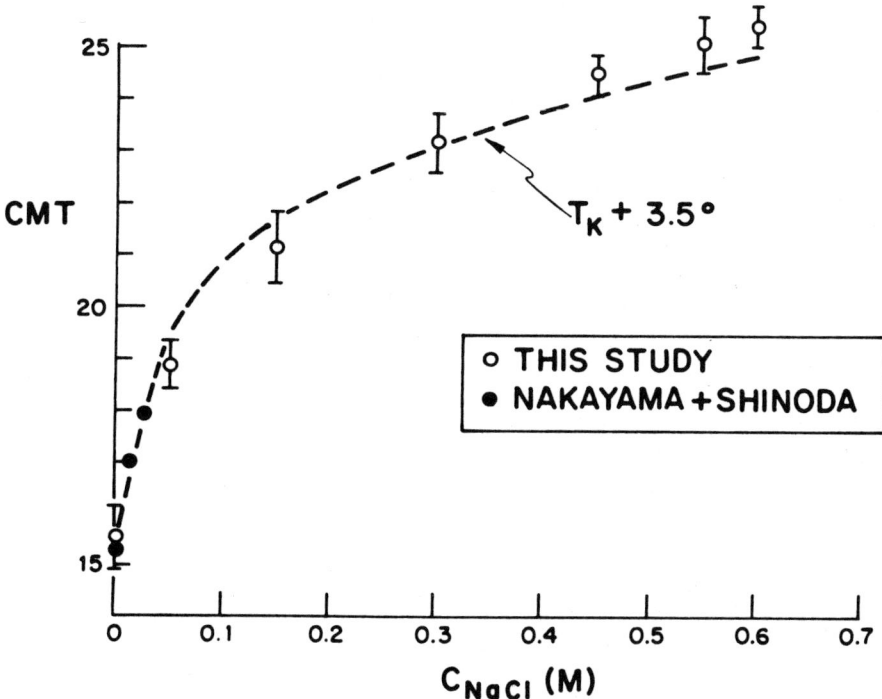

Figure 10. Critical micellar temperature (CMT) as a function of NaCl concentration. Dashed curve represents theoretical prediction based on Murray-Hartley theory. Open and closed circles represent observed CMT values together with maximum uncertainties in brackets.

From our previous thermodynamic analysis of micellar growth and the discussion of the Murray-Hartley theory, it would appear that there is no direct relationship between micellar growth and the existence of the CMT. That is, the SDS micelles do not grow in size because the temperature is approaching the CMT, but rather they grow in size because at low temperatures, the large rod-shaped micelles require less free energy to form than the spherical micelles. This conclusion is important as other authors[18] have recently suggested an analogy between the temperature dependence of micellar size near the cloud point of Triton X-100 and the behavior of the correlation range near a critical point. While this analogy may be correct for the Triton system, it does not seem to be justified for the SDS system.

## CONCLUSIONS

We have employed quasielastic light scattering spectroscopy to study the formation of SDS micelles at concentrations well above the CMC. Our investigation has demonstrated dramatic effects of temperature, NaCl concentration and SDS concentration on the size and shape of SDS micelles. We have interpreted these effects in terms of a simple thermodynamic model of micellar growth and have shown that the effect of NaCl concentration is consistent with theoretical calculations based on the Gouy-Chapman theory of the electric double layer. From the temperature dependence of micellar size we have derived estimates of the enthalpic and entropic contributions to the free energy of hydrophobic interaction. Lastly we have shown that the Murray-Hartley theory of detergent solubility can quantitatively account for the dependence of the CMT on NaCl conentration observed for the SDS system.

## ACKNOWLEDGEMENTS

Supported in part by U.S. Public Health Service Grant AM 18559 and Academic Career Development Award AM 00195 (M.C. Carey). Supported by the National Science Foundation Grant DMR 72-03027A05 to the Center for Materials Science and Engineering, M.I.T; by the National Institute of Health Interdisciplinary Program in Biomaterials Science under Grant #1P01-HL-14322-04 (G. B. Benedek and N. A. Mazer).

## REFERENCES

1. N. A. Mazer, G. B. Benedek and M. C. Carey, J. Phys. Chem. 80, 1075 (1976).
2. N. A. Clark, J. H. Lunacek and G. B. Benedek, Am. J. Phys. 38, 575 (1970).
3. P. W. Pusey, in "Photon Correlation and Light Beating Spectroscopy", H. Z. Cummins and E. R. Pike, Editors, Plenum Press, New York, N.Y., 1974.
4. G. D. J. Phillies, J. Chem. Phys. 60, 976 (1974).
5. G. D. J. Phillies, G. B. Benedek and N. A. Mazer, J. Chem. Phys. (accepted for publication).
6. G. S. Hartley, "Aqueous Solutions of Paraffin-Chain Salts", Hermann et Cie., Paris (1936).
7. K. J. Mysels and L. H. Princen, J. Phys Chem. 63, 1696 (1959).
8. D. Stigter, J. Coll. and Int. Sci. 47, 473 (1974).
9. P. Mukerjee, J. Phys. Chem. 76, 565 (1972).
10. R. J. M. Tausk and J. Th. G. Overbeek: Biophys. Chem. 2, 175 (1974).
11. B. Chu in "Laser Light Scattering", Academic Press, New York, N.Y., 1974.
12. P. Debye and E. W. Anacker, J. Phys. Colloid Chem. 55, 644 (1951).

13. E. J. W. Verwey and J. Th. G. Overbeek, "Theory of the Stability of Lyophobic Colloids", Elsevier, Amsterdam (1968).
14. J. A. Reynolds, D. B. Gilbert and C. Tanford, Proc. Nat. Acad. Sci. 71, 2925 (1974).
15. R. C. Murray and G. S. Hartley, Trans. Faraday Soc. 31, 183 (1935).
16. H. Nakayama and K. Shinoda, Bull. Chem. Soc. Japan 40, 1797 (1967).
17. R. J. Williams, J. N. Phillips and K. M. Mysels, Trans. Faraday Soc. 51, 728 (1955).
18. M. Corti and V. Degiorgio, Opt. Commun. 14, 358 (1975).

# QUASIELASTIC LIGHT SCATTERING SPECTROSCOPIC STUDIES OF AQUEOUS BILE SALT, BILE SALT-LECITHIN AND BILE SALT-LECITHIN-CHOLESTEROL SOLUTIONS

N. A. Mazer, R. F. Kwasnick, M. C. Carey
and G. B. Benedek
Peter Bent Brigham Hospital, Harvard Medical School
Boston, Massachusetts 02115 and
M.I.T., Cambridge, Massachusetts 02139

Using the technique of quasielastic light scattering spectroscopy we have investigated the micellar aggregation of two bile salts (BS) taurodeoxycholate (TDC) and taurocholate (TC) alone and with lecithin (L) and cholesterol (Ch) in aqueous solution. Our measurements of the mean hydrodynamic radii ($\bar{R}$) of simple BS micelles suggest the formation of small primary micelles ($\bar{R}$=10-15 Å) which can polymerize to form large rod-like secondary micelles ($\bar{R}$=15-60...Å). The degree of polymerization increases with the concentration of BS in a manner consistent with a stepwise addition model ($P_{n-1} + P_1 \xleftrightarrow{K} P_n$ n=2,3,...). The binding constant K for TDC is ~100 times greater than for TC and for both species K increases with added NaCl, and decreases with temperature elevation and added urea. For TDC-L and TC-L mixed micellar systems the dependence of $\bar{R}$ on L/BS molar ratio suggests the coexistence of simple and mixed micelles at low L/BS ratios, whereas at high molar ratios only mixed micelles are present. For the mixed micelles, $\bar{R}$ increases from 21 Å to greater than 75 Å, and appears to diverge as the L/BS molar ratio approaches the phase limit. The $\bar{R}$ values are consistent with a <u>mixed disc</u> micellar model in which both L and BS exist within the interior of the disc in the same molar ratio as that of the phase limit, while BS alone constitutes the disc's perimeter. The progressive addition of Ch to TC-L-Ch mixed micelles at a constant TC mole fraction of 75% leads to a slight increase in $\bar{R}$, whereas addition of Ch at a TC mole fraction of 50% causes a decrease in $\bar{R}$. Finally, in solutions that are supersaturated with Ch, we have observed the precipitation of cholesterol as microcrystals whose $\bar{R}$ values are ~1200 Å.

## INTRODUCTION

The aggregative behavior of bile salt (BS), lecithin (L) and cholesterol (Ch) in aqueous solution plays a central role in the solubility of cholesterol in bile and the formation of cholesterol gallstones.[1] To review what is known about this behavior let us first consider micellar aggregation in pure bile salt solutions, and then mixed micelle formation in bile salt-lecithin and bile salt-lecithin-cholesterol systems.

Bile salts are steroid amphiphiles which form simple* micelles in aqueous solution.[2] The size, shape and aggregation numbers of these micelles have been estimated previously (see Ref. 2 for a comprehensive review); however, in most of these studies the techniques employed have required extrapolation to the CMC in order to eliminate the sizeable effects of micellar interactions. Hence the micellar properties obtained are applicable only to concentrations in the vicinity of CMC. It is difficult therefore to infer from these studies the structure and properties of micelles formed at high BS concentrations such as those found physiologically. As conventional techniques have not facilitated a systematic investigation of the influence of BS concentration on micellar size, our present knowledge of bile salt aggregation is incomplete.

There is little experimental data on the size, shape or aggregation numbers of BS-L and BS-L-Ch mixed micelles. However, there is a body of indirect information which led Small[3] and Dervichian[4] to propose structure models for these mixed micelles. These models, while attractive, have not been subjected to adequate experimental verification.

On account of its importance to cholesterol gallstone formation the limits of cholesterol solubility in BS-L-Ch solutions have been studied in considerable detail.[5] In particular the solubility limits are known as functions of total and relative lipid compositions, temperature and ionic strength.[6,7] In the case when a BS-L-Ch solution is supersaturated with cholesterol, the excess Ch will precipitate from solution if sufficient time is allowed.[7] The occurrence of such precipitation in vivo is thought to be a prerequisite for the formation of cholesterol gallstones.[3] At present we know very

---

*Simple implies that the micelles contain no other lipid additives.

little about the size, composition and structure of these precipitates or of the rate at which they form from supersaturated solutions. Such information is vitally needed to further our understanding of cholesterol gallstone disease.

In order to obtain direct quantitative information on the size, shape and structure of simple and mixed BS micelles and precipitated Ch aggregates, we have employed quasielastic light scattering spectroscopy[8] (QLS), a technique well suited to providing this information even at concentrations well above the CMC.[9] Furthermore, under such conditions of high concentration, the variations of micellar size with temperature, ionic strength and concentration can provide important quantitative information on the thermodynamics of micelle formation.[10]

In this paper we have combined the salient features of three separate studies.[11] First we will present measurements of the mean hydrodynamic radii ($\bar{R}$) of simple BS micelles formed by two different bile salt species: taurodeoxycholate (TDC) and taurocholate (TC) together with a thermodynamic model of BS micelle formation. Second, the $\bar{R}$ values of TDC-L and TC-L mixed micelles will be given and these results will be shown to be incompatible with current models of mixed micellar structure. The new results, however, have led us to propose a new model of micellar structure.[11] Lastly we shall present physiologically important results on mixed micelle formation and cholesterol precipitation in TC-L-Ch solutions.

## EXPERIMENTAL

### Materials

The sodium salts of $3\alpha$, $7\alpha$, $12\alpha$ trihydroxy $5\beta$ cholanoyltaurine and $3\alpha$, $12\alpha$ dihydroxy $5\beta$ cholanoyltaurine, known trivially as sodium taurocholate and sodium taurodeoxycholate respectively, were A grade products (Calbiochem, San Diego, Calif., U.S.A.) and were chromatographically (t.l.c.) and potentiometrically pure. Grade I egg yolk lecithin was obtained from Lipid Products, South Nutfield, U.K., and was chromatographically pure (t.l.c.) in a number of solvent systems. NaCl and urea were of A.C.S. analytical quality. Water was doubly distilled from an all-pyrex still and filtered to exclude dust.

### Solutions

Simple BS solutions were made on a w/v basis in volumetric flasks with the appropriate aqueous solvent. Solutions of BS-L and BS-L-Ch (saturated or less than saturated with Ch) were made by dissolving the appropriate amounts of each lipid in a mutual organic solvent, drying the mixture thoroughly first under a stream of purified $N_2$ and then in vacuo (24-48 hours), followed by addition of the appropriate amount of aqueous-NaCl solvent to give the desired concentration of mixed micellar lipids (10 g/dl).[12] The mixed micellar solutions were flushed with purified $N_2$, sealed and equilibrated for two days at 4°C with gentle shaking. In order to prepare supersaturated BS-L-Ch solutions the lipids were mixed in proportions to give two phase systems at room temperature; however, prior to studying these solutions, they were heated to 85°C for 3-5 minutes in order to dissolve the precipitated Ch microcrystals and then injected into pre-warmed scattering cells. By rapidly cooling the scattering cells to lower temperatures, metastable or labile supersaturated solutions were produced.[6,7] Because the acidic pK'a values[2] of TC, TDC and the zwitterionic L are less than 2, no attempt was made to adjust the pH, which was 6.8 ±.2 for all solutions.

## QLS Methods

The apparatus and data analysis techniques used in the QLS experiments have been described by us in detail elsewhere.[9] The definitions of the experimentally measured parameters (such as the mean hydrodynamic radius $\bar{R}$, and mean scattered intensity) have been summarized in our SDS paper appearing elsewhere in this symposium.[10]

## SIMPLE BS MICELLAR SOLUTIONS

### Experimental Results

In Figure 1, the $\bar{R}$ values for TDC and TC micelles as a function of temperature in 0.15 and 0.6 M NaCl are shown. The BS concentration in these experiments was 10 g/dl, the typical total lipid concentration of gallbladder bile.[2] For TDC micelles, increases in temperature cause an appreciable decrease in micellar size whereas increases in NaCl concentration result in significant micellar growth. The TC micelles, on the other hand, are much smaller ($\bar{R} \simeq$ 10 Å in 0.15 M NaCl) and show only a weak dependence on temperature and NaCl concentration. However, when TC micelles are investigated in the presence of high NaCl concentrations (3 molar) the $\bar{R}$ values (data not shown) grow to 38 Å at 20°C and decrease to 31 Å when the temperature is increased to 60°C.

In Figure 2, the effect of BS concentration on micellar size is illustrated for TDC solutions (20-60°C) and TC solutions (20°C) in 0.6 M NaCl. In the case of TDC, $\bar{R}$ decreases dramatically with dilution, whereas the $\bar{R}$ values for TC decrease only slightly. We have found that my extrapolating the concentration dependence of $\bar{R}$ measured at a fixed temperature to zero BS concentration (dashed curves in Figure 2), a minimum value for $\bar{R}$ of ~15 Å is obtained which is essentially independent of temperature and BS species. We estimate from the molecular dimensions of the BS monomers[2] that this minimum micelle could contain between 4 and 10 BS molecules.

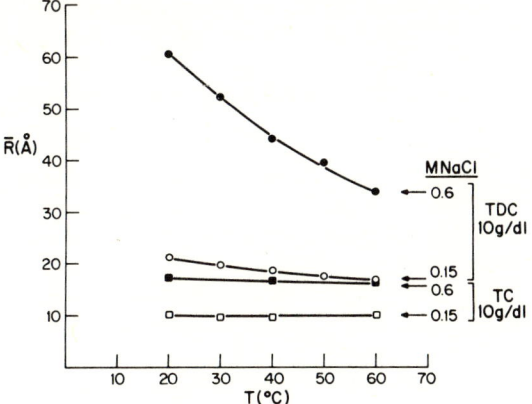

Figure 1. $\bar{R}$ values for TDC and TC micelles at 10 g/dl in 0.15 and 0.6 M NaCl as a function of temperature.

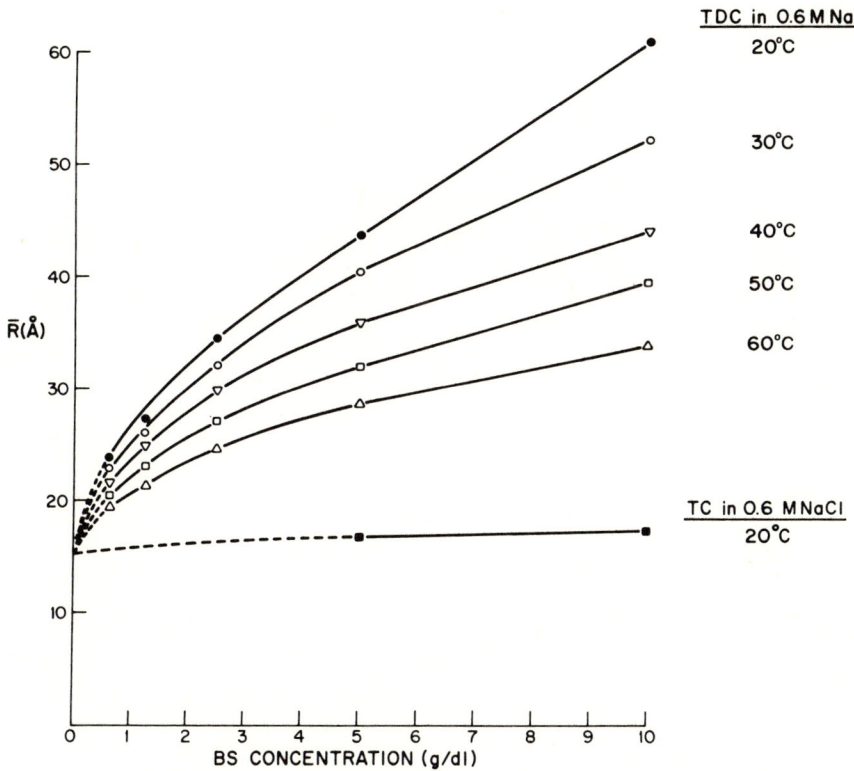

Figure 2. $\bar{R}$ values for TDC and TC micelles in 0.6 M NaCl as a function of BS concentration.

The dependence of $\bar{R}$ on urea concentration for 10 g/dl TDC and TC solutions in 0.6 M NaCl at 20°C is shown in Figure 3. Urea, which modifies water-solute interactions [13], causes a large decrease in the size of TDC micelles, but has little effect on the size of the small TC micelles.

Information on the shape of the BS micelles was obtained using measurements of the temperature dependence of the mean light intensity scattered from TDC solutions in 0.6 M NaCl in conjuction with the $\bar{R}$ values. Our analysis of micellar shape is analogous to that given in our work on SDS micelles[10]. The intensity data will be presented in a future paper, however, our conclusion is that the large BS micelles have a rod-like shape which can be modeled as a prolate elliposid with a semi-minor axis of 15 Å (the size of the minimum BS micelle in 0.6 M NaCl) and a semi-major axis that varies from 15 Å to greater than 200 Å depending on the temperature, BS species and BS concentration.

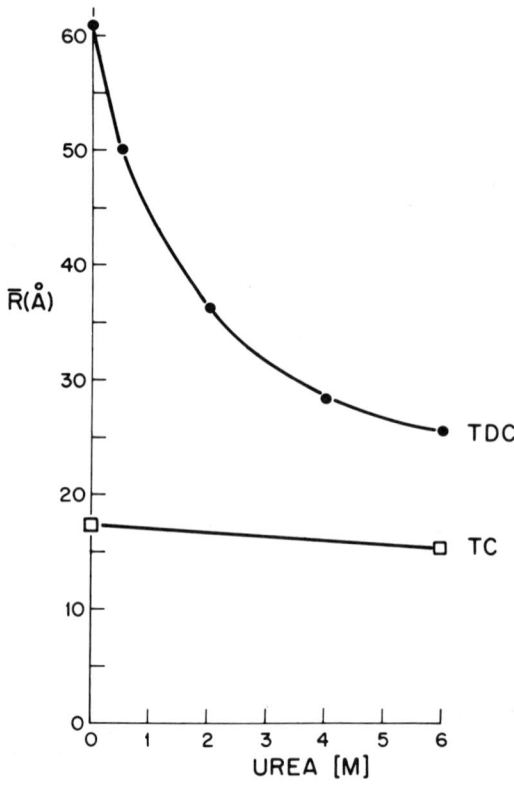

Figure 3. $\bar{R}$ values for TDC and TC micelles (10 g/dl, 20°C, 0.6 M NaCl) as a function of added urea concentration.

## Interpretation of Data on Simple BS Micelle Formation

An attractive model that has been presented to account for simple BS micelle formation is the primary-secondary micelle hypothesis[2]. This hypothesis proposes that hydrophobic interactions induce the BS molecules to form aggregates containing between 2 and 10 molecules called primary micelles. The structure of the primary micelle is such that the hydrocarbon backs of the constituent BS molecules face the interior and the OH groups of the BS molecules are located on the surface of the micelles. It is then suggested that these OH groups, by hydrogen bonding*, permit the self-association of the primary micelles into polymers called secondary micelles. Our deduction of micellar shape indicates that these secondary micelles are linear, rod-shaped polymers.

It is reasonable to suppose that in aqueous solution, the primary micelles are in dynamic equilibrium with the secondary micelles. Our $\bar{R}$ measurements thus reflect the mean micellar size of the distribution of primary and secondary micelles, which we have seen depends on BS species, BS concentration, temperature, NaCl concentration, and urea. We will now interpret these dependences in terms of a simple quantitative model of primary-secondary micelle equilibrium.

Let us consider the primary micelles as "monomers", (not to be confused with the bile salt monomers) denoted by $P_1$, and the secondary micelles, as polymers, denoted by $P_n$ where $n=2,3,4,...$ represents the number of "monomers" in the polymer.

We describe the stepwise growth of the secondary micelles by the chemical equilibria $P_{n-1} + P_1 \xrightarrow{K} P_n$  $n=2,3,...$ where K is the binding constant for the polymerization reaction expressed in inverse concentration units. Letting $C_n$ represent the molar concentration of the n-mer, (i.e. polymer with n "monomer" units) the law of mass action yields the following equation:

$$C_n = K C_{n-1} C_1 \qquad (1)$$

---

*We wish to emphasize that no attempts have been made to directly verify the proposed hydrogen bonding between primary BS micelles by conventional methods[14] appropriate for the detection of intermicellar hydrogen bonds in aqueous systems. Furthermore, it is difficult to explain why intermicellar hydrogen bonds should be favored over hydrogen bonds between micelles and the aqueous solvent. For these reasons we do not wish to exclude the possibility that other types of interactions (i.e. hydrophobic [15]) may play a significant if not predominant role in the formation of secondary micelles.

From this equation it follows that $C_n$ will be given by:

$$C_n = C_1 (KC_1)^{n-1} \tag{2}$$

If $C_o$ represents the total concentration of BS in the system expressed as an equivalent concentration of primary micelles (moles/liter) and we assume that the BS concentration is much greater than the CMC[2] (a valid assumption for our experiments), then the conservation of mass implies:

$$C_o = \sum_{n=1}^{\infty} nC_n = \sum_{n=1}^{\infty} nC_1 (KC_1)^{n-1} \tag{3}$$

Equation (3) enables us to calculate $C_1$ from $C_o$ and $K$. Knowing $C_1$ and $K$, the concentration $C_n$, of each species can be obtained from Equation (2). Given these concentrations we can then calculate the $\bar{R}$ value that would be obtained for such a system of linear polymers in a QLS experiment (see ref. 10 for relevant equations). The results of these calculations is that $\bar{R}$ depends solely on the dimensionless product $KC_o$ as shown in Figure 4*.

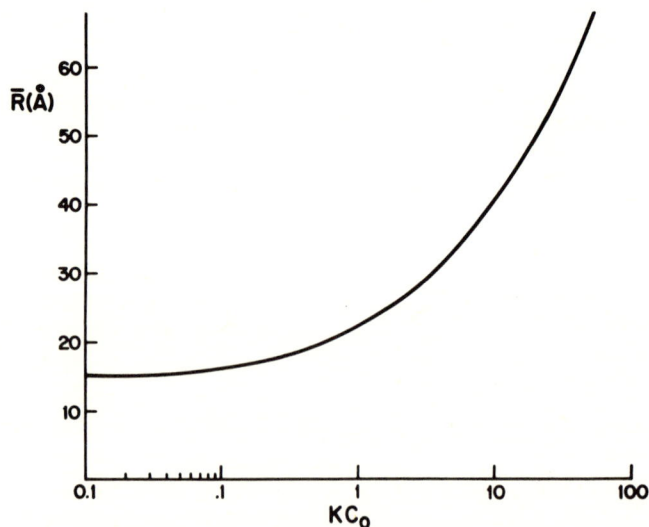

Figure 4. Theoretical dependence of $\bar{R}$ as a function of $KC_o$ (see text for derivation).

---

*In generating the values of $\bar{R}$ we have assumed for the purpose of calculating the hydrodynamic radius of each micellar species that the primary micelle is a sphere of radius 15 Å and that the secondary micelle containing n "monomer" units is represented by a prolate ellipsoid with a semi-minor axis of 15 Å and a volume equal to n times the volume of a primary micelle.

Our theory can account for the concentration dependence of $\bar{R}$ observed for TDC micelles in 0.6 M NaCl at 20°, 30°, 40°, 50° and 60°C. To show this, we have used the theoretical curve in Figure 4 to fit the observed concentration dependence at each temperature. The fit requires finding a value of K for each temperature that allows the best agreement between theory and experiment. The excellent agreement shown in Figure 5 provides compelling support for our model of primary-secondary micelle equilibrium. In the Table we list the value of K (in units of liters/mole of primary micelles) found for each temperature. To express K in these units we have assumed that the number of BS molecules in a primary micelle is 7 in order to calculate the value of $C_o$ from the BS concentration (in g/dl). Variations in this estimate (i.e. 4-10 BS molecules/primary micelle) will affect the absolute magnitude of K, but will not affect our deductions based on the temperature dependence of K.

TABLE: Temperature dependence of K values for TDC Micelles in 0.6 M NaCl

| T(°C) | 20° | 30° | 40° | 50° | 60° | |
|---|---|---|---|---|---|---|
| K | 910 | 680 | 490 | 340 | 220 | (liters/mole of primary micelles) |

By plotting ln K vs $T^{-1}$ we have deduced the enthalpy of bonding ($\Delta H$) between primary micelles[16]. This plot was approximately linear, with a slope that yielded a $\Delta H$ of -6.9 kCal/mole. This value is of the order of magnitude of typical hydrogen-bonds[14] but is also consistent with our estimates of the enthalpy change involved in a purely hydrophobic association of the primary micelles (4-12 kCal/mole)*. Our observation that added urea decreases mean micellar size is also difficult to interpret unequivocally as urea is thought to interfere with hydrophobic interactions[13] as well as hydrogen bonding[2]. Our present data thus do not permit us to resolve the precise nature of the interactions responsible for secondary micelle formation.

In summary, the essential result of our model is that $\bar{R}$ depends only on the dimensionless product $KC_o$. BS concentration affects the equilibrium between primary and secondary micelles by determining the value of $C_o$, whereas the parameters of bile salt species, temperature, NaCl concentration and urea concentration affect the value of K. With regard to the effect of bile salt species on K we conclude from our data that the value of K for TDC micelles is ~100 times greater than

---

*These estimates were calculated assuming that an enthalpy loss of 22.4 cal/mole occurs when a $Å^2$ of hydrocarbon looses contact with water (see ref. 10). In the case of the self-association of primary BS micelles we assume that between 5-15% of the total hydrophobic component of the external surface of each micelle is eliminated from water contact when polymerization occurs.

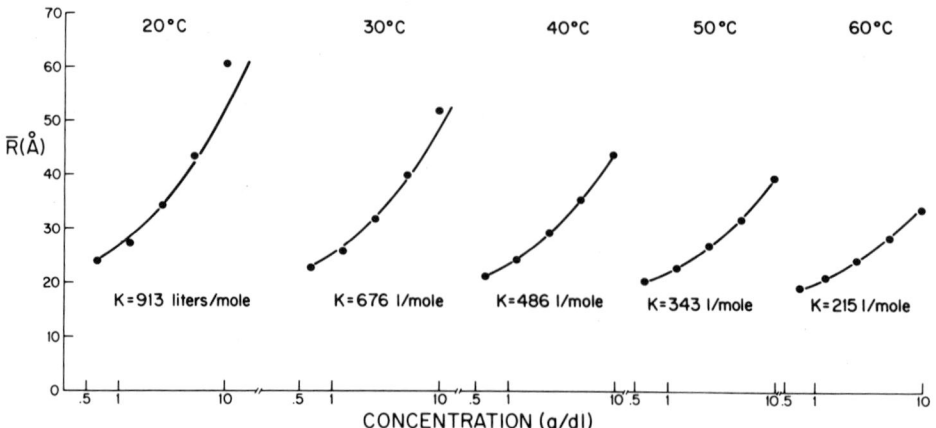

Figure 5. The fit of theoretical concentration dependence of $\bar{R}$ (solid curve) to the experimental data for TDC micelles in 0.6 M NaCl at 20, 30, 40, 50 and 60°C.

the value for TC micelles in 0.6 M NaCl at 20°C. We are currently investigating the physical basis for this great difference in K between bile salt species.

## BS-L MIXED MICELLES

### Experimental Results

The $\bar{R}$ values of 10 g/dl TDC-L and TC-L solutions in 0.15 M NaCl at 20°C are plotted in Figure 6 as a function of the L/BS molar ratio. For L/TDC ratios between 0 and 0.45 and L/TC ratios between 0 and 0.65 the two systems differ markedly. We have both experimental[17] and theoretical evidence[18] which suggests that at these low L/BS ratios simple and mixed micelles coexist in solution, and we are conducting further experiments in order to clarify this interpretation. For L/TDC ratios between 0.45 and the micellar phase limit of ∼1.5* and for L/TC ratios between 0.65 and the phase limit of ∼2* the two systems behave similarly. We believe that only mixed micelles exist at these L/BS ratios. The $\bar{R}$ values of the mixed micelles are observed to increase dramatically with increases in the L/BS ratio and appear to diverge as the phase limit is approached.

---

*These values were determined macroscopically from 10 g/dl BS-L solutions in 0.15 M NaCl at 20°C. In general the phase limit depends on lipid concentration, temperature, ionic strength and bile salt species[6,7].

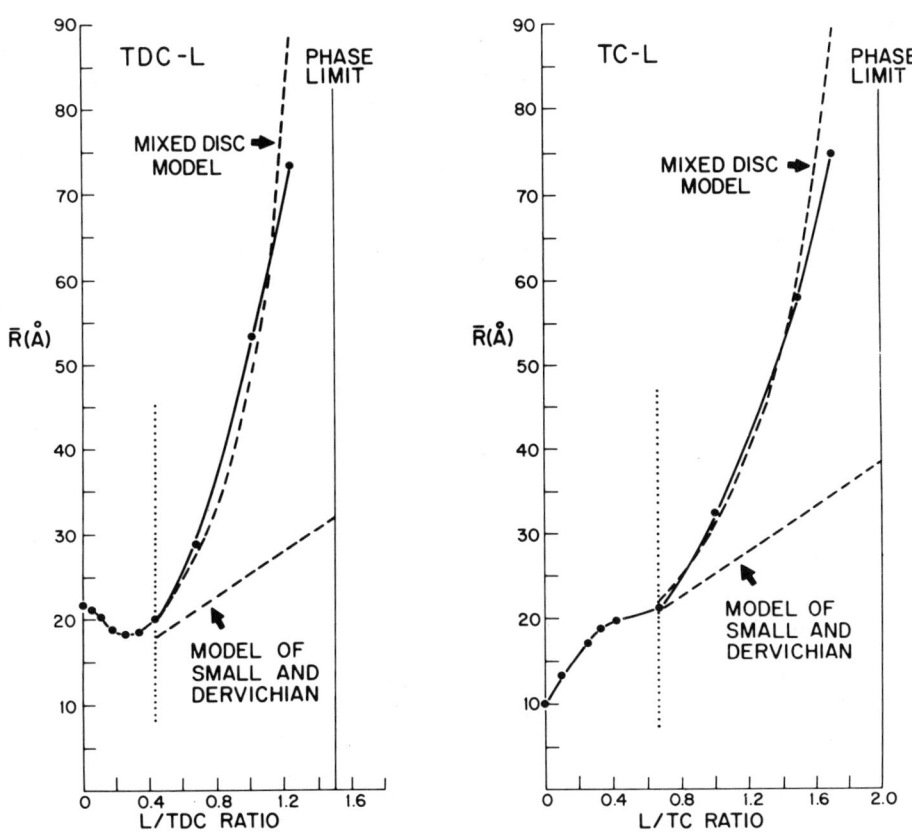

Figure 6. $\bar{R}$ values for TDC-L and TC-L solutions (10 g/dl, 20°C, 0.15 M NaCl) as a function of L/BS molar ratio. Dotted vertical lines separate regions where simple and mixed micelles coexist (to the left) from regions where only mixed micelles are present (to the right). Dashed oblique lines represent theoretical predictions of $\bar{R}$ based on the Small-Dervichian model. Dashed curves represent the fitting of our "mixed disc" model to the experimental data. Solid vertical lines on the right of each graph indicate macroscopically determined micellar phase limits.

It should be mentioned that we have observed other physical properties of the BS-L solutions to undergo dramatic changes as the micellar phase limits are approached. Most impressive is an extraordinary increase in solution viscosity. Indeed, our measurements indicate that the viscosities are in effect diverging at the phase limit[19]. The mean intensity of the scattered light also appears to diverge at the phase limit, furthermore, solutions with L/BS ratio greater than the phase limit appear grossly turbid and viscous.

It is apparent from Figure 6 that our experimental data does not agree with theoretical predictions of $\bar{R}$ based on the simple disc model of the BS-L mixed micelle proposed by Small[3] and Dervichian[4]. However, we have developed a new model for the molecular structure of BS-L micelles, which we call the "mixed disc" model and this new model can account for the divergence of $\bar{R}$ as seen by the close agreement between the fit of the theory to the data.

### Interpretation of BS-L Mixed Micellar Structure

Here we present the "mixed disc" model for the structure of the BS-L micelle. Although our $\bar{R}$ values are not consistent with the model of micellar structure proposed by Small and Dervichian, it is useful, nevertheless, to describe the Small-Dervichian model, as it will facilitate our presentation of the new "mixed disc" model.

In the Small-Dervichian model, the mixed micelle, depicted in Figure (7), consists of a disc of a lecithin bilayer surrounded on its perimeter by bile salt molecules which are oriented so that their hydrophilic OH groups can interact with the aqueous solvent and their hydrophobic surfaces can interact with the lecithin chains. It is helpful to envision the formation of these micelles in the following way. In a solution containing $N_L$ molecules of lecithin and $N_{BS}$ molecules of bile salt, suppose that all of the L molecules first form a bilayer, while all of the BS molecules* form a ribbon-like structure two molecules in width having hydrophobic and hydrophilic surfaces. Now divide both the bilayer and "ribbon" structure into n equal portions so that each piece of the bilayer is a disc of radius r that can be surrounded on its perimeter by the corresponding strip of BS "ribbon" which has a length $2\pi r$. It can be readily seen that the radius of the disc, r, will depend on the ratio $N_L/N_{BS}$. If σ is the number of L molecules per unit area of bilayer and ρ is the number of BS molecules per unit length of "ribbon" then n and r will be related to $N_{BS}$ and $N_L$ by the

---

*The low intermicellar concentration of BS monomers is justifiably neglected in these considerations[17].

following relations:

$$N_L = n\pi r^2 \sigma \qquad (4)$$

$$N_{BS} = n 2\pi r \rho \qquad (5)$$

Dividing equation (4) by equation (5) and solving for r we obtain:

$$r = \left(\frac{2\rho}{\sigma}\right) \frac{N_L}{N_{BS}} \qquad (6)$$

Thus, in the Small-Dervichian model, r increases <u>linearly</u> with the ratio $N_L/N_{BS}$, which is identical to the L/BS molar ratio. It is clear from our data that $\bar{R}$, and consequently r have a much stronger dependence on the L/BS ratio than is predicted by equation (6). In addition, the Small-Dervichian model does not readily explain why a micellar phase limit should exist.

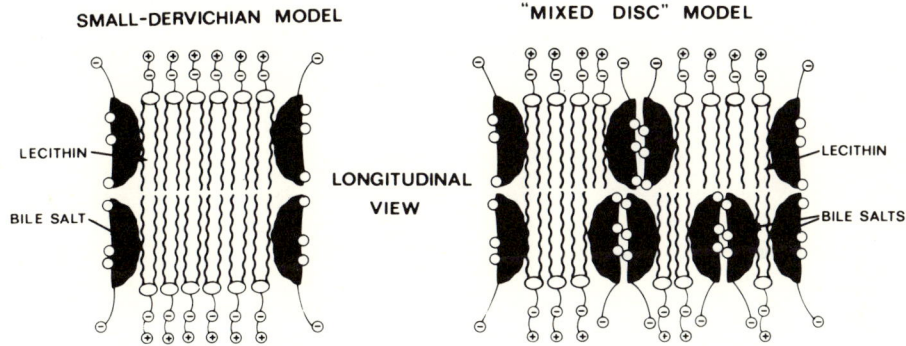

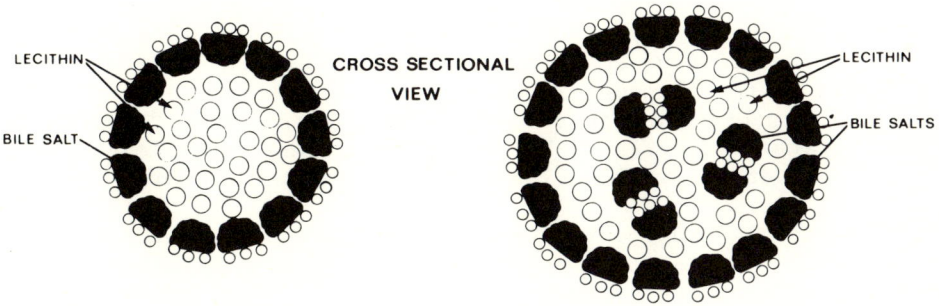

Figure 7. Schematic models for the structure of the Bile Salt-Lecithin mixed micelle.

In our "mixed disc" model also depicted in Figure 7, we make only one assumption which is different from the Small-Dervichian model. We assume that initially the lecithin and bile salt molecules interact to form a mixed bilayer in which the ratio of lecithin to bile salt molecules in the bilayer is equal to the value $\alpha$, regardless of the values of $N_L$ and $N_{BS}$. We shall see shortly that $\alpha$ corresponds to the L/BS ratio at the micellar phase limit. As all of the $N_L$ lecithin molecules are contained in the mixed bilayer, the number of BS molecules incorporated in the mixed bilayer must equal $N_L/\alpha$. This leaves $N_{BS}-N_L/\alpha$ BS molecules available to form BS "ribbon". As before we shall divide both the mixed bilayer and BS "ribbon" into n equal portions, so that each piece of the mixed bilayer is a disc of radius r, surrounded by the corresponding strip of BS "ribbon" which has length $2\pi r$. If $\sigma'$ represents the number of L molecules per unit area of mixed bilayer and $\rho$ has the same meaning as before, then n and r will be related to $N_L$ and $N_{BS}$ by two new relations:

$$N_L = n\pi r^2 \sigma' \quad (7)$$

$$N_{BS} - N_L/\alpha = n 2\pi r \rho \quad (8)$$

Dividing equation (7) by equation (8) and solving for r, we obtain:

$$r = \frac{2\rho}{\sigma'} \frac{N_L}{N_{BS}-N_L/\alpha} = \frac{2\rho}{\sigma'} \frac{N_L/N_{BS}}{1-\alpha^{-1}N_L/N_{BS}} \quad (9)$$

Analysis of equation (9) reveals that r <u>diverges</u> as the ratio $N_L/N_{BS}$ approaches the value $\alpha$. The physical basis for this divergence is quite simple. As the L/BS molar ratio (i.e., $N_L/N_{BS}$) approaches the value $\alpha$, less and less bile salt molecules remain to form "ribbon". Since r is proportional to the amount of bilayer divided by the amount of "ribbon", r must diverge as the amount of "ribbon" approaches zero, (i.e. as L/BS approaches $\alpha$). For L/BS ratios greater than $\alpha$, there exists no bile salt "ribbon" with which to form mixed micelles, thus $\alpha$ must correspond to the micellar phase limit. We have suggested in Figure 7 that the bile salts within the interior of the disc are hydrogen bonded as dimers so that their hydrophobic surfaces have maximal contact with the lecithin chains. Hydrogen bonding between the bile salt molecules within lecithin bilayers would be consistent with the behavior of di- and trihydroxy methylcholanoates in nonpolar solvents[20].

In order to quantitatively test the mixed disc model we have estimated from our $\bar{R}$ data the dependence of the apparent disc radius r on the L/BS ratio assuming the thickness of the mixed micellar disc was 50 Å*. Equation (9) predicts that a plot of 1/r versus

---
*This value is based on the dimensions of the lamellar liquid crystalline phase of egg lecithin[12].

$N_{BS}/N_L$ (the inverse of the L/BS ratio) should obey the following linear equation.

$$\frac{1}{r} = \left(\frac{\sigma'}{2\rho}\right)\left(\frac{N_{BS}}{N_L} - \frac{1}{\alpha}\right) \qquad (10)$$

In Figure 8 a plot of $1/r$ versus $N_{BS}/N_L$ is given for both the TDC-L and TC-L systems at 20°C. Both systems show the linear dependence predicted by equation (10). For the TDC system the value of $\alpha$ deduced from the BS/L ratio at which $1/r$ extrapolates to 0, is 1.54 while for the TC system $\alpha$ is found to be 2.12. These values of $\alpha$ are nearly the same as the macroscopically observed L/BS phase limits, which are ~1.5 and ~2.0 respectively for the L-TDC and L-TC systems at 20°C. This close correspondence between $\alpha$ and the actual phase limit provides strong support for the model. In addition, the values of $\frac{2\rho}{\sigma'}$, 17.4 Å for TDC-L and 13.3 Å for TC-L derived from the slopes of the curves in Figure 8, are in reasonable agreement with estimates for this fraction (15-20 Å) derived from the molecular dimensions of the bile salt and lecithin molecules[2,12]. This interesting finding adds additional support for the "mixed disc" model.

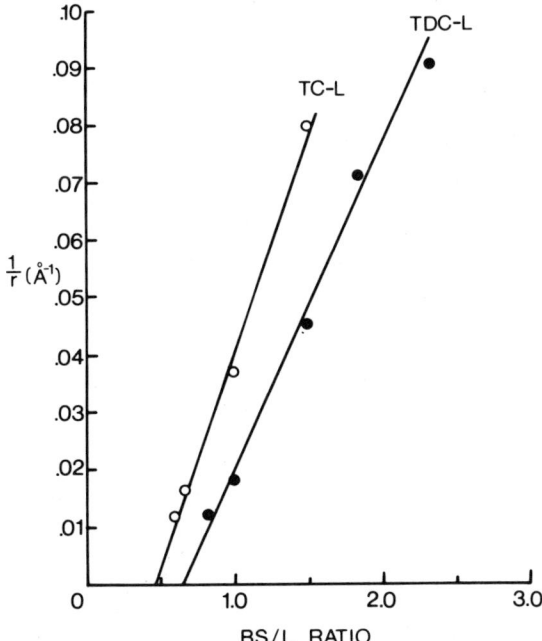

Figure 8. The inverse disc radius ($1/r$) versus $N_{BS}/N_L$ molar ratio for TDC-L and TC-L mixed micelles.

Employing the values of $\alpha$ and $\frac{2\rho}{\sigma}$ derived from Figure 8 for both systems, we used equation (9) and a formula for calculating the hydrodynamic radii of disc-like structures[21] to obtain the theoretical $\bar{R}$ values for the "mixed disc" model which appear in Figure 6. We see from this figure that the "mixed disc" model can clearly account for the observed divergence of $\bar{R}$ which occurs as the phase limit is approached. It is remarkable that this slight modification of the Small-Dervichian model can explain the extraordinary growth of the L-BS micelle and can in addition, provide insight into the physical basis of the micellar phase limit.

## TC-L-Ch MIXED MICELLAR SOLUTION

### Mixed Micelles

We have measured $\bar{R}$ values for mixed micellar solutions containing TC-L-Ch at 10 g/dl total lipid concentration in 0.15 M NaCl at 20°, 40° and 60°C (Figure 9). When Ch is progressively added so that the mole fraction of TC remains fixed at 75%, $\bar{R}$ increases slightly with increases in Ch, and is essentially independent of temperature. When the mole fraction of TC is fixed at 50%, added Ch progressively decreases $\bar{R}$ and a slight temperature dependence is observed. We are presently extending the "mixed disc" model to account for the effects of Ch on the size of the BS-L-Ch mixed micelle.

### Precipitation of Ch Microcrystals

A 10 g/dl solution of TC-L-Ch with molar proportions of 75%/17.5%/7.50% respectively in 0.15 M NaCl has a composition similar to gallbladder bile. This system is unsaturated with Ch at 60°C, supersaturated with Ch but in a metastable state at 40°C ($\bar{R}$ value shown in Figure 7) and supersaturated with Ch in a labile state at 10°C[7]. Plotted in Figure 10 are measurements of the intensity of light scattered from such a solution as a function of time after the temperature was lowered from 60° to 10°C. After a short time delay,* the intensity increased rapidly but after 15 minutes it reached an equilibrium value which was nearly 250 times the initial value. At equilibrium, the solution was observed to be turbid and contained precipitated Ch which remained suspended in solution. The $\bar{R}$ value of this solution was measured to be about 1200 Å, roughly 60 times greater than the $\bar{R}$ value for the mixed micellar solution at 60°C (21 Å). We believe that the value of 1200 Å represents the size of cholesterol microcrystals precipitated under these conditions. The QLS technique has thus provided us with a method for observing the embryonic stages of Ch gallstone formation _in vitro_.

---

*The scattering cell required about 3 minutes to change temperature from 60°C to 10°C and therefore contributed to this delay.

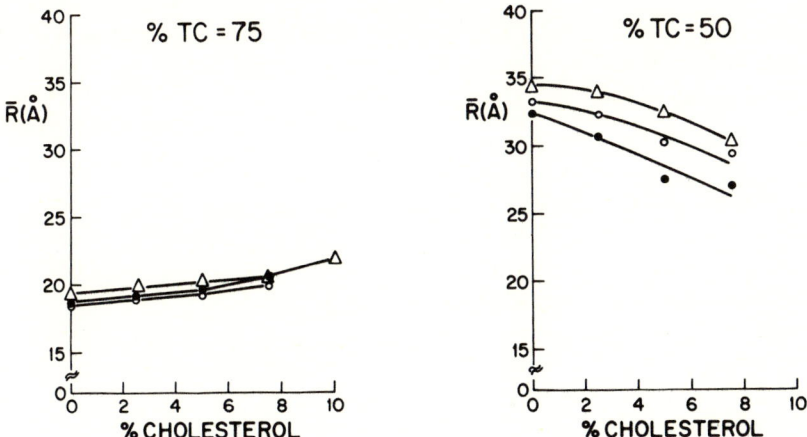

Figure 9. $\bar{R}$ values for TC-L-Ch mixed micelles (10 g/dl, 0.15 M NaCl, ●=20°C, O=40°C, △=60°C) as a function of added Ch expressed as % of total moles. On the left graph, the mole fraction of BS remains fixed at 75% and on the right graph, the mole fraction of BS remains fixed at 50%.

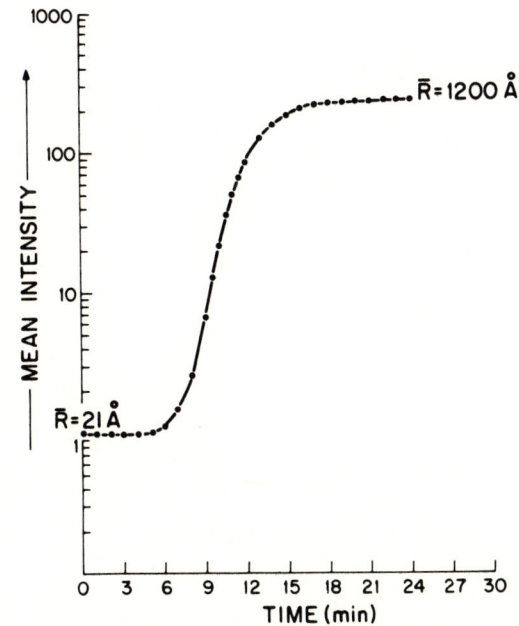

Figure 10. Mean light intensity scattered from a supersaturated TC-L-Ch solution as a function of time after lowering temperature from 60°C to 10°C.

## CONCLUSIONS

Using the technique of quasielastic light scattering spectroscopy we have measured the mean hydrodynamic radii of simple BS micelles, BS-L and BS-L-Ch mixed micelles and precipitated Ch microcrystals over a wide range of conditions including those which are physiologically relevant.

Our study of simple BS micelles at concentrations well above the CMC, has shown that the mean micellar size is affected by BS species, temperature, ionic strength, BS concentration and added urea. Our most important findings are that under the conditions of low temperatures and high NaCl concentration the micelles are rod-shaped and that their mean size grows appreciably as the BS concentration increases. These results on simple BS micelles are found to be consistent with the primary-secondary micelle hypothesis and we have developed a simple quantitative model of primary-secondary micelle equilibrium. This model relates the mean hydrodynamic radius of the BS micelles ($\bar{R}$) to the product $KC_o$, where $C_o$ is the BS concentration and K is the binding constant for the self association of primary micelles. From the dependence of K on temperature we have deduced the enthalpy of bonding between primary micelles to be - 6.9 kCal/mole.

For mixed BS-L micelles we find that at low L/BS molar ratios the TDC-L and TC-L systems are not alike, probably reflecting the coexistence of simple and mixed micelles. At high L/BS ratio the two systems behave similarly suggesting that only mixed micelles are present. The $\bar{R}$ values of the mixed micelles appear to diverge as the L/BS ratio approaches the micellar phase limits for the L-TDC and L-TC systems. This divergence is consistent with a new model for the structure of the L-BS micelle called the "mixed disc" model. This model proposes that the L-BS micelle is a disc in which L and BS exist within the interior of the disc in the ratio of the phase limit, while BS alone constitutes the disc's perimeter.

Lastly we have studied mixed micellar solutions containing TC, L and Ch. At high mole fractions of TC, addition of Ch causes a slight increase in $\bar{R}$, whereas at low mole fractions of TC, $\bar{R}$ decreases with added Ch. In solutions which are supersaturated with Ch, we have observed the precipitation of Ch microcrystals whose $\bar{R}$ values are $\sim$1200 Å.

The results of our investigation question previously held beliefs concerning the structure and homogeneity of simple and mixed micelles and have enabled us to develop detailed alternative models.

## ACKNOWLEDGEMENTS

Supported in part by U.S. Public Health Service Grant AM 18559 (M.C. Carey, P.I.), Academic Career Development Award AM 00195 (M.C. Carey), National Science Foundation Grant DMR 72-03027A05 to the Center for Materials Science and Engineering, M.I.T. and the National Institutes of Health Interdisciplinary Program in Biomaterials Science under Grant #1P01-HL-14322-04 (G.B. Benedek, P.I.). N.A. Mazer wishes to thank the American Gastroenterological Association for a Summer Research Fellowship. The authors would also like to thank Drs. Mukerjee, Cardinal, Birdi and Muller for their constructive comments following the oral presentation of this paper.

## REFERENCES

1. M. C. Carey and D. M. Small, Am. J. Med. 49, 590 (1970).
2. M. C. Carey and D. M. Small, Arch. Intern. Med. 130, 506 (1972).
3. D. M. Small, Gastroenterology 51, 607 (1967).
4. D. G. Dervichian, Advan. Chem. Ser. 84, 78 (1968).
5. M. C. Carey and D. M. Small in "Advances in Bile Acid Research III", S. Matern, P. Back, J. Hackenschmidt and W. Gerok, eds., New York, K. Schattauer Verlag, 1975, p. 277.
6. M. C. Carey and D. M. Small, Gastroenterology 58, 1057 (1970).
7. M. C. Carey and D. M. Small (submitted for publication).
8. N. A. Clark, J. J. Lunacek and G. B. Benedek, Am. J. Phys. 38, 575 (1970).
9. N. A. Mazer, G. B. Benedek and M. C. Carey, J. Phys. Chem. 80, 1075 (1976).
10. N. A. Mazer, M. C. Carey and G. B. Benedek, "The Size, Shape and Thermodynamics of Sodium Dodecyl Sulfate (SDS) Micelles Using Quasielastic Light Scattering Spectroscopy (this symposium).
11. N. A. Mazer, G. B. Benedek and M. C. Carey, Gastroenterology 70, 998 (1976).
12. M. Bourgés, D. M. Small and D. G. Dervichian, Biochem. Biophys. Acta 144, 189 (1967).
13. F. Franks ed., "Aqueous Solutions of Amphiphiles and Macromolecules" (Water -- A Comprehensive Treatise Series: Vol. 4), Plenum Press, New York, 1975.
14. G. C. Pimentel and A. L. McClellan, "The Hydrogen Bond", W. H. Freeman and Co., San Francisco, California, 1960.
15. C. Tanford, "The Hydrophobic Effect", J. Wiley and Sons, New York, 1973.
16. W. J. Moore, "Physical Chemistry", Third Edition, Prentice-Hall, Inc., Englewood Cliffs, New Jersey, 1962.
17. M. C. Carey, Proceedings of the VIIth International Congress on Surface Active Substances, Moscow, September 1976 (in press.)

18. N. A. Mazer, M. C. Carey and G. B. Benedek (unpublished observations).
19. R. McCabe, N. A. Mazer and M. C. Carey (unpublished observations).
20. W. S. Bennet, G. Eglinton and S. Kovac, Nature 214, 776 (1967).
21. R. J. Cohen and G. B. Benedek (manuscript in preparation).

# QUASIELASTIC LASER SPECTROMETRY STUDIES OF PURE BILE SALT AND BILE SALT-MIXED LIPID MICELLAR SYSTEMS

R. T. Holzbach, S. Y. Oh, M. E. McDonnell, and
A. M. Jamieson
Cleveland Clinic Foundation and Case Western Reserve University
Cleveland, Ohio

Quasielastic laser light scattering offers a rapid accurate method for determining translational diffusion coefficients which has certain advantages for characterizing micellar systems. We have used the technique to measure diffusion coefficients in a variety of single bile salt and bile salt-mixed lipid micellar solutions as a function of total solute concentration at physiological temperatures. The evidence suggests that the validity of a Stokes-Einstein relation for micellar size determination at higher concentrations is questionable. Preliminary evidence indicates that addition of phosphatidylcholine and cholesterol to form unsaturated, saturated and metastably supersaturated systems respectively does not alter micellar size.

## INTRODUCTION

A variety of experimental techniques have been applied to characterize the physical parameters of bile salt micelle systems, but reported measurements of size, micellar molecular weight (MMW), and aggregation numbers show considerable disagreement. Among the thermodynamic methods[1,2,3,4], both classical light scattering and sedimentation equilibrium require difficult experimental and mathematical corrections to obtain a true (anhydrous) aggregate molecular weight in typical multicomponent systems of bile salt, solvent and supporting electrolyte. Among the hydrodynamic methods[4,5,6,7,8,9], which detect a solvated particle, ultracentrifuge measurements are subject to experimental difficulties in maintaining integrity of the micelle near the critical micellar concentration (CMC)[4]. Free diffusion measurements[6] or gel filtration[7,8,9], among other classical hydrodynamic techniques, are also prone to measurement artifact. X-ray small-angle scattering studies (XSAS)[10,11], and observations from transmission electron microscopy (TEM) have not been conclusive. XSAS measurements in these diffusely scattering systems have been hindered because of the low intensity of the energy sources and the mathematical approximations in the data analysis[10]. TEM work has been afflicted with artifact from sample dehydration because of the high vacuum[12,13].

The sodium taurodeoxycholate (STDC) and sodium taurocholate (STC)[4] systems appear to have been the most rigorously studied of the single bile salt systems, especially by classical light scattering[3]. In many other single bile salt solutions information on apparent aggregation numbers (Ag#) as a function of pH, ionic strength, solute concentration, and temperature is available[1]. The numerical values of the true micellar weights determined in the STC and STDC systems[3,4] when the effects of preferential interactions are properly accounted for, differ widely from the apparent weights obtained by simpler analysis[1]. It appears, therefore, that the effect of bile salt structure on even the simplest correlated micelle properties (MMW or aggregation number, shape, and degree of hydration) has not yet been completely resolved. Such difficulties are compounded when one attempts to study more complex systems such as the bile salt-mixed lipid micellar solutions. These mixed micelles assume greater importance because they are biologically and physiologically the most relevant.

## BACKGROUND

The classical experimental methods for obtaining micelle molecular weight and size measurements have been reviewed comprehensively by Kratohvil and Dellicolli[3]. Total intensity light scattering and sedimentation equilibrium have been the thermodynamic techniques most widely applied to micelle systems. Careful correction for the effects of preferential thermodynamic interaction between components are required with either method. Among the problems are the fact that charge interactions may make concentration dependence non-linear at low ionic strengths. In addition, estimates of micellar dimensions cannot be made directly by light scattering, because of the small sizes of the micelles. With x-ray small-angle scattering, measurements of MMW and radius of gyration can be obtained[10,11]. These experiments are, however, time-consuming and, when employed for solutions at lower solute concentrations, require both angular resolutions and high intensity anodal sources not generally available.

The general principles involved in obtaining hydrodynamic measurements in simplified form may be summarized as follows. Particle hydrodynamic size may be determined from a measurement of intrinsic viscosity $[\eta]$, translational diffusion coefficient at zero concentration, $D_t^o$, or sedimentation coefficient at zero concentration, $s^{o3}$. A weight average MMW can be obtained by a combination of a weight average $S^o$ and a z-average diffusion coefficient. Alternatively, a value of MMW can be obtained by a combination of $S^o$ or $D_t^o$ with $[\eta]$[14,15]. In systems characterized by significant polydispersity, as recent information strongly suggests applies to the case of bile salts, the statistical average measured is complex[16]. Size, as measured by hydrodynamic methods, is dependent upon the hydrodynamic model, e.g. it is deduced from $D_t^o$ by application of Stokes-Einstein relation

$$D_t^o = \frac{KT}{6 \pi \eta_s R_h} \quad (1)$$

where $R_h$ is the hydrodynamic size and represents the effective radius of the particle viewed as a sphere and $\eta_s$ represents solvent viscosity. This size estimate includes all the tightly bound solvent and ions which move with the micelle and may be either larger or smaller than the radius of gyration measured by light or X-ray scattering depending upon the particle structure. In the case of spherical, relatively compact particles like micelles, it appears most likely that $R_g < R_h$.

Intrinsic viscosity, diffusion coefficients $D_t^o$ and sedimentation coefficients $S^o$, extrapolated to the CMC have been reported for STDC and STC by several authors[3,4,11]. In both systems, problems are encountered in correlating viscosity and frictional coefficient data. This has led to the suggestion[3,4] that each technique 'sees' a different hydrodynamic particle. Obviously more frictional coefficient measurements are needed to resolve such questions. Sedimentation and diffusion coefficient measurements are traditionally measured by either ultracentrifuge or free diffusion techniques, both of which require the establishment of a macroscopic concentration gradient and are therefore somewhat imprecise especially in micellar systems where special precautions are required to ensure micellar integrity near the CMC. Also, since these methods are time-consuming they are not adaptable for kinetic studies in, for example, thermodynamically metastable micellar systems. The classical light scattering technique usually adopted for such experiments cannot directly detect size changes in the important range 10-600Å where nucleation events occur, and must be applied with caution to mixed micellar systems at physiological concentrations because of interparticle effects.

Described in this paper is a simple, rapid, and accurate method of estimating hydrodynamic sizes of bile salt micelles utilizing quasielastic laser light scattering (QLLS) which can be applied in principle to both pure and mixed micelle systems. Compared with the previously applied conventional methods for estimating hydrodynamic size, the recently developed laser light scattering technique enables the measurement of $D_t$ rapidly (in minutes) with accuracies in the range of 1 to 5 per cent. Further advantages with this approach are that it does not involve the induction of microscopic concentration gradients nor subject the system to external perturbation. The technique has been recently employed to measure sizes of mixed sphingomyelin micelles, using as the detergent Triton X-100 which, however, produces micelle sizes much larger than those observed in bile salt systems[16]. We report preliminary experiments designed to test the usefulness of the QLLS technique for micellar size measurement in bile salt systems at concentrations up to the physiological range.

A simplified illustration of the principle involved in quasielastic laser spectroscopy is shown in Figure 1. Coherent laser

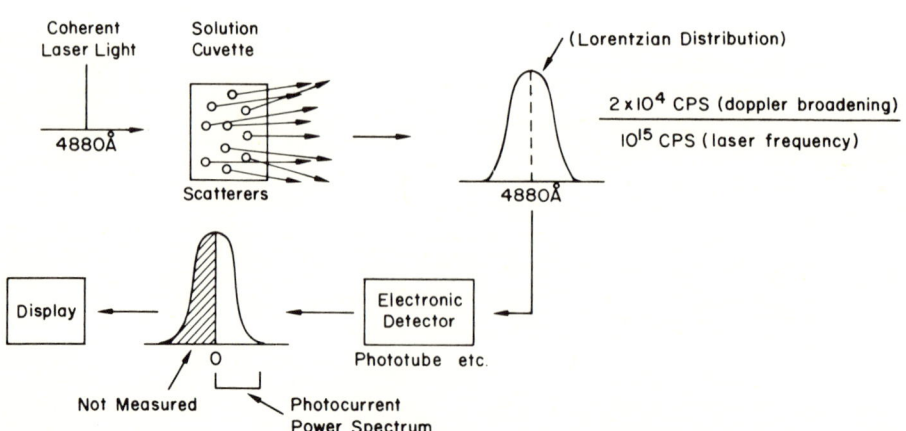

Figure 1. Principle for measurement of translational diffusion ($D_t$) by quasielastic laser spectroscopy.

radiation at 4880 Å is focused into the thermo-regulated cuvette containing the scattering particles. The light scattered by the moving particles will experience small Doppler frequency shifts proportional to particle velocity. When the particles are reasonably monodisperse, on the basis of theoretical and experimental observations, the resulting Rayleigh spectrum is a single Lorentzian distribution centered at the laser frequency and with half-width $\Gamma_D$ given by

$$\Gamma_D = D\kappa^2 \quad (2)$$

where $\kappa$ is the scattering vector. In the quasielastic scattering regime the scattering vector may be written as

$$|\kappa| = \frac{4\pi n}{\lambda_o} \sin\theta/2 \quad (3)$$

where n is the refractive index, $\lambda_o$ is the laser wavelength and $\Theta$ is the scattering angle. By optical mixing techniques[17], the center of this Rayleigh spectrum is shifted from the laser frequency to zero. The resulting homodyne power spectrum from light scattered by the monodisperse suspension is a Lorentzian of half-width $2\Gamma_D$. Depending upon the intensity of scattering, this power spectrum can be obtained with high quality in as brief a period as a few minutes, using a real time spectrum analyzer and digital integrator. As an alternative to current measurements, the same information can be obtained from the auto-correlation function of the photocurrent[17]. For monodisperse particles this approach yields a single exponential decay with characteristic time $2\Gamma_D$.

The spectrometer used in the experiments detailed below has been essentially described before[18]. It consists of a Coherent Radiation Model 52 Ar$^+$ laser, custom-built thermoregulated scattering cell enclosure, EMI 9656 KR phototube, Keithley Model Wide band amplifier and Saicor 42 real time spectrum analyzer and digital integrator. The dc level, proportional to scattering intensity, is monitored by a Digitec digital millivoltimeter.

For completeness, it must be pointed out that all of the modes of molecular fluctuation, e.g., thermodynamic fluctuations of the micellar mass action equilibrium can in principle cause spectral broadening[19], but these modes do not contribute to the scattering here which is found to obey Equation (2). A more important consideration arises from the fact that the micellar systems we wish to study are certainly polydisperse, although thermodynamic considerations of simple amphiphile micellar systems would indicate that this polydispersity might be small[20]. Suspensions of polydisperse micelles will give non-Lorentzian Rayleigh spectra. In principle, a complete spectral analysis can be carried out to give z-average $D_t$ and parameters related to the width and shape of the distribution[21]. In this report we have estimated average diffusion coefficients from single-Lorentzian fits to the full spectral distribution.

## METHODS AND MATERIALS

The source and preparation of sodium conjugates of bile salts and the stock solution preparation have been previously given[22]. Likewise, the purification of ovalecithin and preparation of stock solutions for lecithin and cholesterol solutions have been described elsewhere[22]. In ternary systems, the previously described physiologically balanced mixture of conjugated bile salts were used[23]. Utilizing procedures essentially similar to that in previous work[22], binary mixtures containing bile salt-lecithin, or ternary micellar systems with three components (bile salts-lecithin-cholesterol) for use as bile analog solutions were prepared. To ensure optimal coprecipitation or "close molecular association" of the dried lipids, careful procedural detail was followed to ensure, among other things, mixing and complete drying prior to addition of sufficient diluent to produce the desired W/V solution.

The present study of three-component bile analog solutions was designed to include systems, unsaturated, saturated, and supersaturated with respect to cholesterol. The three different bile analog solutions were designed to fall on a single phase diagram tie-line, to provide a useful comparison of bile salt-mixed-lipid micellar size as a function of degree of cholesterol saturation. The tie-line selected was that representative of the physiologic concentration range for these lipids as found in human bile with the bile salt:lecithin ratio equalling 3:1. For these systems, the thermodynamic definition of equilibrium saturation of Holzbach et al[22] was used to determine the molar ratio of cholesterol added to each bile salt-lecithin mixture (lecithin = phosphatidylcholine).

## PREPARATION OF SOLUTION SAMPLES FOR LIGHT SCATTERING STUDIES

All micellar solutions of any solute concentration used in this study were passed through a 0.22 µ micropore filter into a clean cylindrical scattering cell. For concentration dependence studies, a solution was diluted 1:1 using the purified buffered solution (0.15M NaCl) and passed again through the micropore filter. A final check for cleanliness was made by placing the scattering cell in the spectrometer sample holder and viewing both the dc voltage developed by the phototube and the uniformity of the laser beam in the solution. When the dc voltage across the photomultiplier tube, registered by a digital millivoltimeter, was constant to within 1%, the solutions were assumed clean enough to give homodyne spectra. All measurements were made at 37°C. after pre-equilibrating the sample cell in the thermal jacket sample holder of the laser spectrometer for at least 5 minutes.

## POWER SPECTRUM ANALYSIS

The voltage spectra of the systems studied usually represented the accumulation of 1024 individual spectral records on the Saicor analyzer. This involved less than 3 minutes averaging time. Following acquisition of the integrated voltage spectrum, determination of the shot noise spectrum was obtained as follows: A white light source of appropriate intensity was inserted to reproduce the dc phototube voltage, following which a number of records were accumulated equivalent to the experimental spectrum.

Spectra were analyzed by measuring the voltage amplitude at fifty equally spaced intervals from zero frequency. This voltage squared minus the shot noise voltage squared is taken as proportional to the signal power P. This may be represented, assuming a small uncertainty in the shot noise level, by the equation

$$P = \frac{A}{B + f^2} - C \quad (4)$$

where f is frequency. Since, by design

$$\frac{C}{A - BC} \ll 1,$$

we may write

$$\frac{1}{P} = \alpha + \beta f^2 + \gamma f^4 \quad (5)$$

where

$$A = (1 + \alpha\gamma/\beta^2)^2/\beta$$

$$B = \alpha (A/B)^{1/2}$$

$$C = \gamma A/B$$

Equation (5) is a parabola which we fit by least squares procedure using the weighting coefficient $\sigma^2 = P^{-4}$, which assumes the same accuracy for each P value. A detailed description of procedures has been given elsewhere[16]. Examples of experimental voltage spectra of

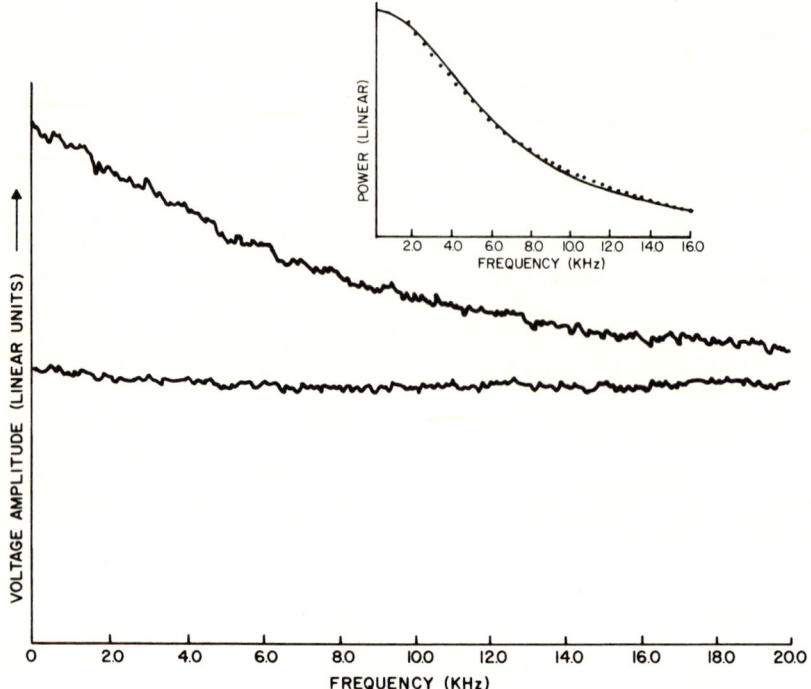

Figure 2. Experimental voltage spectrum of light scattered by a 10% W/V solution of sodium glycodeoxycholate (SGDC) micelles, 0.15 M Nacl at 37°C. Scattering angle is 38.2°. Inset is the single Lorentzian fit to points on the power spectrum with half-width at half-height 6.08 $KH_z$.

single and mixed micelle solutions are shown in Figures 2 and 3, together with their power spectral fits.

## RESULTS

To establish that our spectra were only the result of translational diffusion of the micelles and did not have contributions from fluctuations about the monomer-micelle equilibrium[24] or from rotational diffusion, we studied the angular dependence of a ternary micellar system. Excellent quadratic dependence of the spectral line-width on κ-vector was noted, proving that only a single diffusion relaxation mechanism [Equation (2)], was involved. The concentration dependence of the diffusion coefficient was studied in a number of systems. The diversity of results are illustrated in Figure 4. Panel A gives data for the pure sodium taurocholate system and is

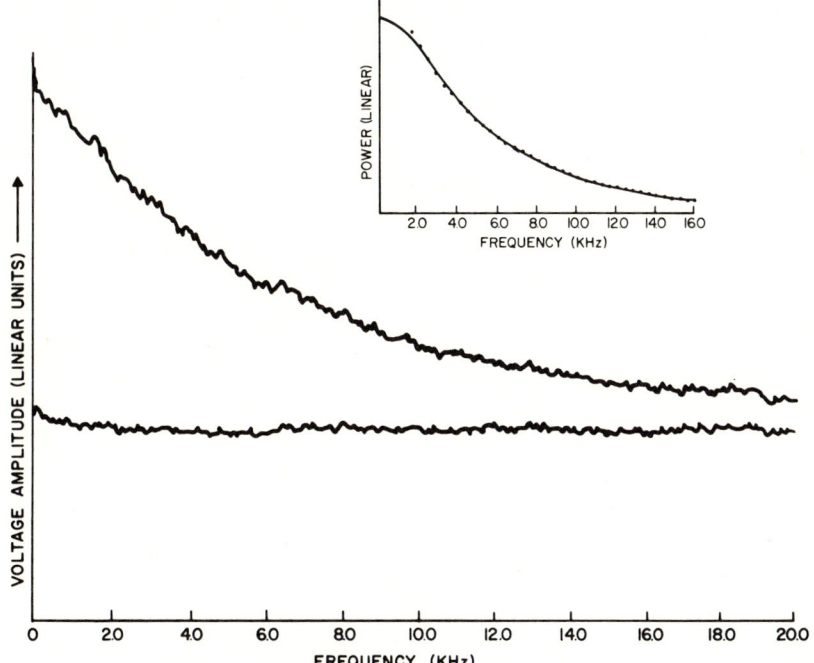

Figure 3. Experimental voltage spectrum of light scattered by a 10% W/V ternary bile analog (bile salts-phosphatidylcholine-cholesterol) in .15 M NaCl at thermodynamically defined saturation, 37°C. Scattering angle is 38.2°. Inset is the single Lorentzian fit to points on the power spectrum with half-width at half-height 4.64 KHz.

representative of many pure and mixed binary bile salt systems where there is negligible concentration dependence from values slightly above the critical micellar concentration through bile component concentrations characteristic of physiological fluids (∿ 10%). Panels B and C show positive and slightly negative concentration dependence, respectively. The former solution (B) is sodium taurodeoxycholate (STDC) which was earlier studied by Laurent and Persson[5] who obtained similar results. The latter (C) is a 3:1 molar mixture of sodium glycocholate and phosphatidylcholine (SGC + PC).

The linear concentration dependence of the diffusion coefficient can be represented by

$$D_t(c) = D_t^o [1 + MA_2 - k_f - \bar{v})c] \quad (6)$$

where M is the micellar molecular weight, $A_2$ is the second osmotic virial coefficient, $k_f$ is the relative concentration dependence of the frictional coefficient, and $\bar{v}$ is the partial specific volume of the solute in the solvent. It is apparent that the relative magnitudes of the terms can lead to positive, negative, or zero concentration dependence in different systems. For the case of non-interacting hard spheres, the magnitude of the positive and negative contributions to the concentration dependence are equal to within better than 2%[25]. The predicted residual concentration dependence is therefore too small to be detected from the accuracy of these measurements.

When micelles can be modeled as hard, neutral spheres, negligible concentration dependence is predicted and the experimentalist can use diffusion coefficient measurements from any concentration and the Stokes-Einstein equation (Equation 1) to determine the micellar radius at that concentration. Viewed in this light, Panel A of Figure 4 would indicate that above the CMC the degree of association is independent of concentration; Panel B would suggest a somewhat preposterous uniform decrease in aggregation with increasing concentration, and Panel C would denote the opposite.

Experimental data consistent with panels A and B respectively have been reported for the STC system by Vitello[4] and for the STDC system by Laurent and Persson[5] using the ultracentrifuge. Behavior consistent with Panel C is implicated in experiments of Mukerjee and Cardinal[26] on the system naphthalene-sodium cholate. These authors measured the change in concentration of sodium cholate. This deriva-

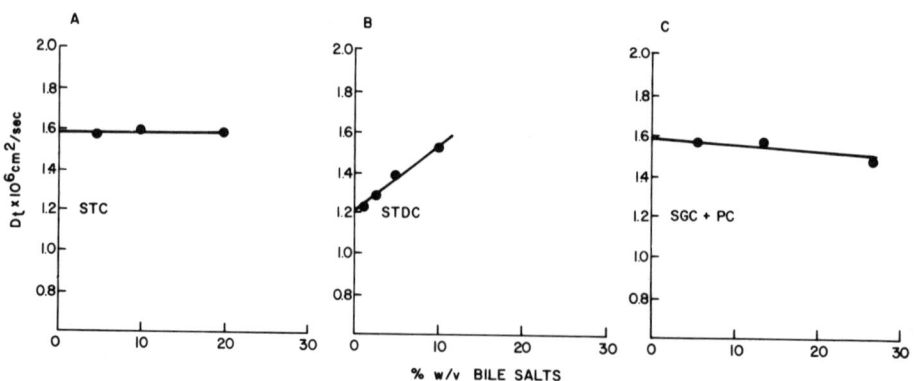

Figure 4. Concentration dependence plots showing different trends with differing systems: A. Sodium taurocholate (STC), B. Sodium taurodeoxycholate (STDC) and C. Sodium glycocholate plus lecithin (3:1) (SGC + PC).

tive does not exhibit a sharp transition at the CMC and indeed is still increasing at the largest concentrations studied (1.5% W/V). This indicates that the amount of naphthalene solubilized per unit of sodium cholate is still increasing and in turn demonstrates that the micellar size is also increasing at least up to 1.5% W/V. It should be noted that these data are derived from the pure bile salt system repeatedly shown to have the smallest aggregation number (Ag#) and that comparatively sharper transitions could be observed for micellar systems characterized by a greater tendency to self-association i.e. the dihydroxy bile salts.

Returning to the interpretation of the concentration dependence predicted by Equation (6), it is probably not safe to conclude that there is complete cancellation of the thermodynamic and hydrodynamic terms in all cases. There is no theoretical framework to assess the effects of the charge or anisometric shape of the micelle on the relative concentration dependence of the frictional coefficient. If such effects are ignored, however, it is difficult to arrive at a physical interpretation of the data shown in Panel B of Figure 4. Work is currently in progress to use X-ray small angle scattering (XSAS) to independently determine micellar size as a function of concentration[27,28]. Combining these results with the present information will permit the unambiguous interpretation of size from measurements of bile salt micelles under physiological conditions. Until this has been done we feel that diffusion coefficient measurements of micelles should be extrapolated to zero concentration where the hydrodynamic and thermodynamic concentration contributions are irrelevant. Within the 5% accuracy of these measurements, these extrapolated zero concentration values are indistinguishable from the diffusion coefficient at the critical micelle concentration.

## DISCUSSION

Our findings indicate, in keeping with the findings of earlier observers, that the molecular structure of a bile salt does indeed affect the hydrodynamic size of the micelles in pure systems. Our observations indicate that loss of the 7-Alpha or 12-Alpha hydroxyl moiety leads to a size increase (diffusion coefficient decrease). The concentration dependence of the chenodeoxycholate systems was not studied[29] but, if this can be assumed the same as observed in the deoxycholate micelles as shown in Panel B, Figure 4 (i.e., a measurement at higher concentration underestimates hydrodynamic size in the STDC system), it may be concluded that isomerization from the 12-Alpha to the 7-Alpha hydroxyl on the cyclopentenophenanthrene ring causes an increase in micellar size of taurodeoxycholate micelles within experimental error. In summary, our data[29] indicate that micellar size extrapolated to CMC varies with bile salt structure in the order SGC < SGDC $\lesssim$ SGCDC, and STC < STDC < STCDC. An increase in hydrodynamic size, it is recognized, may be the result of an increase in anisometry, or of micelle number, or a combination of both. Additional information such as viscometry, rotational measurements or total intensity scattering would be necessary to distinguish between these alternatives. For comparison, we note that the sedimentation equilibrium experiments of Small[1] indicate that the apparent Ag# varies in the order SGC < SGDC < SGCDC, and STC < STCDC ≲ STDC, respectively.

Addition of a known amount of phosphatidylcholine (PC) to the micelles in the proportions 1:3 had a different effect on different systems[29]. Incorporation of all the PC in the existing micelles should increase the micelle aggregation number by 33% and the radius by about 10%. This is observed in some cases such as SGDC where hydrodynamic radius extrapolated to CMC changes from 22Å to 26Å. However, other bile salts, for example STDC, do not demonstrate a change in hydrodynamic radius within experimental error upon the incorporation of the PC, both radii being measured as 27Å. This could be the result, in the STDC system, of a change in micelle shape (transition to a less anisometric shape) accompanying the increase in micellar volume due to the solubilization of lecithin which masks the expected change in hydrodynamic radius. An alternative explanation would be an increase in the number of mixed micelles that have smaller amounts of bile salt per micelle and therefore the same radius.

Hydrodynamic radii were also measured for ternary bile salt-mixed lipid micelles in the unsaturated, saturated and metastably supersaturated cholesterol regimes[29]. The hydrodynamic radii of these ternary micelles extrapolated to the CMC are much larger than those of single bile salt and salt-PC micelles. The diffusion coefficients of these systems were very different when measured at physiological concentrations, however the diffusion coefficients decreased rapidly with decreasing concentrations to give indistinguishable values when extrapolated to zero concentration. This implies that the mixed micelles have the same size at the CMC regardless of the cholesterol content. It is not clear whether the differences in diffusion coefficient at physiological concentrations indicate different radii or incomplete cancellation of hydrodynamic and thermodynamic concentration dependences. X-ray small angle scatter (XSAS) from these same ternary systems can distinguish between these possibilities. Preliminary XSAS spectra that were collected in collaboration with Dr. Peter Laggner[27] suggest that at physiological concentrations the unrefined spectra of unsaturated and saturated micelles may have a nearly similar scattering distribution.

If this observation is supported by subsequent more detailed experiments, we would conclude that a substantial contribution to the concentration dependence of the diffusion coefficient in these systems arises from the second virial coefficient in Equation (6).

Further, the implication is that the micellar sizes do not change as a function of saturation. This is also supported by the concentration extrapolation of the diffusion coefficient in these systems[29].

Additional XSAS experiments[28] in which the scattering curves have been 'desmeared', apparently indicate: (i) STDC micelles in deionized water have radii of gyration of 15Å, which is independent of concentration in the range 2.9% (W/V) to 12.8% (W/V); (ii) STDC + PC micelles in 0.15M NaCl have a radius of gyration $\simeq$ 25Å; (iii) all micelles studied (single bile salt, binary, and ternary systems) have an overall shape which cannot be distinguished from a sphere, since the value of $R_g$ determined from the data assuming spherical symmetry is indistinguishable from the exact value in all cases. If the $R_g$ data for the STDC system at low ionic strength shows similar behavior at higher ionic strengths, this would indicate that the concentration dependence of $D_t$ in the STDC system shown in Figure 4B must be due to the effect of $A_2$ and/or $k_f$, and not to changes in micellar size. It is also interesting to note that our value of $R_h$ = 27Å in the STDC + PC system extrapolated to the CMC compares closely with that of $R_g$ = 25Å measured by XSAS[28] at 10% and 5% W/V.

Table I. Comparative Diffusion Measurements on the Systems STDC and STC Corrected to 20°C

| System | Method | Temp.°C. | $D_t^o \times 10^6$ cm$^2$ sec$^{-1}$ | Reference |
|--------|--------|----------|---------------------------------------|-----------|
| STDC | Ultra-centrifuge | 20 | 0.89 | 5 |
| | QLLS | 37 | 0.82 | This work |
| STC | Ultra-centrifuge | 20 | 1.8 | 4 |
| | Free Diffusion | * 37 | 2.2 | 6 |
| | Free Diffusion | 37 | 2.0 | 30 |
| | QLLS | 37 | 1.04 | This work |

* Note that Woodford's free diffusion experiments[6] involve an extrapolation to infinite concentration and hence cannot rigorously be compared with our data at infinite dilution.

Diffusion coefficients reduced to 20°C for the systems STC and STDC have been reported by other workers and are compared with our data in Table I. The STDC values represent good agreement, but our STC value is substantially lower than the others, although closer to the Vitello[4] result. The discrepancy could be due to changes in MMW with temperature. Unfortunately, the apparent MMW values of Small[1] indicate that it remains constant between 5°C and 37°C for STC but decreases by some 30% for STDC micelles. Thus, one ought to expect good agreement for the STC and poor agreement for the STDC systems, the opposite of that observed. These discrepancies need to be investigated by further experiments. The data are consistent with the interpretation that techniques employing macroscopic concentration gradients measure consistently larger diffusion coefficients.

It is interesting to note that, assuming Small's observation[1] that the MMW of STDC micelles decreases by 25% from T = 20°C to T = 37°C while STC micelles remain constant, and using the true micellar weights deduced by Kratohvil[3] and Vitello[4] to evaluate MMW at 37°C for STDC = 13,200 and MMW 37°C (STC) = 4700 with the specific volume value 0.76, our diffusion coefficients lead to an equivalent hydrodynamic ellipsoid whose values of the maximum axial ratio J and maximum solvation number W (gm $H_2O$/gm bile salt) shown in Table II lead one to conclude that a combination of high asymmetry and high solvation are the only reasonable structural deductions. For example, J values of 5 (prolate) or 6 (oblate) together with w values of 0.5 for STDC and 1.0 for the more hydrated trihydroxy species STC would satisfactorily explain our diffusion data.

It should be observed that implicit in the above discussion, has been the assumption that micellar polydispersity does not change significantly from system to system. That this is not an entirely reasonable assumption has now, for the first time, been most clearly shown by the data given by Mukerjee and Cardinal[26] and by other measurements using different techniques[31]. The concentration dependence seen for sodium cholate cannot be explained by the formation

TABLE II. Derived Maximal Hydrodynamic Parameters for STC and STDC

|  | $f/f_{min}$ | $J_{max}$ | $w_{max}$ |
|---|---|---|---|
| STC | 1.82 | 16.5 (prolate)<br>22 (oblate) | 3.82 |
| STDC | 1.67 | 13.0 (prolate)<br>16.5 (oblate) | 2.78 |

of micelles of a single aggregation number. They conclude that single bile salt micelles have both high polydispersity and a mean micellar size that is concentration dependent. The polydispersity of the samples we studied is evident from small departures from a complete match between the data and the single Lorentzian power spectral fits (Figures 2 and 3). The average diffusion coefficients estimated by our approach are smaller than the z-average values $<D_t>_z$. This means that the sizes we estimate will be larger than those obtained by z-average or weight-average measurements.

## CONCLUSIONS

Micellar size for pure bile salt species is strongly affected by bile salt molecular structure, and we have observed some minor differences in trends from previous workers[1]. Phosphatidylcholine causes small changes in micellar size in most systems and seems most plausibly interpreted as simple swelling of the pure micelles. In some systems, when lecithin is added, size or shape changes are indicated. Preliminary evidence indicates that addition of phosphatidylcholine and cholesterol to form unsaturated, saturated and metastably supersaturated systems respectively may not alter micellar size. Our experiments also indicate that substantial errors may arise in applying the Stokes-Einstein equation for size determination from diffusion data at higher concentrations in micellar systems without properly accounting for thermodynamic interactions.

## REFERENCES

1. D. M. Small, Adv. Chem. Series, "Molecular Associations in Biological and Related Systems", 84, 31 (1968).
2. J. P. Kratohvil and H. T. Dellicolli, Can. J. Biochem. 46, 945 (1968).
3. J. P. Kratohvil and H. T. Dellicolli, Fed. Proc., 29, 1335 (1970).
4. L. B. Vitello, Ph.D. Thesis, Clarkson College of Technology, Potsdam, New York, 1972.
5. T. C. Laurent and H. Persson, Biochim. Biophys. Acta, 106, 616 (1965).
6. F. P. Woodford, J. Lipid Res., 10, 539 (1969).
7. A. Norman, Proc. Soc. Exp. Biol. Med. 115, 936 (1964).
8. A. Norman, Proc. Soc. Exp. Biol. Med. 116, 902 (1964).
9. B. Borgstrom, Biochim. Biophys. Acta, 106, 171 (1965).
10. K. Fontell, in "Surface Chemistry", P. Ekwall, K. Groth and V. Runnström-Reio, Editors, 253-267, Academic Press, New York, 1965.
11. K. Fontell, Kolloid Z. U. Z. Polym., 246, 710 (1971).

12  J. I. Howell, J. A. Lucy, R. C. Pirola and I. A. D. Bouchier, Biochim. Biophys. Acta, 210, 1 (1970).
13  S. Y. Oh and R. T. Holzbach, Biochim. Biophys. Acta, 441, 498 (1976).
14  L. Mandelkern and P. J. Flory, J. Chem. Phys., 20, 212 (1952).
15  M. E. McDonnell and A. M. Jamieson, Biopolymers 15, 1283 (1976).
16  V. G. Cooper, S. Yedgar and Y. Barenholz, Biochim. Biophys. Acta, 363, 86 (1974).
17  B. Chu, "Laser Light Scattering", Academic Press, New York, 1974.
18  A. M. Jamieson and A. G. Walton, J. Chem. Phys., 58, 1054 (1973).
19  B. J. Berne and R. Pecora, Ann. Rev. Phys. Chem., 25, 233 (1974).
20  C. Tanford, J. Phys. Chem., 78, 2469 (1974).
21  D. E. Koppel, J. Chem. Phys., 57, 4814 (1972).
22  R. T. Holzbach, M. Marsh, M. Olszewski, and K. Holan, J. Clin. Invest., 52, 1467 (1973).
23  D. H. Neiderhiser and H. P. Roth, Proc. Soc. Exp. Biol. Med., 128, 221 (1968).
24  D. H. McQueen and J. J. Hermans, J. Colloid Interface Sci., 39, 387 (1972).
25  C. W. Pyun and M. Fixman, J. Chem. Phys., 41, 937 (1964).
26  P. Mukerjee and J. R. Cardinal, J. Pharm. Sci., 65, 882 (1976).
27  P. Laggner, (1976), personal communication.
28  K. Müller, (1976), personal communication.
29  S. Y. Oh, R. T. Holzbach, M. E. McDonnell and A. M. Jamieson, (1976), unpublished data. Submitted for publication. *
30  R. C. Sehlin, E. L. Cussler and D. F. Evans, Biochim. Biophys. Acta, 388, 385 (1975).
31  P. Mukerjee, (1976), unpublished data.

\* This manuscript contains data which are referred to and briefly summarized in Discussion section of the present paper.

# DISCUSSION

## On the paper by C. Tanford

**L. E. Scriven,** *University of Minnesota*: It is not clear to me what requirements demand a disk-like shape for the micelles you are considering. Only under fairly severe constraints do disk-like forms have minimal surface area. Thermal fluctuations have stronger effects the smaller the number of molecules in an equilibrium structure.

If the micelles are disk-like, is their planform necessarily circular, or elliptical, or otherwise, and what controls it? If they are edged with half-torus-like closures, what causes some molecules to make up the closures, some to form into the disk faces, and many fewer to line up into the seams? Might not a biconcave disk, well-rounded all over, be as likely? And even a full torus, above a certain aggregation number?

On physical grounds, I would expect that all of these forms and many more are possible; that prolate ellipsoids are not improbable; that oblate ellipsoids may be most probable; that thermal fluctuations carry a liquid micelle continually through all of these states, shape changes propagating over the micelle surface like random waves on some turbulent globe; that this mild chaos renders every molecular location in the structure nearly equivalent; and that the most reasonable first approximation where the average shape is wanted is an ellipsoid. Would you care to comment?

**Charles Tanford:** With at least one dimension more or less fixed, all possible micellar shapes fall into one of two categories: rod-like (prolate ellipsoids, cylinders with caps, etc.) and disk-like (oblate ellipsoids, biconcave disks, etc.). The area per head group decreases only slowly with aggregation number for rod-like micelles, but decreases rapidly for disk-like micelles, regardless of the precise shape chosen for area calculations. The requirements of the hydrophobic driving force are therefore more easily met by disk-like micelles, especially as the cratic factor, $[Z]^m$ in equation 1, favors smaller micelles when other factors are equally well satisfied. Only when the head group repulsion factor is extremely large, as in ionic micelles in the absence of added salt, would I expect rod-like micelles to be favored, and even then only at high amphiphile concentrations.

I agree fully with your final paragraph, except that I would expect prolate ellipsoids to make only a very small contribution to the equilibrium distribution.

K. Shinoda, *Yokohama National University*:
 1. Can you explain the finite aggregation of nonionic surfactants?
 2. If we add co-surfactants, such as hexanol or heptanol, the aggregation number of micelle increases and finally becomes infinite. Mole fraction of alcohol in this mixture is not very big. 0.5~0.2 in mole fraction. Can you explain this fact by your theory?

C. Tanford:
 1. The repulsion factor for nonionic surfactants is presumably steric, a requirement for a relatively large free volume for the head group. The cratic factor, $[Z]^m$ in equation 1, also becomes more unfavorable as micelle size increases.
 2. The theory is easily modified to allow for co-micellization with molecules such as alcohols, but I have not yet made any calculations on the effect this has on the optimal micelle size.

## On the paper by K. S. Birdi

J. H. Fendler, *Texas A & M University*: What is precisely the "micellar interior"? What is neglected in the calculation?

K. S. Birdi: Any property of the micelle which is related to the alkyl chain of the amphiphile molecule can be safely regarded as being related to the micellar interior. The free energy of solubilization, $\Delta G_s^o$, is mostly determined by the hydrophobic interactions between the alkyl chains inside the micellar interior, and hence it is related to the micellar interior.

In the calculation we have not neglected anything, since the aim was merely to determine how RT ln CMC $- \Delta G_\phi^o$ quantity varied with the area per polar head group available at the micelle-water interface. The only assumption made was that $\Delta G_\phi^0 \sim \Delta G_\phi^\emptyset$. Since we found linear relation between RT ln CMC - $\Delta G_\phi^0$ and area per polar group (or $1/N^{1/3}$) shows that our assumption is acceptable within the experimental accuracy. This indicates that the opposing forces are related to the surface free energy of the micelle.

## On the paper by N. Muller

K. S. Birdi, *Fysisk-Kemisk Institut, Lyngby, Denmark*: Without exception, in nonionic aqueous solutions, as one increases the temperature, the CMC changes little with temperature (T), while at the same T, the aggregation number ($\nu$) changes drastically. Since your model is mainly based on any arbitrary value for S, then it is clear that one cannot easily apply it to these systems.

Further, I would like to bring to your attention some more examples where CMC-T method gives incorrect enthalpies, as compared to calorimetric methods: See K. S. Birdi in "Colloidal Dispersions and Micellar Behavior", K. L. Mittal, Editor, ACS Symposium Series No. 9, American Chemical Society, Washington, D.C., 1975.

N. Muller: The model is not set up to deal with the concentration dependence of $\nu$ at constant T, but for data at nearly constant concentration it can accommodate any combination of values of s, $\nu$, $d\nu/dT$, and $d(\ln CMC)/dT$ that might be suggested by other experimental results. Your interesting paper in the reference cited provides additional illustrations of my statement that "Equation (6) often, but not always, yields results in good agreement with those of calorimetric determinations," and I thank you for calling it to my attention.

J. R. Mendel, *Eastman Kodak Co.*: I would like to know more of your interpretation of the dependence of $\Delta G$ on chain length. I want to compare some of Mukerjee's ideas with those of Tanford. I'm interested in knowing how you have found free energy to be related to chain length of surfactant. Do you have a paper, or could you suggest a reference?

N. Muller: I have not dealt with the dependence of $\Delta G^O$ on chain length in this paper, and so I am unprepared to answer your question. A really satisfactory answer should include an explanation on how the contributory $\Delta H^O$ and $\Delta S^O$ values vary with chain length, and this is still an unsolved problem.

## On the paper by P. Mukerjee, J. R. Cardinal, and N. R. Desai

J. H. Fendler, *Texas A & M University*: We agree with the solubilization sites you suggest (see J. Phys. Chem., 75, 3907, 1971).

L. E. Scriven, *University of Minnesota*: Calculations which bear out your surmises about 'areas' and 'humps' were made by Professor Bidner while a Fulbright visitor in my department in January-March 1974. Model calculations established what is intuitively obvious once intuition is developed:

    1. If the amphiphile inventory in premicelles is low, successive addition equilibrium constants must fall rapidly with increasing aggregation number of small complexes and may then level off at a low value.

    2. Successive addition equilibrium constants must then rise, peak, and fall again if the size distribution of completed micelles is narrow. The rise and peak are consequences of 'closing the package'. When the micelle is completed the amphiphile zone defines a closed surface with an inside and an outside where neither existed before. The change is a topological one. The topological change may be a more significant feature than the cooperativity of the process.

    3. If micellar aggregation numbers are large and total premicellar inventory is small, there can be relatively very few premicelles of any given aggregation number, and therefore substantial rate limitations on establishing micellization equilibrium.

    4. The Bury-Hartley model, with only amphiphile monomers and monodisperse micelles, is a singular limit of multiple equilibria.

These calculations and additional material appear in a paper the printing of which was unfortunately delayed until January by economic and political troubles abroad: Latin American Journal of Chemical Engineering and Applied Chemistry, 6, 1-32 (1976), published by SAICIQQA, Avenida 1, No. 867, La Plata, Argentina. I believe the results both confirm and complement yours, and I commend the paper to all who are interested in the subject.

P. Mukerjee: I am happy to learn about this very important paper.

## On the paper by R. Zana, J. Lang, S. H. Yiv, A. Djavanbakht, and C. Abad

C. Tanford, *Duke University*: How secure can you be in the interpretation of $\tau_1$ and $\tau_2$? Is it possible that these kinetic parameters refer to processes quite different from those to which you assign them?

## DISCUSSION

R. Zana: Of course I cannot say that the interpretation given to $\tau_1$ and $\tau_2$ is 100% sure. However, all of the experimental results presently available for the fast and the slow relaxation processes agree with the model presented to explain the relaxation behavior of fairly dilute micellar solutions. Moreover, from a quantitative point of view this model yields data (rate constants, micelle polydispersity, rate of change of the rate constants with chain length) which are physically reasonable and/or quantitatively consistent with the postulated model. I must also add that other interpretations, such as that presented by Professor Yasunaga during this symposium, do not permit the explanation of all of the experimental results and/or lead to rate constants not consistent with the model. Therefore, until the finding of new results which would contradict the model of Aniansson et al. (J. Phys. Chem., 80, 905, 1976), this model must be considered as good as a model can be.

T. J. Gilligan, *Carnegie-Mellon University*: You interpret the presence of only one distribution of relaxation times to mean that micellization is absent. Given the small aggregation numbers could this result because the amplitude of this micellization step is small and/or the rate for this step is close to that for the exchange step.

R. Zana: The word micellization in its present sense refers to any process by which compounds associate to form aggregates containing anything between 2 and a very large number of associating molecules. In thisrespect we do not claim the presence of only one distribution of relaxation times to mean that micellization is absent. On the basis of the Aniansson and Wall theory we interpret such a result to mean that the micelle size distribution curve does not present a maximum and a minimum and we correlate this result with the absence of a CMC (and thus of a "true" micellization) on our ultrasonic absorption versus concentration curve. As pointed out in the paper this agrees with the conclusions of the recent thermodynamic study of Ruckenstein and Nagarajan. In case of "true" micellization with a small aggregation number, I cannot say much about the amplitude of the micellization step. It is likely, indeed, that the micellization rate will then be close to the exchange rate but I do not know if this will result in only one distribution of relaxation times (the ratio of these rates would have to be well below 10). However, even in case of small aggregation numbers a choice can be made between "true" micellization and continuous association by looking at the shape of the ultrasonic absorption versus concentration curve. Indeed such a plot presents a marked maximum in case of "true" micellization, and a monotonous increase for a continuous association.

## On the paper by K. Takeda, T. Yasunaga, and H. Uehara

T. Gilligan, *Carnegie-Mellon University*: In your analysis of the amplitude data of your ultrasonic measurements, you appear to neglect the contribution of the enthalpy to amplitude. Given the variation of $\Delta H_i$ for different size aggregates from exothermic (small) to endothermic (large), would you not expect $\Delta V$ to be affected and perhaps be best interpreted in terms of series of $\Delta V_i$ that are a function of aggregate size?

T. Yasunaga: As pointed out by you, the contribution of $\Delta H$ was neglected to the amplitude. It may be possible that these data are interpreted in terms of a series of $\Delta V_i$. In such a case, however, one needs more bold assumptions or simplifications. On the other hand, if $\ldots < \Delta V_i < \ldots \Delta V_n$ and $\ldots < [A_i]_e < \ldots [A_n]_e$, the contribution of $\Delta \bar{V}_i$ to the relaxation amplitude may be small.

G. Aniansson, *Chalmers University of Technology*; H. Hoffmann, *University of Bayreuth*; and R. Zana, *Centre de Recherches sur les Macromolecules, Strasbourg*: The model that is used in this paper is one of the two that were considered by Graber et al. (Kolloid Zeitschrift 1970). This phenomenological model was shown to yield volume changes upon micelle formation which did not agree with values obtained by other methods. In addition the model presented by Dr. Yasunaga assumes a slow exchange of monomers between the micellar phase and the bulk solution, which is in contradiction with nmr measurements made by a number of workers. Furthermore, the model does not give the concentration dependence of both relaxation times that have been obtained experimentally (see G. Aniansson, J. Phys. Chem. **80**, 905, 1976). Finally, it should be pointed out that models such as the one presented in this paper yield overall rate constants which cannot be correlated with elementary processes involved in micelle formation.

T. Yasunaga: Please see with care especially the discussions on $\Delta V^o$, nmr, elementary processes, conclusive part, and so on.

## On the paper by M. Almgren, E. A. G. Aniansson, S. N. Wall, and K. Holmaker

L. E. Scriven, *University of Minnesota*: Is the case for the exclusively <u>stepwise</u> dissociation kinetics of micelles a strong one? When the relaxation spectrum is approximated by just two processes, one fast and the other slow, how certain is it that the slow process is dominated by readjustment of the <u>entire</u> distribution through 'flux' of amphiphile across the entire premicellar range? One wonders if perturbations may not promote thermal fluctuations that lead to fission processes, i.e. changes in topology. Examples are the breaking apart of a globular micelle into premicellar fragments, or the puncturing of a disk-like micelle to give a torus, or the pinching off of part of a cylindrical micelle. The reverse processes of fusion would seem to be rather less probable, and the restoration of equilibrium by stepwise association is surely relatively slow. Have the experimental data been examined with an eye to fission processes as alternatives to purely stepwise dissociation? The appropriate equations one expects are differential-summation equations, and the analogy is with flow in a branched, highly interconnected network of tubes.

E. A. G. Aniansson: The fact that only relaxation times, or perhaps, two groups of closely similar relaxation times are found in surfactant micellar systems is mainly due to the fact that the signal in relaxation measurements on these systems is essentially proportional to the amount of surfactant material in monomer, or equivalently, micellar form. If the concentration of individual micellar sizes could be followed, a larger set of relaxation times would in principle be obtained.

The basic assumptions underlying our treatment are:

1. The <u>attainment of equilibrium</u> occurs mainly through the entrance and exit of one monomer at a time into or out of the aggregates. (If the entrance or exit of dimers gave a noticeable contribution to the rate the theory would be the same except for the replacement of $k_s^-$ with a closely similar quantity.)

2. The minimum in the distribution curve at intermediate-sized aggregates is <u>deep.</u>

3. The product $k_s A_s$, also, has a deep minimum among the rare intermediate-sized aggregates.

4. The initial relative deviations from equilibrium are much smaller than unity.

We believe that on the basis of existing knowledge about surfactant micellar systems these assumptions constitute a least-biased or least-peculiar set.

In essence, Dr. Scriven's question concerns the assumption (4) and is a particular case of a general one pertaining to all jump relaxation experiments: Do the strongly non-equilibrium physical conditions during the formation of the jump, the perturbation, bring about extremely fast changes in chemical composition so that notwithstanding their very short time span measurable changes in composition results?

This question has been contemplated before but to our knowledge there exists so far no clear evidence for such effects. One therefore assumes that the concentrations of the various species immediately after the perturbation are the same as before, i.e. the concentrations are those of the equilibrium conditions before the jump.

If the perturbation, as Dr. Scriven suggests, causes a measurable amount of fission the fragments will either fall in the region of rare intermediate-sized aggregates or in the micellar region itself. Only if measurable amounts of fragments fall in the former region will an initial situation result that in effect differs from that of point 4.

The stepwise motion out of the intermediate region would be characterized by a time constant of the order of $\ell^2/k_m$, where $\ell$ is a measure of the half width of the minimum in $\bar{A}_s$ similar to $\sigma$ for the maximum and $k_m$ is an average value of $k_s^-$ around the minimum. Reasonable estimates of these quantities would yield for the time constant a value intermediate between those of the fast and the slow processes or closer to the former. If fusion contributed to this process to a considerable extent due to the abnormally high concentrations of intermediate aggregates the time required would be even smaller. Also, it would even in effect be a second order reaction since the relative deviations from the equilibrium concentration caused by the initial fission process would be much larger than unity in the intermediate region if the process is of measurable extent. Considerable deviations from exponentiality would result.

At the end of such a process, whether it occurs by fusion or stepwise or both, the number of micelles would not in general be the equilibrium one. Since the concentrations are now close to the equilibrium ones fusion is now improvable at least for ionized aggregates and the final restoration of equilibrium would occur through a flow over the whole range of aggregation space.

In conclusion, then, the initial perturbation process invoked would result in three distinct processes and one would expect to find experimentally three distinct relaxation times (in addition to purely physical or counterion relaxations).

Editor: The same question was addressed by Dr. Scriven to Dr. Zana and Dr. Hoffmann but these authors asked Dr. Aniansson to answer this question for them also.


On the paper by N. A. Mazer, M. C. Carey and G. B. Benedek

G. S. Hartley, *Hamilton, New Zealand*: You are confident that light-scattering evidence for "super-spherical" size in sodium dodecyl sulfate solutions combined with no great increase of viscosity means that the micelles must be oblate spheroids. Are you confident that the viscosity data are correctly interpreted with due allowance for effect of shear rate which, as I pointed out in my paper, can increase viscosity over a low range (by creating rods?) and decrease it over a high range (by breaking them).

DISCUSSION

C. Tanford: I can see no basis for questioning viscosity and diffusion data obtained in numerous laboratories which demonstrate that (at lower ionic strength than Mazer was using) sodium dodecyl sulfate micelles must approximate an oblate ellipsoidal shape. A transition to rod-shaped micelles has been observed in the absence of added salt when the repulsion between head groups can be expected to become very strong. These results are in accord with theoretical expectation because the separation between head groups is considerably larger in rod-shaped micelles than in disk-like micelles, so that the former should be favored only when repulsion is strong. Mazer claims to find rod-shaped micelles at <u>very high ionic strength</u>; clearly an unexpected result. My remarks at the end of his paper were, however, not intended to dispute his result. I was objecting to the presentation of his results without any reference to previous work on the subject and without an explanation to suggest why the unexpected result, inconsistent with previous observations on the effect of ionic strength might have been obtained.

Editor: Dr. Hartley has directed this question to Dr. Tanford as it arose from Dr. Tanford's comment on the above paper.


On the paper by N. A. Mazer, R. F. Kwasnick,
M. C. Carey and G. B. Benedek

T. J. Gilligan, *Carnegie-Mellon University*: Your model for the aggregation of bile salts to form large aggregates is isodesmic. This model does not result in critical behavior (cf. Mukerjee), but rather gives a decreasing number of aggregates with increasing aggregate size; yet you refer to the transition to this larger aggregate as a second CMC (cf. Small). This does not appear consistent. Would you like to comment on this point in regard to the controversy of continuous association vs. multiple-step micelle formation for bile salts?

M. C. Carey: We thank Dr. Gilligan for asking this question as it offers an opportunity to clarify an important point regarding bile salt aggregation. Our model for simple bile salt micelles assumes first that the bile salt monomers form small aggregates containing 2-10 monomers which we call primary micelles. The <u>primary micelles</u> then aggregate isodesmically (i.e. single polymeration constant model) to form <u>secondary micelles</u>. If the formation of the primary micelles involves some degree of cooperativity (i.e. aggregation number larger than 2) then there will be a CMC associated with the formation of the primary micelle. This is how we have consistently used CMC in the present paper. We do not, nor does Small, refer to a "second CMC" in the bile salt system to describe the formation of secondary micelles. Our model thus suggests that the aggregation of bile salts involves both cooperativity (bile salt monomers→primary micelles) and continuous association (primary micelles→secondary micelles).

J. K. Thomas, *University of Notre Dame*: I would like to add that we have shown a distinct decrease in the Chapman transition temperature of distearoyl phosphotidylcholine bilayers on addition of cholesterol. This indicates strong interaction of materials with structures similar to cholesterol, *viz.* bile acids with lecithins.

M. C. Carey: We thank Dr. Thomas for making his comment. Bile salts can interact with the lecithin bilayers to form a single phase lamellar system. Phase equilibria, and x-ray diffraction experiments have been carried out on these mixed bilayers[*] but calorimetric studies have not yet been done. One would expect that the phase transition temperature of lecithins would be depressed if the bile salts are intercalated between the fatty acid chains of the bilayer.

---

[*]D. M. Small, M. Bourgés and D. G. Dervichian, Biochim. Biophys. Acta 125, 563, 1966.

# Part III

# Micelles in Nonaqueous Media

# MICELLES IN APOLAR MEDIA

H. F. Eicke
Physikalisch-Chemisches Institut
der Universität Basel
Klingelbergstr. 80
CH-4056 Basel, Switzerland

The investigation of lipophilic micelles has been neglected for a long time and now proved to be of considerable scientific interest. Also many phenomena which are common to "normal" and "inverted" micelles are much more easily investigated in nonaqueous, apolar surfactant solutions. It now appears that the essential properties of micelles in apolar solvents are understood, namely their stability, the slight variation of micelle size in various dispersion media, the observed dipole moments, their dissociation behavior, and solubilizing capacities. Furthermore, in connection with the formation of micelles in apolar media there seems to be general agreement with regard to the existence of premicellar aggregates which may adopt the role of nuclei in the frame of the micellar pseudo-phase model, whose presence was again demonstrated by recent dielectric increment and kinetic measurements applying the dielectric field effect. Particularly, these kinetic investigations revealed information in favor of a hypothesis proposed in order to explain the transition from an initially linear arrangement of premicellar aggregates to a closed aggregate. The experimental results conform satisfactorily with a model which predicts a conformational transformation between two premicellar states.

## INTRODUCTION

The remarkable lack of interest which prevailed until recently concerning colloid science in non-aqueous media is probably due to the outstanding importance of aqueous systems, particularly with respect to the biological implications of aqueous colloidal solutions. Thus micelle formation in non-aqueous colloidal solutions, although recognized for some time, has been investigated systematically only in recent years, both experimentally and theoretically. It should be noted, actually, that much information on micelles is probably more easily accessible from a study of apolar surfactant solutions, i.e. avoiding the complications imposed by electrical double layer and dissociation phenomena.

The apparent segregation of considerations on aqueous and non-aqueous (in particular non-polar organic) surfactant solutions is not based primarily on purely formal or historical reasons. The interactions governing the formation of surfactant aggregates in apolar media are, indeed, quite different from those in aqueous solutions, in spite of the apparent similar building principle of lipophilic and hydrophilic micelles.

With regard to the term "micelle" there appears to exist still no universally accepted definition: It should designate any soluble aggregate, spontaneously and reversibly formed from ampiphilic molecules or ions.[1]

The micellization processes according to the commonly used equilibrium thermodynamical descriptions, namely
        a) the multiple equilibrium model, and
        b) the pseudo phase model,
are - like the micelle definition - equally well applicable to aqueous and non-polar surfactant solutions.[2]

The second model (b) would better conform to the above definition of a micelle. Both models describe limiting experimental situations and there should not be much discussion as to which has to be applied. If the pseudo-phase model is appropriate it has the advantage of a rather limited number of parameters.

In order to realize the above mentioned remarkable differences between interactions encountered in aqueous and non-polar surfactant solutions, considerations on a molecular level are necessary. For this purpose stability considerations as to the existence of micelles in apolar media are appropriate.

## STABILITY OF LIPOPHILIC MICELLES

It is essential to note that the enthalpic and entropic contributions to the formation and stability of micelles in apolar media are qualitatively and quantitatively different from their corresponding values in aqueous systems. The *entropic* part appears to be primarily determined by the properties of the particular surfactant (i.e. the solute-solvent interactions are in general less pronounced except that in some cases a weak solvent dependence of the micelle size has been observed)[3] and differs in this respect significantly from the situation in aqueous systems. The entropic contribution has been determined according to the Hildebrand model[4] considering the different molal and intrinsic van der Waals volumes of mixing components[5] (see Figure 1).

Considering the *enthalpic* contribution with regard to the micelle stability which is composed in principle of electrostatic and dispersion interactions, it appears necessary to realize from experimental observations that the monomers within the micelle are

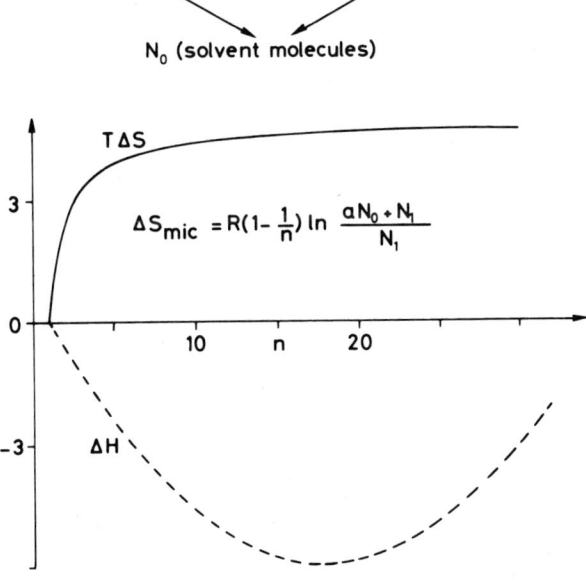

Figure 1. Entropic energy contribution to $\Delta G_{mic}$ according to the Hildebrand model: $N_1$ monomers and $N_2 = N_1/n$ micelles are mixed with $N_0$ solvent molecules: Entropies of mixing: $\Delta S_1$ and $\Delta S_2$. $\Delta S_{mic} = \Delta S_2 - \Delta S_1^o$. Units of ordinate e.g. kcal/mol (see (5)).

still dispersed in the solvent and may, in this respect, be compared with a "solvent penetrated random polymer coil". This would imply that the dispersion interactions should yield only minor contributions compared with the electrostatic terms. In order to acquire better judgment a detailed examination[3] of electrostatic and dispersion interaction energies is necessary, the latter in terms of the so-called microscopic theory, i.e. additivity of molecular interaction forces which may be improved by correction terms according to Renne and Nijboer,[6] thus conforming with Lifshitz's theory.[7] The choice of a geometric model corresponding to the shape and structure of micelles in apolar media is not too critical with respect to the approximative character of the calculation. Generally, spheroidal or ellipsoidal shapes have been proposed based on various experimental determinations.[8,9] A closed surface structure is to be expected, however, in view of energetic considerations (shielding of polar groups, dispersion interactions of apolar tails, etc.). In general, the free surface energy (enthalpy) of the particular aggregate has not to be considered, since this is usually negligibly small in apolar media. Thus, the enthalpic contribution to the micellization is thought to be composed of $\Delta H_{mic} = \Delta W^D + \Delta W^C$ where

$$\Delta W^D = W^D_{11} + W^D_{22} - 2W^D_{12}$$

and
(1)

$$\Delta W^C = W^C_{dd} + W^C_{cd} + W^C_{cc} + W^C_{aa} - W^C_{ca} + W^C_I$$

where D denotes dispersion interactions: subscripts 11 = intramicellar, 22 = solvent-solvent, 12 = solute-solvent and C Coulombic interactions: subscripts: a = anion, c = cation, d = dipole, I = ionization.

The ionic groups of the monomers will be considered as a mixture of $(1-\beta)$ dipoles continuously distributed over the surface of the polar core assumed (without specific assumptions) to be spherical and $\beta$ ionic pairs which are dissociated where $\beta$ is the degree of dissociation.

A combination of the above discussed enthalpic and entropic energy contributions as well as of the free enthalpy due to thermal fluctuations (to be discussed in the next paragraph) yields the free enthalpy change of micellization $\Delta G_{mic}(n)$ as a measure of the stability of lipophilic micelles.

## DYNAMIC ASPECTS OF LIPOPHILIC MICELLES

The difficulty of explaining the dipole moments of spheroidally or ellipsoidally shaped micelles may be overcome by considering the thermal fluctuations of the polar surfactant molecules within the micelle. Without any specific assumption it is possible by means of this consideration to calculate dipole moments of symmetrically shaped aggregates and these are found to be in close agreement with the experimentally determined values.

The basic idea is to assume that due to these thermal fluctuations each of the dipolar surfactants building up the micellar phase alternates between being a constituent part of the proper micelle and a solvated (solubilized) dipole molecule in the polar core of the micelle with a mean dielectric constant $\varepsilon$.

Recently, fluorine-NMR measurements have been made in different solvents (including AOT micelles in $C_6H_{12}$) with $CF_3CH_2OH$ as an indicator assuming the perturbations of the polar micellar core by the $CF_3CH_2OH$ to be negligible. This assumption appears to be justified by the high pk-value of fluoroethanol in water which is approximately 12.30, thus ruling out a possible proton exchange between $CF_3CH_2OH$ and AOT. From this preliminary investigation a reasonable value for $\varepsilon_{mic}$ has been deduced[10] (see Figure 2). In order to

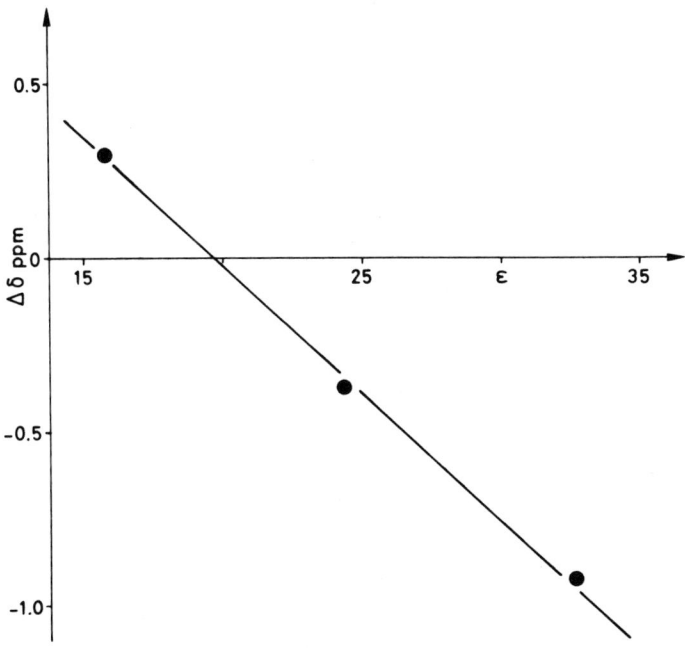

Figure 2. Correlation between chemical shifts ($\Delta\delta$) and dielectric constant ($\varepsilon$) using $CF_3CH_2OH$ in 2-butanol, AOT/cyclohexane, ethanol and methanol. Order of experimental points from left to right corresponds to that of the solvents given in the legend.

obtain an estimate of the order of magnitude of the dipole moment produced by these thermal fluctuations, the differential free energy of solvation of a dipole molecule ($\mu$) with radius (a) fluctuating in the gradient of the dielectric constant $\varepsilon$ within the micelle is calculated assuming that the dielectric constant is decreasing radially outwards (Figure 3). In this way it is found that

$$dG_{solv} = - \frac{\mu^2}{3a^3} \frac{d\varepsilon}{\varepsilon^2} \qquad (2)$$

if only small fluctuations, i.e. small variations of $\varepsilon$ are considered. Calculating approximately the dielectric constant attributed to the polar region in terms of the dipole moment of the constituent dipolar surfactant molecules within the micelle according to an expression derived from Onsager's theory

$$\varepsilon \simeq \frac{32 \pi n \mu^2}{9 V_{mic} kT} \qquad (3)$$

$dG_{solv}$ can be represented as a function of $\mu$. Thus, according to the fluctuation theory relation[11]

$$\frac{d^2 G}{d\mu^2} = \frac{1}{n} (kT/<\Delta\mu^2>) \qquad (4)$$

(where n is the degree of association per micelle) one obtains

$$\frac{<\Delta\mu^2>}{\mu^2} = \frac{16\pi}{3} \frac{1}{n} \qquad (5)$$

and the corresponding contribution to the micelle stability

$$dG_{mic} = - \frac{kT}{2} \qquad (6)$$

kT refers to the micelle as a kinetic unit.[12]

The existence of considerable dipole moments of the monomeric surfactant molecules and the lipophilic micelles, respectively, open the possibility to investigate molecular rearrangements and thus, rotational motions of surfactant molecules within the aggregational state. This implies, however, that the mean dielectric increment of the surfactant solution or the overall observed dipole moments are changing during the aggregational process. Since this is the case (according to Figure 4 and the above proposed fluctuation model), relaxation kinetic investigations have been carried out with the help of the dielectric field effect.[13] This can be utilized to derive kinetic data if the chemical transformation is accompanied by a finite change of the electrical moment of reaction $\Delta M$, which is the case with a number of anionic surfactants as was demonstrated

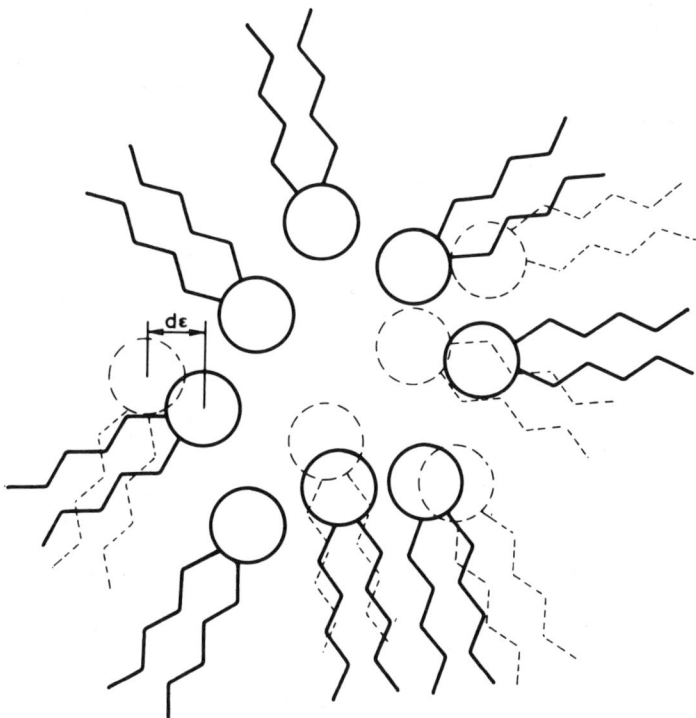

Figure 3. Fluctuating ionic surfactant molecules in the gradient of the dielectric constant within the micellar aggregate.

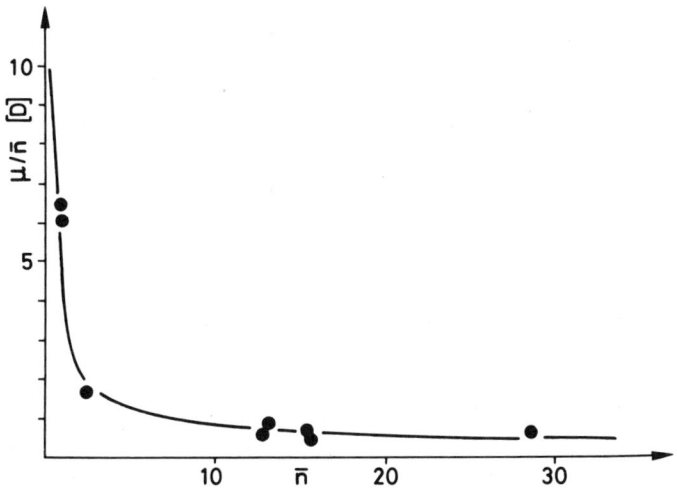

Figure 4. Apparent decrease of dipole moment/surfactant monomer during micellization (see (5)).

experimentally by dielectric increment measurements.[5] With respect to the dependence of the equilibrium constant K on the electric field strength E described by a van't Hoff type equation, i.e. $\delta \ln K/\delta E = \Delta M/RT$ (which yields a nonlinear relation between ln K and E for the usual case that $\Delta M$ is proportional to E) high static fields modulated by a low amplitude a.c. field must be used to observe finite shifts of the equilibrium. Consequently the dielectric excess loss, i.e. the loss difference with and without high field, is the experimentally determined quantity. From the characteristic frequencies and the associated amplitude factors the chemical relaxation time and the physical parameters of the reacting chemical system have been derived.[14] The results verified again the cooperative feature of the micelle formation for the system AOT/cyclohexane as shown in Figure 5 where the amplitude factors of the orientational and chemical relaxation process have been plotted versus the weighed-in concentration of AOT. The most remarkable result is believed to be shown by Figure 6. It presents a graph of the characteristic frequency $f_c$ plotted versus the weighed-in concentration of AOT. This kinetic behavior with decreasing relaxation times in a particular concentration range re-

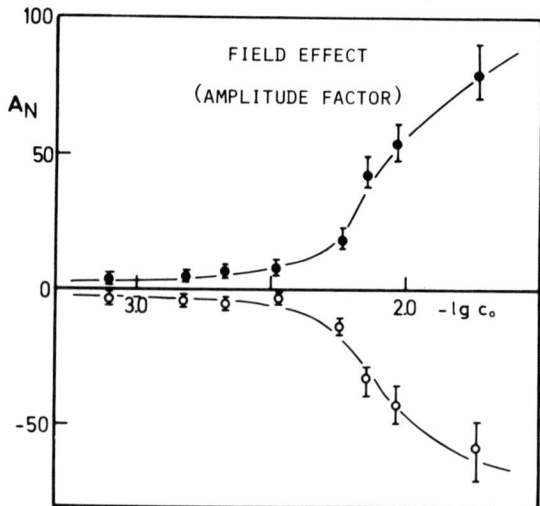

Figure 5. Amplitude factors of field effect measurements normalized with respect to applied DC-field:AOT/cyclohexane. Chemical excess loss (solid circles), orientation field effect (open circles). From Reference 14, Berichte der Bunsengesellschaft für Physikalische Chemie.

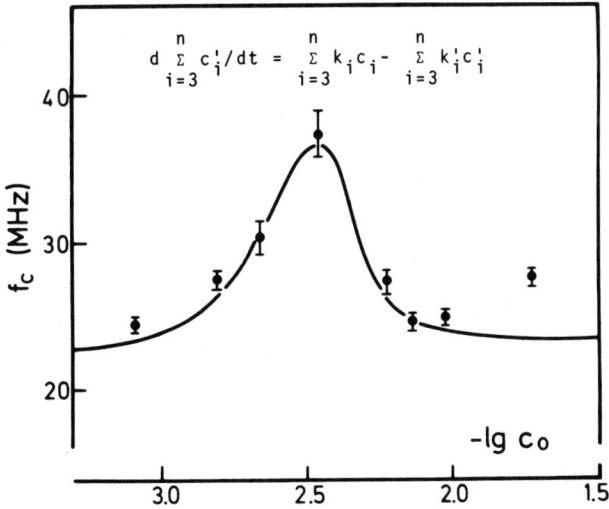

Figure 6. Characteristic frequency due to the chemical field effect versus AOT concentration. Curve through the data points is calculated according to the reaction scheme (see Table I). From Reference 14, Berichte der Bunsengesellschaft für Physikalische Chemie.

sembles peculiarities which have been ascribed to Eigens' treatment of the Monod model,[15] conformational changes induced by ligand enzyme interactions. In order to explain this result the following reaction scheme has been proposed where each association step may undergo a conformational change $C_i \rightleftarrows C_i'$, with an equilibrium constant $K_i$. The conformational transformation is assumed not to influence the binding energy of a monomer to a premicellar aggregate. Starting from Equation (7),

$$d \sum_{i=3}^{n} c_i'/dt = \sum_{i=3}^{n} k_i c_i - \sum_{i=3}^{n} k_i' c_i' \qquad (7)$$

(with $k_i$ and $k_i'$ as rate constants of the isomerization) describing the kinetics of this transformation one obtains finally - considering only small perturbations of the equilibrium - an expression for the concentration dependence of $1/\tau_{chem}$, the reciprocal chemical relaxation time (which is proportional to the characteristic frequency $f_c$) (Figure 6). The initial steep increase of $f_c$ followed by its rapid decrease after passing a maximal value of the relaxation time can be described by the above proposed model only if the rate constants $k_i$ and $k_i'$ at the beginning and the end of the

Table I. Reaction Scheme Suggested for Conformational Transformation.

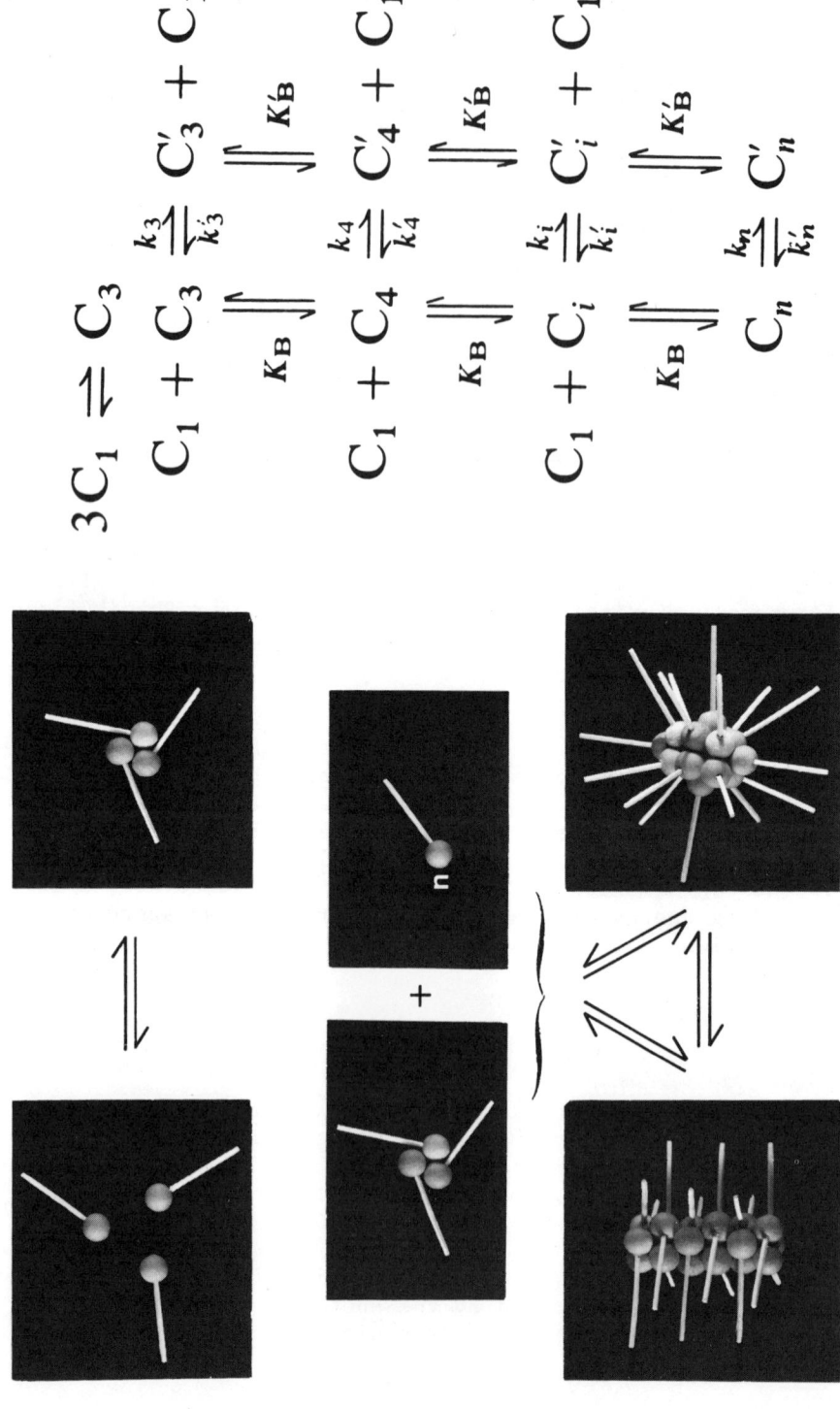

micellization process are smaller than for premicellar aggregates of medium size. Under this condition it could be inferred from the parameters chosen to fit the theoretical curve to the experimental points that the transformation is to be expected between the sixth and seventh association step of the premicellar subunit. The rate constant for this process is found to be $3.2 \times 10^8$ sec$^{-1}$. Since $f_c$ is related with the chemical relaxation time $\tau_{chem}$ by $1/\tau_{chem} = 2\pi f_c$ which in turn determines the rate constant of this isomerization process, $f_c = 5 \times 10^7$ Hz could be considered as a lower limit of the fluctuation frequency of AOT molecules within the (almost) built up micelle.

## PROPERTIES OF LIPOPHILIC MICELLES AND THEIR SOLUTIONS

With regard to the above proposed micellar model in apolar surfactant solutions it appears essential to realize that rather general assumptions with respect to the enthalpic and entropic energy contributions to $\Delta G_{mic}$ are evidently sufficient to describe the characteristic features of typical micelle forming compounds as, for example, AOT, AY, Di-2-ethylhexylphosphate[3] etc. The experimentally determined data used for the calculation of the following graphs refer to AOT. It is then found that

1. The micelle size varies only slightly with the solvent used (Figure 7).
2. The temperature dependence of the cmc can be described satisfactorily.
3. The observed dipole moments can be derived - independently of the micellar shape and structure.
4. The degree of dissociation ($\beta$) can only be varied within narrow limits to avoid unrealistically large Coulombic repulsion terms. Thus, $\beta$ was calculated to be approximately 10%.
5. The maximum of the $\Delta G(n)$-curve (Figure 8) may be interpreted in terms of a "potential" barrier which exactly simulates the expected behavior of a micellar nucleus. The existence of these "nuclei", i.e. of premicellar aggregates, which are thought to be necessary for the micellization process was shown with the help of various experimental methods, including, for example, conductivity,[16] dielectric increment,[5] vapor pressure osmometric,[15] and x-ray measurements.[17] The concentration of these aggregates, however, is small which corresponds to the model shown in Figure 8, i.e. attributing to them an essential feature of a "classical" nucleus. From the existence of the above mentioned barrier which is now assumed to be attributed to the nucleation step, it may be concluded that it is more favorable to attach monomers to a nucleus than to assume aggregation of subunits. Quite generally, nuclei

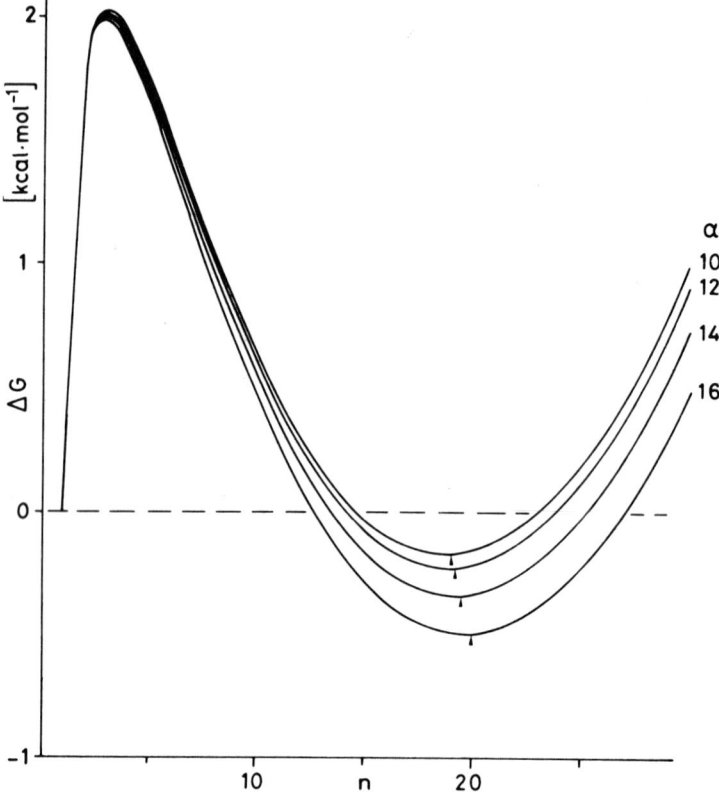

Figure 7. Dependence of micelle size on solvent expressed by polarizability $\alpha_{solvent}$. $\alpha$ in units of $(.1 \text{ nm})^3$ (see (3)).

(as well as other aggregates) formed by polar surfactants in apolar media are expected to possess preferred geometrical structures due to the "vectorial" character of the electric dipole moment.

Finally, it appears essential to realize that this type of treatment is in no way restricted to a special geometric model. The main feature is that the dispersion interaction term is proportional to the number n of the constituent monomers, whereas the electrostatic repulsion of like charged ions is proportional to a higher power of n. Thus it is to be expected - since all substantial interactions have been considered - that, for example, long narrow cylindrical micelles (which are believed to exist in water)[1] could not be stabilized in apolar media since their corresponding electrostatic interactions would be proportional to n. These have

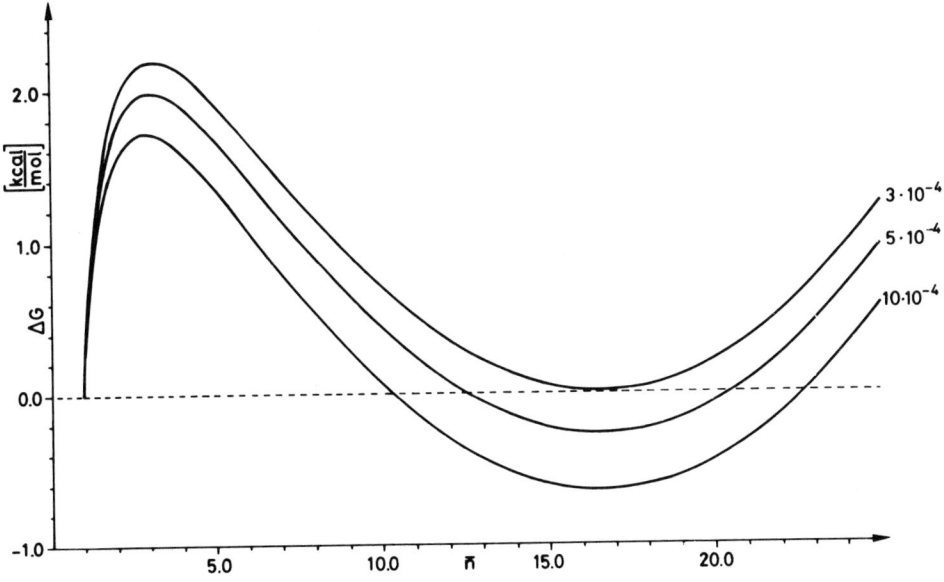

Figure 8. Free enthalpy change of micellization. Parameter: $c_{surfactant}$  CMC = $3 \times 10^{-4}$ mol dm$^{-3}$.

in fact not been experimentally observed in apolar surfactant solutions at moderate concentrations.

On the other hand, however, an initially linear arrangement of premicellar subunits has to be concluded, i.e. starting from a nucleus (triangular shaped as in the case of AOT) which is verified by the above mentioned X-ray measurements of Ekwall. The linear arrangement of premicellar subunits corresponds to a dipolar chain association (→→) with increasing dipole moment.

This type of association offers, furthermore, a minimum of sterical hindrance with respect to the apolar tails of the surfactant molecules. The growth of the premicelle starting from the nucleus would then proceed via an addition of surfactant molecules such that each dipole within the aggregate is compensated by way of a polypole association(⇅). Since the polar ends of the subunits formed in this way can hardly be shielded from the solvent molecules the need to attain a closed aggregate appears easily understandable. Such a transformation should thus occur at an intermediate aggregation number as is actually observed with the help of the dielectric field method.

A particularly interesting feature of these lipophilic micelles is closely related to the above mentioned fluctuation determined dielectric property of the polar core, namely their solubilizing power. This phenomenon is inherently connected with the existence of micelles, i.e. the tendency of surfactant molecules to enrich themselves in interfaces or even to produce an interface by the micellization. The possibility of stabilizing these micelles by solubilizing ions (e.g. $H^+$, $Li^+$, $Na^+$ etc.) and polar liquids (e.g. water, aqueous electrolytes, glycerol, etc.) has been extensively investigated, both experimentally [16,18,19,20,21] and theoretically.[22] The solubilization phenomena have also been the subject of many investigations with respect to both fundamental and technical aspects. Examples are the interpretation of the action of antistatic additives in hydrocarbon fuels[16] and considerations on structure and properties of the so-called microemulsions in apolar media.[23,24]

In conclusion it appears that the above presented experimental observations and theoretical considerations offer already a comprehensive picture of the structure and stability of lipophilic micelles and some of their characteristic properties.

## ACKNOWLEDGMENT

A grant from the Schweizerischer Nationalfonds zur Förderung der Wissenschaften under the project Number 2.314.75 is gratefully acknowledged.

## REFERENCES

1. C. Tanford, "The Hydrophobic Effect", John Wiley and Sons, New York, 1973.
2. P. Mukerjee, J. Pharm. Sci. 63, 972 (1974).
3. H. F. Eicke and H. Christen, J. Colloid Interface Sci. 46, 417 (1974).
4. J. H. Hildebrand, J. M. Prausnitz and R. L. Scott, "Regular and Related Solutions", Van Nostrand Reinhold Co., New York, 1970.
5. H. F. Eicke and H. Christen, J. Colloid Interface Sci. 48, 281 (1974).
6. M. J. Renne and B. R. A. Nijboer, Chem. Phys. Letters 1, 317 (1967).
7. E. M. Lifshitz, J. Exp. Theor. Phys. USSR, 29, 94 (1955).
8. J. P. Peri, J. Amer. Oil Chem. Soc. 35, 110 (1958).
9. P. A. Winsor, Chem. Rev. 68, 1 (1968).
10. E. M. Haidegger, Dissertation, Universität Basel, 1976.
11. J. C. Slater, "Introduction to Chemical Physics", Dover Repr., New York, 1970.

12. H. F. Eicke and H. Christen, (1976) J. Phys. Chem. submitted for publication.
13. K. Bergmann, M. Eigen and L. DeMaeyer, Ber. Bunsenges. Phys. Chem. 67, 819 (1963).
14. H. F. Eicke, R. Hopmann and H. Christen, Ber. Bunsenges. Phys. Chem. 79, 667 (1975).
15. M. Eigen, Quarterly Rev. Biophysics 1, 3 (1968).
16. H. F. Eicke and V. Arnold, J. Colloid Interface Sci. 46, 101 (1974).
17. P. Ekwall, L. Mandell and K. Fontell, J. Colloid Interface Sci. 33, 215 (1970).
18. H.F. Eicke and J. Rehak, Helv. Chim. Acta (1976) in press.
19. H. F. Eicke and J. C. W. Shepherd, Helv. Chim. Acta 57, 1951 (1974).
20. H. F. Eicke, J. C. W. Shepherd and A. Steinemann, (1976) J. Colloid Interface Sci., 56, 168 (1976).
21. J. H. Fendler and E. J. Fendler, "Catalysis in Micellar and Macromolecular Systems, Chapter 10, Academic Press, New York, 1975.
22. H. F. Eicke, J. Colloid Interface Sci. 52, 65 (1975).
23. H. F. Eicke, (1976) J. Colloid Interface Sci., in press.
24. G. Gillberg, H. Lehtinen and S. Friberg, J. Colloid Interface Sci. 33, 40 (1970).

AGGREGATION OF SURFACTANTS IN HYDROCARBONS. INCOMPATIBILITY OF THE CRITICAL MICELLE CONCENTRATION CONCEPT WITH EXPERIMENTAL DATA

A. S. Kertes

Institute of Chemistry
The Hebrew University
Jerusalem, Israel

An increasing volume of new experimental data on the physical chemistry of aggregation of surfactants in hydrocarbons and their nonpolar derivatives indicates that the process of aggregation cannot be satisfactorily explained by the conventional monomer $\rightleftarrows$ micelle equilibrium known to exist in aqueous media. Instead, both these new data and a reassessment of earlier ones suggest that the process of self-association of surfactants in nonionizing hydrocarbon media leads to the formation of oligomers (dimers, trimers, tetramers, etc.), all in a dynamic equilibrium. Such a process of stepwise aggregation of amphiphiles formed reversibly by a progressive molecular association is satisfactorily presented in terms of a series of mass action law equilibria. The smooth continuity as a function of surfactant concentration observed in most conventional physical properties measured in these systems (as contrasted to the phenomenon of great sharpness of transition in aqueous media) is the result of such a stepwise aggregation tendency. The existence of CMC in aqueous solutions is clearly a consequence of the monomer-micelle equilibrium. It is demonstrated that once that equilibrium ceases to be operative and is replaced by a stepwise aggregation equilibrium, the concept of CMC is rendered inapplicable. The two concepts are thus mutually exclusive.

## INTRODUCTION

In <u>aqueous</u> solutions the aggregation of amphiphilic molecules is largely due to hydrophobic interactions between the oleophilic parts of the molecules. As a consequence, there is a substantial free-energy change involved in the process of micellization which is due to the transfer of the oleophilic alkyl-chain part of the surfactant from an aqueous to a quasi-hydrocarbon environment.

In contrast to earlier concepts aiming to explain the Hartley picture of the inverted or reversed micelle in <u>organic</u> solutions, we believe we have shown[1] that the driving force behind the process of aggregation and micellization of surfactants in hydrocarbon media is essentially due to dipole-dipole interactions between polar heads of the amphiphilic molecules.

Obviously, the two types of interactions - the hydrophobic one in water and the dipole-dipole one in nonpolar organic solvents have thus little in common. This in turn poses the question whether the fundamental concepts which appear to explain the behavior of aqueous surfactant solutions can blindly be adopted to interpret the phenomena which surfactants exhibit in hydrocarbons.

We submit that the answer to this question is no. For the purpose of the present discussion we propose that the existence or nonexistence of a clearly detectable critical micelle concentration is the key factor in describing and defining the process of aggregation in solution.

## CRITICAL MICELLE CONCENTRATION: AQUEOUS VERSUS HYDROCARBON SOLUTIONS

At sufficiently low concentrations both in aqueous and non-aqueous media, surfactants are in nonassociated monomeric form. At such concentration levels, in water and aqueous solutions, the monomers of ionic amphiphiles are dissociated into ions, while in non-ionizing hydrocarbons they form ion pairs, when the extent of ion pairing depends essentially on the polarity of the solvent and the factors affecting it. As a rule, the concentration of the surfactants at which association of the molecules sets in is two or more orders of magnitude lower in hydrocarbons[1] than in water. This is true even if considering the conflicting interpretation of a possible premicellar association in aqueous system.[2]

In <u>aqueous</u> solutions the cmc is conventionally defined as the narrow range of concentrations at which the surfactant molecules aggregate to form micelles according to the equilibrium

$$mS \rightleftarrows S_m \qquad \beta_m = a_{S_m} a_S^{-m} \qquad (1)$$

where $m$ stands for the number of monomeric surfactant molecules S in the micelle, and a represents the activities of the species indicated by the subscript. At concentrations above cmc additional surfactant molecules are aggregated into micelles. By conventional thermodynamic concepts, for a finite value of $m$, the activities of both the monomer and the micelle will increase with increasing concentration of the solute. However, when $m$ is large the changes in activities tend to level off and may be assumed to be negligible.

It is, of course, a well-established fact that cmc is exhibited in the form of an abrupt change in the physical, colligative or spectral properties of the aqueous solution. Operatively, the cmc is then determined from the intersection of the extrapolated branches of two straight lines of different slopes. In terms of Equation (1), the values of $m$ and $\beta_m$ govern the sharpness of the transition at cmc of the monomer $\rightleftarrows$ micelle equilibrium. A recent critical evaluation of cmc data[2] for a large variety of surfactants leaves little doubt that indeed equilibrium (1) represents the predominating reaction of amphiphilic compounds in water and aqueous solutions. It appears that cmc is a quantity which can be determined quantitatively with a precision and accuracy equal to that of nearly any other property which characterizes aqueous solutions of surface-active agents. It is clearly detectable and, most frequently, differences between various methods for determining cmc are within experimental errors.[2]

Turning now to the aggregation phenomena in nonpolar <u>organic</u> solvents, one should emphasize that in recent years an increasing number of experimental data tend to indicate that at concentrations as low as $10^{-7} - 10^{-6}$ mole dm$^{-3}$ ionic surface active agents are not entirely monomeric[3-10]. At such low solute concentrations, of course, few experimental tools are sufficiently sensitive to provide reliable and indisputable quantitative data on the self-association of surfactants. Even at significantly higher concentrations, in the $10^{-4} - 10^{-2}$ mole dm$^{-3}$ range, quantitative studies are hampered by the fact that most of the conventional methods for cmc determination in aqueous media become inadequate in hydrocarbon solvents because of lack of ionization of the ampiphile in low-dielectric constant media, on which most methods are based.

These systems have been reviewed recently,[1] and a reassessment of the experimental data leads us to question the reliability of the evidence on which the existence of cmc in nonpolar organic solvents has been claimed.[11-24] The main criticism voiced concerned the applicability of the conventional method consisting of a plot of gross colligative, thermodynamic or spectroscopic data of a solution versus the solute concentration, to hydrocarbon system. As long as the difference in the slope between two straight lines, one representing the partial molar property of the monomer and the other that of the micellar aggregate, is large, the concentration at which the break occurs reflects the cmc. However, in less favorable cases when such a plot does not exhibit a sharp enough break, no well-defined concentration to which the cmc is to be assigned can be ascertained with any degree of precision. There is clearly the hazard of bias creeping into the interpretation of such curves where the slopes change only slightly, and/or the data are not of high enough precision. Several examples have been reproduced[1] to show that there is a reasonable doubt that breaks representing cmc values have been read into the shape of such plots. It thus appears that in these systems the concept of cmc becomes an elusive matter. An elusive matter, we submit, for the simple reason that its existence is doubtful. The search for it is likely to be the result, at least in some cases, of preconceived concepts originating through imitation of the behavior of surfactants in aqueous media.

As a possible alternative to the monomer $\rightleftarrows$ micelle equilibrium represented by Equation (1) and operative in aqueous solutions, an increasing number of reports in recent years suggest that the experimental data on the properties of nonaqueous solutions of amphiphilic compounds can be interpreted in terms of a continuous aggregation process.[25-40] We have formulated such a multiple equilibrium model for cationic surfactants a decade ago[41,42] and have recently reexamined most of the existing literature data.[1] The stepwise aggregation model expressed in terms of a set of mass-action law equilibria (vide infra) involving overall formation constants of oligomers of increasing sizes accounts for the smooth continuity observed in conventional physical properties of solutions measured in these systems, in contrast to the sharp transitions observed in aqueous systems.

We thus submit that it is unjustified to speak of one concentration, or even a narrow range of concentrations, describing cmc of surfactants in hydrocarbon solutions. In other words, there is no unique concentration at which the existence of micelles becomes detectable, thus rendering even the definition of cmc in such systems rather ambiguous.

In order to make the clear distinction (vide infra) between the stepwise aggregation process of amphiphiles in hydrocarbons, and that assumed to take place in water,[43,44] one should emphasize that the former is due to the behavior of a highly polar solute being dissolved in a nonpolar solvent. This in turn results in a dipole-dipole type interaction[31] between the surfactant molecules when the characteristic features of such systems are (i) the presence of aggregates at low solute concentration, and (ii) the stability of aggregates with a small number of monomers. Both features are in contrast to the behavior of surfactants in aqueous solutions.

## STEPWISE AGGREGATION EQUILIBRIA

The general procedure adopted by us[1,29,41,42] for dealing with the multiple equilibria in a stepwise aggregation process of amphiphilic molecules in hydrocarbon solutions where oligomers of various sizes are in a dynamic equilibrium, is essentially a modification of the Bjerrum method[45] for the evaluation of stepwise formation constants of metal complexes in solution, extended to binary systems of undissociated solutes in nonionizing solvents. The procedure has since been applied to explain the aggregation behavior of a number of surfactants in organic solvents.[25-40] The various calculation procedures including those mentioned by Adams et al.[39] depend on the nature and quality of the available experimental data, and have been critically reviewed by Rossotti and Rossotti.[45]

A straightforward mathematical definition requires that a stepwise aggregation of monomers involves the formation of every and all oligomers starting with the dimer

$$S_1 + S_1 \rightleftarrows S_2 \qquad k_{211} = a_{S_2} a_{S_1}^{-2} \tag{2}$$

$$S_2 + S_1 \rightleftarrows S_3 \qquad k_{321} = a_{S_3} a_{S_2}^{-1} a_{S_1}^{-1} \tag{3}$$

$$S_3 + S_1 \rightleftarrows S_4 \qquad k_{431} = a_{S_4} a_{S_3}^{-1} a_{S_1}^{-1} \tag{4}$$

$$S_{n-1} + S_1 \rightleftarrows S_n \qquad k_{n(n-1)1} = a_{S_n} a_{S_{n-1}}^{-1} a_{S_1}^{-1} \tag{5}$$

where $S_1$ is the undissociated monomeric surfactant molecule, $S_2$ the dimer, $S_3$ the trimer, etc., and $a$ denotes activities.

Simultaneously, however, a semi-stepwise aggregation can proceed along several additional routes by interactions between lower aggregates of different sizes. Thus, for example, a tetramer can be formed by reaction shown in Equation (4) and/or either one or both of the interactions

$$S_2 + 2S_1 \rightleftarrows S_4 \qquad k_{4211} = a_{S_4} a_{S_2}^{-1} a_{S_1}^{-2} \tag{6}$$

$$2S_2 \rightleftarrows S_4 \qquad k_{422} = a_{S_4} a_{S_2}^{-2} \tag{7}$$

The stepwise aggregation constant and the constants representing interactions between oligomers are interrelated through the overall aggregation constant

$$nS_1 \rightleftarrows S_n \qquad \beta_n = a_{S_n} a_{S_1}^{-n} \tag{8}$$

being the product of all stepwise aggregation constants up to $k_n$

$$\beta_n = k_1 k_2 k_3 \ldots k_n \tag{9}$$

For the tetramer example

$$4S_1 \rightleftarrows S_4 \tag{10}$$

$$\beta_4 = a_{S_4} a_{S_1}^{-4} = k_{4211} k_{211} = k_{422} k_{211}^2 = k_{431} k_{321} k_{211} \tag{11}$$

The corresponding thermodynamic functions are thus rigorously defined for any one of the oligomers formed as

$$\Delta G_n^\circ = -RT \ln \beta_n \tag{12}$$

$$\Delta H_n^\circ = RT \frac{d \ln \beta_n}{dT} \tag{13}$$

$$\Delta S_n^\circ = R \ln \beta_n + \frac{\Delta H_n^\circ}{T} \tag{14}$$

The numerical values of $\beta_n$ depend on the concentration scale defining activities. Since it is usually rather difficult to distinguish between specific and nonspecific nonidealities in the type of systems under consideration, it is reasonable to assume that the deviation of the solution from ideal behavior is essentially due to the formation of aggregates. This in turn implies that the monomer $S_1$ and the individual oligomers $S_n$ behave ideally in solution, their activity coefficients being unity. Thus, in terms of concentration

$$\beta_n^c = [S_n][S_1]^{-n} \tag{15}$$

is the concentration overall-aggregation constant. The ratio between the analytical concentration of the surfactant in solution

$$c_S = [S_1] + 2[S_2] + 3[S_3] + \ldots + (n-1)[S_{n-1}] + n[S_n] =$$
$$= \sum_1^N n[S_n] = \sum_1^N n\beta_n^c [S_1]^n \tag{16}$$

and the molar concentration of the various aggregates as determined experimentally

$$c_{S(exp)} = [S_1] + [S_2] + [S_3] + \ldots + [S_{n-1}] + [S_n] =$$
$$= \sum_1^N [S_n] = \beta_n^c [S_1]^n \tag{17}$$

Is the number-average aggregation number

$$\tilde{n} = \frac{\sum_1^N n\beta_n^c [S]^n}{\sum_1^N \beta_n^c [S]^n} \tag{18}$$

This set of mass-action law equilibria implies that with increasing solute concentration, the concentration of aggregates increases. Thus, regardless of whether the extent (number of aggregated units) or the degree (size of aggregated units) of aggregation increases, or both, $\tilde{n}$ will increase.

The determination of the various $n$ and $\beta_n^c$ values can be done both graphically and numerically, as described in detail elsewhere,[45] when the concentration of the free monomer in equilibrium is derived from a graphical extrapolation of the Bjerrum relationship discussed previously.[1,29,45]

Computational problems in solving equations of type (16) and (17) can be largely overcome by a successive approximation method employing a suitable digital computer program, for values of $n$ and $\beta_n^c$. The equations may be inconsistent on account of small experimental errors, and two or more sets of combinations of $n$ and $\beta_n^c$ values may satisfy the same experimental data. The most desirable model, out of a large number of aggregation models tested to fit the experimental data, is considered the one which involves the smallest number of aggregates and yields an error-squares sum which cannot be significantly lowered by a second model which involves one or two additional aggregates. It should be emphasized that for the evaluation of $n$ and the corresponding $\beta_n^c$ values with any degree of accuracy, a large volume of experimental data must be used. It is our experience[46-49] that a large number of models must be tested against experimental data before adopting a given model which describes the oligomerization of the amphiphilic compound throughout the concentration range employed in the experimental work. It is thus an unsound practice to force[39] any preconceived model of oligomerization upon a system when the advantage of the use of computers for the least-squares method is the ease of weighting the variables with their appropriate statistical weight.

Though the concept of stepwise aggregation implicitly assumes that the distribution of aggregate sizes is that of the series of oligomers starting with the dimer, some aggregate sizes ($n$ values) may be present in very small concentrations or not be present at all. Such is the case, for example, when aggregation in a particular system proceeds preferentially by clustering of dimers only. As a consequence, the fraction of aggregates with odd $n$ values will be negligibly small.

The highest values of $n$ in a set of oligomers as derived from a computer-fit calculation procedure should not be taken too literally. The fraction of such a large aggregate is usually small even at the highest solute concentrations at which meaningful physicochemical determinations are still feasible. As a result, some of the evaluated $n$ values may actually be those of $\tilde{n}'$ - an average aggregation number for several $n$ values actually present. For example, if the best fit indicates that the predominating oligomers in a system are the dimer, tetramer and the 16-mer, this does not necessarily mean that the large aggregate contains indeed 16 monomers.

It may well be that $\tilde{n}' = 16$, rather than $n = 16$, representing thus an average aggregation number between aggregates having 12, 14, 16, 18 and 20 monomers. Indeed, there appears to be a unique distribution of aggregate sizes for any given binary surfactant-solvent system at a given temperature. A number of examples have been discussed recently,[1] for which various experimental methods yield aggregation data with a characteristic set of $n$ values.

Some rationalization has been offered for such a uniqueness of aggregate sizes of surfactants in hydrocarbon media. Muller[38] believes that small oligomers with odd number monomers per aggregate tend to disproportionate into the more stable even-order aggregates. The suggestion is based on estimates of electrostatic potential energies which indicate that the even-numbered aggregates are slightly favored. Eicke and Christen[40] found that the total energy of micelles of some ionic surfactants in nonpolar solvents as derived from calculations based on a geometrical model involving coulombic and London-Van der Waals interactions, shows a marked minimum at $n \simeq 16$. It is rather surprising, however, that this $n$ value of the aggregate with the lowest energy is not much affected by the solvent as long as its dipole moment is low or zero. Further estimates[50] of energies and frequencies of nucleation in aggregate formation based on a Monte-Carlo model using a two-dimensional liquid lattice suggest that the redistribution or disproportionation of the number of monomers per oligomer might in reality be considerably more complex than suggested by Muller.[38]

## REFERENCES

1. A. S. Kertes and H. Gutman, in "Surface and Colloid Science", E. Matijevic, Editor, Vol. 8, pp. 193-295, Wiley, New York, 1976.
2. P. Mukerjee and K. J. Mysels, "Critical Micelle Concentration of Aqueous Surfactant Systems", National Bureau of Standards, NSRDS-NBS 36, Washington, D.C., 1971.
3. P. Debye and H. Coll, J. Colloid Sci., 17, 220 (1962).
4. S. M. Nelson and R. C. Pink, J. Chem. Soc., 1952, 1744.
5. N. Pilpel, Chem. Rev., 63, 221 (1963).
6. C. R. Singleterry and L. A. Weinberger, J. Amer. Chem. Soc., 73, 4574 (1952).
7. S. Kaufman and C. R. Singleterry, J. Colloid Sci., 10, 139 (1955) and 12, 465 (1957).
8. S. Kaufman, J. Colloid Sci., 17, 231 (1962).
9. P. H. Elworthy, J. Chem. Soc., 1959, 813.
10. J. B. Peri, J. Amer. Oil Chem. Soc., 35, 110 (1958).
11. P. Debye and W. Prins, J. Colloid Sci., 13, 86 (1958).
12. A. Ray, J. Amer. Chem. Soc., 91, 6511 (1969).
13. K. Shinoda and H. Arai, J. Colloid Sci., 20, 93 (1965).
14. C. W. Brown, D. Copper and J. C. S. Moore, J. Colloid Interface Sci., 32, 584 (1970).

15. S. Kaufman and C. R. Singleterry, J. Colloid Sci., $\underline{7}$, 453 (1952).
16. A. Kitahara, in "Cationic Surfactants", E. Jungermann, Editor, Chapter 8, Marcel Dekker, New York, 1970.
17. A. Kitahara, Bull. Chem. Soc. Japan, $\underline{28}$, 234 (1955); $\underline{29}$, 15 (1956); $\underline{30}$, 586 (1957) and $\underline{31}$, 288 (1958).
18. A. Kitahara, J. Colloid Sci., $\underline{12}$, 342 (1957).
19. A. F. Sirianni and B. A. Gingras, Can. J. Chem., $\underline{39}$, 331 (1961).
20. J. F. Yan and M. B. Palmer, J. Colloid Interface Sci., $\underline{30}$, 177 (1969).
21. N. Sata and H. Sasaki, Kolloid-Z., $\underline{152}$, 76 (1957).
22. P. Becher, J. Phys. Chem., $\underline{64}$, 1221 (1960).
23. S. Ross and J. P. Olivier, J. Phys. Chem., $\underline{63}$, 1671 (1959).
24. J. H. Fendler, E. J. Fendler, R. T. Medary and O. A. Elseoud, J. Chem. Soc. Faraday Trans. I., $\underline{69}$, 280 (1973).
25. S. R. C. Hughes and D. H. Price, J. Chem. Soc., $\underline{1967}$, 1093 and $\underline{1968}$, 1464.
26. D. F. Evans and P. Gardam, J. Phys. Chem., $\underline{72}$, 3281 (1968).
27. J. J. Bucher and R. M. Diamond, J. Phys. Chem., $\underline{69}$, 1565 (1965).
28. W. Muller and R. M. Diamond, J. Phys. Chem., $\underline{70}$, 3469 (1966).
29. A. S. Kertes and G. Markovits, J. Phys. Chem., $\underline{72}$, 4202 (1968).
30. A. S. Kertes, O. Levy and G. Markovits, J. Phys. Chem., $\underline{74}$, 3568 (1970).
31. O. Levy, G. Markovits and A. S. Kertes, J. Phys. Chem., $\underline{75}$, 542 (1971).
32. E. Hogfeldt, P. R. Danesi and F. Fredlund, Acta Chim. Scand., $\underline{25}$, 1338 (1971).
33. A. S. Kertes, H. Gutmann, O. Levy and G. Markovits, in "Chemie, physikalische Chemie und Anwendungstechnick der grenzflachenaktiven Stoffe", pp. 1023-1034, Carl Hanser Verlag, Munich, 1973.
34. G. Y. Markovits, O. Levy and A. S. Kertes, J. Colloid Interface Sci., $\underline{47}$, 424 (1974).
35. H. Gutmann and A. S. Kertes, J. Colloid Interface Sci., $\underline{51}$, 406 (1975).
36. A. S. Kertes, O. Levy and G. Y. Markovits, in "Proc. Intern. Conf. Colloid and Surface Science", E. Wolfram, Editor, Vol. 1, p. 497, Akademiai Viado, Budapest, 1975.
37. O. Levy, G. Y. Markovits and I. Perry, J. Phys. Chem., $\underline{79}$, 239 (1975).
38. N. Muller, J. Phys. Chem., $\underline{79}$, 287 (1975).
39. F. Yun-Fat Lo, B. M. Escott, E. J. Fendler, E. T. Adams, R. D. Larsen and P. W. Smith, J. Phys. Chem., $\underline{79}$, 2609 (1975).
40. H. F. Eicke and H. Christen, J. Colloid Interface Sci., $\underline{46}$, 417 (1974).
41. G. Markovits and A. S. Kertes, in "Solvent Extraction Chemistry", D. Dyrssen, J. O. Liljenzin and J. Rydberg, Editors, p. 390, North-Holland Publ., Amsterdam, 1967.

42. A. S. Kertes and G. Markovits, in "Thermodynamics of Nuclear Materials, 1967", p. 227, Intern. Atomic Energy Agency, Vienna, 1968.
43. P. Mukerjee, J. Pharm. Sci., $\underline{63}$, 972 (1974).
44. P. Mukerjee, in "Physical Chemistry: Enriching Topics from Colloid and Surface Science", H. van Olphen and K. J. Mysels, Editors, pp. 135-153, Theorex, La Jolla, 1975.
45. F. J. C. Rossotti and H. Rossotti, "The Determination of Stability Constants", McGraw-Hill, New York, 1961.
46. G. Markovits, Ph.D. Thesis, The Hebrew University, Jerusalem, 1968.
47. O. Levy, Ph.D. Thesis, The Hebrew University, Jerusalem, 1970.
48. J. David-Auslaender, Ph.D. Thesis, The Hebrew University, Jerusalem, 1972.
49. H. Gutmann, Ph.D. Thesis, The Hebrew University, Jerusalem, 1974.
50. H. Christen and H. F. Eicke, J. Phys. Chem., $\underline{78}$, 1423 (1974).

MIXED NON-IONIC DETERGENT SYSTEMS IN AQUEOUS AND NON-AQUEOUS SOLVENTS

Issa Lo, Florence Madsen, Alexander T. Florence, Jean-Paul Treguier, Monique Seiller and Francis Puisieux

Laboratoire de Pharmacie Galénique, Université de Paris-Sud, 92290 Châtenay-Malabry, Paris, France and The School of Pharmaceutical Sciences, University of Strathclyde, Glasgow, G1 1XW, United Kingdom

Mixtures of surfactants are known to be more efficient stabilisers of emulsion systems than single surfactants. Little work has been reported on the properties of mixed non-ionic detergents in aqueous and non-aqueous systems, yet the behaviour of such mixtures is important in the understanding of the mechanisms of emulsion and suspension stabilisation. Ternary phase diagrams of mixed non-ionic surfactant-oil-water systems have been constructed for dodecane and mineral oil and mixtures of a water soluble and a water insoluble oleyl polyoxyethylene ether with HLB values of 12.4 and 4.9 respectively. Mixtures of the two surfactants solubilise water in isotropic dodecane and mineral oil systems more efficiently than the single surfactants. Inverse mixed micelles are favoured probably because of the more efficient packing of the short and long chain oxyethylene groups together in the aqueous core of the micelles which appear from viscosity measurements to be nearly spherical. In aqueous solution the addition of low amounts of the less soluble non-ionic to the water soluble surfactant results in the formation of asymmetric micelles. Addition of further quantities of the low HLB compound results in an apparent breakdown of the large micelles and a decreased viability of the neat phase. The question of the validity of the calculation of HLB values for surfactant mixtures is discussed.

## INTRODUCTION

Although surfactant mixtures rather than single entities are commonly used in commercial emulsion formulations, relatively little attention has been paid to the properties of mixtures of non-ionic surfactants in either aqueous or non-aqueous solvent systems. The few papers[1-3] on non-ionic surfactant mixtures deal with the critical micelle concentration and surface activity of the mixtures rather than with the bulk solution properties, although Shinoda and Kunieda[4] have investigated the requirements for non-ionic surfactant mixtures to produce so-called microemulsions.

Non-ionic surfactant mixtures are more efficient emulsion stabilisers than are single surfactants. The reasons for this are several, but may in part be due to the ability of two species to interact in some unique manner at phase interfaces or to their ability to alter the properties of the bulk phases, for example by increasing the viscosity of the continuous phase. Work in our laboratories on the stabilisation of emulsions had led to an investigation of phase diagrams of oil water non-ionic surfactant systems and to a study of some bulk solution properties of non-ionic surfactant mixtures. The preparation of triangular phase diagrams allows the definition of the whole spectrum of phases of the three-component mixtures and their conditions of existence and stability[5,6]. The complexity of oil-water-surfactant systems, however, often precludes an adequate analysis of conditions in the individual phases; this suggests that investigation of the isolated aqueous and non-aqueous surfactant-solvent system is advantageous although it is acknowledged that the lack of the third component phase oversimplifies the system. However, when non-ionic mixtures are being studied, little progress can be made in understanding the mechanisms of the interactions between the component surfactants without separate investigation of the individual solvent systems.

This paper summarises work on dodecane-water-non-ionic surfactant mixtures, consisting of Brij 92[a], a lipophilic non-ionic oleyl alcohol with a hydrophile-lipophile balance (HLB) of 4.9, and Brij 96, a polyoxyethylated oleyl alcohol derivative with an HLB of 12.4. Measurements on aqueous micellar solutions of Brij 96 in water in the presence of small quantities of the low solubility homologue Brij 92 are also reported.

---

(a) The name Brij is a trademark of ICI, United States Inc., and is also used by Honeywell-Atlas, U.K. and by Atlas Chemical Industries N.V. in Europe.

Because of the miscibility of the oxyethylene chain with both aqueous and non-aqueous solvents, polyoxyethylated non-ionic surfactants based on long chain alcohols while readily forming micelles in aqueous systems do not do so in anhydrous solvent systems such as benzene or cyclohexane. However, on the addition of small quantities of water to a solution of such a surfactant in benzene or cyclohexane, the water molecules convert the polyoxyethylene chains of the surfactant into lipophobic units and inverse micelles form[7]. The importance of polyoxyethylene chain length in this process, first reported by Schulman, Cohen and his collaborators[8], is made clear in the present work with mixtures.

In practice when surfactant mixtures are used in the formulation of emulsions, HLB values for the surfactant mixtures are calculated algebraically. The validity of calculation of HLB values of surfactant blends has been questioned by Boyd et al.[9] in view of the possibility of intermolecular associations occurring between surfactant species. Supporting this, we have previously shown[10] that the critical HLB for maximum emulsion stability for a given oil depends on the blend of non-ionic surfactant used, and in addition is dependent on the concentration of surfactant[11]. As electrolytes added to the aqueous phase can shift the critical HLB[12] (i.e. the HLB for maximum stability) it is feasible that polar non-electrolytes may also. The results in this work are analysed from the point of view of HLB and of deviations from the additivity of the properties of non-ionic mixtures.

## EXPERIMENTAL

### Materials

Surfactants were commercial samples from Honeywill-Atlas Limited, U.K. or Atlas Chemical Industrial N.V. through Atlas-Seppic, Paris, France. Brij 92, an oleyl alcohol derivative with an average of 2 ethylene oxide residues, and Brij 96 with an average of 10 ethylene oxide residues were used as received and were the samples used in previous work[4,10,12]. Mineral oil was of pharmacopoeial quality and dodecane a specially purified sample. Water was demineralised water. Salts were analytical grade. The critical HLB for maximum stability of dodecane emulsions has previously been determined as 9 - 9.5 and as 8.5 for mineral oil-water emulsions both stabilised by Brij 92-96 mixtures[12].

### Phase Diagrams

Ternary phase diagrams were constructed with dodecane or mineral oil and mixtures of Brij 92 and Brij 96 to produce surfactant systems with calculated HLB values of 4.9 (Brij 92 alone), 7, 9.5, 10.5, 11.5 and 12.5 (Brij 96 alone). The phase diagrams were prepared by varying the concentrations of the components by intervals of 10% to cover the whole area of the equilateral triangle and near phase boundaries this was reduced to 5% to define the limits of the phases. In the regions where emulsions were found, 5% increments in concentration were used.

The components were added to a beaker (150 ml) and equilibrated at 70°C (20 minutes) before homogenisation (10 minutes) with a turbine blade stirrer. The resultant preparations were then maintained at ambient temperatures for one week during which time their phase structure was identified. Transmitted light microscopy and phase-contrast microscopy (Reichert-Zetopan) was used in addition to conductivity determinations and examination under polarising microscope (Leitz, Dialux-Pol) between crossed Nicol prisms.

## Viscosity

A suspended level dilution viscometer of about 10 ml capacity was used for investigation of solutions of Brij 96. Some measurements were made with a Haacke (Eraus) viscometer. Relative viscosities, $\eta_{rel}$, were measured using the time of flow of water or salt solution as solvent value, and the reduced specific viscosities $\eta_{sp}/c$ (c being the concentration of the surfactant in g ml$^{-1}$) were plotted against c. Extrapolation to $c \rightarrow 0$ gave the intrinsic viscosity, $[\eta]$, for each system.

## RESULTS

The phase diagrams obtained with the dodecane-water-surfactant mixtures are reproduced in Figure 1 and those obtained with mineral oil in Figure 2. The following regions can be identified; an area of single aqueous isotropic liquid $L_1$; an area of organic isotropic liquid $L_2$; a region where the two phases co-exist as an emulsion phase, $E(L_1 + L_2)$, an aqueous neat phase (G), a middle phase (M) and in some of the systems an inverse middle phase ($M_2$) appears. The phase diagram for the polyoxyethylene (10) oleyl ether (Brij 96) mineral oil-water system is very similar to the diagram obtained for the same components by Lachampt and Vila[14] allowing for the differences in sample composition for both oil and surfactant. At approximately a 5:1 ratio of Brij 92 to Brij 96 an inverse middle phase ($M_2$) appears between 75-80% surfactant, a phase according to Winsor[14] that has not been encountered in other systems of polyoxyethylated non-ionic surfactants.

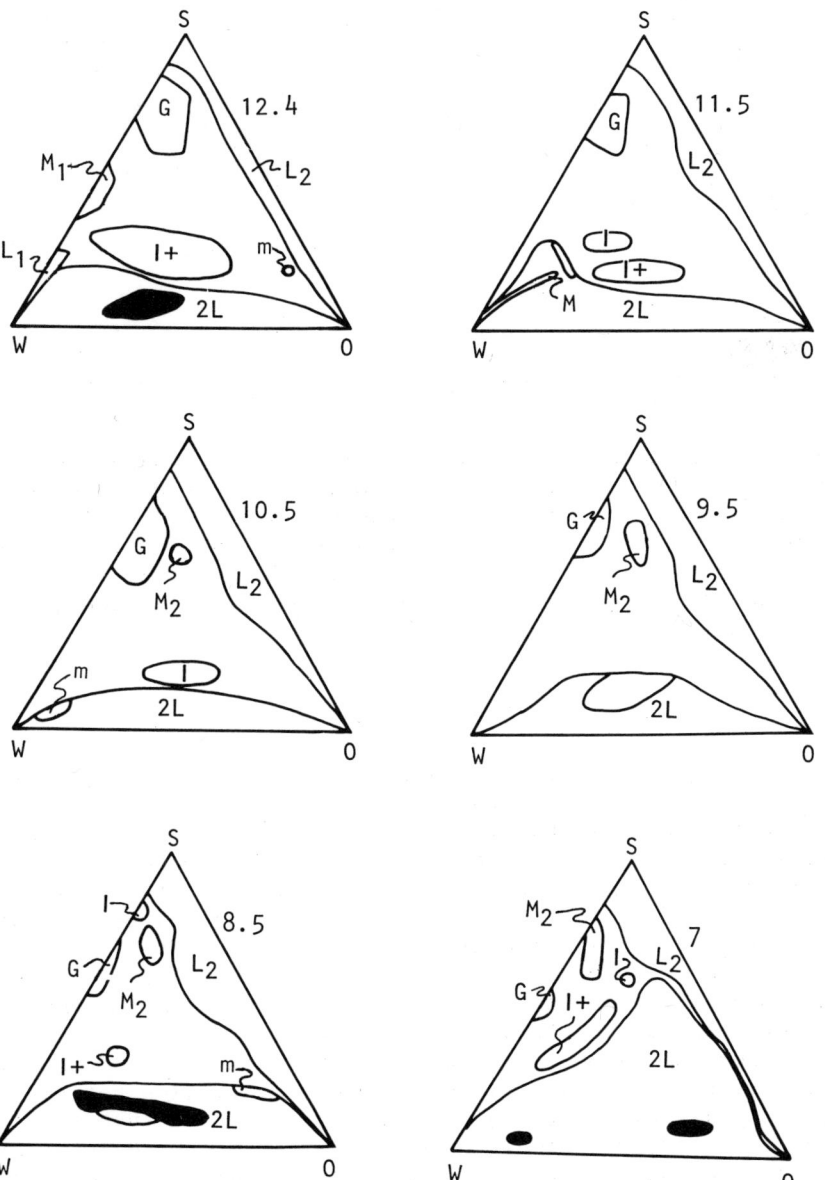

Figure 1. Phase diagrams of dodecane-water-Brij 92/96 mixtures. O,W,S = 100% oil, water, surfactant, respectively. HLB values of the surfactant mixtures are shown to the right of each diagram. Phase boundaries for $L_1$, $L_2$, G, $M_1$ and $M_2$ phases shown. Other symbols represent: 2L - two phase oil-water system (emulsions of varying stability), M - micro emulsion, I - isotropic elastic, I+ - isotropic elastic + disperse oil phase. Filled black regions - stable emulsion zones.

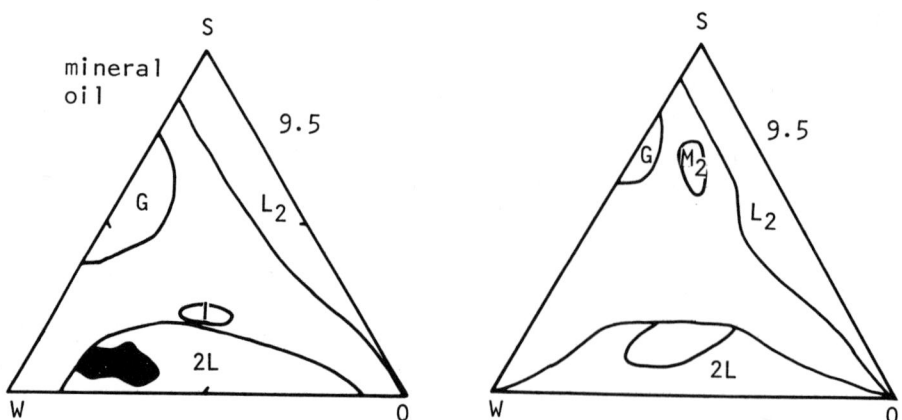

Figure 2. Comparative phase diagrams for mineral oil and dodecane systems at a surfactant HLB of 9.5 showing the extreme sensitivity of the phases to the properties of the oil phase. Symbols represent: 2L - two phase oil-water system (emulsions of varying stability), M - micro emulsion, I - isotropic elastic, I+ - isotropic elastic + disperse oil phase. Filled black regions - stable emulsion zones.

## Non-Aqueous Phases

The non-aqueous phases of interest are the $L_2$ and $M_2$ phases: From Figure 1 it is readily apparent that the boundaries of the $L_2$ phase extend as the HLB rises and at an HLB of about 8 the area is maximised. The isotropic dodecane-surfactant-water system must result from the formation of inverse micelles with the polyoxyethylene chains directed away from the non-polar solvent into the micelle interior. The solubility of water in dodecane is negligibly small. In the presence of the polyoxyethylene chains of the surfactant molecules, the water becomes associated with the chains and converts the chains into more hydrophilic species. The micelles therefore grow. We would therefore expect that the $L_2$ phase is greater in pure polyoxyethylene (10) oleyl ether systems than in polyoxyethylene (2) oleyl ether systems and that addition of water to the former should cause a greater increase in micellar uptake than to the latter. Both expectations are borne out in Figure 1. Maximum water uptake in the two systems can be obtained from the data in figures 1 and 2 at several fixed surfactant concentrations. Figure 3 summarises the results of water uptake in 25%, 40% and 50% surfactant solutions in oil. Maximal solubilisation in dodecane occurs at an HLB of 8 where the molar ratio of polyoxyethylene (2) oleyl ether to polyoxyethylene (10) oleyl ether is approximately 3:1. At surfactant levels below 25% there is negligible uptake of water by polyoxyethylene (2) oleyl ether in

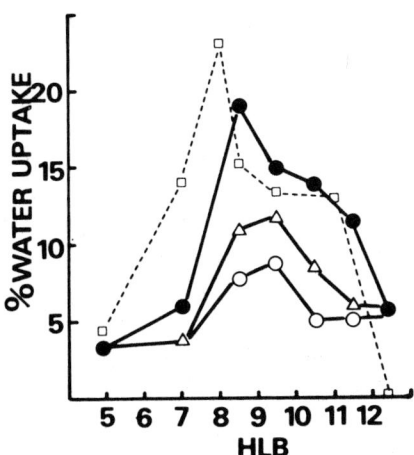

Figure 3. Water uptake into non-aqueous $L_2$ — dodecane and ---- mineral oil phases as a function of the calculated HLB of Brij 92/96 mixtures. The dodecane results obtained at ○ 25%, △ 40% and ● 50% w/w surfactant levels; the mineral oil results at 50% w/w levels.

dodecane, and in mineral oil neither of the surfactants allow measurable solubilisation of water at levels below 50% surfactant.

As far as it is possible to calculate, at the point of maximum water solubilisation there are 3 water molecules associated with each ethylene oxide residue, this corresponding to an average of 10 water molecules for each surfactant molecule. In the single surfactant system of 50% polyoxyethylene (10) oleyl ether in dodecane there are about 0.4 molecules of water for each ethylene oxide residue (6% water). As the potential binding capacity in the absence of micelle formation is 2 water molecules for each ether oxygen, this suggests that micelle formation is limited in the single surfactant system. Addition of the short chain molecule encourages micelle formation as is apparent from the results in Figures 1 - 3 and further uptake is possible by formation of a hydrophilic pool in the micelle core. Preliminary measurements of intrinsic viscosity indicate that the micelles are spherical.

The slight difference in the dielectric constants of mineral oil and dodecane are reflected in differences in the behaviour of the surfactant mixtures in the two oils. Figure 4 demonstrates these differences exhibited in the concentrations of the surfactant mixtures required to solubilise a fixed percentage (5%) of water in the non-aqueous isotropic phase. More hydrophilic mixtures of the

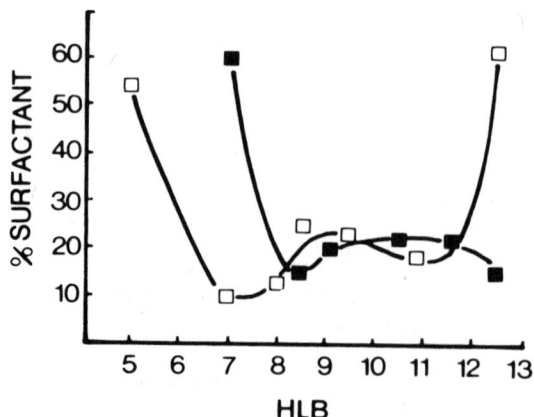

Figure 4. Percentage of various surfactant mixtures required to solubilise 5% water in isotropic systems of □ mineral oil and ■ dodecane.

surfactant solubilise more water in dodecane ($\varepsilon = 2.01$) than in mineral oil ($\varepsilon = 2.10$) but with hydrophobic mixtures the situation is reversed, water uptake being greater in mineral oil. The hydrophilic mixture would form larger aggregates in the less polar solvent and hence solubilise more water; this tendency is reversed when hydrophobic surfactant mixtures, being less polar, form larger aggregates and hence solubilise more water molecules. (Contrary to general trends noted with Brij 96 and oils of increasing dielectric constant[6]). Both Figure 3 and Figure 4 demonstrate that there is some synergistic quality of some mixtures with HLB values of around 8-9.

## Simple Aqueous Systems of the Surfactant Mixtures

Uptake of small quantities of polyoxyethylene (2) oleyl ether by polyoxyethylene (10) oleyl ether micelles markedly affects the hydrodynamic properties of the micelles. Viscosity measurements show this effect. The maximum additive concentration of the hydrophobic surfactant is of the order of 20% at a molar rate of 0.54 moles of polyoxyethylene (2) ether to 1 mole of polyoxyethylene (10) oleyl ether but the behaviour is relevant to an understanding of emulsion stability and of the formation of mixed liquid crystalline states.

Table I. Intrinsic Viscosities (ml g$^{-1}$) of Polyoxyethylene (2) oleyl Ether - Polyoxyethylene (10) oleyl Ether Mixtures (Brij 92 - Brij 96 mixtures).

| Calculated HLB of mixture | % Brij 92 | $[\eta]$ (ml g$^{-1}$) |
|---|---|---|
| 12.4 | 0 | 10 |
| 12.0 | 6% | 19 |
| 11.5 | 12% | 23 |
| 11.0 | 19% | 42 |

Table 1 lists the intrinsic viscosities of mixtures of the non-ionic surfactants. The rapidly increasing asymmetry is obvious. Addition of mœ than 20% of the insoluble surfactant results in loss of the isotropic $L_1$ phase. The high bulk viscosity of these mixtures has several consequences. The thinning of aqueous films between the oil droplets is decreased[11] with a concomitant increase in stability, and droplet creaming or sedimentation is retarded.

The middle ($M_1$) phase disappears on addition of more than 20% of the hydrophobic surfactant. The neat (G) phase which, because of its more linear structure, one would expect to be more dependent on surfactant size and shape, diminishes in size on addition of the short chain surfactant and indeed does not form when Brij 92 is present as sole surfactant. Nevertheless addition of Brij 92 to the system allows formation of the neat phase at lower surfactant concentrations than in solutions of polyoxyethylene (10) oleyl ether (Figure 5).

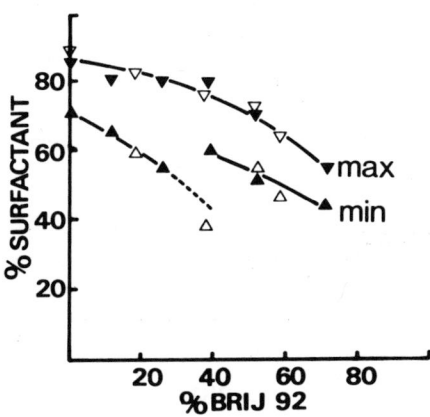

Figure 5. Upper and lower concentration limits for the existence of the neat G phase in surfactant-water mixtures from two series of experiments (open and filled symbols) plotted as a function of increasing Brij 92 content.

The capacity for solubilisation of dodecane, however, is greater the higher the concentration of the Brij 96 in the mixture, i.e. Brij 92 depresses solubilisation in the neat phase. Preliminary results suggest that Orange OT uptake in micellar mixtures of the surfactants is little affected by the incorporation of the hydrophobic surfactant into the Brij 96 micelles.

As might be expected, the mixed micelles are sensitive to the presence of electrolyte in the aqueous phase. Electrolytes which salt out the non-ionic detergents increase micellar size and electrolytes which salt in the detergents decrease the size. The micelles for Brij 96 in fact appear to be sensitised to the effect of electrolyte by the incorporation of the hydrophobic surfactant.

Figure 6 shows the effect of sodium chloride and sodium iodide on the viscosity of solutions of mixtures at an HLB of 12. While the polyoxyethylene (10) oleyl ether is not markedly affected by

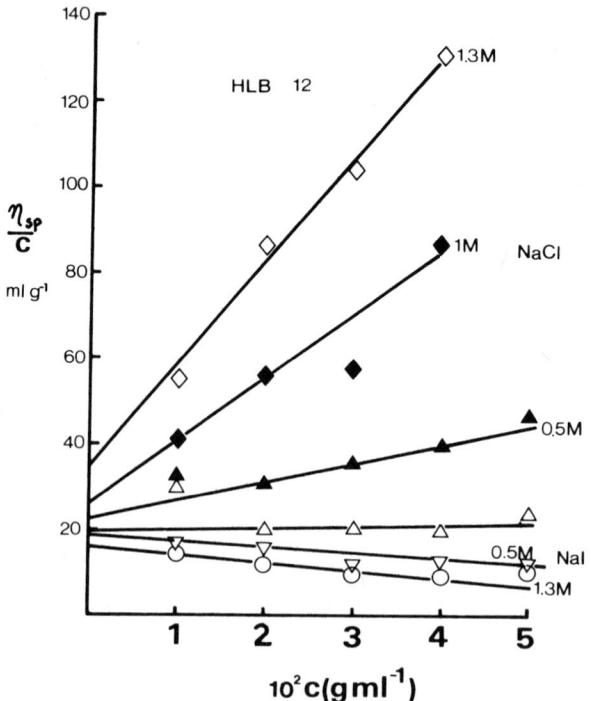

Figure 6. Reduced specific viscosity as a function of surfactant concentrations for a mixture of Brij 92 and Brij 96 with a calculated HLB of 12 in the presence of sodium chloride and sodium iodide.

these additives, their addition to the mixed systems is reflected in marked changes in the concentration dependence of the reduced specific viscosity and intrinsic viscosity. NaI makes the complex more hydrophilic - that is it decreases the effect of the short-chain surfactant, while NaCl increases the effect of Brij 92 on the asymmetry of Brij 96 micelles. This perhaps explains the results obtained previously[12] on the equilibrium thicknesses of free films of these mixed systems. In the presence of NaCl the equilibrium thickness increases significantly from 14 nm to over 25 nm.

## CONCLUSION

The properties of systems containing mixtures of non-ionic surfactants are by no means simple functions of the concentrations of each component. While the HLB number is a useful index for cataloguing mixtures, the concentration-dependence of optimal HLB values for solubilisation (Figures 3 and 4) and for emulsion stabilisation[13], the influence of the oil phase and the effect of additives on the HLB number[12], decreases its theoretical value. While most of the changes and influences can be rationalised, there is need for another index to aid formulators of surfactant systems.

## REFERENCES

1. H. Lange and K-H. Beck, Kolloid Z.Z. Polym., 251, 424 (1973)
2. N. Nishikido, Y. Moroi and R. Matuura, Bull. Chem. Soc. Japan, 48, 1387 (1975)
3. J.H. Clint, JCS Faraday Trans. I., 71, 1327 (1975)
4. K. Shinoda and H. Kuneida, J. Colloid Interface Sci., 42, 381 (1973)
5. J-P. Treguier, I. Lo, M. Seiller and F. Puisieux, Pharm. Acta Helv., 50, 421 (1975)
6. I. Lo, A.T. Florence, J-P. Treguier, M. Seiller and F. Puisieux, J. Colloid Interface Sci. In Press
7. A. Kitahara, J. Phys. Chem., 69, 2788 (1965)
8. Schulman, J.H., R. Matalon and M. Cohen, Discuss. Faraday Soc., 11, 117 (1951)
9. J. Boyd, C. Parkinson and P. Sherman, J. Colloid Interface Sci., 41, 359 (1972)
10. M. Seiller, T. Legras, F. Puisieux and A. le Hir, Ann. Pharm. Franc., 28, 425 (1970)
11. M. Seiller, C. Arguillere and P. David, F. Puisieux and A. le Hir, Ann. Pharm. Franc., 26, 587 (1968)
12. A.T. Florence, F. Madsen and F. Puisieux, J. Pharm. Pharmac., 27, 385 (1975)
13. I. Lo, T. Legras, M. Seiller, M. Choix and F. Puisieux, Ann. Pharm. Franc., 30, 211 (1972)
14. F. Lachampt and R.M. Vila, Amer. Perfum. Cosmet., 82, 29 (1967)

15. P.A. Winsor in "Liquid Crystals and Plastic Crystals", G.W. Gray and P.A. Winsor, Editors, Vol. 1, John Wiley: London (1974)

CALORIMETRIC STUDIES OF THE MICELLE FORMATION IN SOLUTIONS OF

SODIUM OCTANOATE AND WATER IN ALIPHATIC ALCOHOLS

>Jarl B. Rosenholm, Per J. Stenius
>and Marja-Riitta Hakala
>Department of Physical Chemistry
>Åbo Akademi Porthansgatan 3-5
>SF-20500 Åbo (Turku) 50, Finland

The relative molar enthalpies of the three components in the isotropic pentanol-rich solution phase in the system sodium N-octanoate-N-pentanol-water have been measured calorimetrically at 25°C. The results are compared with similar measurements in the system sodium N-octanoate-N-decanol-water. It is found that the sodium octanoate is of dominating importance for the thermal interactions in the whole phase studied. At high concentrations of alcohol there is a strong increase in the partial molar enthalpy of the sodium octanoate as well as in the enthalpy of water. When the concentration of the alcohol is reduced a fast but continuous decrease of these enthalpies to a fairly constant value is observed. The partial molar enthalpies of the alcohols show only minor variations with the concentration. This is interpreted as being indicative of a continuous aggregation process towards aggregates with predominantly constant properties.

## INTRODUCTION

Several investigations of the aggregation in solutions of water and alkyl carboxylates in aliphatic alcohols have been published.[1-3] They have been mainly aimed at elucidating the structure of the aggregates and few experiments give quantitative information about the factors that stabilize them. In this report we describe a calorimetric investigation of the partial molar enthalpies in the isotropic alcohol-rich solution phase L2 in the system sodium N-octanoate-N-pentanol-water ($NaC_8$-$C_5OH$-$H_2O$) (see Figure 1). The results are compared with similar measurements in the system $NaC_8$-decanol ($C_{10}OH$)-$H_2O$.[4]

The solubilities of water and $NaC_8$ in decanol at 20°C are only 3.8 and 0.35 weight %, respectively.[5] When both are dissolved simultaneously, however, the solubility increases strongly: The isotropic solution phase L2 extends to quite high concentrations (see Figure 1b). The right-hand phase boundary is defined by the lowest amount of water required to hydrate the counterion of the carboxylate; it has been shown to be relatively independent of the chain length of either the alcohol (from $C_1OH$ to $C_{10}OH$) or the carboxylate.[6] For sodium octanoate, 5-6 moles of water per mole of sodium is required.

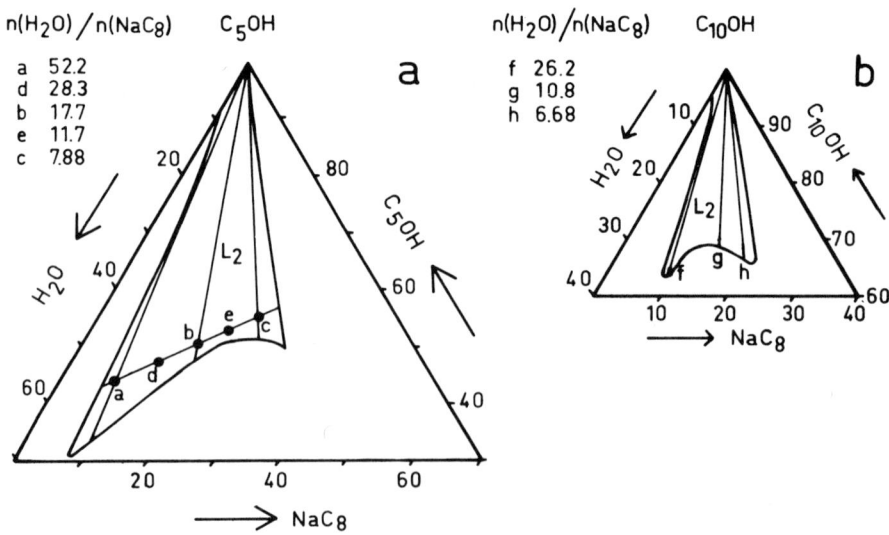

Figure 1. The isotropic alcohol-rich phase, L2, for the systems sodium octanoate ($NaC_8$)-pentanol ($C_5OH$)-water ($H_2O$) (a) and sodium octanoate-decanol ($C_{10}OH$)-water (b). The phase diagram (in weight per cent) pertains to 20°C.[17] The measured enthalpies correspond to compositions determined by the straight lines.

The location of the left-hand border depends on the maximum amount of water that can be bound by the polar groups and the counter ions. The phase boundary towards low alcohol contents depends on the alcohol, the carboxylate and the temperature.[7] The temperature has virtually no effect on the water-rich boundary of the phase and a quite small effect on the right-hand boundary.[5]

As is seen in Figure 1, the exchange of decanol for pentanol causes an extension of the $L_2$ region towards the aqueous corner due to the increased mutual solubility of water and alcohol.

## THERMODYNAMIC EQUATIONS

<u>Notation</u>. We give the component and its state within parentheses after the symbol for each quantity referring to this component. Thus, for example, $x(C_5OH, L2)$ means the mole fraction of pentanol in an L2 solution. Different solutions in the same phase are denoted by Roman numerals, for example, $x(C_5OH, L2, I)$.

<u>Standard states</u>. We choose as standard states for the components pure water, pure pentanol and infinitely dilute $NaC_8$ in water, respectively, at 25°C and 101.3 k Pa.

We wish to determine the relative molar enthalpy $L_m(L2)$ of solutions of $NaC_8$ and $H_2O$ in $C_5OH$, that is, the molar enthalpy for the process

$$C_5OH(1) + H_2O(1) + NaC_8(aq) \rightarrow L2\text{-phase} \qquad (1)$$

We measure $L_m(L2)$ for constant molar ratios $r = n(H_2O)/n(NaC_8)$ (see below). Once $L_m(L2)$ is known for several r's the partial molar enthalpies of each component may be calculated using the equations

$$l(H_2O, L2) = L_m(L2) - (1-x(H_2O,L2)) \frac{\partial L_m(L2)}{\partial x(H_2O,L2)} - \qquad (2a)$$

$$- x(NaC_8,L2) \frac{\partial L_m(L2)}{\partial x(NaC_8,L2)}$$

$$l(NaC_8,L2) = L_m(L2) - x(H_2O,L2) \frac{\partial L_m(L2)}{\partial x(H_2O,L2)} + \qquad (2b)$$

$$+ (1 - x(NaC_8,L2)) \frac{\partial L_m(L2)}{\partial x(NaC_8,L2)}$$

$$l(C_5OH, L2) = L_m(L2) - x(H_2O, L2) \frac{\partial L_m(L2)}{\partial x(H_2O, L2)} \qquad (2c)$$

$$- x(NaC_8, L2) \frac{\partial L_m(L2)}{\partial x(NaC_8, L2)}$$

Details of the method of calculation will be published elsewhere.[4]

$L_m(L2)$ has been determined for three r's, corresponding to the lines a, b, c and for the line through the points a to e in Figure 1a. The former determination is made in four steps:

(i) We mix solutions (L2, I) of water and $NaC_8$ in $C_5OH$ with pure pentanol to form another L2 solution (L2, II):

$$n(H_2O, L2, I) + n(C_5OH, L2, I) + n(NaC_8, L2, I) +$$
$$n(C_5OH, 1) \rightarrow n(H_2O, L2, II) + n(C_5OH, L2, II) + n(NaC_8, L2, II) \qquad (3)$$

The enthalpy of this process, $\Delta H_3$, is given by

$$\Delta H_3 = n(L2, II) L_m(L2, II) - n(L2, I) L_m(L2, I) \qquad (4)$$

Since $L_m$ for $C_5OH$, by definition, = 0. Here, n(L2, I) and n(L2, II) are the total amounts of matter in solutions I and II, respectively. Figure 2 illustrates the resulting dependence of $L_m(L2, II)$ on $x(C_5OH)$ when $L_m(L2, I)$ is set equal to zero.

(ii) To calculate $L_m(L2, II)$ from (4) we need to know $L_m(L2, I)$. We determine this enthalpy by dissolving $L_m(L2, I)$ in an aqueous solution of $NaC_8$ (L1, I):

$$n(L2, I) + n(L1, I) \rightarrow n(L1, II) \qquad (5)$$

where $n(L1, I) = n(H_2O, L1, I) + n(NaC_8, L1, I)$ and n(L1, II) is the total amount of matter in solution (L1, II) which contains $C_5OH$ solubilized in an aqueous solution of $NaC_8$. The enthalpy of this process is

$$\Delta H_5 = n(L1, II) L_m(L1, II) - n(L2, I) L_m(L2, I) - n(L1, I) L_m(L1, I) \qquad (6)$$

(iii) $L_m(L1, I)$ is determined by dilution of the solution with a large amount of water:

$$n(L1, I) + n(H_2O, I) \rightarrow n(NaC_8, aq) \qquad (7)$$

The enthalpy of this process is

$$\Delta H_7 = -n(L1,I) L_m(L1,I) \qquad (8)$$

since, by definition, $L_m(H_2O,I) = L_m(NaC_8,aq) = 0$

(iv) $L_m(L1,II)$ is determined by dissolving pure $C_5OH$ in an aqueous solution of $NaC_8(L1,III)$ having the same ratio $n(H_2O)/n(NaC_8)$ as solution (L1,II):

$$n(C_5OH,1) + n(NaC_8,L1,III) + n(H_2O,L1,III) \rightarrow n(L1,II) \qquad (9)$$

where $n(NaC_8,L1,III) = n(NaC_8,L1,II)$ and $n(H_2O,L1,III) = n(H_2O,L1,II)$. The enthalpy of this process is

$$\Delta H_9 = n(L1,II) L_m(L1,II) - n(L1,III) L_m(L1,III) \qquad (10)$$

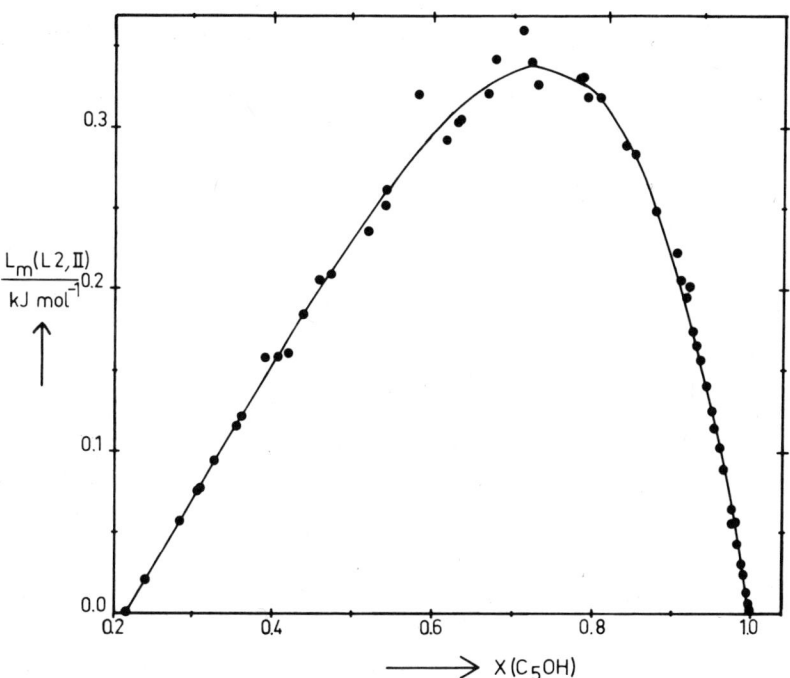

Figure 2. The molar enthalpy for the solutions with the constant ratio $n(H_2O)/n(NaC_8) = 17.7$ (line c in Figure 1) as a function of the mole fraction of pentanol, at 25°C. The points correspond to experimentally derived values, the full drawn line corresponds to the polynomial describing them. $L_m(L2,II)$ pertains to Equation (4) in the text with $L_m(L2,I)$ set equal to zero.

(v) $L_m(L1, III)$ is determined in the same way as $L_m(L1, I)$:

$$n(L1,III) + n(H_2O,I) \rightarrow n(NaC_8,aq) \tag{11}$$

$$\Delta H_{11} = -n(L1,III) \, L_m(L1,III) \tag{12}$$

Figure 3 illustrates steps (ii)-(v).

From Equations (4), (6), (8), (10) and (12) we find

$$L_m(L2,II) = \frac{1}{n(L2,II)} (\Delta H_3 - \Delta H_5 + \Delta H_7 + \Delta H_9 - \Delta H_{11}) \tag{13}$$

As the solutions (L2, I) we use solutions a, b, and c for each r, as shown in Figure 1.

## EXPERIMENTAL

<u>Chemicals</u>. The sodium octanoate was prepared by neutralization of octanoic acid (Fluka AG puriss.) with NaOH (Merck 1 M "Titrisol"). The octanoate was dried in vacuum at 110°C. The purity was checked by photometric titration with percloric acid in glacial acetic acid, using crystal violet as indicator. The maximum deviation allowed

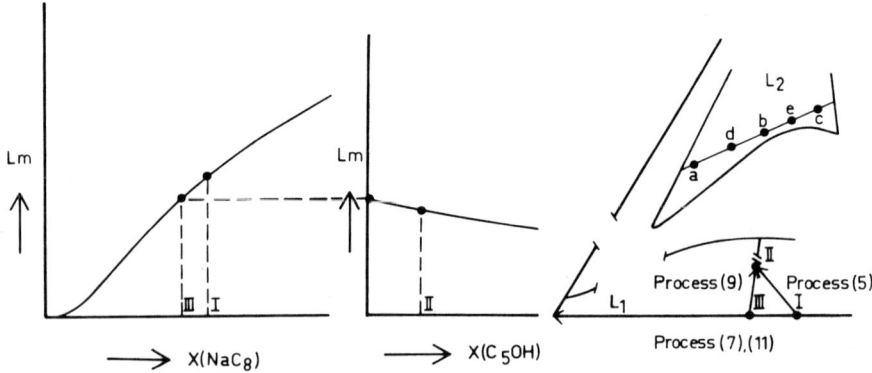

Figure 3. Schematic illustration of the steps (ii)-(v) in the text: Process (5) - mixing of the solutions a to e, respectively, of the pentanol-rich solution phase (L2) with a binary solution of $NaC_8/H_2O$ of the composition I, giving the ternary solution of $NaC_8/H_2O/C_5OH$ of the composition II. Process (9) - mixing of the binary solution of $NaC_8/H_2O$ of the composition III with pure pentanol, giving the ternary solution of $NaC_8/H_2O/C_5OH$ of the composition II. Processes (7) and (11) - mixing of the binary solutions (I) and (III) with a large amount of pure water, giving infinitely diluted sodium octanoate in water, $NaC_8$ (aq). All the solutions I to III are of the isotropic water-rich solution phase, L1.

from the theoretical molecular weight was 0.5%. The pentanol was Merck pro analysi grade. The water was distilled and ion exchanged.

Apparatus. The batch and flow versions of the LKB 10700 microcalorimeter system and the ampuole and titration calorimeter version of the LKB 8700 precision calorimetry system were used.

Experimental methods. The enthalpies for Process (3) were determined by mixing solutions at constant r to give evenly spaced enthalpies along each line corresponding to a given r over the whole L2 region (see Figure 2). For the largest values, the flow calorimeter was used, while the high viscosity of solutions with low r made it necessary to use the batch calorimeter. A few batch experiments were performed for each one of the lines otherwise studied by flow calorimetry. The agreement between batch and flow values was found satisfactory.

To calculate the final concentrations in the flow experiments, it is necessary to know the densities of each solution. These were calculated by interpolation from a few separate pycnometric density determinations. Since agreement was good with the batch experiments, in which the concentrations are determined by direct weighing, it was concluded that significant errors were not introduced by this method.

For Process (5), ampuole calorimetry was used. For Process (7), (9) and (11) the titration calorimeter was utilized. Details of the results of these last three experiments have been published elsewhere.[8]

## RESULTS AND DISCUSSION

Figure 4 shows the enthalpies for lines a, b and c. The relative partial molar enthalpies were calculated from these curves and the curve through a to e by fitting regressional polynomial through the points and are shown in Figure 5. As a result of the uncertainties introduced by the additions and subtractions in steps (ii)-(v), not too much significance should be attached to the absolute values of the enthalpies in Figure 4. The large number of experimental points in phase L2, however, implies that the curves for the partial molar enthalpies are well defined by the experiments.

It is obvious that the sodium octanoate is of dominating importance for the thermal interactions in all regions of the L2 phase. The partial molar enthalpies of $NaC_8$ are larger than the other by one order of magnitude. This agrees well with other investigations, which show that ionic effects are of dominating importance, in particular at concentrations where there is a dissociation to smaller

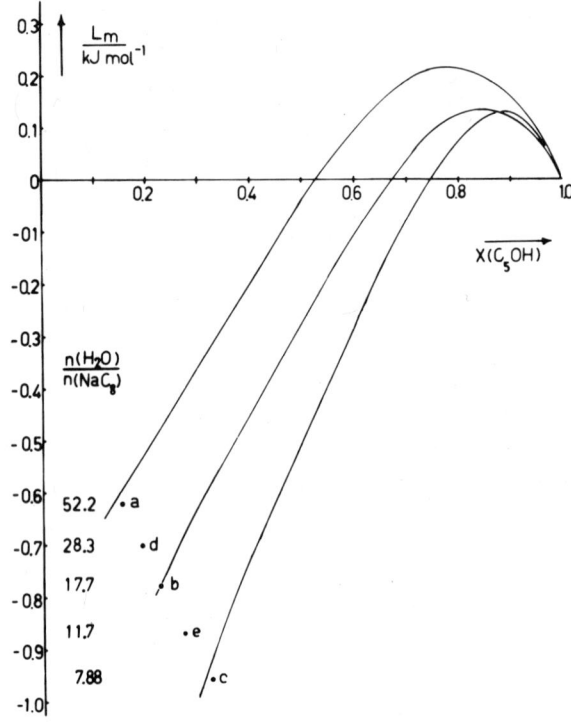

Figure 4. The relative molar enthalpy, $L_m$, as a function of the mole fraction of pentanol at 25°C. Series with constant molar ratio water/sodium octanoate.

aggregates.[9-11] As the concentration of water becomes low there is a strong increase in the partial molar enthalpy of the $NaC_8$ as well as in the enthalpy of the water. The simplest explanation for this behavior is that the counter ions are strongly hydrolyzed even by small amounts of water; the water in these low concentrations is strongly polarized. The pentanol is not much affected but there is a small continuous increase in its relative molar enthalpy as the concentration of $NaC_8 + H_2O$ increases. The rapid decrease in the enthalpies for $NaC_8$ and $H_2O$ at low concentrations becomes continuously slower at intermediate concentrations and at high concentrations the enthalpies are almost constant. The enthalpy of $C_5OH$, on the other hand, undergoes changes almost up to the phase boundary.

These observations agree well with similar measurements for the system water-$NaC_8$-$C_{10}OH$.[4] The general appearance of the relative molar enthalpies in this system, shown in Figure 6, is the same.

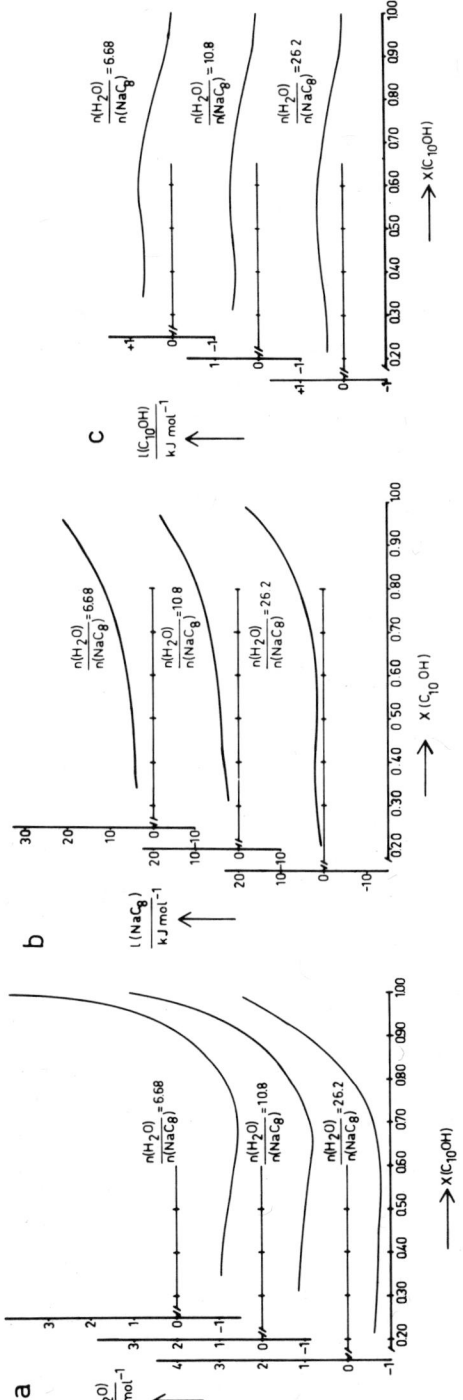

Figure 5. The partial molar enthalpies of water ($H_2O$) (a), sodium octanoate ($NaC_8$) (b) and pentanol ($C_5OH$) (c) as a function of the mole fraction of pentanol, at 25°C. Series with constant molar ratio water/sodium octanoate.

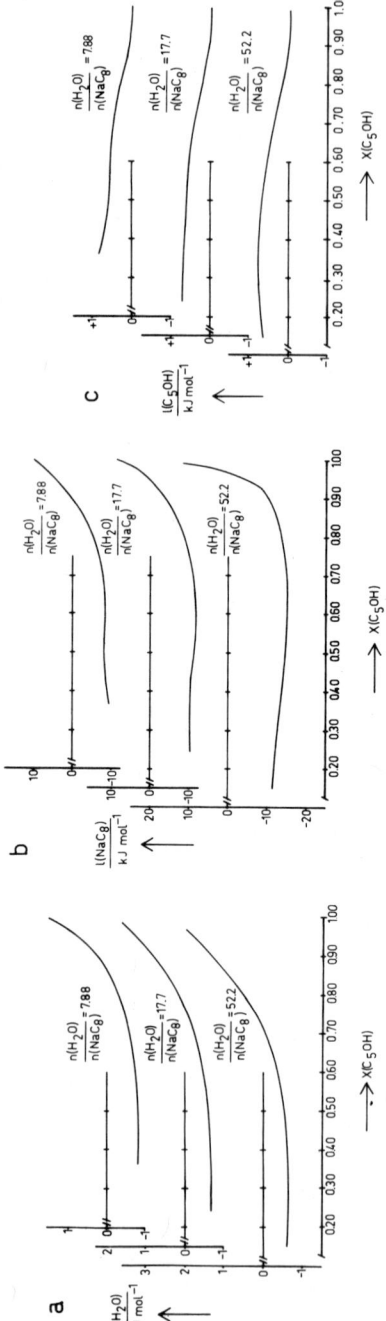

Figure 6. The partial molar enthalpies of water ($H_2O$) (a) sodium octanoate ($NaC_8$) and decanol ($C_{10}OH$) (c) as a function of the mole fraction of decanol, at 25°C. Series with constant molar ratio water/sodium octanoate.

It has been stated that the association in this type of system passes from lower aggregates to large structures ("inverted micelles").[1-3] The calorimetric investigations do not indicate separate steps in the aggregation process; one gets the impression of a continuous change. The invariant enthalpies at high concentrations, however, indicate that aggregates with relatively constant properties predominate.[12] We will not attempt to give a definite picture of the structure of these aggregates. Low angle X-ray investigations, however, indicate aggregates of colloidal dimensions.[13] A light scattering study on the other hand resulted in an estimated molecular weight of about 800 g mole$^{-1}$.[14] Ekwall and co-workers[15] state on the basis of density and viscosity measurements that the aggregates contain not only octanoate and water but also alcohol. They also assert that the interaction between the polar end group, the counterions and the water is of predominating importance in the aggregation.[16] Our results do not contradict this model, which can well be reconciled with a slow increase in the enthalpy of the alcohols while the enthalpies of the $H_2O$ and the $NaC_8$ decrease rapidly to a relatively constant value.

Our results, however, do not confirm all the differences between the aggregations processes in regions with low and high water concentrations, respectively. Thus, Gillberg and Ekwall[16] have found a sudden shift towards higher energies in the pmr signals of the $CH_2$ protons in the $NaC_8$ as the concentration of decanol decreases below 75 weight % in solutions with low concentrations of water. In the water-rich regions a continuous downfield shift is observed. We have recently found similar differences by ir spectroscopy.[11] The enthalpies show no analogous difference. They are well reconcilable, however, with the observation (Ekwall[17]) that the vapor pressure of water is not much dependent on the concentration of $C_{10}OH$ at high concentrations of $H_2O$ and $NaC_8$.

When the concentration of water is decreased at constant concentration of $C_{10}OH$ or $C_5OH$, the enthalpy of the water decreases continuously while that of the $NaC_8$ increases. Again, this indicates continuous changes towards aggregates with constant properties. Along the line through points a-e all the enthalpies are fairly constant.

We conclude that the calorimetric investigations, which we consider as having only just passed the stage of developing methods, are capable of giving important new information on the aggregation processes in solutions of surfactants and water in apolar media. It is obvious that the aggregation does not lead to aggregates as large or as stable as in water. The details of the aggregate structure, however, can only be elucidated by a combination of results from investigations by many different methods, spectroscopic as well as thermodynamic. There is still too much room for speculation to allow the setup of a model with reasonably predictive capabilities.

ACKNOWLEDGMENTS

We wish to thank Professor I. Danielsson for discussions and Mr. R. Slätis for invaluable help with the experiments. Financial support from the Finnish Research Council for Natural Sciences is gratefully acknowledged.

REFERENCES

1. P. Ekwall, J. Colloid Interface Sci., 29, 16 (1969).
2. P. Ekwall, I. Danielsson and P. Stenius, in "Surface Chemistry and Colloids", M. Kerker, Editor, MTP Intern. Rev. Sci., Phys. Chem. Ser. No. 1, Vol. 7, pp. 129-145, Butterworths, London, 1972.
3. P. Ekwall and P. Stenius, ibid., Phys. Chem. Ser. 2, Vol. 7, pp. 239-248, 1975.
4. Zs. Bedö, K. Blomqvist and J. B. Rosenholm, (1976), to be published.
5. P. Ekwall and L. Mandell, Acta Chem. Scand., 21, 1612 (1967).
6. P. Ekwall and L. Mandell, ibid., 22, 699 (1968).
7. P. Ekwall, "Advances in Liquid Crystals", G. H. Brown, Editor, Vol. 1, pp. 1-142, Academic Press, New York, 1975.
8. P. J. Stenius, J. B. Rosenholm and M.-R. Hakala, presented at the Conf. Colloids and Surfaces, San Juan, Puerto Rico, June 1976.
9. B. Lindman and P. Ekwall, Mol. Cryst. 5, 79 (1968).
10. J. B. Rosenholm and B. Lindman, J. Colloid Interface Sci., in press.
11. J. B. Rosenholm (1976), to be published.
12. N. Kamenka, H. Fabre and B. Lindman, C. R. Acad. Sci., Ser. C, 281, 1045 (1975).
13. B. Svens and M. Turpeinen, Progr. Colloid & Polymer Sci., 56, 30 (1975).
14. R. Friman (1976), to be published.
15. P. Ekwall and P. Solyom, Acta Chem. Scand., 21, 1619 (1967).
16. G. Gillberg and P. Ekwall, ibid., 21, 1630 (1967).
17. P. Ekwall, "Liquid Crystals and Ordered Fluids", J. F. Johnson and R. S. Porter, Editors, Vol. 2, pp. 177-197, Plenum, New York, 1974.

DISCUSSION

On the paper by H. F. Eicke

L. E. Scriven, *University of Minnesota*: Dr. Kertes showed data on cationic detergents in apolar solvent which indicate that dipole moment increases with aggregation number, for small aggregates at least, and this does indeed suggest that the aggregates are linear. But his further argument against the existence of closed micellar aggregates does not seem to accord with Dr. Eicke's results.
  The critical step in the formation of closed micelles in apolar media, if I understand Dr. Eicke's presentation, is a conformational change at constant aggregation number, from a linear aggregate of comparatively high dipole moment, to a rounded, closed package of much lower polarity, as measured by multipole moments as well as dipole moment. Evidently this transformation (which I would term topological) is analogous to closure of premicelles to give protomicelles in aqueous detergent solutions, and likewise can give rise to easily detected critical micelle phenomena. Can both authors comment?

H. F. Eicke: Among ionic surfactants in apolar solvents a number of compounds, in particular those with sulfonate groups, are definitely known to form micelles which can be identified by methods common with respect to aqueous surfactant systems, i.e. conductivity, light scattering, vapor pressure osmometry, and ultracentrifuge measurements. Many detergents in apolar media, however, do not form micelles - probably most of the cationic amphiphilic compounds investigated so far - but exhibit multiple association behavior typical for a large class of polar compounds in apolar solvents. In the latter case the shape of the molecules and the particular charge distribution determine whether the minimal energy is obtained for chain association of the dipolar molecules or for "polypole" association (anti-parallel dipoles). Dr. Kertes observed one of these association patterns for small cationic detergent aggregates which, however, can tell us nothing about micelle formation in general in non-polar solvents. Moreover, it is impossible to draw any conclusion from these observations as to the possible existence of micelles in apolar media.

If micelles are formed, as for example in the case of AOT in several nonpolar solvents, an initially linear aggregation of triangular subunits of AOT molecules with partially compensated dipole moments should be assumed according to conclusions drawn from Ekwall based on small angle x-ray scattering. The instability of such aggregate with increasing length requires a transformation to a more rounded package whereby the polarity again is reduced compared with the linear arrangement. This "topological" transformation is derived from the model proposed to explain experimental data resulting from the dipole chemical field effect measurements.

Editor: Dr. Scriven has addressed this question to Dr. Kertes also. Dr. Kertes' comments are given under discussion of his paper.

K. Kalyanasundaram, *University of Notre Dame*: In your model, the reverse micelle formation proceeds through trimers, tetramers, etc. to the n-mer where n is the aggregation number. I wonder whether you have any information on the time-scales involves in each of these steps or on rate limiting steps in the multiple step process.

H. F. Eicke: The proposed model assumes according to the experimental results that the main observable relaxation process is determined by the isomerization process with an estimated rate constant of $5 \times 10^7$ sec$^{-1}$ for the transformation from a linear to a "closed aggregate.

## On the paper by A. S. Kertes

J. H. Fendler, *Texas A & M University*: What extent is the issue of semantics or terminology? You are talking of trimeric, tetrameric and pentameric of surfactants in nonpolar solvents. Some people refer to them as reversed micelles. Did you consider surfactants such as aerosols and aluminum soaps which form large micelles? Finally would you agree that the sudden appearance of large rate enhancements in the three component system is due to some change of aggregations.

A. S. Kertes: I believe that micelles exist in hydrocarbon media. I oppose, however, the misleading term of inverted or reversed

## DISCUSSION

micelles, and even more the implied spherical structure which goes with it. To the best of my knowledge there is no experimental evidence that spherical inverted micelles exist. The fact that the concept of reversed micelles is frequently used is no evidence yet.

I don't think that small aggregates should be termed micelles, because usually they do not exhibit the characteristic behavior of micellar solutions, such as solubilization, for example. On the other hand, at what size-distribution will a solution start to exhibit this and other micellar characteristics depends both on the surfactant and the solvent. Thus, the possibility that a solution in which the predominating aggregate is a tetramer will exhibit micellar behavior cannot a priori be excluded. The important difference between hydrocarbon and aqueous solutions of surfactant remains the fact that in the former case aggregation sets in before the bulk solution starts to exhibit micellar characteristics, and of course, that the process of aggregation is a stepwise one.

This applies to heavy-metal salts of fatty acids, and I would expect that true and pure aluminum tricarboxylates (and pure dialkyl sodiumsulfosuccinates-aerosols) behave similarly. Unfortunately, existing literature data refer to aluminum soaps which do not analyze as 1 aluminum to 3 carboxylates (neither are the commercially available aerosols well defined compounds). The presence of hydroxy group "impurities" in the aluminum soaps is bound to affect the process of aggregation.

The above sets the stage for the answer to the last part of your question. Indeed, the presence of a third component - whether additive, solubilizate or simple impurity - is bound to affect the aggregation equilibrium, and it is conceivable that in some systems this effect may be dramatic.

E. Ruckenstein, *State University of New York, Buffalo*; and R. Zana, *Centre de Recherches sur les Macromolecules, Strasbourg*: We fully agree with the point of view that for surfactants in non-polar solvents a critical micelle concentration (CMC) cannot be identified. In support of this view we report that ultrasonic absorption measurements performed on various surfactants in cyclohexane [1] have failed to reveal a CMC. It is our opinion, however, that a unitary thermodynamic treatment can be used both in aqueous and non-polar solvents. In non-polar solvents the nature of the interaction forces results in general in a size distribution function of the aggregates that decreases monotonically with the size of the aggregate and, therefore, no critical concentration is found. On the other hand, in water the size distribution function of aggregates of amphiphiles with a straight alkyl chain decreases monotonically with the size of the aggregate only below a certain critical concentration of surfactant, but presents a minimum and a maximum above

this concentration [2,3]. This assures the existence just above
this critical concentration of a CMC where a sharp change of physicochemical properties occurs. This does not mean, however, that all
amphipathic compounds present a CMC in aqueous solutions. Indeed,
the critical concentration can be larger than the solubility of the
compound as happens, for instance, for long chain alcohols or
amines with six or more carbon atoms. We note further that some
amphiphiles with a double paraffinic chain or with a particular
structure (bile salts) do not generally aggregate in water as
micelles but can organize in other structures [3,4]. The long
double paraffinic chains organize, for instance, as vesicles and
have a critical vesicle concentration [3].

References:
  [1] R. Zana and J. Lang (unpublished results).
  [2] E. Ruckenstein and R. Nagarajan, J. Phys. Chem. $\underline{79}$, 2622 (1975).
  [3] E. Ruckenstein and R. Nagarajan, this symposium.
  [4] R. Zana, et al., this symposium.

A. S. Kertes: Thank you for the interesting comments, and the
unpublished information in support of the doubts expressed in my
talk concerning the existence of CMC in hydrocarbon solvents.

Indeed, Mukerjee (references 43 and 44) has suggested some
theoretically possible models of stepwise self-association of surfactants in water and aqueous solutions. The sets of mass-action
law equilibria are similar to those given in our earlier works
(references 1, 41 and 42). It remains, however, to examine experimental data in terms of the models suggested.

One word of caution might be in place in view of your comment:
Not all amphiphiles are surfactants. Long-chain alcohols, alkyl-
amines and fatty acids are amphiphiles but their hydrocarbon solutions do not exhibit surface active properties in spite of the fact
that some aggregate, though usually much less than surfactants
derived from them. Thus, for example, tridodecylamine in benzene
behaves ideally, at least up to 0.2 m concentration, whereas several
tridodecylammonium halides aggregate in benzene at concentrations as
low as 0.005 m (reference 29).

A. Kitahara, *Science University of Tokyo*: I myself have some doubts
about the exact meaning of CMC in non-aqueous systems, and am
willing to accept that a stepwise aggregation process is operative
in the formation of very small aggregates.

However, we have many data, obtained by vapor pressure measurements, which indicate that the apparent average aggregation number
reaches a constant value. To these cases the concept of stepwise
aggregation does not apply.

I also believe that ionic interactions between the ionic parts
of surfactants are operative, though you have been talking of
dipole-dipole interactions only.

# DISCUSSION

A. S. Kertes: Thank you for your partial support of my thesis.
Expressing aggregation of ionic surfactants in hydrocarbons in terms of apparent average aggregation number, $\tilde{n}$, is bound to be misleading, frequently very seriously. $\tilde{n}$ does not distinguish between the extent (number of aggregated units) and degree (size of aggregated units) of aggregation as I tried to emphasize in my talk. As a consequence of the definition of $\tilde{n}$ (equation 18 in my text) a constant value for $\tilde{n}$ with increasing solute concentration can result from a decreased extent of aggregation, compensated by an increased degree of aggregation. This is an artifact arising from size-distribution, and is no evidence against the concept advocated here.

As to your second question, I don't believe that in low dielectric constant and zero dipole moment hydrocarbons ionic interactions can significantly contribute to the stability of the aggregates (micelles) formed. I do allow, however, for hydrogen bonding and/or metal-coordination bonds, as discussed in more detail elsewhere (reference 1).

F. M. Fowkes, *Lehigh University*: I agree that a careful study of the CMC in hydrocarbon media has yet to be made, but there is absolutely no doubt that micelles of appreciable aggregation number exist over wide concentration ranges. In some cases (see the work of Reerink) the aggregation number increases to large values (>100) with increase in concentration. In others, such as the salts of dinonylnaphthalene sulfonic acid, constant micelle size of about 25 sulfonates is observed from $10^{-6}$ M to $10^{-1}$ M (J. Phys. Chem. **66**, 1843, 1962). Dr. Kertes' equilibrium constants are perfectly straightforward and correct, the same as used in aqueous systems, but they do not predict much with respect to aggregation number.

A. S. Kertes: Your question appears to be a continuation of our discussion last year in Budapest (Proc. Intern. Conf. on Colloid and Surface Science, Akademiai Kiado, Budapest, 1976, Vol. I, p. 497 and Vol. II, p. 126). As I have emphasized in my reply to the question of Dr. Fendler earlier, I do not question the existence of micelles in hydrocarbon solutions. All I am trying to advocate is that these micelles are formed by a stepwise process of aggregation rather than by a simple monomer⇌micelle equilibrium characteristic to aqueous systems. The difference in the process of aggregation, as I have tried to emphasize in my talk, is due to the very different nature of the driving forces which are operative in the two types of systems: hydrophobic interactions in water, versus dipole-dipole interactions in hydrocarbon.

As to your last comment, I believe that we have successfully interpreted experimental data by explaining through our set of mass-action law equilibria, why no CMC can be expected and observed in hydrocarbon solutions. Using equation 18 in my paper, one can derive the average aggregation number from the $\beta$ values calculated from experimental data.

L. E. Scriven, *University of Minnesota*: Dr. Kertes showed data on cationic detergents in apolar solvent which indicate that dipole moment increases with aggregation number, for small aggregates at least, and this does indeed suggest that the aggregates are linear. But his further argument against the existence of closed micellar aggregates does not seem to accord with Dr. Eicke's results.

The critical step in the formation of closed micelles in apolar media, if I understand Dr. Eicke's presentation, is a conformational change at a constant aggregation number, from a linear aggregate of comparatively high dipole moment, to a rounded, closed package of much lower polarity, as measured by multipole moments as well as dipole moment. Evidently this transformation (which I would term topological) is analogous to closure of premicelles to give protomicelles in aqueous detergent solutions, and likewise can give rise to easily detected critical micelle phenomena. Can both authors comment?

A. S. Kertes: We have been very careful in discussing the possible structure of aggregates in hydrocarbon solutions. Our dielectric constant measurements and dipole moment calculations (reference 31) appear to be in line with what could be expected for small linear aggregates. We don't claim that this should be taken as evidence.

My comment Dr. Scriven is referring to, which does not appear in the written version of my talk, concerns earlier thermodynamic arguments (reference 29) and some new preliminary calorimetric data which also favor the linear structure, but again do not prove it.

I don't think there is necessarily a contradiction between our and Dr. Eicke's interpretation. Our dipole moments refer to dimeric or trimeric species (reference 31) for which a "closed micellar" structure, as Dr. Scriven calls it, is rather unlikely. The dipole moment, the distance of closest approach, the polarizability and the angle between the molecules in the aggregate are affected by the size of the aggregate. Consequently, the dipole moment of Dr. Eicke's higher aggregates is expected to be different. One should emphasize that a low dipole moment of the higher aggregate is responsible, at least partly, for limiting the growth of the aggregate in solutions.

CONCLUDING REMARKS

G. S. Hartley

57 Aurora Terrace, Hillcrest

Hamilton, New Zealand

This has been an extremely interesting and lively symposium and I am sure you would first wish me to express our great thanks to Dr. Kashmiri Mittal for the hard work and time he has put into organizing it. Most of us know how difficult it is to get contributions in and questions replied to on time but we still need a driving force. Thank you, Dr. Mittal.

As a winding-up speaker, I have a wonderful opportunity to be provocative without incurring riposte. I will try not to take advantage but confine myself to non-controversial points. It is useful, I think, to remind ourselves what a very small fraction of the whole diagram - water, amphiphile, hydrophobe - has been examined with precision and interpreted with clarity. These extreme corners are easy - the aqueous corner in which I spent some time, and the oil corner where Dr. Kertes finds no evidence for a critical concentration. He does well to remind us that the behavior dictated by water aggregation when this solvent is the continuum can be very different when it is absent or contained only in the discrete aggregates.

The most important part of the whole field for future development, particularly for its physiological importance, is the labyrinthine complex of separate phases, some anisotropic, occupying the large central area. Like the impressive and elegant SUNY campus in which we have been priviliged to meet, the correlations are often obscure and the direction confusing. I found Dr. Scriven's description of interpenetrating continuous phases in sponge-type structures most stimulating.

It is obviously desirable to understand the simple before proceeding to the complex but I feel that I was very unadventurous in retreating into my simple corner as soon as the complexities appeared.

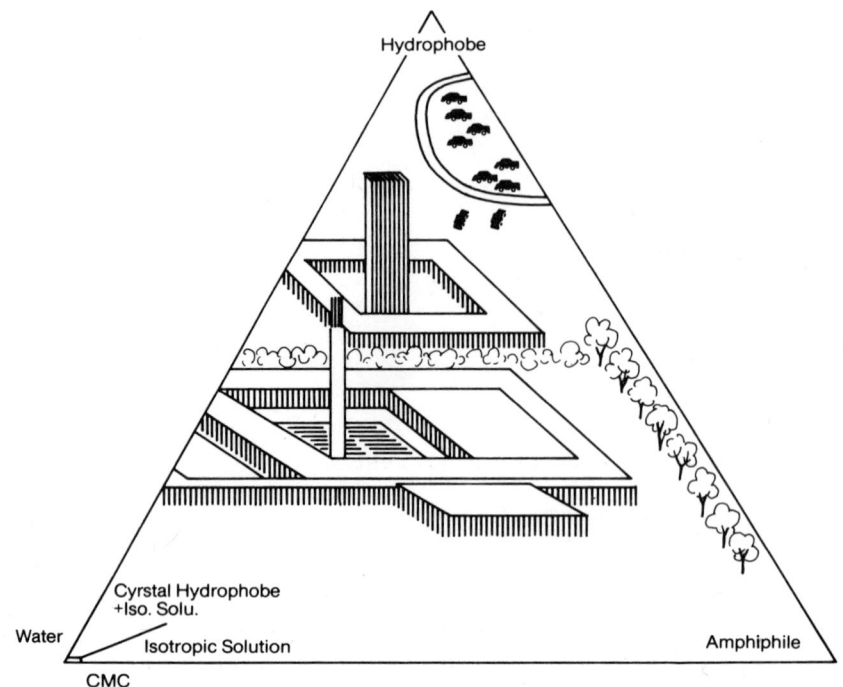

Diagram - Water, Amphiphile, Hydrophobe

I pay my respects to those who are actively investigating the spontaneous formation of laminae or rods which are, so to speak, equilibrium systems in one or two dimensions but not in all. I hope the work in the corners has been of some help to them.

They have had the work in the corners to build on. They also have sophisticated tools at their disposal which were not dreamed of at the time of my own contribution - the use of photochemical probes and relaxation methods, discussed in this symposium by Dr. Turro and Dr. Aniansson. Light scattering technique has become much more advanced but Dr. Mazer draws conclusions from its use which Dr. Tanford finds at variance with those from a combination of less sophisticated measurements.

Unquestionably, these new tools have great power, but I wonder whether as much effort is put into the interpretation of the results they provide as is put into the design of the tools themselves. I raise this question with some hesitation because I know that an old man can become bewildered with innovations and feel jealous of those who can understand them, but I recall the errors of the first interpretation of X-ray diffraction data in this field and could illustrate my point with historical examples from other fields. I would have liked to have heard more argument over the analysis of just how certain and unique is the interpretation of some of the new data.

The micelle state makes possible some interesting possibilities in chemical reaction through promoting molecular contacts which would otherwise occur relatively infrequently. This has importance both in industry and in vital chemistry. We have had some illustrations of these reactions. They are necessarily tied up with the complexities of structure and make the unravelling of these complexities difficult, though it is all the more worthwhile. This symposium has helped considerably the mutual understanding of the many interests in micellization, solubilization and micro-emulsions.

ABOUT THE CONTRIBUTORS

Here are included biodata of only those authors who have contributed to this volume. Biodata of contributors to Volume 2 are included in that volume.

*Conception Abad* is a postdoctoral investigator at the Centre de Recherches sur les Macromolecules, C.N.R.S., Strasbourg, France. He obtained his D.Sc. in 1974 from the University of Madrid, Spain.

*Mats Almgren* is at the Chalmers University of Technology, Sweden. He received his Doctor of Technology degree in 1974 from the Chalmers University of Technology. His research interests are photochemical and micellar kinetics.

*E. A. Gunnar Aniansson* is Professor of Physical Chemistry at the University of Gothenburg, Sweden. He received his Doctor of Technology degree in 1961 from the Royal Institute of Technology in Stockholm. His research interests are surface chemistry, radioactivity methods, molecular collisions, and micellar kinetics.

*Vinod Kumar Bansal* is a Research Associate in the Chemical Engineering Department, University of Florida, Gainesville. He obtained his Ph.D. degree in Metallurgical Engineering from the Indian Institute of Technology, Kanpur, India. He has published in the areas of collector-frother interaction in flotation, microemulsions and micellar solutions for improved oil recovery.

*George B. Benedek* is Professor of Physics at the Massachusetts Institute of Technology and is a lecturer in physics on the Faculty of Medicine at Harvard University. He has published over 50 research papers in a variety of fields including quasielastic light scattering spectroscopy. He is a Fellow of the American Physical Society, a member of the Governing Board of the American Institute of Physics, and a member of the Editing Board of the Biophysical Journal.

*Kulbir S. Birdi* is Assistant Professor ("lektor" in Danish) at the Technical University of Denmark, Lyngby, Denmark. His research interests are monolayers, biological macromolecules, and micelles.

*John R. Cardinal* has been an Assistant Professor, College of Pharmacy, University of Utah, since 1972. He received his Ph.D. degree in 1973 from the School of Pharmacy, University of Wisconsin.

## ABOUT THE CONTRIBUTORS

*Martin C. Carey* is Assistant Professor of Medicine at Harvard Medical School and Associate in Medicine at the Peter Bent Brigham Hospital, Boston. He received his M.D. degree from the National University of Ireland in 1962. His current research interests are concerned with the physical chemistry and biophysics of alimentary tract lipids in health and disease. In 1975 he was appointed a Fellow of the J. S. Guggenheim Memorial Foundation. He serves on the Editorial Board of the American Journal of Physiology and on the Steering Committee of the Gastroenterology Research Group.

*Narendra R. Desai* received his M.S. degree from the School of Pharmacy, University of Wisconsin, Madison, in 1968.

*Abassali Djavanbakht* is presently at the University of Meched, Iran. He obtained his M.S. in 1973 from the University of Strasbourg.

*Hans Friedrich Eicke* is Professor of Physical Chemistry at the University of Basel, Switzerland. He carried out his Ph.D. thesis research (1959-1961) at the Max-Planck-Institute for Physical Chemistry in Gottingen. He has published about 40 papers, mostly on the properties and behavior of surface active (weak) electrolytes in hydrocarbon solutions. He is president of the Chemical Society of Basel.

*Alexander T. Florence* is J. P. Todd Professor of Pharmaceutical Technology and Chairman of the Department of Pharmaceutical Technology, University of Strathclyde, Glasgow, Scotland. He received his Ph.D. degree in 1965. In 1972 he was awarded the British Pharmaceutical Conference Science Award for his published work in the Pharmaceutical Sciences. He is co-author with Professor P. H. Elworthy and Dr. L. B. MacFarlane of the book "Solubilization by Surface Active Agents" published in 1968.

*Margaret W. Geiger* is presently a postdoctoral Research Associate at Yale University. She completed her Ph.D. degree in chemistry in the summer of 1976 at Columbia University.

*William John Gettins* is Research Technician at the University of Salford.

*Hans Gustavsson* is a graduate student in physical chemistry at the Lund Institute of Technology.

*Marja-Riitta Hakala* is a Research Assistant at the Finnish Research Council for Natural Sciences.

*Gilbert Spencer Hartley* has been living in Hamilton, New Zealand since his retirement in 1970 as Research Director from the Agrochemical Division of Fisons Limited which he joined in 1955.

## ABOUT THE CONTRIBUTORS

He had his education at University College, London from 1924-1930, and remained in the same laboratories until 1944. The war took him into Chemical "Defence" Research until 1945. His main work on paraffin chain salts was done during 1927-1939. Presently, he retains an informal interest in pest problems and is collaborating with a U.K. author on a new textbook on the physics of pesticide use.

*Richard R. Hautala* is Assistant Professor at the University of Georgia in Athens. He received his Ph.D. degree in 1969 from Northwestern University. His specialty is organic photochemistry involving reactions in micelles and other organized molecular aggregates.

*Heinz Hoffmann* is Professor at the University of Bayreuth. He had his university education at Würzburg and Karlsruhe. He has published in the area of electrochemistry, relaxation methods, fast reactions, metal complexation, and micelle formation.

*Kjell Holmaker* is Lecturer in Mathematics at the Gothenburg Center for Applied Mathematics. He received his Doctor of Technology degree in 1976 from the Chalmers University of Technology. He has published on optimal control theory.

*R. Thomas Holzbach* is Head of the Gastrointestinal Research Unit, Department of Medicine, The Cleveland Clinic Foundation, Cleveland, Ohio. He received his M.D. degree in 1955 from Case Western Reserve University School of Medicine. His research interests include investigation of bile salt and bile salt mixed lipid micelles.

*Alexander M. Jamieson* is in the faculty of the Department of Macromolecular Science, Case Western Reserve University. He received his D. Phil. degree in 1969 from the University of Oxford, England. His research interests include quasielastic laser light scattering.

*A. Steven Kertes* is in the faculty of the Hebrew University, Jerusalem. He obtained his Ph.D. degree in 1954 from the Hebrew University. He has over 100 publications to his credit, and is the co-author of two monographs in the field of solvent extraction chemistry. He is on the Editorial Board of the J. Inorganic Nuclear Chemistry, and the CRC Critical Reviews in Analytical Chemistry. Since 1973, he is chairman of the Subcommittee on Solubility Data of IUPAC.

*Robert F. Kwasnick* is an undergraduate physics major in the Honors Program at Swarthmore College in Pennsylvania. During the past two years he has been a research assistant at M.I.T., and at the Peter Bent Brigham Hospital and Harvard Medical School.

*Jacques Lang* is Charge de Recherches in the C.N.R.S. Centre de Recherches sur les Macromolecules, Strasbourg, France. He received his Ph.D. degree in 1968 from Strasbourg. His research interests are kinetic studies of micellar equilibria of ionic and nonionic detergents, and reactions in biomolecules.

*Göran G. Lindblom* is a Research Associate at the Swedish Natural Science Research Council. He received his Doctor of Technology in 1974 from the Lund Institute of Technology. He has published in the field of NMR studies of amphiphile and biological membrane systems.

*Björn Lindman* is Lecturer in Physical Chemistry at the Lund Institute of Technology. He obtained his Doctor of Technology degree in 1971 from the Lund Institute of Technology. He has published in the field of NMR studies of protein solutions, amphiphile systems, and electrolyte solutions. He is the coauthor (with S. Forsen) of a monograph "Chlorine, Bromine, and Iodine NMR, Physicochemical and Biological Applications" soon to be published by Springer-Verlag, Berlin.

*Issa Lo* has recently been appointed to a position in the Faculty of Pharmacy at Dakar. He holds the degree of Docteur d'Etat en Pharmacy of the Faculty of Pharmacy of Paris. He is interested in the surfactant-water-oil systems.

*Florence Madsen* recently completed her doctoral thesis at the Universite de Paris-Sud at Châtenay-Malabry. Her research interests are in the area of emulsion stability.

*Norman A. Mazer* is presently a candidate for the Ph.D. in Physics at the Massachusetts Institute of Technology and also a candidate for the M.D. degree. He has been interested in the technique of quasielastic light scattering spectroscopy for studying the size, shape and polydispersity of biological macromolecules, and also for studying micelle formation.

*Milton E. McDonnell* is a Research Associate in the Department of Macromolecular Science, Case Western Reserve University. He received his Ph.D. degree in 1975 from the University of Delaware. His research concerns characterizing the hydrodynamic properties of macromolecules, primarily by quasielastic laser light scattering.

*Kashmiri Lal Mittal* is presently at the IBM Corporation, Poughkeepsie, New York. He received his Ph.D. degree in Physical Chemistry (Surface and Colloid) from the University of Southern California in 1970. He has organized and chaired a number of international symposia. In addition to these volumes, he is also the editor of "Adsorption at Interfaces" and "Colloidal Dispersions and

# ABOUT THE CONTRIBUTORS

Micellar Behavior" (both published by the ACS), and of "Adhesion Measurement of Thin Films, Thick Films, and Bulk Coatings" (to be published by the ASTM). His research interests include: Colloid science, adhesion science and technology, corrosion and passivation of metals, and surface properties of polymers. He has published about 35 scientific or technical papers, and has given many invited talks on the various facets of adhesion on the invitation of various societies or organizations.

*Pasupati Mukerjee* is Professor, School of Pharmacy, University of Wisconsin, Madison. He obtained his Ph.D. degree in Colloid and Physical Chemistry from the University of Southern California in 1957. He has served as a member of the Executive Committee of the Division of Colloid and Surface Chemistry, ACS, and is currently a member of the Commission on Colloid and Surface Chemistry of IUPAC. He is a Fellow of AAAS.

*Norbert Muller* is in the faculty of the Chemistry Department at Purdue University. He obtained his Ph.D. degree in 1953 from Harvard University. His major area of research has been nuclear magnetic resonance spectroscopy. In about 1965, he became interested in micellar solutions.

*R. Nagarajan* is currently pursuing research in the area of membranes toward his Ph.D. degree at SUNY-Buffalo. He received his M.S. degree in Chemical Engineering from the Indian Institute of Technology, Kanpur.

*H. Nüsslein* is a graduate student at the Universität Erlangen - Nürnberg, West Germany.

*Suk Yon Oh* received his Ph.D. in 1974 from Colorado State University.

*Leon M. Prince* is a Consultant in Surface Chemistry. In January 1976, he retired from the Lever Brothers Company, where he was a Development Scientist at the Research Center. Mr. Prince is the Co-Editor (with Dr. Sears) of "Biological Horizons in Surface Science", published in 1973, and he is editing "Microemulsions: Theory and Practice" which will be published soon.

*Francis Puisieux* is Professor in Pharmaceutical Technology at the Centre d'Etudes Pharmaceutiques, Université Paris-Sud at Châtenay-Malabry. Since 1964, he has investigated the chemical composition of ointment excipients, aspects of the physics of tablet compression, emulsions, and surfactant systems.

*Jorgen Eilir Rassing* is Professor of Chemistry, Roskilde University Centre, Denmark. He is currently Visiting Professor at the University of Salford. He has published approximately 40 papers.

*Jarl B. Rosenholm* is a Research Assistant at the Finnish Research Council for Natural Sciences. He obtained his M.Sc. degree in 1972 from the University Åbo Akademi (Turku), Finland. He has published about 15 papers in the field of association colloids.

*Eli Ruckenstein* is Faculty Professor of Engineering and Applied Sciences, State University of New York, Buffalo. He was Professor at the Polytechnic Institute, Bucharest, Romania, 1949-1969. He received the Gheorghe Spacu Award of the Romanian Academy of Sciences (1963) for research in surface phenomena, and two awards for research of the Romanian Department of Education (1958, 1964). He has published extensively dealing with many subjects. His research interests are catalysis, colloids and interfaces, and biophysics.

*Neil E. Schore* is currently teaching at the University of California at Davis. He received his Ph.D. degree in 1973 from Columbia University.

*Monique Seiller* is Maître de Conférences in the Faculty of Pharmaceutical Sciences at Caen. Her research interests are in the area of emulsion stability.

*Dinesh O. Shah* is Professor of Anesthesiology, Biophysics, and Chemical Engineering, University of Florida, Gainesville. He received his Ph.D. degree in Biophysics from Columbia University in 1965. He received the University of Florida's "Excellence in Teaching Award" in 1972-73, "President's Scholar Award" in 1975-76, and "Outstanding Service Award" in 1975-76. He has published in theareas of biological and model membranes, chemical evolution, monomolecular films, foams, microemulsions, boundary lubrication, and surface chemical aspects of lungs, vision, and anesthesia.

*Per J. Stenius* is Head of the Department of Polymer Science and Docent at the Department of Physical Chemistry, Åbo Akademi, Åbo (Turku), Finland. He obtained his Phil. Dr. degree in 1973 from the University Åbo Akademi. He has published about 40 papers in the area of association colloids, emulsion stability, and emulsion polymerization.

*Kunio Takeda* is a Research Fellow of the Japan Society for the Promotion of Science. He received his Ph.D. degree in 1976 from Hiroshima University.

# ABOUT THE CONTRIBUTORS

*Charles Tanford* is James B. Duke Professor of Physical Biochemistry at Duke University, Durham, North Carolina. He is a member of the American Academy of Arts and Sciences, and of National Academy of Sciences. He is author of an advanced textbook, "Physical Chemistry of Macromolecules", and about 15 research papers. He has been on the editorial boards of the Journal of the American Chemical Society and the American Chemical Society and the Journal of Biological Chemistry.

*Jean Paul Treguier* is a pharmacist and Maître-Assistant in the Centre d'Etudes Pharamceutiques at the Université de Paris-Sud. He has just completed his thesis for the Doctorat D'Etat in pharmacy.

*Nicholas J. Turro* is Professor of Chemistry, Columbia University, New York. He received his Ph.D. from the California Institute of Technology. He has served as editor of "Spectroscopic Methods and Organic Techniques" for the W. A. Benjamin Company and on the editorial boards of several journals. He has received many diverse awards. He has authored "Molecular Photochemistry", published in 1969, and is finishing a new book tentatively called "Modern Molecular Photochemistry". His bibliography includes over 160 publications.

*Hiromoto Uehara* is in the faculty at Kinki University, Japan. He received his Ph.D. degree in 1976 from Hiroshima University.

*Werner Ulbricht* is Akademischer Rat at the University of Bayreuth. He had his university education in Erlangen-Nürnberg. He has published in the area of metal complexation and micellization.

*Staffan N. Wall* is a graduate student at the University of Gothenburg, Gothenburg, Sweden. He has published on micellar kinetics.

*Håkan Wennerstrom* is a Research Associate at the Lund Institute of Technology. He received Doctor of Technology in 1974 from the Lund Institute of Technology. His publications deal with applications of MNR to amphiphilic systems and applications of quantum chemistry to chemical problems.

*Evan Wyn-Jones* is Reader in Chemistry, University of Salford, United Kingdom. He has published approximately 70 papers.

*Tatsuya Yasunaga* is Professor of Physical Chemistry at Hiroshima University, Japan. He obtained his Ph.D. degree in 1955 from Hiroshima University. He has published more than 60 papers dealing with kinetics of fast reactions in solutions by means of relaxation techniques.

*Seang H. Yiv* is at the Centre de Recherches sur les Macromolecules, Strasbourg, France. Completed Master's thesis in 1975.

*Raoul Zana* is Maître de Recherches at the Centre de Recherches sur les Macromolécules, C.N.R.S., Strasbourg, France. He received his D.Sc. in 1964 from the University of Strasbourg. His current research interests are ion-solvent interactions and micellar solutions.

# SUBJECT INDEX

Pages 1-488 appear in Volume 1
Pages 489-930 appear in Volume 2

Acrylamide, 628
Aerosol OT, 649, 667, 676
Aggregates (see also Aggregation and Micelles)
  in surfactant solutions, 90-91
  size distribution of, 302
Aggregation or Association
  (see also Aggregates, Micelles and Micellization)
  in hydrocarbons, 445-452
  model (for large micelles) of, 183-186
  model (general) of, 188-193
  of amphiphiles, 199-200
  of amphiphiles with single hydrocarbon tails, 140-143
  of amphiphiles with two hydrocarbon tails, 143-148
  of drug molecules, 70-71
  of naturally occurring molecules, 58
  structures, 797
Alkyl Sulfates
  aggregation number, 160-162
  chemical shifts, 205
  c.m.c., 158-160, 244
  enthalpy changes, 232
  reciprocal relaxation times, 311-312
Alpha-pinene, 46
Alpha-terpineol, 46
Amphipath, 3
Amphipathic, 24
Amphipathicity, 3
Amphiphiles
  importance of, 25
  phase diagrams of, 879-880

Amphiphilic, 24
Amphiphilicity, 3
Amphiphilic Systems
  ionic interactions in, studied by NMR, 195-224
Antihistamines, 229, 355-356
Applications
  of micelles, 11-16
  of microemulsions, 16-17
Aqueous Ionic Micelle
  schematic representation of, 8
ATP, 62
Azo Dye Sulfonate, 821

Bicontinuous Structures, 877-892
Bifunctional Micellar Catalysis, 603-613
Bilayers, 917-919
Bile Salts, 64-65, 289-302, 384-391, 405-417, 851
Binding Constants, 495-497
Biphenyl Anion, 649
Bolaform Detergents, 293

Catalysis
  micellar (see Micellar Catalysis)
  microemulsions, 806-814
Calorimetric Studies of Micelle Formation, 467-477
Cellulose Ethers, 835-844
Chemical Relaxation Methods, 291
Chlorophyll, 63
Conductivity, 267-268, 294
Continuous Association, 291

Cooperativity and Anticooperativity, 186-188
Cosurfactant, 47, 803
Counterion Binding, 196
Critical Dodecyl Content, 897
Critical Micelle Temperature, 378-379
Critical Micellization Concentration (c.m.c.), 45, 121, 158-166
(see also Micelles of)
effect of alkyl chain length, 244
in aqueous vs. hydrocarbon solutions, 446-448
in hydrocarbons, 445-452
of a number of amphiphiles as determined by chemical shifts (table), 206
of a number of amphiphiles as determined by quadrupole relaxation (table), 211
of chlorpromazine, 70
of docosane-1,22-bis(trimethylammonium bromide), 298
of dodecyl pyridinium with various counterions, 268
of dodecyl trimethyl ammonium bromide, 307
of lysoplasmalogen, 917
of sodium alkyl sulfates, 158, 160, 244, 307
of sodium dodecyl sulfate, 84, 128, 837
of sodium octanoate, 220
of sodium perfluorooctanoate, 7
of sodium tetradecylsulfate, 307
of surfactants containing octyl chains, 176
of tetradecane trimethylammonium bromide, 298
of tetradecylpyridinium with various counterions, 268, 278
of trifluoropromazine, 7
of various ionic surfactants as determined by conductivity method (table), 309

Critical Micellization Concentration (c.m.c.), contd.
second, 308, 427
temperature dependence of, 229-237
Critical Vesicle Concentration, 133

Dielectric Constants at Interfaces of Micelles, 249
Diethyl Malonate, 622
Diffusion Coefficients, 405-406
Diffusion Measurements, 415
Dipalmitoyl Lecithin, 914
Diphenyl Acetonitrile, 620
Duroquinone, 534
Dye Pairing, 822-833
Dye-surfactant Mixed Aggregate, 829-832

Effective Polarity, 248-250
Electron Transfer Reactions, 531-546, 549-567
Electron Transport across the Interface, 559
Electron Tunneling, 550-552
Electroorganic Synthesis, 129-137
Electrophoretic Mobility, 97-98
Electromotive Force, 294
Emphysema, 851
Emulsion, 50, 621, 756
Equivalent Conductivity, 267-268, 294

Fluid Theories, 780-788
Fluorescent Probes, 76-85
Fluorescent Quenching, 916-924
Fluorinated Surfactants, 245
Functionalized Micelles, 603-613
hydroxyl, 603-613
imidazolyl, 603-613
Functionalized Surfactants, 603-613

Hard Sphere Repulsion, 780-789
Hemin, 697
Hydration
of amphiphilic aggregates, 196
of ionic micelles, 247-248

# INDEX

Hydrophile-Lipophile Balance, 457, 716
HLB-Temperature Range, 903-908
Hydrodynamic Parameters, 416
Hydrophobic Counterions, 274-286
Hydrophobic Interactions, 156-157
Hydrophobic Polyelectrolytes, 895-899

Improved Oil Recovery, 87
Indicator Dyes, 250-252
Indole Detergent, 84
Interfacial Curvature, 730
Interfacial Tension Minima, 856
Interphase, 50
Intramacromolecular Micelles, 895-899
Intramicellar Electron Transfer, schematic illustration of, 545
Inverse Micelles, 9, 12, 794
Inverse Micellar Solution, 797-798
Invert Micelle, 666, 695
Ionic Interactions, Studied by NMR, 195-224
Ionic Micelles, 263-287
  kinetics of, 263-287, 305-325, 329-344, 347-355

Kinetic Equation
  computer program for, 516
  first order, 523
  second order, 490-492, 515
  third order, 523
Kinetic Method for Determination of Binding Constants, 495-497
Kinetics of Ionic Micelles, 263-287, 305-325, 329-344, 347-355
Kinetic Theory of Rate Enhancements, 509-528

Lecithins, 914
Light Scattering, 408, 781-784
Lipids, 59
Lipophilic Micelles (see also Inverse Micelle, Invert Micelle, and Nonaqueous Media, Micelle Formation in)

Lipophilic Micelles, contd.
  dipole moments of, 434
  dynamic aspects of, 432-439
  properties of, 439-442
  solubilizing power of, 442
  stability of, 431-432
Liquid Crystalline Phases, 49, 200, 878
Lumophores, 76
Lung Gas Transference, 847-852
Lung Surfactants, 847-852
Lysoplasmalogen, 914

Membrane Transport, 847-852
Mesophases, 10, 217, 222
Mesomorphous States, 878
Methyl Orange Dye, 822
Micellar Adsorption, 15
Micellar Aggregation Number from Kinetic Data, 279-280
Micellar Catalysis (see Micelle Catalyzed Reactions)
Micellar Catalysis and Inhibition, 14
Micellar Cations, 640-642
Micellar Concentration, 5
Micellar Effects
  kinetic theory of, 489-506, 509-528
  mechanisms of, 502-506
  of nonionic micelles, 569-586
  on kinetics and equilibria of chemical reactions, 489-506
Micellar Emulsion, 802
Micellar Nucleus, 280-286
Micellar Slug, 94
Micellar Solution, 52, 87-110, 714, 797
  adsorption from, 107-108
  interfacial properties of, 95-99
  light scattering from, 99-102
Micelle, Definition, 2, 4, 53, 430
Micelles
  dielectric constant at interfaces of, 249
  functionalized, 603-613
  in nonaqueous media, 9, 429-442, 445-452

Micelles, contd.
  ionic, 263-287
  intramacromolecular, 895-899
  inverse, 9, 12, 794
  invert, 666, 695
  mixed, 353, 819-833, 855-875
  molecular weight of, 178-181
  nonionic (see Nonionic Micelles)
  polymerized, 69
  shape of, 124, 181-183, 359-379
  size and size distribution of, 120-121, 171-193, 329-344, 359-379
  small vs. large, 181-186
  structure of, 7-10
Micelles as Competitors, 26
Micelles as Model Systems, 11
Micelle Catalyzed Reactions in Aqueous Media (same as Micellar Catalysis) (see also Reversed Micellar Catalysis)
  acid catalyzed hydrolysis of p-nitrobenzaldehyde diethyl acetal, 521
  acrylamide polymerization, 635-637
  anodic dimerization, 621-622
  anodic oxidation, 620
  basic hydrolysis of p-nitrophenyl acetate-hexanoate-laurate, 517-519
  benzidine rearrangement, 524-526
  bisulfite redox reactions, 627-643
  carbonate oxidation, 544
  cyanide addition to pyridinium ions, 527-528
  dismutation reaction of $Br_2^-$ radical, 563
  ester hydrolysis, 517-518
  hydrolysis of p-nitrophenyl acetate, 605-613
  hydrolysis of p-nitrophenyl hexanoate, 605-613
  interaction of hemin with cyanide ion, 703
  ketal hydrolysis, 522

Micelle Catalyzed Reactions, contd.
  laser photolysis of duroquinone, 546
  ligand exchange reactions of vitam $B_{12a}$, 695-706
  methyl acrylate polymerization, 638-639
  photoredox processes, 531-547
  radiation-induced peroxidation, 589-600
  radiation-induced processes in nonionic micelles, 569-586
  radiation-induced redox reactions, 549-567
  reduction of $Eu^{+3}$, 542
  vinyl polymerization, 627-643
Micelles Formation (see also Aggregation)
  biological implication of, 55-74
  calorimetric studies of, 467-477
  general theory of, 152-154
  in nonaqueous media, 9, 429-442, 445-452
  surface activity and, 3-6
Micellization
  enthalpies of, 229-237
  free energy of, 122, 125-126, 137-140
  kinetic studies on, by means of relaxation methods, 305-325
  mechanism of, 312-325
  membranes and, 58-61
  multiphase, 855-875
  thermodynamics of, 119-131, 151-166, 229-237, 263-287, 368-379, 467-477
  true, 291
Micelles of (see also c.m.c. of)
  Aerosol OT, 676
  alkyl polyoxyethylene ethers, 69
  bifunctional surfactants, 603-613
  bile salts, 64-65, 383-400, 403-417
  bolaform detergents, 293-297
  Brij-35, 570

# INDEX

Micelles of, contd.
  Brij-96, 456, 464-465
  k-casein, 63
  cetyl trimethylammonium
    bromide, 497, 550, 553, 557,
    629-631, 828
  chlorpromazine, 70
  didodecyl dimethylammonium
    halides, 677
  dodecyl ammonium propionate,
    677
  dodecyl-N-betaine, 37
  dodecylpyridinium iodide, 36
  dodecylpyridinium with various
    counterions, 263-288
  dodecyl trimethyl ammonium
    chloride, 543, 550-557
  fatty acids, 592
  Igepal CO-530, 700
  Igepal CO-630, 570
  indole detergent, 84
  lysoplasmalogen, 917
  nonionic detergents, 455-465
  β-D-octyl glucoside, 249
  sodium alkyl sulfates, 158
  sodium dodecyl sulfate, 128-
    130, 181, 359-380, 497, 512,
    539, 550, 557, 837
  tetradecyl pyridinium with
    various counterions, 263-288
  tridodecyl ammonium halides,
    677
  Triton X-100, 570
Micelle-water Interface, 246-247
Microemulsion, Definition, 53
Microemulsions, 46-58, 713-754,
  755-774, 779-789, 791-798,
  801-814, 877-892
  anionic, 728
  applications of, 16
  containing ionic surfactants,
    791-798
  effect of electrolytes on
    stability of, 796
  interfacial free energy of,
    767-769
  light scattering by, 779-788
  nonionic, 720-726
  phase behavior of, 719, 806

Microemulsions, contd.
  phase inversion of, 760-762
  reactions in, 806-814
  stability of, 757-767, 804-806
  ternary diagrams for, 717
  theory of, 730-747
  thermodynamic stability of,
    757-766
Microenvironments in Aqueous
  Micellar Systems, 241-259
Mixed Micelles, 353, 819-833,
  855-875
  of bile salt-lecithin, 392-398
  of bile salt-lecithin-choles-
    terol, 398-400
  of decyltrimethyl ammonium
    bromide - cetyl trimethyl
    ammonium bromide, 353
  of linoleate-laurate, 597
  of linoleate-oleate, 597
  of linolenate-oleate, 597
Molecular Exciton Model, 825
Monomer Concentration, 5
Monomer-Micelle Equilibria, 173,
  347-353
Monomer-Micelle Exchange Processes, 347-355
Multiphase Systems, 757

Naphthalene, 80-82, 256-257
Negative Interfacial Tension,
  730
Nonaqueous Media, Micelle Formation in (see also Lipophilic
  Micelles), 9, 429-442, 445-
  452
NMR Chemical Shifts, 7-8, 205,
  685-688
NMR Properties of Alkali and
  Halide Ions, 201
Nonionic Micelles
  C-13 NMR studies of, 575-577
  energy transfer within, 583
  fluorescence probing of, 577
  fluorescence quenching effi-
    ciency in, 579
  hydrated electron reaction
    in, 586
  laser photoionization and
    radiolysis, studies in, 584-586

Nonionic Micelles, contd.
 model of, 571
 proton NMR studies of, 572
Nonionic Surfactant Mixtures, 455-465
Nonionic Surfactants, 455-465, 901-911
 in aqueous media, 455-465
 in nonaqueous media, 901-910

Oil Blue A, 836
Oil-Water Interface, 802
Orange OT Dye, 836
Organized Structures, 4

Paraffin Chain Salt Solutions, 30-36
Particulate Matter, 850-851
Partition Coefficient, 489
Permeability Coefficient, 71
Petroleum Sulfonates, 95-99
Phase Diagram of Ternary Systems, 48, 200, 458-460
Phase Inversion Temperature, 902
Phase Transfer Catalysts, 129-137
Pinocytosis, 60
P-jump Method, 296
Polycations, 640-642
Polymerized Micelles, 69
Polymer-Sodium Dodecyl Sulfate Systems, 835-844
 surface tension of, 839-841
 viscosity of, 841-842
Polyoxyethylene Oleyl Alcohols, 456
Polysoaps, 896-899
Polysorbate 80,
Protomorphyrin, 63
Pyrene, 82-84, 552, 555, 648, 914
Pyrene-3-aldehyde, 923
Pyrene Butyric Acid, 655
Pyrene Butyric Acid, 648

Quasielastic Light Scattering Spectroscopy Study
 of bile salt and bile salt-mixed lipid micellar system, 403-416

Quasielastic Light Scattering Spectroscopy Study, contd.
 of bile salt, bile salt-lecithin-cholesterol solutions, 383-400
 of sodium dodecyl sulfate micelles, 359-379

Reaction Rate Constant, Determination, 497-498
Relaxation Methods, 268-273, 305-325
Reverse Micelle (see also Lipophilic Micelles), 9, 12, 695
 of Aerosol OT, 649, 667
 in nonpolar media, schematic representation, 12
Reverted Micelle, 9
Reversed Micellar Catalysis
 acid catalyzed ester decarboxylation, 666-670
 acid catalyzed Diels-Alder condensation, 671
 electron transfer reactions, 654
 energy transfer reactions, 655
 ester hydrolysis, effect of solubilizate, 688-690
 interaction of hemin with cyanide ion, 703
 laser photolysis, 656
 ligand exchange reactions, 695-706
 pulse radiolysis, 649
 reactions of biphenyl anions, 650
 reactions of biphenyl triplet, 661
 with mixed cations, 672
 Ziegler catalyzed diene polymerization, 671

Salt Effect on Reaction Rates, 526-527
Second c.m.c., 308, 427
Self Association, 5, 7
Self Association Equilibria, 172-173

# INDEX

Shear-induced Structures, 38-39
Size Distribution Model, 135-137
Small Micelles vs. Large Micelles, 181-182
Sodium Bisulfite, 628
Sodium-N-Ocatanoate-N-Pentanol-Water System, 467-477
Sodium Perfluorooctanoate, 8
Solubilization, 11, 28-30, 102-105, 154-156, 909-910, 675-692, 837-839
  and catalysis, 691-692
  and phase equilibria, 102-105
  by nonionic surfactants, 901-910
  by polysoaps, 896-898
  of acetic acid, 681
  of diethyl malonate, 622
  of diphenyl acetonitrile, 619
  of dyes, 835-844
  of hydrocarbons by nonionic surfactants, 902
  of indicator dye, 249
  of methanol, 679
  of nonpolar gases, 847-852
  of Oil Blue A, 837
  of Orange OT, 837
  of n-propylamine, 680
  of pyrene, 919
  of tetramethylbenzidine, 533
  of Vitamin $B_{12a}$, 705
  of water, 460-462
  schematic representation of, 12
Solubilization, Definition, 53
Solubilization Isotherms, 682-685
Solubilized Probe Molecules, 531-546
Solubilized Species
  microenvironments of, 252-258
  surface activity of, 258
Solubilized Systems, 714
Sphere-Rod Transition, 188-193
Stepwise Aggregation Equilibria in Hydrocarbons, 448-452
Structure Formation in Surfactant Solution, 91
Structure of Micelles, 7-10

Surface Active Drugs, 56-71
Surface Activity and Micelle Formation, 3-6
Surface Tension Minima, 856
Surfactant-Polymer Interaction, 108-111

Ternary Systems, 467-477
Tetramethylbenzidine, 533
Thermodynamics of Amphiphilar Aggregation, 133-148
Thermodynamics of Micellization, 119-131, 151-166, 229-237, 263-287, 368-379, 467-477
T-jump Method, 296
α-Tocopherol
  as antioxidant in soap micelles, 598-600
  as prooxidant in soap micelles, 598-600
Transient Irregular Growth, 34-37
Trifluoropromazine Hydrochloride, 8
True Micellization, 291

Ultrasonic Absorption, 299, 307
Ultrasonic Relaxation Method, 347-355

Vesicle Formation, 143-148, 922
Vesicle Size Distribution, 143-145
Vitamin $B_{12a}$, 697

Water-Amphiphile Binary Phase Diagram, 897
Water-Amphiphile-Hydrophobe Diagram, 486, 880, 930